# Advances in Mushroom Research

Bhoopander Giri • Isha Gunwal
Payal Mago • Rupam Kapoor
Siva Raseetha
Editors

# Advances in Mushroom Research

## Breakthroughs and Future Directions

Springer

*Editors*
Bhoopander Giri
Department of Botany
University of Delhi
Delhi, Delhi, India

Isha Gunwal
Swami Shraddhanand College
University of Delhi
Delhi, Delhi, India

Payal Mago
Shaheed Rajguru College of Applied Sciences for Women
University of Delhi
Delhi, Delhi, India

Rupam Kapoor
Botany
University of Delhi
Delhi, Delhi, India

Siva Raseetha
Faculty of Applied Sciences
Universiti Teknologi MARA (UiTM)
Shah Alam
Shah Alam, Selangor, Malaysia

ISBN 978-981-95-7691-3 ISBN 978-981-95-7692-0 (eBook)
https://doi.org/10.1007/978-981-95-7692-0



This Springer imprint is published by the registered company Springer Nature Singapore Pte Ltd.
The registered company address is: 152 Beach Road, #21-01/04 Gateway East, Singapore 189721, Singapore

# Foreword

Fungi, since long, have been known for their unique features like eukaryotic, nonmotile, heterotrophy, and chitin in their cell walls. Ecologically these living organisms have been recognized as decomposers (saprophytes), parasites, symbiotic partners (mycorrhizal and lichens), and disease agents (pathogens). However, their direct human applications have also been well recognized in food and drink, medicines, agriculture, bioremediation, enzyme productions, etc. Their vast biodiversity is highly appreciated as they help manage plant diseases producing beneficial products, treat human fungal infections, and have been integral to life. Mushrooms constitute the very important part of kingdom fungi. The term mushrooms were popularly used to identify the edible sporophores.

For centuries, mushrooms were regarded with a mixture of respect, suspicion, fear, agents of poison, and provider of food and medicines (in some cultures). The edible mushrooms are free of cholesterol and contain small amounts of essential amino acids and B Vitamins. Nowadays mushrooms are emerging as central players in science, medicine, and sustainability. They play an essential role in human health and environmental sustainability. Mushrooms and their mycelial networks are recognized as essential associates in addressing some of the most demanding issues, such as healing degraded ecosystems, developing sustainable materials and foods, and discovering new medicines. They act as nature's recyclers, relentless workers to decompose organic matters, thereby maintaining soil health. Besides, mushrooms with symbiotic partnership with plants through mycorrhizal networks facilitate nutrient exchange and sustain entire ecosystems. To study mushrooms is to uncover the hidden systems that reinforce biodiversity and planetary stability.

The book *Advances in Mushroom Research*, edited by Bhoopander Giri, Isha Gunwal, Payal Mago, Rupam Kapoor, and Siva Raseetha, provides the latest insights into the ecological significance of mushrooms, the remarkable therapeutic potential against neurodegeneration, immune dysfunction, cancer, and mental illness. Besides, readers will learn mushrooms as bio-remediators of polluted materials, potential sources of leather-like textiles and packaging materials, protein-rich foods for an increasing population, and partners in intensifying agriculture. This book comprises 16 authoritative chapters written by the experts in their fields. The opening chapter of this book expressively captures the ecological significance of mushrooms, and determines their role in ecological restoration, focusing on their contribution to nutrient cycling, enhancement of soil structure, suppression of plant pathogens, regulation of water dynamics, and formation of beneficial symbiotic relationships with plants. Chapter 2 explores diversity, ecology, distribution, and applications of edible and medicinal fungi, highlighting their multifunctional role in ecosystems and society. Chapter 3 reveals the basic biology of mushrooms and biodiversity of wild edible and medicinal mushrooms. Chapter 4 addresses the socioeconomic dimensions of mushroom cultivation and its role in rural livelihoods, waste repurposing, and sustainable development. Chapter 5 highlights biological, nutritional, medicinal, and economic importance of mushrooms and their role in sustainable agricultural practices, whereas Chap. 6 addresses the transformative role of Internet of Things (IoT) technologies in advancing mushroom production.

The promise of these fungi extends far beyond, as this book explores the profound therapeutic potential of several mushroom species. Certain mushroom species contain potent bioactive compounds that can modulate our immune systems, fight disease, and support cognitive health. The exploration of these compounds for their nutraceutical and pharmaceutical applications represents one of the most exciting frontiers in bioprospection. Chapters 7, 8, 9, 10, 11, and 12 address these issues and discuss role of mushrooms in coping up with such problems. Besides, the roles of mushrooms in agriculture, developing sustainable packaging materials, leather alternatives, and redefining biofabrication have been described in the last four chapters of this book. This book, as well, weaves together ecological, medicinal, and biotechnological approaches to present a holistic portrait of a field undergoing a resurgence. This volume brings together cutting-edge research and compels us to move beyond viewing mushrooms as mere food or folklore but to recognize them as indispensable partners in addressing global challenges from environmental degradation to human wellness. I hope this volume will serve as an invaluable resource for researchers, students, botanists, mycologists, industrialists, and environmentalists, and for anyone fascinated by the fungal kingdom.

I congratulate the contributors of different chapters and editors of the book for bringing out this volume on the very fascinating subject.

Former Vice-Chancellor, Chairperson,
Haryana State Higher Education Council
Panchkula, Haryana, India

Kailash Chandra Sharma

# Preface

Since ancient times, mushrooms have been valued both as a culinary delicacy and for their therapeutic properties. They are rich in bioactive compounds such as polysaccharides, terpenoids, and phenolic compounds, which contribute to a wide range of health benefits. In addition, mushrooms provide an excellent source of vitamins, proteins, and essential nutrients. Many mushroom varieties contain potent antioxidants that help combat oxidative stress, a key factor in aging and the development of chronic diseases. They also exhibit anti-inflammatory properties, which could alleviate arthritis inflammation and inflammatory bowel diseases. Certain mushroom varieties such as Reishi and Shiitake are known to enhance immune functions and improve ability of an individual to control infections and related illnesses. Certain polysaccharides present in mushrooms demonstrate potential against tumor growth and metastasis, possibly by activating immune response against cancer cells, inducing apoptosis, or inhibiting tumor-related blood vessel formation. Mushrooms such as Lion's Mane have been reported to potentially protect against neurodegenerative disorders like Alzheimer's and Parkinson's diseases. They support the production of nerve growth factor, aiding in the repair and regeneration of neurons. Besides, several mushrooms possess antiviral and antibacterial properties, positioning them as promising candidates for the development of novel antimicrobial agents.

The book *Advances in Mushroom Research* presents a compelling exploration of the latest breakthroughs in mycology. This volume brings together contributions from leading experts across 16 insightful chapters, covering a wide range of topics with a particular emphasis on mushrooms diversity and ecology, the discovery of promising therapeutic compounds and their nutraceutical applications, and highlights their environmental, industrial, and agricultural applications. We hope that this extensive volume will rouse interest and generate innovative ideas among young researchers, fostering the development of new approaches and methodologies that could drive further advances in our understanding of mushroom ecology and its implications for pharmacology, environmental science, industry, and agriculture. Serving as a valuable resource, it is deliberated for medical researchers, botanists, mycologists, environmentalists, agriculturists, biotechnologists, and scholars interested in pharmacognosy.

We are grateful to our contributing authors, whose unwavering support and expertise were instrumental in creating these authoritative and invaluable chapters. Our sincere appreciation also extends to our colleagues for their invaluable assistance in the preparation and compilation of this volume. We are indebted to the dedicated team at Springer, particularly William Achauer, Anil Chandy, Sean Eickhoff, Momoko Asawa, Aishwarya Thyagarajan, and Swati Sharma for their dedicated commitment and diligent efforts in bringing this project to realization. Finally, a special and heartfelt thanks must go to our families for their unwavering support and endless encouragement throughout this endeavor.

Delhi, India — Bhoopander Giri
Delhi, India — Isha Gunwal
Delhi, India — Rupam Kapoor
Delhi, India — Payal Mango
Shah Alam, Selangor, Malaysia — Siva Raseetha

# Contents

# Editors and Contributors

## About the Editors

**Bhoopander Giri** is a Professor in the Department of Botany, University of Delhi, Delhi, India. He has worked as a Post-Doctoral Fellow at the International Center for Genetic Engineering and Biotechnology and as a Raman Post-Doctoral Fellow at the Department of Plant Pathology, North Carolina State University, USA. He has served as an Associate Fellow at The Energy and Resources Institute, New Delhi. He has published more than 60 papers in peer-reviewed journals and books. He has served as the General Secretary for International Symbiosis Society, USA, and an editorial board member of the journal *Scientific Reports*. He has presented his research work at several international conferences.

**Isha Gunwal** is an Assistant Professor in the Department of Botany, Swami Shraddhanand College, University of Delhi, Delhi, India. She has published more than 20 papers in peer-reviewed journals. She is a life member of several prestigious academic societies. Her research focuses on phytoremediation using mycorrhiza and the effects of biological factors on edible mushrooms. She has made significant contributions to bio-waste management through the cultivation of oyster mushrooms and the phytoremediation of heavy metals using Vetiver grass. She is committed to teaching and mentoring undergraduate students.

**Payal Mago** is a distinguished academician and administrator. She is the Principal of Shaheed Rajguru College of Applied Sciences for Women and holds prestigious positions including Director of the Campus of Open Learning, Director of Udhmodya foundation, Dean of the Faculty of Open Learning, and Director of the Women's Studies & Development Centre, at the University of Delhi. She has authored over 37 research papers in peer-reviewed journals and 8 chapters and edited 11 books. She played a pivotal role in developing more than 137 innovative skill courses to enhance student's employability. She is a consultant for mushroom cultivation, showcasing her diverse expertise to both education and practical applications in her field.

**Rupam Kapoor** is a Professor in the Department of Botany, University of Delhi, India. She specializes in the field of plant-fungus interactions. She has contributed immensely to understanding the role of AMF in alleviation of abiotic stress. Recently she has completed project on mushrooms of Kargil—the cold dessert of India. She has successfully completed several research projects sponsored by various funding agencies such as DBT, DST, and Ministry of Environment, Forest and Climate Change. Her name has been listed in the global ranking of top 2% scientists of the world for their contribution in their respective fields by Stanford University for 4 consecutive years (2020–2023). She has more than 100 publications in peer-reviewed journals and edited 4 books. She is an Adjunct Professor at Department of Environmental Science Hamdard University.

**Siva Raseetha** is a distinguished faculty member at the Faculty of Applied Sciences, UiTM Shah Alam, Malaysia. She has completed her postdoctoral research at the Mushroom Research Centre, University of Malaya. Her research expertise encompasses mushroom studies, including fresh mushrooms, postharvest treatments, and downstream products. She serves as the Vice Treasurer for the Malaysian Mushroom Researchers and Entrepreneurs Association. She has organized international symposium in collaboration with PERTEMA, AIMS partners, and the Malaysian Mushroom Researchers and Entrepreneurs Association, supported by the Malaysian Societies Registration Department and served as a prelude to the 10th International Conference on Mushroom Biology and Mushroom Products held at University Malaya (2023).

## Contributors

**K. N. Abhini** Department of Microbiology, St. Pius X College, Kasaragod, Kerala, India

**Abzar** Faculty of Applied Sciences, Universiti Teknologi MARA, Shah Alam, Selangor, Malaysia

Department of Crop Science, Faculty of Agriculture, University Putra Malaysia, Serdang, Selangor, Malaysia

**F. M. N. A. Aida** Faculty of Applied Sciences, Universiti Teknologi MARA, Shah Alam, Selangor, Malaysia

**Maria José Alves** AquaValor – Centro de Valorização e Transferência de Tecnologia da Água – Associação, Chaves, Portugal

LiveWell – Research Centre for Active Living & Wellbeing, Instituto Politécnico de Bragança, Bragança, Portugal

CIMO – Centro de Investigação de Montanha, Instituto Politécnico de Bragança, Bragança, Portugal

**Suresh Balasubramanian** Division of Microbiology, School of Life Sciences, Ooty Campus, JSS Academy of Higher Education and Research, Longwood, Ooty, The Nilgiris, Tamil Nadu, India

**Pooja Bhadoriya** Shaheed Rajguru College of Applied Sciences for Women, University of Delhi, Vasundhara Enclave, Delhi, India

**Bhawana** Department of Botany, Institute of Home Economics, University of Delhi, New Delhi, India

**Fernando C. Lidon** Faculdade de Ciências e Tecnologia, Universidade Nova de Lisboa, Campus da Caparica, Caparica, Portugal

GeoBiotec—GeoBioTec Research Institute, Faculdade de Ciências e Tecnologia, Universidade Nova de Lisboa, Campus da Caparica, Caparica, Portugal

**Indranil Chakraborty** Department of Chemistry, Kharagpur College, Kharagpur, Paschim Medinipur, West Bengal, India

**José Manuel Costa** Agrarian School of Viseu, Polytechnic University of Viseu, Viseu, Portugal

CERNAS—IPV Research Centre, Polytechnic University of Viseu, Viseu, Portugal

**Rizki Rabeca Elfirta** Research Center for Applied Microbiology, National Research and Innovation Agency, The Republic of Indonesia, Bogor, Indonesia

**Evan Febriansyah** Department of Biology, Faculty of Mathematics and Natural Sciences, Universitas Indonesia, Depok, Indonesia

**Nur Azalina Suzianti Feisal** Department of Diagnostic and Allied Health Science, Faculty of Health and Life Sciences, Management and Science University, Shah Alam, Selangor, Malaysia

**Juliana Garcia** AquaValor – Centro de Valorização e Transferência de Tecnologia da Água – Associação, Chaves, Portugal

LiveWell – Research Centre for Active Living & Wellbeing, Instituto Politécnico de Bragança, Bragança, Portugal

CITAB – Centre for the Research and Technology of Agro-Environment and Biological Sciences, Inov4Agro, University of Trás-os-Montes e Alto Douro, Vila Real, Portugal

**Elsa M. Gonçalves** UEIS-TSA - Unidade Estratégica de Investigação e Serviços de Tecnologia e Segurança Alimentar, Instituto Nacional de Investigação Agrária e Veterinária, Oeiras, Portugal

GeoBiotec—GeoBioTec Research Institute, Faculdade de Ciências e Tecnologia, Universidade Nova de Lisboa, Campus da Caparica, Caparica, Portugal

**Irene Gouvinhas** Animal and Veterinary Research Center (CECAV), University of Trás-os-Montes and Alto Douro, Vila Real, Portugal, Vila Real, Portugal

Associate Laboratory of Animal and Veterinary Sciences (AL4AnimalS), Vila Real, Portugal, Vila Real, Portugal

**Raquel P. F. Guiné** Agrarian School of Viseu, Polytechnic University of Viseu, Viseu, Portugal

CERNAS—IPV Research Centre, Polytechnic University of Viseu, Viseu, Portugal

**Bhoomika Rajeev Gupta** Shaheed Rajguru College of Applied Sciences for Women, University of Delhi, Vasundhara Enclave, Delhi, India

**Charu Khosla Gupta** Acharya Narendra Dev College, University of Delhi, New Delhi, India

**Rajni Gupta** Department of Botany, Kirori Mal College, University of Delhi, New Delhi, India

**Anggita Rahmi Hafsari** Functional Omics and Bioprocess Development Laboratory, Institute of Biological Sciences, Faculty of Science, Universiti Malaya, Kuala Lumpur, Malaysia

Department of Biology, Faculty of Science, Universitas Islam Negeri (UIN) Sunan Gunung Djati Bandung, Bandung, West Java, Indonesia

**Zarimah Mohd Hanafiah** Department of Civil Engineering, Faculty of Engineering and Build Environment, Universiti Kebangsaan Malaysia (UKM), Bangi, Malaysia

Functional Omics and Bioprocess Development Laboratory, Institute of Biological Sciences, Faculty of Science, Universiti Malaya, Kuala Lumpur, Malaysia

**Tengku Nilam Baizura Tengku Ibrahim** Department of Environmental Health, Faculty of Health Sciences, Universiti Teknologi MARA Cawangan Pulau Pinang, Kampus Bertam, Penang, Malaysia

Occupational Health and Safety Risk Management (OHSeRM) Research Initiative Group, Universiti Teknologi MARA, Cawangan Pulau Pinang, Kampus Bertam, Kepala Batas, Pulau Pinang, Malaysia

**Damanpreet Kaur** Department of Microbiology, Institute of Home Economics, University of Delhi, New Delhi, India

**Gurpreet Kaur** Department of Botany, SGTB Khalsa College, University of Delhi, Delhi, India

**Kohinoor Kaur** Department of Microbiology, Institute of Home Economics, University of Delhi, New Delhi, India

**Surinder Kaur** Department of Botany, SGTB Khalsa College, University of Delhi, Delhi, India

**Manuela Lageiro** UEIS-TSA - Unidade Estratégica de Investigação e Serviços de Tecnologia e Segurança Alimentar, Instituto Nacional de Investigação Agrária e Veterinária, Oeiras, Portugal

GeoBiotec—GeoBioTec Research Institute, Faculdade de Ciências e Tecnologia, Universidade Nova de Lisboa, Campus da Caparica, Caparica, Portugal

**Hariprasath Lakshmanan** Division of Biochemistry, School of Life Sciences, Ooty Campus, JSS Academy of Higher Education and Research, Longwood, Ooty, The Nilgiris, Tamil Nadu, India

**Payal Mago** Shaheed Rajguru College of Applied Sciences for Women, University of Delhi, Delhi, India

**Prasenjit Maity** Department of Chemistry, Sabang Sajanikanta Mahavidyalaya, Lutunia, Paschim Medinipur, West Bengal, India

**Shibdas Maity** Department of Botany, Sabang Sajanikanta Mahavidyalaya, Lutunia, Midnapore, Paschim Medinipur, West Bengal, India

**Avrajit Manna** Department of Chemistry, Raja Narendra Lal Khan Women's College (Autonomous), Gope Palace, Paschim Medinipur, West Bengal, India

**Hom-Singli Mayirnao** Department of Botany, University of Delhi, Delhi, India

**Rekha Mehrotra** Department of Microbiology, Shaheed Rajguru College of Applied Sciences for Women, University of Delhi, Delhi, India

**R. Mohammad-Syaril** Johor Bahru Public Health Laboratory, Tampoi, Johor Bahru, Johor, Malaysia

**Wan Hanna Melini Wan Mohtar** Department of Civil Engineering, Faculty of Engineering and Build Environment, Universiti Kebangsaan Malaysia (UKM), Bangi, Malaysia

**Eva Olo-Fontinha** AquaValor – Centro de Valorização e Transferência de Tecnologia da Água – Associação, Chaves, Portugal

LiveWell – Research Centre for Active Living & Wellbeing, Instituto Politécnico de Bragança, Bragança, Portugal

Universidade Católica Portuguesa, CBQF – Centro de Biotecnologia e Química Fina, Laboratório Associado, Escola Superior de Biotecnologia, Porto, Portugal

**Tamilarasi Sambu Periyasamy** Division of Biochemistry, School of Life Sciences, Ooty Campus, JSS Academy of Higher Education and Research, Longwood, Ooty, The Nilgiris, Tamil Nadu, India

**Jegadeesh Raman** Agricultural Microbiology Division, National Institute of Agricultural Sciences, Rural Development Administration, Wanju, Jeollabuk, Republic of Korea

**Ana Cristina Ramos** UEIS-TSA - Unidade Estratégica de Investigação e Serviços de Tecnologia e Segurança Alimentar, Instituto Nacional de Investigação Agrária e Veterinária, Oeiras, Portugal

GeoBiotec—GeoBioTec Research Institute, Faculdade de Ciências e Tecnologia, Universidade Nova de Lisboa, Campus da Caparica, Caparica, Portugal

**Siva Raseetha** Faculty of Applied Sciences, Universiti Teknologi MARA, Shah Alam, Selangor, Malaysia

**R. Rasyiqah-Nabihah** Faculty of Applied Sciences, Universiti Teknologi MARA, Shah Alam, Selangor, Malaysia

**Fernando H. Reboredo** Faculdade de Ciências e Tecnologia, Universidade Nova de Lisboa, Campus da Caparica, Caparica, Portugal

GeoBiotec—GeoBioTec Research Institute, Faculdade de Ciências e Tecnologia, Universidade Nova de Lisboa, Campus da Caparica, Caparica, Portugal

**Atik Retnowati** Research Center for Biosystematics and Evolution, National Research and Innovation Agency, The Republic of Indonesia, Bogor, Indonesia

**Rini Riffiani** Research Center for Applied Microbiology, National Research and Innovation Agency, The Republic of Indonesia, Bogor, Indonesia

**Luisa Cristina Roseiro** UEIS-TSA - Unidade Estratégica de Investigação e Serviços de Tecnologia e Segurança Alimentar, Instituto Nacional de Investigação Agrária e Veterinária, Oeiras, Portugal

GeoBiotec—GeoBioTec Research Institute, Faculdade de Ciências e Tecnologia, Universidade Nova de Lisboa, Campus da Caparica, Caparica, Portugal

**Maria Isabel Santos** Agrarian School of Viseu, Polytechnic University of Viseu, Viseu, Portugal

**Iwan Saskiawan** Research Center for Applied Microbiology, National Research and Innovation Agency, The Republic of Indonesia, Bogor, Indonesia

**Ryan Haryo Setyawan** Research Center for Applied Microbiology, National Research and Innovation Agency, The Republic of Indonesia, Bogor, Indonesia

**Manisha Sharma** Department of Botany, Kirori Mal College, University of Delhi, New Delhi, India

**Samiksha Sharma** Department of Microbiology, University of Delhi, South Campus, New Delhi, India

**Jani Silva** AquaValor – Centro de Valorização e Transferência de Tecnologia da Água – Associação, Chaves, Portugal

LiveWell – Research Centre for Active Living & Wellbeing, Instituto Politécnico de Bragança, Bragança, Portugal

**Mafalda Silva** Faculdade de Ciências e Tecnologia, Universidade Nova de Lisboa, Campus da Caparica, Caparica, Portugal

**Yashika Singh** Department of Botany, Kirori Mal College, University of Delhi, New Delhi, India

**Sinosh Skariyachan** Department of Microbiology, St. Pius X College, Kasaragod, Kerala, India

**N. V. Vinod** Department of Microbiology, St. Pius X College, Kasaragod, Kerala, India

**Aditi Virmani** Shaheed Rajguru College of Applied Sciences for Women, University of Delhi, Vasundhara Enclave, Delhi, India

**Wan Abd Al Qadr Imad Wan-Mohtar** Functional Omics and Bioprocess Development Laboratory, Institute of Biological Sciences, Faculty of Science, Universiti Malaya, Kuala Lumpur, Malaysia

**Leena Catherine Anthonimuthu Francis Xavier** Division of Microbiology, School of Life Sciences, Ooty Campus, JSS Academy of Higher Education and Research, Longwood, Ooty, The Nilgiris, Tamil Nadu, India

**Preeti Yadav** Acharya Narendra Dev College, University of Delhi, New Delhi, India

**Merina Yasmin** Department of Human Physiology, Vidyasagar University, Midnapore, Paschim Medinipur, West Bengal, India

# Mushrooms: The Humble Warrior of Eco-Restoration

1

Preeti Yadav and Charu Khosla Gupta

**Abstract**

Anthropogenic activities, compounded by climate change, are rapidly accelerating the degradation of natural ecosystems, making ecological restoration an urgent global priority. Conventional approaches have largely focused on reintroducing native species, reforestation, wetland rehabilitation, and erosion control. However, emerging scientific insights highlight the significant ecological role of mushrooms and the reproductive structures of fungi in ecosystem recovery. Once overlooked, mushrooms are now increasingly recognized as vital natural allies in environmental restoration. This chapter examines the multifaceted role of mushrooms in ecological restoration, with a focus on their contribution to nutrient cycling, enhancement of soil structure, suppression of plant pathogens, regulation of water dynamics, and formation of beneficial symbiotic relationships with plants. A particular emphasis is placed on mycorestoration-a fungal-based restoration method involving the enzymatic and physical breakdown of both organic and inorganic pollutants by different fungi. Field experiments and case studies not only demonstrate the ability of fungi to facilitate vegetation growth in degraded landscapes but also indicate their role in improving the diversity and function of soil microbiomes. These humble species aid in the detoxification of pollutants such as petroleum hydrocarbons and heavy metals. Spent mushroom substrate (SMS) and spent mushroom compost (SMC), the leftover materials from the cultivation of mushrooms, are widely employed for their effectiveness as biofertilizers, soil conditioners, and tools for carbon sequestration. Economi-

P. Yadav · C. K. Gupta (✉)
Acharya Narendra Dev College, University of Delhi, New Delhi, India
e-mail: charukhoslagupta@andc.du.ac.in



B. Giri et al. (eds.), *Advances in Mushroom Research*,
https://doi.org/10.1007/978-981-95-7692-0_1

cally, mushroom-based restoration is a superior practice, which offers not only a low cost but also an environmentally sustainable alternative to traditional remediation approaches. Besides reducing dependency on chemical inputs, these processes create value-added by-products, in addition to promoting rural employment. Despite the plentiful advantages, challenges remain in selecting appropriate fungal species, scaling up implementation, integrating fungal approaches into policy, and overcoming cultural barriers to acceptance. This chapter highlights mushrooms as vital ecological engineers and calls for their greater integration into restoration ecology practices and policies.

**Keywords**

Eco-restoration · Fungi · Mushroom · Spent mushroom substrate (SMS)

## 1.1 Introduction

Human-driven activities such as deforestation, pollution, urban development, and climate change have increasingly disrupted natural ecosystems, resulting in widespread soil degradation, biodiversity loss, and the collapse of vital habitats. In response to these challenges, ecological restoration (eco-restoration) has emerged as a vital strategy to rehabilitate damaged environments by reinstating ecological integrity and enhancing essential ecosystem services such as water purification, air filtration, and carbon sequestration (Calabi-Floody et al. 2024).

Traditional approaches to restoration have focused primarily on above-ground interventions, reintroducing native species, rehabilitating degraded areas, restoring wetlands, and stabilizing erosion-prone landscapes. However, in the subterranean world, the fungal kingdom is gaining attention for its critical role in ecosystem recovery. Until recently, mushrooms were primarily recognized as culinary delights or sources of medical compounds (Lindahl and Tunlid 2015; Ma et al. 2025), but nowadays they are considered nature's secret weapon to heal Mother Earth. While mushrooms are the visible fruiting bodies of fungi, the true biological engine lies in the extensive underground mycelial network, which permeates soil and organic debris to drive numerous essential ecological functions.

This hidden fungal web plays a central role in nutrient cycling, with fungi being the primary decomposers, breaking down complex organic materials like lignin-rich woody debris and dead leaves into simpler, bio-available nutrients. This decomposition not only sustains plant life but also supports overall soil health and productivity. Recent studies show that spent mushroom substrate (SMS) can significantly increase microbial diversity, enzyme activity, and organic carbon turnover, directly enhancing the biological quality of soil (Calabi-Floody et al. 2024; Igual et al. 2024). Without this fungal activity, organic material would keep accumulating in the soil, and all critical nutrients would remain inaccessible to plants, effectively stalling ecosystem regeneration (He et al. 2024).

Mushrooms serve as a gateway into the broader ecological potential of fungi, giving rise to a concept known as mycorestoration, a fungi-based restoration strategy that includes the usage of fungal enzymes to degrade or detoxify environmental pollutants. It has been reported that mushrooms such as *Pleurotus* and *Ganoderma* species produce extracellular enzymes capable of breaking down petroleum hydrocarbons, pesticides, dyes, and even heavy metals (Ali et al. 2025; Mishra et al. 2024). As an eco-friendly alternative, mycoremediation reduces dependence on chemical and mechanical remediation techniques while delivering comparable, if not superior, results. Interestingly, these results are sustainable too.

Incorporating fungal spores or mycelium into restoration practices has shown promising results, with improved soil structure and increased organic matter decomposition when mycelium was applied to degraded soils (He et al. 2024). Wang et al. (2025) demonstrated that fungal inoculation not only boosts nutrient cycling but also accelerates vegetation establishment in degraded grasslands and wetlands.

Although evidence in favor of fungal applications continues to grow, substantial obstacles to their widespread adoption still persist. Presently, there is not only a limited understanding of fungal diversity across different ecosystems but also logistical constraints in large-scale inoculum production, along with the risk of disrupting native microbial communities (Hawkes and Keitt 2015). Furthermore, long-term monitoring and adaptive management are essential to ensure the ecological stability of fungal interventions.

## 1.2 Roles and Applications of Mushrooms in Eco-Restoration

### 1.2.1 Role in Soil Structure Improvement

The extensive network of fungal mycelium plays a crucial role in improving the structure of soil by binding soil particles together, increasing soil aggregation, and the overall stability of the soil. This structural improvement significantly contributes to increasing soil aeration, water retention, and resistance to erosion, these being the essential factors in soil rehabilitation (Rillig and Mummey 2006). Glomalin-a sticky glycoprotein secreted by fungal hyphae, acts as a natural adhesive, facilitating the aggregation of fine soil particles, forming larger and more stable soil aggregates (Matos et al. 2022; Anandakumar et al. 2025), resulting in a porous, sponge-like soil texture capable of holding moisture efficiently.

A well-textured soil not only supports root development but also promotes diverse and active microbial communities, thereby creating optimal conditions for eco-restoration. Additionally, SMS is considered a viable and sustainable alternative to peat in formulations of microbial inoculants, helping in mineral solubilization (Hu et al. 202[illegible]). Further research indicates that SMS enhances soil porosity along with water-holding capacity more effectively than traditionally used peat-based carriers, thus promoting plant growth and soil microbial health (Igual et al. 2024).

### 1.2.2 Formation of Symbiotic Relationships (Mycorrhizae)

Several mushroom-forming fungi show beneficial symbiotic relationships with plant roots, commonly referred to as mycorrhizae. These interactions play a significant role in ecological restoration, owing to their capacity to improve plant health, growth, and survival. By this association, the fungi facilitate the uptake of water and minerals from the soil and, in return, are provided with carbohydrates produced through photosynthesis (Pérez-Moreno et al. 2021). This reciprocal exchange enhances plant resilience to environmental stresses like drought, soil-borne pathogens, and the toxicity of heavy metals (Rillig and Mummey 2006). Furthermore, recent studies indicate that these mycorrhizae can increase the level of ergothioneine, a potent antioxidant with nutritional and medicinal relevance in crops like wheat and asparagus, highlighting the ecological and health benefits of mycorrhizal symbiosis (Carrara et al. 2023).

### 1.2.3 Bioremediation and Detoxification

Nowadays, many degraded or disturbed ecosystems are often loaded with environmental contaminants like heavy metals, pesticides, and petroleum residues. Mushrooms possess specialized enzymes that enable them to break down these toxic compounds. Mycoremediation, the application of fungi for detoxification of these toxic compounds, is rapidly gaining importance within the field of ecological restoration, as certain mushroom species have a remarkable ability to metabolize toxic compounds through enzymatic activity capable of degrading hydrocarbons, synthetic chemicals, and heavy metals (Akpasi et al. 2023).

Species within the *Pleurotus* genus (oyster mushrooms) produce certain enzymes that efficiently degrade petroleum hydrocarbons, making them effective agents for cleaning up oil-contaminated sites (Roshandel et al. 2021). Similarly, *Phanerochaete* (a white-rot fungus) utilizes lignin-degrading enzymes to break down a wide range of persistent pollutants, industrial dyes, and polychlorinated biphenyls (Vaksmaa et al. 2023). Further, some field-based studies support the efficacy of native fungal strains in reducing concentrations of heavy metals in urban brownfield soils. Many mushrooms are applicable for the mobilization of various toxic metals in soil, like *Agaricus, Cortinarius, Boletus, Leccinum, Suillus,* and *Phellinus* (Malik et al. 2021). *Pleurotus platypus, Agaricus bisporus,* and *Calocybe indica* are well-known as biosorbent agents for the removal of Cu, Zn, Fe, Cd, Pb, and Ni from aqueous solutions (Kulshreshtha et al. 2014). These findings show that fungi are an environmentally sustainable and cost-effective solution for bioremediation (Ali et al. 2025). Mycoremediation further creates favorable conditions for the re-establishment of native plant and animal communities by restoring soil health and reducing the pollutant content (Mishra et al. 2024) (Fig. 1.1).

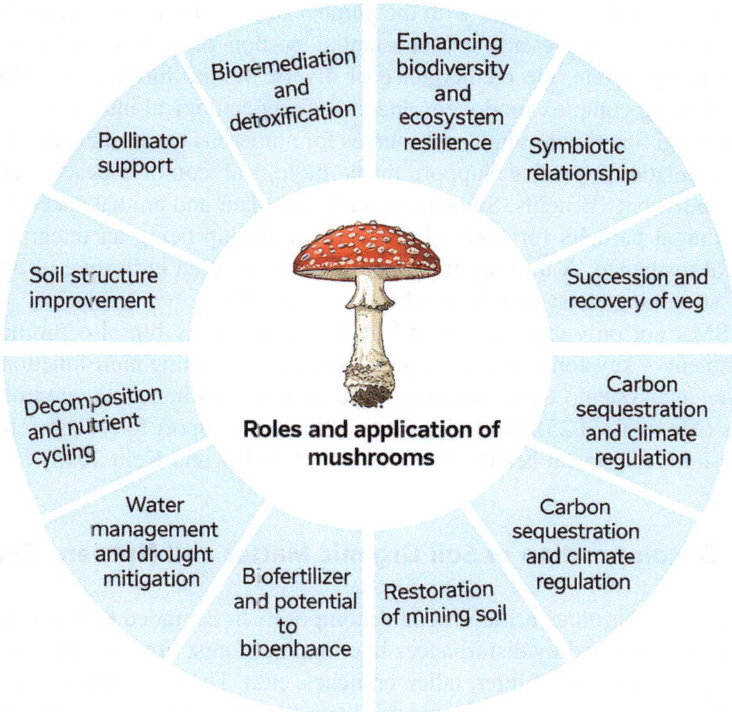


**Fig. 1.1** Various roles and applications of mushrooms

### 1.2.4 Facilitating Succession and Vegetation Recovery

As noted previously, numerous mushroom species contribute significantly to the initial phases of ecosystem restoration by enhancing soil quality and supporting plant establishment (Neuenkamp et al. 2019). Mycorrhizal associations enhance the growth of pioneer species, initiating natural ecological succession and paving the way for the development of more complex plant species (Smith and Read 2008). Experimental studies have shown that using SMS instead of conventional nursery soil can significantly enhance seedling growth, nutrient uptake, and microbial activity in degraded soils (He et al. 2024). Moreover, combining SMS with plant growth-promoting rhizobacteria has been found to facilitate the survival and productivity of certain crops like blueberries (Wang et al. 2025).

### 1.2.5 Enhancing Biodiversity and Ecosystem Resilience

Mushrooms not only facilitate the restoration of soil structure and chemical balance but also help in the revitalization of biological diversity. In most ecosystems, fungal diversity exceeds that of plants, with the number of fungal species ranging between 1.5 and 5 million, representing a substantial portion of global biodiversity. As ecosystems regenerate, the reemergence of diverse fungal forms drastically helps in reestablishing complex food webs and enhancing ecological intricacy.

These fungi function as vital food sources for numerous organisms and engage in symbiotic relationships that support many biological communities, resulting in several biodiversity benefits. Since many endemic plant and animal species rely on specific fungal partners for survival, these fungi end up being an integral part of ecological resilience. A thriving fungal community is often indicative of a healthy and self-sustaining ecosystem (Pérez-Moreno et al. 2021).

The SMS not only enhances overall microbial diversity but also promotes the establishment of keystone microbial species. Besides imparting multifunctionality to any given ecosystem, these microbes also provide resilience to environmental stressors (Ma et al. 2025). Such dynamics ultimately support food web robustness and long-term ecological health of ecosystems (Hawkes and Keitt 2015).

### 1.2.6 Decomposition of Soil Organic Matter and Nutrient Cycling

Saprotrophic mushrooms act as crucial decomposers in degraded ecosystems, especially the ones affected by disturbances like logging, forest fires, and the accumulation of organic debris (leaf litter, fallen branches, etc.). These mushrooms accelerate the breakdown of organic matter and facilitate the recycling of essential nutrients back into the soil (Fig. 1.2). In terrestrial ecosystems, these fungi are among the most efficient decomposers in degrading complex plant polymers such as lignin and cellulose using fungi-specific enzymes like cellulases and ligninases. Though mushrooms are primarily known to be efficient decomposers, they also play an important role in mobilizing essential nutrients like carbon, nitrogen, and phosphorus in soil and eventually support new plant growth.

The extensive mycelial networks of mushrooms not only enhance the decomposition of organic matter by transforming resistant material into available forms but also contribute to soil nutrient recycling. Recent studies show that the use of SMS can enhance soil microbial diversity and enzymatic activity, thereby accelerating nutrient cycling and improving overall levels of organic carbon (Calabi-Floody et al. 2024). As a consequence, fungi contribute substantially to soil fertility—the primary requirement for the successful restoration of flora in disturbed or degraded ecosystems (Lindahl and Tunlid 2015).

### 1.2.7 Carbon Sequestration and Climate Regulation

Fungal hyphal networks play a very important role in the storage of carbon within terrestrial ecosystems. These fungi contribute to long-term carbon retention in the soil by transforming organic material into stable, recalcitrant carbon compounds. Ectomycorrhizal fungi are particularly efficient in facilitating this process, especially

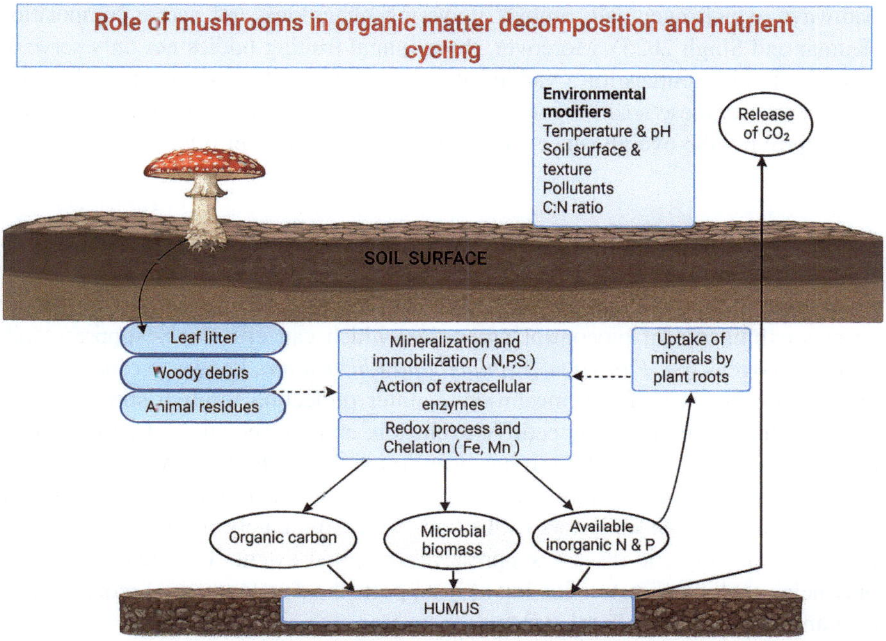


**Fig. 1.2** Organic matter decomposition and nutrient cycling by mushrooms

within forest ecosystems, where they form thick mycelial networks, augmenting the transfer of host photosynthates belowground and also producing oxidative enzymes (e.g., laccases and peroxidases). In many ECM-dominated forests, competition with saprotrophic decomposers can slow litter breakdown (the "Gadgil effect"), further enhancing soil carbon retention (Shao et al. 2023). Even after the death of these fungi, their tough cell walls containing chitin and melanin degrade after a long time and stick to the soil minerals, which lock carbon in some places. Additionally, the application of SMS to degraded soils has shown an increase in the organic carbon levels of the soil, highlighting its potential as an effective tool for carbon sequestration in degraded landscapes (Clemmensen et al. 2013; Calabi-Floody et al. 2024).

### 1.2.8 Pollinator Support and Aboveground Biodiversity

Pollinators are crucial for the sustainability and well-being of an ecosystem. Certain mushroom species, such as *Laccaria, Boletus, Russula, and Cortinarius* spp., indirectly support the well-being of pollinators and some aboveground plants by modulating their physiology through mycorrhizal associations (van der Heijden et al. 2015). They are proficient in altering nutrient availability and hormone-signaling pathways, which eventually impact flowering phenology and nectar composition (Kumar and Singh 2025). Moreover, these fungal fruiting bodies not only serve as essential food resources for a variety of invertebrates and small mammals but also offer habitat to these small organisms, thereby enriching aboveground biodiversity and improving the overall success of ecological restoration efforts.

### 1.2.9 Suppression of Soil-Borne Diseases

In addition to their roles in decomposition and nutrient cycling, certain mushroom species exhibit potent biocontrol properties, which can effectively suppress soil-borne pathogens like *Fusarium, Pythium, and Rhizoctonia*, known to cause root rot and wilting diseases. These mushrooms confer protection through stimulation of plant immune responses, competitive exclusion, and the secretion of antimicrobial compounds (Bello et al. 2025). It has been found that SMC and SMS enhance soil microbial communities by introducing some beneficial microorganisms along with a few natural pathogen inhibitors to the soil, which ultimately makes the soil more active biologically and disease-suppressive to a good extent. This dual benefit not only helps ecological restoration but also supports the development of more sustainable and resilient agricultural systems.

### 1.2.10 Biofertilizer and Bioenhancer Potential

Mushrooms are well-known for their potential as biofertilizers and soil enhancers within sustainable land use systems. Besides improving soil pH, these ecosystem warriors also enhance nutrient retention and support microbial equilibrium in soil, eventually leading to enhanced crop productivity. Such properties of mushrooms drastically reduce the reliance on synthetic fertilizers (Igual et al. 2024; He et al. 2024). Also, some AMF and endophytic fungi function as bio-stimulants, enhancing drought tolerance and boosting yields in major crops, such as cereals and legumes.

### 1.2.11 Water Management and Drought Mitigation

Mushrooms, with their numerous ecological restoration benefits, also raise the efficiency of plant water use and increase the ability of the soil to store and supply water. These mycorrhizal fungi significantly boost plant water-use efficiency by increasing root systems and accessing moisture from the soil. Their function is particularly valuable in arid and drought-prone environments or in reclaimed mining sites, where fungal inoculants have been shown to improve plant hydration and facilitate vegetation by increasing soil water-holding capacity (He et al. 2024).

### 1.2.12 Restoration of Post-mining and Urban Soils

Mushroom species like *Pisolithus* and *Rhizopogon* exhibit strong potential for postindustrial land restoration, including the rehabilitation of mine tailings, urban brownfields, and compacted soils (Frutos et al. 2017). These fungi contribute toward the restructuring of soil, mitigating toxicity and creating favorable conditions for plant reestablishment. Recent research from various regions in Asia and Europe has documented accelerated vegetation recovery in post-mining landscapes treated with SMS (Ali et al. 2025).

## 1.3 Range of Mushrooms Used in Ecological Restoration

### 1.3.1 Mycorrhizal Mushrooms

These fungi play a vital role in ecological restoration by forming symbiotic relationships with plants, especially trees. This association significantly enhances nutrient uptake, water absorption, and overall plant resilience in degraded or nutrient-poor soils. These fungi are integral to forest regeneration and are commonly used in reforestation and ecological rehabilitation programs.

Among the ectomycorrhizal species, *Pisolithus tinctorius*, *Laccaria bicolor*, *Amanita muscaria,* and some species of *Rhizopogon* have been widely documented for their beneficial associations with the roots of common trees like *Pinus, Quercus,* and *Betula.* These mutual relationships not only enhance seedling recruitment but also improve environmental stress tolerance by increasing access to nitrogen and water, which are particularly valuable in disturbed or nutrient-depleted landscapes (Smith and Read 2008).

Inoculating soils or nursery-grown seedlings with these fungi is a widely adopted approach throughout the world. *Pisolithus tinctorius* is especially effective in colonizing seedlings grown in acidic and nutrient-deficient soils. Its application has resulted in boosting survival and promoting growth under challenging post-disturbance conditions created by logging, mining, or wildfires.

### 1.3.2 Saprotrophic Mushrooms

Saprotrophic mushrooms are the most common and important decomposers in ecosystems, which break down dead organic matter and return essential nutrients to the soil. As discussed earlier, their activity not only supports nutrient cycling but also improves soil structure and fertility, making them important agents in restoring ecological balance, especially in degraded or contaminated environments.

Species like *Pleurotus ostreatus* (oyster mushroom) and *Trametes versicolor* (turkey tail mushroom) are frequently employed in mycoremediation due to their strong enzymatic systems. These fungi produce extracellular enzymes (laccases and peroxidases) capable of degrading a wide variety of environmental pollutants, including petroleum hydrocarbons, synthetic dyes, pesticides, and other persistent toxic substances (Harman 2006).

In addition to breaking down various pollutants in the soil, saprotrophic fungi contribute to waste management by decomposing agricultural residues and municipal solid waste. The resulting biomass is transformed into a nutrient-rich compost, which enhances soil health and supports vegetation regrowth. This dual function of pollution mitigation and organic waste valorization makes saprotrophic mushrooms indispensable in integrated restoration frameworks.

### 1.3.3 Ligninolytic Mushrooms

Ligninolytic fungi are proficient in breaking down the forest floor litter, composting systems, and in the remediation of contaminated soils, as they specialize in degrading lignin—a complex and recalcitrant component of plant cell walls.

A well-studied example of ligninolytic mushrooms—*Phanerochaete chrysosporium*—is known for its ability to degrade a range of persistent organic pollutants (POPs), including polychlorinated biphenyls (PCBs) and various industrial solvents. This effectiveness primarily stems from the secretion of powerful oxidative enzymes, such as lignin peroxidase and manganese peroxidase, which facilitate the degradation of stable chemical compounds resistant to conventional remediation methods (Akpasi et al. 2023).

Another noteworthy species is *Ganoderma lucidum* (reishi mushroom), one of the common medicinal mushrooms found to have antioxidant, anti-inflammatory, and immune-boosting effects. This fungus has demonstrated the ability to restore ecosystems through its wood-degrading abilities and potential for bioremediation (or mycoremediation). These mushrooms are able to absorb heavy metals and stabilize soil structure, making them particularly valuable in restoring lands degraded by either industrial operations or mining activities.

### 1.3.4 Endophytic and Experimental Mushrooms

These fungi form colonies inside plant tissues and promote their growth, stress resilience, and overall plant health, suggesting a novel symbiotic relationship extending beyond the traditional root–soil interface. *Pleurotus,* one such genus, is shown to enhance the tolerance of host plants to common abiotic stresses like drought, salinity, or heavy metal exposure. When these mushrooms are grown along with restoration species, they extensively support the restoration process, especially in degraded agricultural landscapes and arid zones. Regrettably, our understanding of the restoration potential of endophytic mushrooms across varied landscapes is limited. Continued field experimentation and long-term ecological monitoring are essential in order to integrate these amazing species into ecological restoration practices.

## 1.4 Different Methods for Applying Mushrooms in Eco-Restoration

Mushrooms are well known for using their expansive mycelial networks and fruiting bodies to offer benefits in ecological restoration. The advantages include enhancing soil stability, degrading pollutants, supporting plant establishment, and increasing microbial diversity. Some common methods employing mushrooms for ecological restoration are discussed below and illustrated in Fig. 1.3.

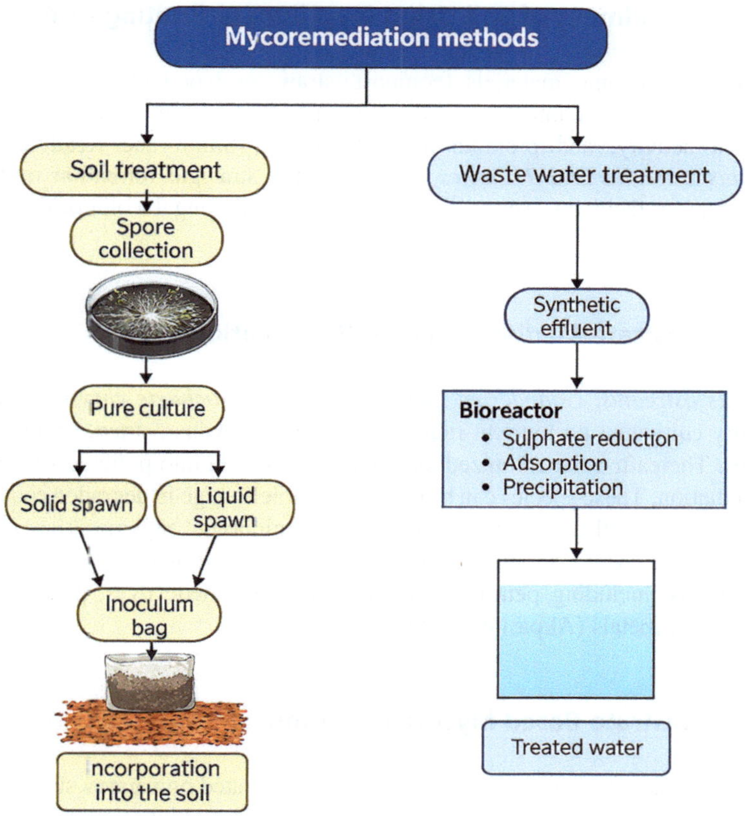


**Fig. 1.3** Mycoremediation methods

### 1.4.1 Inoculation of Seedlings with Mycorrhizal Mushroom Species

Introduction of ectomycorrhizal fungi, such as *Pisolithus tinctorius, Laccaria bicolor, Amanita muscaria,* and some species of *Rhizopogon,* in nursery-grown seedlings during reforestation and afforestation efforts has shown significant improvement in the survival of seedlings along with their nutrient uptake and resilience to abiotic stress (Smith and Read 2008; Himanen et al. 2024). Early establishment of such symbiotic relationships proves particularly beneficial in nutrient-poor or disturbed soils.

### 1.4.2 Amendment of Soil Using Mycelium or Fruiting Bodies

Incorporating organic materials (compost, straw, or wood chips) colonized by mushroom mycelium into degraded soils can enhance soil structure, stimulate microbial activity, and promote nutrient cycling (Smith and Read 2008). In *Pleurotus ostreatus* and *Trametes versicolor*, the strategic formation of fruiting bodies further facilitates spore dispersal and natural fungal colonization (Wang and Tang 2022).

### 1.4.3 Substrate-Mediated Fungal Remediation

*Pleurotus ostreatus*, *Ganoderma lucidum*, and *Phanerochaete chrysosporium* are generally cultivated on nutrient-rich substrates like sawdust, straw, or agricultural residues. Thereafter, the colonized material is introduced into polluted sites in need of remediation. These species can bring considerable change in degraded ecosystems owing to their ability to produce powerful ligninolytic enzymes (laccases and manganese peroxidases), which are able to break down a wide range of contaminants, including petroleum hydrocarbons, synthetic dyes, pesticides, and certain heavy metals (Akpasi et al. 2023).

### 1.4.4 Substrate-Based Mycoremediation

As the name suggests, different mushrooms are cultivated on various substrates like sawdust, straw, or agricultural residues and later introduced into polluted sites. Here too, species like *Pleurotus ostreatus, Ganoderma lucidum,* and *Phanerochaete chrysosporium* are particularly effective due to their production of ligninolytic enzymes, which degrade petroleum hydrocarbons, synthetic dyes, pesticides, and heavy metals (Akpasi et al. 2023).

### 1.4.5 Dispersal of Fruiting Bodies for Passive Colonization

Spreading fruiting bodies of fresh or dried mushrooms in urban areas or in restoration sites facilitates passive spore dispersal. The practice increases fungal diversity and supports long-term ecosystem colonization. This low-input approach not only enhances ecological connectivity and resilience in a given area but also passively augments the rehabilitation and establishment of plant species in that area (Neuenkamp et al. 2019; Wang and Tang 2022).

### 1.4.6 Integrated Application with Vegetative and Soil Restoration Practices

Fungal inoculation is frequently combined with complementary restoration approaches, including native vegetation planting, erosion control, and organic matter enrichment. This integrated method, besides generating synergistic effects that accelerate soil recovery, also supports plant establishment and strengthens the overall ecosystem resilience (Smith and Read 2008; Neuenkamp et al. 2019).

### 1.4.7 Wastewater Treatment Using Mushrooms

Edible mushroom species such as *Pleurotus ostreatus* and *P. eryngii* have demonstrated remarkable potential in treating wastewaters contaminated with heavy metals and synthetic dyes. These fungi exhibit high biosorption capacities, especially absorbing metals like cadmium, lead, and chromium (Goligar et al. 2023). Besides their exceptionally high biosorption potential, these species can degrade synthetic dyes chemically (Chaudhary et al. 2023). Together, these traits make them a cost-effective and sustainable solution for effluent remediation (Johnnie et al. 2023).

## 1.5 Various Mechanisms Employed in Mycoremediation

### 1.5.1 Enzymatic Degradation

This involves the degradation of complex organic pollutants like polycyclic aromatic hydrocarbons (PAHs), petroleum derivatives, synthetic chemicals, and industrial dyes using extracellular enzymes (laccase, lignin peroxidase, and manganese peroxidase) produced by different species of white-rot fungi. These enzymes empower the fungi to degrade pollutants effectively. Chaudhary et al. (2023) reported the ability of SMS derived from *Agaricus bisporus* to remove 68%–90% of pharmaceutical contaminants via enzymatic activity along with Fenton-like oxidative reactions. In a related study, SMS from *Pleurotus eryngii* has been demonstrated to successfully

**Table 1.1** Summary of different mechanisms of mycoremediation

| Mechanisms | Description | Key fungi/ materials | References |
|---|---|---|---|
| Enzymatic degradation | Fungi secrete extracellular enzymes (laccase, lignin peroxidase, and manganese peroxidase) that degrade complex organic pollutants like PAHs, pharmaceuticals, dyes, and synthetic chemicals | *Agaricus bisporus, Pleurotus eryngii* | Chaudhary et al. (2023); Khan (2025) |
| Absorption / biosorption | Heavy metals are passively bound to fungal cell walls and extracellular polysaccharides via functional groups (e.g., –OH, –COOH), resulting in a reduced concentration of metals from soil or water | *Pleurotus tuber-regium, P. eryngii* | Afolabi et al. (2024); Omoregie et al. (2025) |
| Metal immobilization (SMC) | Biochar or compost derived from spent mushroom compost (SMC) or substrate (SMS) alters soil chemistry (pH, porosity) and binds metals through surface complexation and electrostatic interactions, reducing the bioavailability and mobility of heavy metals | *SMC-derived biochar* | Zhang et al. (2020); Madzin et al. (2025) |

degrade pharmaceutical compounds like acetaminophen and sulfonamides in a bioreactor system, with the mechanisms validated through enzymatic assays and metaproteomic profiling (Amin et al. 2018) (Table 1.1).

### 1.5.2 Absorption and Accumulation

Fungal mycelia have a unique capability of sequestering heavy metals (lead, copper, nickel, and zinc) through two primary mechanisms-biosorption and bioaccumulation. Whereas biosorption involves the adherence of metal ions to fungal cell walls and extracellular compounds, bioaccumulation is the uptake and storage of these metal ions by the fungal cell. Research conducted on *Pleurotus tuber-regium* highlights the ability of the mushrooms to remove up to 90% of heavy metals (including Pb, Zn, Cu, and Mn) from contaminated soils (Afolabi et al. 2024; Omoregie et al. 2025). Additionally, *Pleurotus eryngii* exhibited significant accumulation of Pb and Cd in its fruiting bodies (Goligar et al. 2023).

### 1.5.3 Metal Immobilization via Spent Mushroom Compost

Spent mushroom compost (SMC) and biochar derived from spent mushroom substrate (SMS) help in the immobilization of heavy metals in soils through multiple mechanisms, like regulation of pH (influences the solubility of metals), introduction of reactive fungal groups (readily bind with the metal ions), and improvement in porosity and stability of soils. SMC-derived biochar has been shown to effectively adsorb Cu and Pb from mining effluents via electrostatic interactions and surface complexation (Zhang et al. 2020). The biochar produced from SMC at 600 °C exhibits a high adsorption capacity for copper, which can range from 52 to 66 mg/g (Madzin et al. 2025). Further, it has been delineated that composted SMC reduced the bioavailability of Cd and Zn in agricultural soils by approximately 48%, subsequently enhancing the potential for safe and sustainable land reuse (Ramlan and Adam 2022).

## 1.6 Some Recent Case Studies

### 1.6.1 Oil Spill Bioremediation in Estuarine Zones with *Pleurotus ostreatus*

*Pleurotus ostreatus* has an inherent capability of degrading saturate and asphaltene fractions of Louisiana sweet crude oil in estuarine habitats where salinity levels range between 5% and 15%. Both ammonium chloride and potassium phosphate augment the degradation process, and results are visible in just seven days (Crescent et al. 2025).

### 1.6.2 Reforestation of Post-mining Sites Using Some *Pisolithus* and *Rhizopogon* spp.

The reforestation of mining-damaged land is made more effective by inoculating the planted trees with fungi like *Pisolithus* and *Rhizopogon*, as they help the trees to survive and grow in harsh, nutrient-poor soils. The vigor in reforestation is attributed to increased tolerance to heavy metal toxicity and enhanced nutrient uptake by these mushrooms (Smith and Read 2008).

### 1.6.3 Biosorption of Heavy Metals

The biosorption potential of *Pleurotus eryngii* for copper in contaminated wastewater was elucidated by Kan et al. (2015). Their assessment divulged a huge copper uptake of 15.19 mg/g by the mushrooms, highlighting their application in industrial wastewater treatment. *P. eryngii* also demonstrated a strong capacity to survive in aqueous solutions containing heavy metals like lead and cadmium. A substantial reduction in heavy metal concentrations was recorded in wastewaters employing *P. eryngii* (Goligar et al. 2023).

**Table 1.2** Review of studies using different mushrooms for bioremediation

| S. No. | Species | Contaminant | Environment/ context | Key outcome | References |
|---|---|---|---|---|---|
| 1. | *Pleurotus ostreatus* | Crude oil | Estuarine conditions | Effective degradation of hydrocarbons | Crescent et al. (2025) |
| 2. | *P. eryngii* | $Pb^{2+}$, $Cd^{2+}$ | Wastewater | Enzyme activation and biosorption | Goligar et al. (2023) |
| 3. | *P. eryngii* | $Cd^{2+}$ | Contaminated water | Quick and efficient removal of cadmium | Amin et al. (2018) |
| 4. | *P. eryngii* | $Cu^{2+}$ | Industrial effluent | High copper uptake (15.19 mg/g) | Kan et al. (2015) |
| 5. | *Pleurotus* spp. | $Co^{2+}$, $Cu^{2+}$, $Ni^{2+}$ | Contaminated soil/water | High accumulation and tolerance toward heavy metals | Mohamadhasani and Rahimi (2022) |

### 1.6.4 Metal Tolerance and Accumulation Capacity of *Pleurotus* spp.

Apart from biosorption of heavy metals, some species of *Pleurotus* also exhibit elevated antioxidant enzyme activity, which supports their adaptation to metal-stressed environments. *Pleurotus* species cultivated in soils contaminated with cobalt, nickel, and copper have shown robust metal tolerance and significant bioaccumulation ability (Mohamadhasani and Rahimi 2022).

Cadmium in water bodies is a major threat to biodiversity these days. Applying dried biomass of *Pleurotus eryngii* to contaminated water removes up to 99.9% of cadmium ions within 10 minutes under controlled conditions of pH 5 at an optimal temperature of 500 °C (Amin et al. 2018) (Table 1.2).

## 1.7 Overcoming Key Restoration Challenges with Mushrooms in Degraded Ecosystems

### 1.7.1 Post-mining Site Restoration

Post-mining landscapes are typically marked by severely degraded soils with low organic matter, compaction, high acidity from acid mine drainage, and elevated concentrations of toxic heavy metals like aluminum, lead, and cadmium. These conditions are further worsened by the depletion of native microbial populations, making natural recovery extremely slow.

### 1.7.2 Mushroom-Based Approaches

Ectomycorrhizal fungi such as *Pisolithus tinctorius* and *Rhizopogon* spp. play a vital role in overcoming the harsh conditions of degraded ecosystems. By forming mutualistic associations with tree roots, they enhance nutrient uptake and improve plant survival (Bhantana et al. 2021). These fungi secrete organic acids that bind heavy metals and subsequently reduce the solubility and toxicity of these metals. They also sequester metals within their tissues or transform them into less harmful chemical forms, aiding the detoxification of post-mining sites (Yang et al. 2018). Additionally, the application of SMC has been shown to improve soil structure, neutralize acidity, enhance water retention, and stimulate the recovery of microbial communities (Frutos et al. 2017).

### 1.7.3 Deforested Area Restoration

Deforestation leads to the loss of tree canopy and leaf-litter inputs, disruption of soil microbial communities, enhanced soil erosion, and increased susceptibility to invasive species. These disturbances destabilize ecosystem functions and reduce resilience. Introducing mycorrhizal fungi such as *Laccaria* and *Scleroderma* spp. helps rebuild essential belowground symbiotic networks, which enhance seedling establishment, restore nutrient cycling, and contribute to improved ecosystem resilience in degraded forest soils (Patel et al. 2023).

### 1.7.4 Restoration of Nutrient-Poor Soils

Nutrient-deficient soils, especially those lacking nitrogen, phosphorus, and micronutrients (zinc and copper), show poor physical structure, low microbial activity, and reduced water-holding capacity, making them highly unproductive. The addition of SMC further enriches soil fertility by increasing microbial biomass, improving water-holding capacity, and enhancing cation-exchange capacity (Liu et al. 2021). Moreover, saprotrophic fungi decompose plant residues, release immobilized nutrients, and increase soil organic carbon, thereby strengthening soil fertility and long-term ecosystem recovery (Rao and Singh 2021).

## 1.8 Economic Impact of Utilizing Mushrooms in Eco-Restoration

### 1.8.1 Cost-Effective and Environmentally Sustainable

Recent research demonstrates that fungal remediation significantly reduces overall operational expenses while achieving better environmental outcomes (Omoregie et al. 2025). Species like *P. ostreatus* demonstrate a strong capacity for hydrocarbon degradation and heavy metal sequestration, thereby providing a low-impact and eco-friendly solution (Periakaruppan et al. 2025). Hence, we can conclude that mushroom-based soil remediation offers a cost-effective alternative to conventional approaches using chemicals or soil excavation.

### 1.8.2 Resource-Efficient and Circular Waste Management

Mycoremediation aligns with circular economy principles by promoting sustainable resource utilization. SMS, which is a by-product of mushroom cultivation, can be used as a fertilizer, bioenergy feedstock, or even livestock feed, thereby minimizing agricultural waste and promoting nutrient recycling (Leong et al. 2022). Furthermore, using agro-industrial residues as mushroom-growing media reduces organic waste while generating value-added biomass (Dulta et al. 2024).

### 1.8.3 Generation of Marketable By-Products

Mushrooms used in remediation strategies yield commercially valuable lignin-degrading and cellulose-degrading enzymes (ligninase and cellulase) over and above bioethanol and other biochemicals (Viriato et al. 2024). Also, SMC increases agricultural benefits by improving soil fertility and supporting higher crop productivity, ultimately reducing the overall dependency on chemical fertilizers (Mwangi et al. 2024).

### 1.8.4 Environmental and Public Health Benefits

The use of fungi-based strategies of restoration instead of chemical-intensive remediation reduces environmental contamination and human exposure to toxic substances (Mwangi et al. 2024). These practices not only increase the biodiversity, enhance soil structure, and improve water quality but also contribute to long-term agricultural and ecological sustainability besides reinforcing ecosystem services (Viriato et al. 2024).

### 1.8.5 Market Expansion and Employment Generation

Nowadays, incorporating mushroom-based solutions into eco-restoration programs creates employment in mushroom farming, their processing, and product innovation, particularly in rural regions (Leong et al. 2022). Also, a rising demand for sustainable, organic, and environmentally responsible products is creating new markets for innovative products derived from mushrooms (Viriato et al. 2024).

### 1.8.6 Barriers and Implementation Challenges

Although mushrooms serve as modest yet powerful agents of eco-restoration, mushroom-based remediation practices face numerous challenges. At present, investment in infrastructure, resources, and specialized training remains scarce (Martín et al. 2023). Furthermore, policy shortcomings, low public awareness, and cultural resistance continue to hinder the broader adoption of mushroom-based approaches, particularly in developing nations.

## 1.9 Conclusion

Mushrooms, the real crusaders of ecosystem recovery, play many important roles in restoring damaged environments. They not only break down organic waste but also recycle nutrients, improve soil structure, form helpful partnerships with plants, remove harmful pollutants, support wildlife, and even help store carbon in the soil. Unlike many organisms, they can perform multiple restoration tasks at once by connecting with plant roots to share nutrients, breaking down tough substances like wood and chemicals, and helping the soil hold more water. Their underground networks, called mycelium, connect plants, soil, and microbes, eventually creating a living system that helps ecosystems rebuild themselves. Tools like spent mushroom substrate (SMS) and fungal inoculants are now being used successfully in degraded soils to boost plant growth, improve soil quality, and speed up natural recovery. These modest organisms have shown their usefulness in many different environments, from burned forests and abandoned mines to drylands and polluted cities. Their ability to clean soil, restore fertility, and support diverse plant and animal life makes them one of the most valuable tools in modern restoration science. They also help combat climate change by storing carbon in the soil and supporting healthy vegetation that can absorb still more carbon from the atmosphere.

This chapter emphasizes the growing recognition among scientists and restoration practitioners of these unassuming ecosystem builders. In the absence of functional fungal networks, soils remain impoverished, plant establishment is compromised, and ecosystem recovery remains incomplete. Conversely, when fungi act as primary agents of restoration, ecosystems demonstrate accelerated recovery, enhanced biodiversity, improved plant vigor, and greater soil stability. Fungi have been regarded as nature's original healers, facilitating ecological regeneration for millions of years and enabling life to rebound after disturbances like wildfires and climatic fluctuations.

Fungi provide natural, effective, and economically sustainable solutions. Recognizing their ecological value and integrating them as central components of restoration frameworks is, therefore, not optional but essential. Ultimately, the pathway to resilient ecosystems begins underground, driven by the subtle yet powerful influence of fungal networks.

**Acknowledgments** The authors are thankful to the Principal, Acharya Narendra Dev College, for constant motivation and support. Preeti Yadav (Research Scholar) acknowledges the CSIR for the Junior Research Fellowship.

**Declaration** We hereby declare that this chapter is an original review work carried out by the authors and has not been published elsewhere in any form. The authors confirm that there are no conflicts of interest, and the manuscript complies with ethical standards and copyright requirements of Springer.

## References

Afolabi FJ, Babaniyi RB, Obagunwa M, Onile FO, Ajagun EJ, Oyedele OJ, Fasiku SA (2024) Ethnomycology of *Pleurotus tuber regium* and its use in food, medicine and bioremediation, Botanica, pp 183–192

Akpasi SO, Anekwe IMS, Tetteh EK, Amune UO, Shoyiga HO, Mahlangu TP, Kiambi SL (2023) Myco-remediation as a potentially promising technology: current status and prospects—a review. Appl Sci 13:4978

Ali Z, Abdullah M, Hafeez F, Alasmari A, Sardar F, Ahmad J et al (2025) Advances in myco-remediation strategies for emerging contaminants: a comprehensive review. Biorem J:1–37

Amin F, Talpur FN, Balouch A, Samoon MK, Afridi HI, Surhio MA (2018) Utilization of *Pleurotus eryngii* biosorbent as an environmental bioremedy for the decontamination of trace cadmium (II) ions from the water system. Water Sci Technol 78:1148–1158

Anandakumar S, Kalaiselvi T, Kuttimani R, Jeyakumar P, Ravichandran V (2025) Glomalin-related soil protein: unlocking its potential as a soil conditioner and plant biostimulant. In: Management of mycorrhizal symbiosis for mycoremediation and phytostabilization. Elsevier, pp 73–90

Bello AB, Dabban AI, Muhammad WM, Anarachi A, Oko JO, Opeyemi H, Oyewole OA, Adetunji CO (2025) Application of mushrooms in management of soil-borne parasites, nematodes, bacteria and fungi. In: Adetunji CO, Egbuna C, Ijabadeniyi OA, Karunarathna SC (eds) Mushroom biotechnology for improved agriculture and human health. Wiley, pp 85–122

Bhantana P, Rana MS, Sun X, Moussa MG, Saleem MH et al (2021) Arbuscular mycorrhizal fungi and their major role in plant growth, zinc nutrition, phosphorus regulation and phytoremediation. Symbiosis 84:19–37

Calabi-Floody M, Medina J, Richardson AE, Lecaros IZ, Mora-Perez L et al (2024) Enhancing soil carbon sequestration with C-rich carrier materials from spent mushroom substrate and composted wheat straw: implications for smart fertilizer design. J Soil Sci Plant Nutr 24: 6691–6706

Carrara JE, Lehotay SJ, Lightfield AR, Sun D, Smith AH, Richie JP, Heller WP (2023) Linking soil health to human health: Arbuscular mycorrhizae play a key role in plant uptake of the antioxidant ergothioneine from soils. Plants People Planet 5:449–458

Chaudhary N, Kumar P, Kumar A, Dayal S, Rani A, Ahamad F (2023) Efficiency of spent mushroom (*Agaricus bisporus*) waste biomass for the biosorption of basic fuchsin dye from aqueous solution. Environ Conserv J 24:300–309

Clemmensen KE, Bahr A, Ovaskainen O, Dehlberg A, Ekblad A, Wallander H et al (2013) Roots and associated fungi drive Long-term carbon sequestration in Boreal Forest. Science 339:1615–1618

Crescent SS, Pisarski EC, Wirth EF, Long RA (2025) Myco-remediation of Louisiana sweet crude oil with Pleurotus ostreatus and nutrient amendments. Environ Toxicol Chem 44:1154–1163

Dulta K, Awasthi YK, Aman J, Khirwar R, Kulwanshi S, Kumar K (2024) Dual application of waste treatment and fungal cultivation/metabolite production. In: Uppuluri KB, Selvasembian R (eds) Bioprospecting of multi-tasking fungi for a sustainable environment. Springer Nature Singapore, Singapore, pp 361–377

Frutos I, García-Delgado C, Cala V, Gárate A, Eymar E (2017) The use of spent mushroom compost to enhance the ability of *Atriplex halimus* to phytoremediate contaminated mine soils. Environ Technol 38:1075–1084

Goligar N, Saadatmand S, Khavarinejad RA (2023) Myco-remediation of lead and cadmium by lignocellulosic enzymes of *Pleurotus eryngii*. AMB Express 13:127

Harman GE (2006) Overview of mechanisms and uses of *Trichoderma* spp. Phytopathology 96: 190–194

Hawkes CV, Keitt TH (2015) Resilience vs. historical contingency in microbial responses to environmental change (A Classen, Ed.). Ecol Lett 18:612–625

He J, Zeng G, Liu Z, Guo Z, Zhang W, Li Y, Zhou Y, Xu H (2024) Replacing traditional nursery soil with spent mushroom substrate improves rice seedling quality and soil substrate properties. Environ Sci Pollut Res 31:39625–39636

Himanen K, Nygren M, Pennanen T (2024) Mycelial inoculation of containerized Norway spruce seedlings with ectomycorrhizal fungi. New For 55:47–61

Hu Y, Mortimer PE, Hyde KD, Kakumyan P, Thongklang N (2021) Mushroom cultivation for soil amendment and bioremediation. Circ Agric Syst 1:1–14

Igual JM, Andrades MS, Frade C, Carpio MJ, Marin-Benito MJ, Rodriguez-Cruz MS, Valverde A (2024) Impacts of spent mushroom substrate amendment and season on microbial communities in a semiarid vineyard soil. Appl Soil Ecol 203:105689

Johnnie DA, Issac R, Prabha ML, Gomez LA (2023) Biodegradation assay of heavy metals and dyes decolorization in textile industrial effluent using laccase isolated from *Pleurotus ostreatus*. J Pure Appl Microbiol 17:2324–2343

Kan S-H, Sun B-Y, Xu F, Song Q-X, Zhang S-F (2015) Biosorption of aquatic copper (II) by mushroom biomass *Pleurotus eryngii*: kinetic and isotherm studies. Water Sci Technol 71:283–288

Khan MF (2025) Fungi for sustainable pharmaceutical remediation: enzymatic innovations, challenges, and applications—a review. Processes 13:1034

Kulshreshtha S, Mathur N, Bhatnagar P (2014) Mushroom as a product and their role in myco-remediation. AMB Express 4:29

Kumar P, Singh J (2025) Plant-associated fungi. In: Natural bioactives from the endophytes of medicinal plants. CRC Press, Boca Raton, pp 213–228

Leong YK, Ma T-W, Chang J-S, Yang F-C (2022) Recent advances and future directions on the valorization of spent mushroom substrate (SMS): a review. Bioresour Technol 344:126157

Lindahl BD, Tunlid A (2015) Ectomycorrhizal fungi—potential organic matter decomposers, yet not saprotrophs. New Phytol 205:1443–1447

Liu X, Zhang X, Li R, Wang J, Jin Y, Xu W, Wang H, Qu J (2021) Organic amendment improves rhizosphere environment and shapes soil bacterial community in black and red soil under lead stress. J Hazard Mater 416:125805

Ma X, Han F, Yang G, Wu J, Ma Y (2025) Enhanced soil ecosystem multifunctionality and microbial community shifts following spent mushroom substrate application in vineyards. Appl Soil Ecol 213:106230

Madzin Z, Zahidi I, Talei A, Raghunandan ME, Hermawan AA, Karam DS (2025) Optimising spent mushroom compost biochar for heavy metal removal: mechanisms and kinetics in mine water treatment. J Water Process Eng 69:106829

Malik NA, Kumar J, Wani MS, Tantray YR, Ahmad T (2021) Role of mushrooms in the bioremediation of soil. In: Dar GH, Bhat RA, Mehmood MA, Hakeem KR (eds) Microbiota and biofertilizers, vol 2. Springer International Publishing, Cham, pp 77–102

Martín C, Zervakis GI, Xiong S, Koutrotsios G, Straetkvern KO (2023) Spent substrate from mushroom cultivation: exploitation potential toward various applications and value-added products. Bioengineered 14:2252138

Matos PS, Silva CFD, Damian JM, Cerri CEP, Pereira MG, Zonta E (2022) Beneficial services of Glomalin and Arbuscular Mycorrhizal fungi in degraded soils in Brazil. Sci Agric 79: e20210064

Mishra P, Tiwari A, Dubey VK, Pandey GC, Srivastava G (2024) Arbuscular Mycorrhizal fungi (AMF): a natural tool for phytoremediation of heavy metals (HMs). In: Madhav S, Gupta GP, Yadav RK, Mishra R, Hullebusch EV (eds) Phytoremediation. Springer Nature Switzerland, Cham, pp 225–243

Mohamadhasani F, Rahimi M (2022) Assessment of heavy metal tolerance and accumulation in *Pleurotus* spp. for environmental restoration. Environ Health Eng Manag 9(4):263–271

Mwangi JK, Otieno FO, Wamalwa KM (2024) Soil enhancement through spent mushroom compost: implications for sustainable farming. Agric Soil Sci Rev 11(1):54–63

Neuenkamp L, Prober SM, Price JN, Zobel M, Standish RJ (2019) Benefits of mycorrhizal inoculation to ecological restoration depend on plant functional type, restoration context and time. Fungal Ecol 40:140–149

Omoregie EO, Uka CJ, Bello SA (2025) Economic evaluation of myco-remediation versus conventional remediation techniques. Int J Environ Econ 7(1):37–49

Patel S, Nair R, Menon S (2023) Mycorrhizal inoculation enhances seedling survival in Western Ghats reforestation programs. For Ecol and Manag 525:120452

Pérez-Moreno J, Guerin-Laguette A, Rinaldi AC, Yu F, Verbeken A, Hernandez-Santiago F, Martinez-Reyes M (2021) Edible mycorrhizal fungi of the world: what is their role in forest sustainability, food security, biocultural conservation and climate change? Plants People Planet 3:471–490

Periakaruppan R, Vanathi P, Priyanka G (2025) Mycoremediation—a sustainable clear technology for environmental remediation. In: Jeyaseelan A, Murugasen K (eds) Environmental science and engineering. Sustainable environmental remediation: avenues in nano and biotechnology. Springer Nature Switzerland, Cham, pp 321–351

Ramlan NNS, Adam S (2022) The effect of application of spent mushroom waste on cd-polluted soil. IOP Conf Ser Earth Environ Sci 1059:012033

Rao V, Singh M (2021) Role of saprotrophic fungi in nutrient cycling and organic matter decomposition in tropical agroforestry systems. Fungal Ecol 54:101111

Rillig MC, Mummey DL (2006) Mycorrhizas and soil structure. New Phytol 171:41–53

Roshandel F, Saadatmand S, Iranbakhsh A, Ardebili ZO (2021) Mycoremediation of oil contaminant by Pleurotus Florida (P.Kumm) in liquid culture. Fungal Biol 125:667–678

Shao S, Wurzburger N, Sulman B, Hicks Pries C (2023) Ectomycorrhizal effects on decomposition are highly dependent on fungal traits, climate, and litter properties: a model-based assessment. Soil Biol Biochem 184:109073

Smith SE, Read DJ (2008) Mycorrhizal Symbiosis, 3rd edn. Academic Press. Comprehensive resource on mycorrhizal fungi and their ecological roles, including restoration applications

Vaksmaa A, Guerrero-Cruz S, Ghosh P, Zeghal E, Hernando-Morales V, Niemann H (2023) Role of fungi in bioremediation of emerging pollutants. Front Mar Sci 10:1070905

van der Heijden MGA, Martin FM, Selosse M-A, Sanders IR (2015) Mycorrhizal ecology and evolution: the past, the present, and the future. New Phytol 205:1406–1423

Viriato V, Carvalho SADD, Santoro BDL, Bonfim FPG (2024) A business model for circular bioeconomy: edible mushroom production and its alignment with the sustainable development goals (SDGs). Recycling 9:68

Wang L, Tang Z (2022) How do arbuscular mycorrhizas affect the reproductive functional fitness of host plants? Front Plant Sci 13

Wang M, Sun D, Xu Z (2025) Effects of spent mushroom substrate treated with plant growth-promoting Rhizobacteria on blueberry growth and soil quality. Microorganisms 13:932

Yang Z, Liang L, Yang W, Shi W, Tong Y, Chai L, Gao S, Qi L (2018) Simultaneous immobilization of cadmium and lead in contaminated soils by hybrid bio-nanocomposites of fungal hyphae and nano-hydroxyapatites. Environ Sci Pollut Res 25:11970–11980

Zhang G, Liu N, Luo Y, Zhang H, Su L, Oh K, Cheng H (2020) Efficient removal of Cu (II), Zn (II), and Cd (II) from aqueous solutions by a mineral-rich biochar derived from a spent mushroom (*Agaricus bisporus*) substrate. Materials 14:35

# Ecology and Diversity of Edible and Medicinal Mushrooms Across Different Habitats

2

José Manuel Costa, Maria Isabel Santos, and Raquel P. F. Guiné

**Abstract**

Edible and medicinal mushrooms represent a unique intersection of biodiversity, ecology, culture, and biotechnology. This chapter explores their taxonomic and ecological diversity across habitats, emphasizing both their functional ecosystem roles and socioeconomic relevance. The taxonomic spectrum of mushroom-forming fungi spans Basidiomycota and Ascomycota, with major orders such as Agaricales, Boletales, and Russulales encompassing many key edible and medicinal taxa. Methods of identification, ranging from traditional macroscopic observation to molecular phylogenetics and genomics, are discussed as essential tools for reliable classification and sustainable utilization. Ecologically, mushrooms function as mycorrhizal symbionts, saprotrophs, and parasites, driving nutrient cycling, soil fertility, and forest regeneration. Their habitat-specific distribution is illustrated through case studies from temperate, tropical, alpine, grassland, and urban ecosystems, each characterized by emblematic species and unique ecological dynamics. Environmental drivers, including climate, soil properties, host plant associations, and disturbance regimes, are identified as key factors shaping fungal diversity patterns. From a human perspective, mushrooms are valuable resources for nutrition, food security, and healthcare. Cultivated species, such as *Agaricus bisporus*, *Pleurotus ostreatus*, and *Lentinula edodes*, exemplify sustainable production systems, while wild species sustain local economies, cultural traditions, and ethnomedicinal practices.

J. M. Costa (✉) · R. P. F. Guiné
Agrarian School of Viseu, Polytechnic University of Viseu, Viseu, Portugal

CERNAS—IPV Research Centre, Polytechnic University of Viseu, Viseu, Portugal
e-mail: jmcosta@esav.ipv.pt

M. I. Santos
Agrarian School of Viseu, Polytechnic University of Viseu, Viseu, Portugal



B. Giri et al. (eds.), *Advances in Mushroom Research*,
https://doi.org/10.1007/978-981-95-7692-0_2

Medicinal fungi, including *Ganoderma lucidum* and *Cordyceps sinensis*, highlight the pharmacological potential of bioactive compounds with antioxidant, immunomodulatory, and anticancer properties. Finally, this chapter addresses future perspectives in fungal genomics, biotechnology, and integration into agroforestry and ecosystem restoration. Remaining challenges, such as research gaps, threatened species conservation, and risks of toxicity or misidentification, remain crucial for interdisciplinary collaboration. Overall, mushrooms emerge as ecological keystones and socioeconomic assets essential to both planetary health and sustainable development.

**Keywords**

Edible mushrooms · Medicinal fungi · Fungal ecology · Biodiversity · Ethnomycology

## 2.1 Introduction

Fungi are one of the most diverse and ecologically relevant kingdoms of living beings, performing essential functions in terrestrial and aquatic ecosystems (Tedersoo et al. 2014; Hawksworth and Lücking 2017). Despite their importance, they remain comparatively underexplored relative to plants and animals, both in biodiversity assessment and biotechnological potential (Hyde et al. 2019). Recent estimates suggest that fungal diversity may range from 2.2 to 3.8 million species, highlighting the vast unknown potential of this kingdom (Hawksworth and Lücking 2017). This chapter aims to deepen knowledge about taxonomic diversity, ecology, distribution, and potential use of edible and medicinal fungi from an integrated perspective of biology, ecology, ethnomycology, and biotechnology (Agrawal and Dhanasekaran 2019).

The growing recognition of mushrooms as sustainable sources of food and bioactive compounds has driven scientific and industrial interest (Valverde et al. 2015; Hyde et al. 2019). Edible fungi are valued for their high nutritional quality and adaptability to sustainable cultivation, while medicinal species offer bioactive molecules with immunomodulatory, antioxidant, and therapeutic potential (Agrawal and Dhanasekaran 2019). At the same time, fungi sustain ecosystem functioning through decomposition, mycorrhizal symbioses, and the regulation of plant communities (Tedersoo et al. 2014). Understanding the factors that influence the diversity and distribution of fungi—including environmental variables, soil characteristics, biotic interactions, and ecological disturbances—is fundamental to their conservation and valorization (Tedersoo et al. 2014; Sánchez-Hernández et al. 2025). Accurate taxonomic identification, supported by traditional knowledge and genomic tools, provides the foundation for conserving fungal diversity and integrating it into global sustainability strategies (Nilsson et al. 2019; Hyde et al. 2019; Roux and Eyssartier 2024).

Edible and medicinal mushrooms represent a unique interface between biodiversity, ecological functionality, and sustainable economics (Hyde et al. 2019; Pérez-Moreno 2021). Ecologically, fungi are key decomposers, driving nutrient cycling and soil formation (Tedersoo et al. 2014; Heng 2016). Mycorrhizal species enhance plant uptake of water and minerals, while saprotrophic fungi recycle carbon and essential elements (Aerts 2003; Heng 2016). These processes sustain the productivity and resilience of natural and agricultural systems. Economically, edible mushrooms provide nutritious food, rich in protein, fiber, vitamins, and minerals but low in fat (Valverde et al. 2015; López-Hortas et al. 2022), and their cultivation contributes to rural development and food security, expanding globally across diverse production systems (Hyde et al. 2019; Saldanha et al. 2023).

Commonly farmed species include *Agaricus bisporus*, *Pleurotus ostreatus*, and *Lentinula edodes*, while wild taxa such as *Boletus*, *Cantharellus*, and *Tuber* (truffles) hold exceptional gastronomic and market value (Pérez-Moreno 2021). On the other hand, medicinal mushrooms, such as *Ganoderma lucidum*, *Cordyceps sinensis,* and *Hericium erinaceus*, have long histories in traditional medicine and are increasingly recognized in modern therapies for their antioxidant, immunomodulatory, and anticancer compounds (Venturella et al. 2021). The valorization of these resources requires an integrated approach that simultaneously recognizes their ecological role, their economic potential, and the need to conserve fungal diversity (Hyde et al. 2019).

This chapter aims to explore the diversity, ecology, distribution, and applications of edible and medicinal fungi, highlighting their multifunctional role in ecosystems and society. The main taxonomic groups of food and therapeutic interest will be presented, with an emphasis on emblematic species and their morphological and biochemical characteristics. Identification and classification methods, including morphological, molecular, and phylogenetic techniques, will also be addressed. Ecological aspects are explored through the trophic roles of fungi as mycorrhizal symbionts, saprotrophs, and parasites, underscoring their contribution to nutrient cycling, soil fertility, and ecosystem stability. Factors influencing fungal growth and distribution, including climate, soil properties, host plants, and natural or anthropogenic disturbances, are analyzed across temperate, tropical, alpine, agricultural, and urban habitats. Edible mushrooms will also be addressed from the perspective of nutrition and food safety, as well as medicinal mushrooms in terms of their bioactive compounds and pharmacological applications. Regional case studies and ethnomycological knowledge illustrate the cultural and biological diversity of fungus use.

## 2.2 Taxonomic Diversity

### 2.2.1 Edible Fungi

Edible fungi comprise a diverse group of macrofungi, primarily within the phylum Basidiomycota, with some representatives in Ascomycota. Wild species are valued for their culinary tradition, flavor, aroma, and rarity. Both wild and cultivated, their

**Fig. 2.1** *Agaricus bisporus* (**a**), *Pleurotus ostreatus* (**b**), and *Lentinula edodes* (**c**). (Original photos by José Costa)

nutritional profile includes high-quality proteins, essential amino acids, vitamins (B-complex, D), minerals, and dietary fiber (Kalač 2016; Roncero-Ramos and Delgado-Andrade 2017). Additionally, they contain bioactive compounds with antioxidant and immunomodulatory properties (Valverde et al. 2015; Wasser 2017; Rathore et al. 2017; Roux and Eyssartier 2024).

The most widely cultivated edible species belong to the order Agaricales, such as *Agaricus bisporus* (button mushroom), *Lentinula edodes* (shiitake), and *Pleurotus ostreatus* (oyster mushroom) (Fig. 2.1). These species are globally important for food security and functional nutrition (Valverde et al. 2015). Their polysaccharides, particularly β-glucans, have been associated with immune system modulation and reduction of oxidative stress (Chan et al. 2009; Bisen et al. 2010).

Wild edible mushrooms play a significant role in traditional diets and local economies, and key taxonomic groups include the following orders: Boletales, such as *Boletus edulis* (porcini), highly prized for its flavor and antioxidant compounds; Cantharellales, such as *Cantharellus cibarius* (chanterelle), rich in carotenoids and vitamin precursors; Russulales, such as *Lactarius deliciosus* (saffron milk cap), known for unique sesquiterpenes; and Agaricales, such as *Amanita caesarea* (Caesar's mushroom), highly regarded for its flavor and economic value (Heleno et al. 2010; Roux and Eyssartier 2024) (Fig. 2.2).

These species are ecologically associated with forest ecosystems, forming ectomycorrhizal relationships that enhance soil health and biodiversity (Aerts 2003; Tedersoo et al. 2014; Heng 2016; Roux and Eyssartier 2024).

### 2.2.2 Medicinal Fungi

Medicinal fungi include species traditionally used in ethnomedicine and now recognized for their pharmacological potential. They occur mainly in the Basidiomycota and Ascomycota phyla (Wasser 2017; Fang et al. 2025).

**Fig. 2.2** *Boletus edulis* (**a**), *Cantharellus cibarius* (**b**), *Lactarius deliciosus* (**c**), and *Amanita caesarea* (**d**). (Original photos by José Costa)

As examples of Basidiomycota medicinal species, we can name *Ganoderma lucidum* (reishi), which contains triterpenoids and polysaccharides with immunomodulatory, hepatoprotective, and anticancer properties (Jayachandran et al. 2017); *Trametes versicolor* (turkey tail), source of polysaccharide-K (PSK), used as an adjuvant in cancer therapy (Xu et al. 2012; He et al. 2022b); and *Grifola frondosa* (maitake), rich in β-glucans with antitumor and metabolic benefits (Xu et al. 2012; Wasser 2017; Ma et al. 2018) (Fig. 2.3).

Regarding the Ascomycota medicinal species, the most well-known species are *Cordyceps militaris* and *Ophiocordyceps sinensis*, both producing cordycepin and adenosine derivatives, exhibiting anti-inflammatory, anti-fatigue, and immunostimulatory effects (Wasser 2017; Das et al. 2021). These species are widely incorporated into nutraceuticals and functional foods, with ongoing research into their clinical applications (Zhang et al. 2022; Yao et al. 2024).

The health-promoting effects of these fungi are largely attributed to β-glucans, triterpenes, phenolic compounds, and glycoproteins. These molecules modulate immune responses, reduce oxidative stress, and may protect against chronic diseases such as cancer, diabetes, and cardiovascular disorders (Valverde et al. 2015; Wasser 2017; Rathore et al. 2017; Jayachandran et al. 2017).

**Fig. 2.3** *Ganoderma lucidum* (**a**), *Trametes versicolor* (**b**), and *Grifola frondosa* (**c**). (Original photos by José Costa)

### 2.2.3 Core Species and Their Distinguishing Features

Regarding wild edible mushrooms, they exhibit extraordinary morphological and ecological diversity, positioning them as important subjects in taxonomy, ecology, and ethnomycology (Boa 2004; Mishra and Shankar 2025). Reliable field identification integrates macroscopic characteristics, such as cap shape, hymenophore configuration, stipe texture, and spore morphology, with ecological factors such as substrate preference and mycorrhizal associations (Laessøe and Petersen 2019). These fungi represent a wide taxonomic range, including Boletales, Agaricales, Cantharellales, and Russulales, each group defined by distinctive diagnostic traits (Richard et al. 2015; Wang et al. 2015). Beyond their gastronomic appeal, wild mushrooms are vital to rural livelihoods, local markets, and cultural practices, while playing key ecological roles in nutrient cycling and forest resilience through ectomycorrhizal partnerships (Smith et al. 2011; Büntgen et al. 2012). Table 2.1 summarizes the main wild edible mushroom species appreciated worldwide, morphological characteristics, and their typical habitats.

Wild medicinal mushrooms have been central to ethnomedicine for centuries and are increasingly validated by modern pharmacological studies (Baby et al. 2015; Das et al. 2021). These species, often growing as wood-decaying or parasitic fungi, synthesize bioactive compounds such as polysaccharides, triterpenes, and phenolic metabolites that contribute to immune modulation, anti-inflammatory effects, and other health-promoting properties (Zou et al. 2020; Lu et al. 2021). Morphological diversity ranges from the varnished conks of *Ganoderma lucidum* to the coral-like forms of *Hericium erinaceus* (Roncero-Ramos and Delgado-Andrade 2017; Zhang et al. 2024). Table 2.2 compiles the main wild medicinal fungi, outlining their identifying features and natural habitat.

Cultivated mushrooms play a pivotal role in global food security and the nutraceutical industry, representing a controlled and sustainable alternative to wild harvesting (Grimm and Wösten 2018; Shanmugaraj and Biswas 2024). Species such as *Agaricus bisporus*, *Pleurotus* spp., *Lentinula edodes*, and *Ganoderma lucidum* are now produced on an industrial scale, utilizing substrates ranging from composted agricultural waste to sterilized sawdust (Vieira et al. 2023; Geng et al. 2024). Morphological features, combined with specific substrate requirements, determine their suitability for cultivation systems (Jeewanthi et al. 2017; Grimm and Wösten 2018; Das et al. 2021; Shen et al. 2022). Table 2.3 presents the most widely cultivated edible and medicinal mushrooms, emphasizing their distinguishing features and common substrates used in production.

### 2.2.4 Methods of Identification and Classification

The identification and classification of mushroom species rely on a range of techniques integrating classical mycological practices with modern molecular methodologies. Traditionally, taxonomists employed morphological observation, microscopic analysis, and chemical reactions to differentiate taxa. These foundational methods remain relevant, particularly for field mycologists and ecological research, where molecular tools may be limited (Jeewon 2016; Phukhamsakda et al. 2022; Ikechi-Nwogu and Egere 2024).

*Macroscopic examination* of fruiting bodies constitutes the first diagnostic approach. Characteristics such as pileus color and texture, lamellae or pore arrangement, stipe ornamentation, and spore print color provide primary differentiation

**Table 2.1** Core wild edible mushrooms and their distinguishing features

| Order | Species | Morphological features | Habitat/ ecology | References |
|---|---|---|---|---|
| Boletales | *Boletus edulis* (Porcini) | Brown convex cap; whitish pores→greenish; stout stipe with reticulation | Coniferous & deciduous forests | Nuhn et al. (2013) and Roux and Eyssartier (2024) |
| | *Boletus aereus* | Dark brown to blackish cap; dense reticulation; firm context | Mediterranean oak forests | Binder and Hibbett (2006), Nuhn et al. (2013), and Roux and Eyssartier (2024) |
| | *Boletus pinophilus* | Reddish-brown cap; elongated stipe | Montane pine forests | Adamčík et al. (2015) and Roux and Eyssartier (2024) |
| | *Boletus reticulatus* | Pale brown velvety cap; stout stipe with whitish reticulate pattern | Mediterranean oak forests | Binder and Hibbett (2006), Nuhn et al. (2013), and Roux and Eyssartier (2024) |
| | *Leccinum scabrum* (Birch bolete) | Gray-brown cap; scabrous stipe; white pores | Birch forests | Binder and Hibbett (2006) and Roux and Eyssartier (2024) |
| | *Suillus luteus* (Slippery jack) | Glutinous brown cap; ring on stipe; yellow pores | Pine plantations | Binder and Hibbett (2006) and Roux and Eyssartier (2024) |
| | *Xerocomellus chrysenteron* (Red cracking bolete) | Cap cracks showing reddish flesh; slender stipe | Mixed woodlands | Šutara (2008), Frank et al. (2020), and Roux and Eyssartier (2024) |
| Agaricales | *Agaricus campestris* (Field mushroom) | Free gills; pink→brown spore print; smooth cap | Pastures & grasslands | Bashir et al. (2021) and Roux and Eyssartier (2024) |
| | *Agaricus arvensis* (Horse mushroom) | Large, almond odor, yellowing flesh | Meadows & grasslands | Cao et al. (2021) |
| | *Macrolepiota procera* (Parasol) | Tall stipe w/ snakeskin pattern; movable ring; scaly cap | Meadows, woodland edges | Vellinga (2004) |
| | *Coprinopsis comatus* (Shaggy mane) | Cylindrical shaggy white cap; autodigestion | Roadsides, grasslands | Pawlik et al. (2015) and Roux and Eyssartier (2024) |

(continued)

**Table 2.1** (continued)

| Order | Species | Morphological features | Habitat/ ecology | References |
|---|---|---|---|---|
| | *Calocybe gambosa* (St. George's) | Thick white gills; farinaceous odor; spring fruiting | Calcareous grasslands | Roux and Eyssartier (2024) |
| | *Tricholoma matsutake* (Matsutake) | Robust white stipe; brown fibrillose cap; spicy odor | Pine forests | Vaario et al. (2017) |
| | *Clitocybe odora* | Greenish cap; strong anise odor | Deciduous forests | Laessøe and Petersen (2019) and Roux and Eyssartier (2024) |
| Cantharellales | *Cantharellus cibarius* (Chanterelle) | Funnel-shaped yellow cap; blunt ridges (false gills); apricot smell | Mixed forests | Olariaga et al. (2017) |
| | *Cantharellus lateritius* (Smooth chanterelle) | Nearly smooth hymenium; orange cap | Deciduous forests | Buyck (2014) |
| | *Craterellus cornucopioides* (Black trumpet) | Dark, trumpet-shaped, hollow | Beech & oak forests | Adamčík et al. (2015) and Roux and Eyssartier (2024) |
| | *Craterellus tubaeformis* (Yellowfoot) | Hollow stipe; yellow hymenium; brown cap | Mossy coniferous forests | Olariaga et al. (2017) and Roux and Eyssartier (2024) |
| Russulales | *Lactarius deliciosus* (Saffron milk cap) | Orange latex; zonate orange cap; green staining | Pine forests | Adamčík et al. (2015) |
| | *Lactarius indigo* | Striking blue latex; zonate blue cap | Deciduous forests, Americas/Asia | Lee et al. (2019) and Roux and Eyssartier (2024) |
| | *Russula cyanoxantha* | Violet-green cap; flexible gills (non-brittle) | Deciduous forests | Zsigmond et al. (2019) |
| Polyporales & allies | *Laetiporus sulphureus* (Chicken of the woods) | Bright sulfur-yellow shelves | Hardwood trunks | Song et al. (2018) and Roux and Eyssartier (2024) |
| | *Fistulina hepatica* (Beefsteak fungus) | Red, tongue-shaped; exudes reddish juice | Oaks & chestnuts | Laessøe and Petersen (2019) and Roux and Eyssartier (2024) |
| | *Sparassis crispa* (Cauliflower mushroom) | Large, curly, whitish fruit bodies | At the base of conifers | Jorjadze et al. (2024) and Roux and Eyssartier (2024) |

(continued)

**Table 2.1** (continued)

| Order | Species | Morphological features | Habitat/ ecology | References |
|---|---|---|---|---|
| Pezizales (Ascomycota) | *Morchella esculenta* (Morel) | Honeycomb ridges & pits; hollow | Mixed forests | Richard et al. (2015) |
| | *Morchella elata* group (Black morels) | Conical, darker, ridged cap; hollow | Coniferous forests | Sanz-Rocha et al. (2023) |
| | *Tuber melanosporum* (Black truffle) | Warty black ascocarps; marbled gleba | Subterranean, oaks/hazels | Balestrini and Mello (2015), Leonardi et al. (2021), and Roux and Eyssartier (2024) |
| | *Tuber magnatum* (White truffle) | Pale ascocarps; strong aroma; smooth | Hypogeous, hardwoods | Balestrini and Mello (2015), Leonardi et al. (2021), and Roux and Eyssartier (2024) |
| Auriculariales & Hericiales | *Auricularia auricula-judae* (Wood ear) | Ear-shaped, gelatinous brown fruit body | Elder & hardwoods | Looney (2013) and Wu et al. (2021) |
| | *Hericium erinaceus* (Lion's mane) | Cascading white spines in clusters | Hardwood trunks | Hallenberg et al. (2013) and Roux and Eyssartier (2024) |
| Hydnales | *Hydnum repandum* (Hedgehog mushroom) | Pale cap; spines (teeth) instead of gills | Mixed forests | Laessøe and Petersen (2019) and Roux and Eyssartier (2024) |
| Amanitales | *Amanita caesarea* (Caesar's mushroom) | Orange cap; yellow gills & stipe; large white volva | Mediterranean oak forests | Lupala et al. (2025) |

criteria. Spore print color, for instance, distinguishes genera such as *Amanita*, *Lepiota*, and *Agaricus* (Laessøe and Petersen 2019; Roux and Eyssartier 2024; Ikechi-Nwogu and Egere 2024). However, phenotypic variability driven by ecological and developmental factors often complicates identification and may lead to misclassification (Jeewon 2016; Roux and Eyssartier 2024).

*Microscopy* remains central to species delimitation, allowing detailed observation of spores, cystidia, basidia, and hyphal structures. Diagnostic traits include spore size, shape, ornamentation, and the presence of clamp connections (Laessøe and Petersen 2019; Ikechi-Nwogu and Egere 2024). Spore wall morphology, such as amyloid ornamentation or reticulation, is significant in *Russula*, *Lactarius*, and *Boletus* (Pérez-Moreno et al. 2021). Examination of pileipellis and hymenophoral

**Table 2.2** Core wild medicinal mushrooms and their distinguishing features

| Order | Species | Morphological features | Habitat | References |
|---|---|---|---|---|
| Polyporales | *Ganoderma lucidum* (Reishi/ Lingzhi) | Woody, varnished red-brown cap; lateral stipe; tough context | Decaying hardwood logs, stumps | Baby et al. (2015) and Roux and Eyssartier (2024) |
| | *Trametes versicolor* (Turkey tail) | Thin leathery brackets; concentric multicolored zones; white pore surface | Dead hardwood, fallen logs | K et al. (2024) |
| | *Fomes fomentarius* (Tinder fungus) | Hoof-shaped perennial conk; gray to black exterior; hard woody context | Birch and beech trunks | Laessøe and Petersen (2019) |
| | *Inonotus obliquus* (Chaga) | Sterile, black irregular conk with cracked surface | Birch trees (parasitic) | Roux and Eyssartier (2024) and Wang et al. (2025) |
| | *Phellinus linteus* | Large perennial dark basidiocarps; hard woody texture | Hardwood trunks, esp. mulberry | Chen et al. (2019) and Roux and Eyssartier (2024) |
| | *Piptoporus betulinus* (Birch polypore) | White to pale brown hoof-shaped fruit body; smooth surface | Exclusively on birch | Wasser (2017) |
| | *Polyporus umbellatus* | Large underground sclerotium; produces clusters of umbrella-shaped caps | Base of oak and beech trees | Grienke et al. (2014) |
| Russulales | *Grifola frondosa* (Maitake/Hen of the woods) | Large, clustered rosettes of gray-brown fronds | At the base of oak trees | Hobbs (2020) |
| | *Sparassis crispa* (Cauliflower mushroom) | Globose, curly whitish to yellowish fronds | Base of conifer trees | Hobbs (2020) and Jorjadze et al. (2024) |

(continued)

**Table 2.2** (continued)

| Order | Species | Morphological features | Habitat | References |
|---|---|---|---|---|
| Tremellales | *Tremella fuciformis* | Gelatinous, translucent, frond-like fruiting body | Decaying hardwood trunks | Wu et al. (2016, 2019) and Shen et al. (2017) |
| Auriculariales | *Auricularia auricula-judae* (Wood ear) | Ear-shaped gelatinous brown fruit body | Dead elder and hardwood | Looney (2013) and Roux and Eyssartier (2024) |
| Hericiales | *Hericium erinaceus* (Lion's mane) | White cascading spines in clusters | Decaying hardwood trunks | Lai et al. (2013), Hallenberg et al. (2013), and Spelman et al. (2017) |
| | *Hericium coralloides* | Coral-like branched fruit body with pendant spines | Hardwood logs and stumps | Hallenberg et al. (2013) and Laessøe and Petersen (2019) |
| Hymenochaetales | *Phellinus linteus* | Hard, dark fruit body with a yellow-brown pore surface | Mulberry trees | Patel and Goyal (2012), Rahi and Malik (2016), and Khursheed et al. (2020) |
| | *Schizophyllum commune* (Split gill) | Small fan-shaped fruit bodies; split gills that reopen with humidity | Dead wood worldwide | Dasgupta et al. (2025) |
| Agaricales | *Agaricus bisporus* (Button mushroom/ Champignon) | White to brown cap; free pink→brown gills; annulate stipe | Compost (manure + straw-based) | Chang and Wasser (2017) |
| | *Gymnopus fusipes* | Fused clustered stipes; brown caps | Oak stumps and roots | (Pettifor et al. 2022) |
| Pezizales (Ascomycota) | *Cordyceps militaris* | Orange clavate stromata arising from insect pupae | Insect larvae (Lepidoptera pupae) | Sun et al. (2019), Das et al. (2021), and Zhang et al. (2024) |
| | *Ophiocordyceps sinensis* (Caterpillar fungus) | Club-shaped stromata from mummified caterpillars | Alpine meadows, parasitic on *Thitarodes* larvae | Li et al. (2009), Chen et al. (2010b), Jiraungkoorskul and Jiraungkoorskul (2016), |

(continued)

**Table 2.2** (continued)

| Order | Species | Morphological features | Habitat | References |
|---|---|---|---|---|
| | | | | Chellapandi and Saranya (2023), and Yao et al. (2024) |
| | *Morchella* spp. (Morels, used medicinally in some traditions) | Honeycomb ridges and pits; hollow ascocarps | Forest soils, spring fruiting | Richard et al. (2015), Sanz-Rocha et al. (2023), and Roux and Eyssartier (2024) |

trama is also crucial for resolving cryptic taxa (Jeewon 2016; Pérez-Moreno et al. 2021).

*Chemical reagents* represent another enduring yet valuable taxonomic aid. The Melzer's reagent test for amyloid and dextrinoid reactions in spores and asci remains fundamental in Ascomycota (Hongsanan et al. 2025). Potassium hydroxide (KOH) solutions are applied to enhance pigmentation and detect diagnostic color changes in tissues, assisting in the identification of genera such as *Russula* and *Cortinarius* (Wang et al. 2015). Sulfovanillin and guaiac reagents, used to detect secondary metabolites, continue to support systematic studies (Wang et al. 2015; Hongsanan et al. 2025).

In the last two decades, *molecular techniques* have transformed fungal taxonomy. DNA barcoding using the Internal Transcribed Spacer (ITS) region of rDNA serves as the universal marker for species-level identification (Schoch et al. 2012; Zhang et al. 2021). When single-locus data are insufficient, multilocus phylogenetic analyses employing TEF-1α, RPB1, RPB2, and β-tubulin enhance resolution within complex clades (Lambert 2024; Asif et al. 2025). These approaches have clarified evolutionary relationships and resolved taxonomic ambiguities (Nagy and Szöllősi 2017; Keuler et al. 2020).

*Phylogenomics* further strengthens fungal systematics. Genome-scale data enable robust phylogenetic inferences, revision of long-standing classifications, and discovery of new lineages (Nagy and Szöllősi 2017; Keuler et al. 2020; Lambert 2024). Such analyses are particularly valuable for groups exhibiting high morphological convergence, where traditional traits are unreliable (Jeewon 2016).

*Chemotaxonomic profiling* complements morphology and molecular data. Metabolite fingerprinting of sesquiterpenes and polyketides yields species-specific chemical signatures. MALDI-TOF mass spectrometry now enables rapid, high-throughput identification of mycelial cultures and fruiting bodies (Patel 2019; Horisawa and Iwamoto 2022). Such approaches integrate chemistry with systematics and open new avenues for distinguishing species complexes.

Recent *computational* advances have introduced deep learning and computer vision into fungal identification. Convolutional neural networks trained on extensive photographic datasets can distinguish edible and toxic species with high accuracy

**Table 2.3** Main cultivated edible and medicinal mushrooms and their distinguishing features

| Order | Species | Morphological features | Substrate in cultivation | References |
|---|---|---|---|---|
| Agaricales | *Agaricus bisporus* (Button mushroom/ Champignon) and *A. subrufescens (A. blazei)* | White to brown cap; free pink→brown gills; annulate stipe | Compost (manure + straw-based) | Chang and Wasser (2017) |
| | *Pleurotus ostreatus* (Oyster mushroom) | Oyster-shaped cap; decurrent white gills; short lateral stipe | Pasteurized straw, hardwood sawdust, agricultural waste | Fufa et al. (2021) |
| | *Pleurotus eryngii* (King oyster/ Cardoon mushroom) | Thick stipe; small convex cap; decurrent gills | Sawdust, straw, enriched substrates | Iryna et al. (2022) |
| | *Pleurotus pulmonarius* (Phoenix oyster) | Pale cap; smaller, more delicate than *P. ostreatus* | Straw, cotton waste, sawdust | Fufa et al. (2021) |
| | *Lentinula edodes* (Shiitake) | Brown scaly convex cap; white gills; white stipe | Sterilized hardwood logs or sawdust blocks | Bisen et al. (2010) and Song et al. (2025) |
| | *Flammulina velutipes* (Enoki) | Long slender stipes; small convex caps; cultivated white form | Sterilized sawdust or corn cobs | Royse et al. (2017) |
| | *Volvariella volvacea* (Paddy straw mushroom) | Volva at stipe base; gray-brown cap; cultivated in warm climates | Rice straw bundles, cotton waste | Chang and Wasser (2017) |
| Polyporales | *Ganoderma lucidum* (Reishi/ Lingzhi) | Woody basidiocarp; red varnished cap; lateral stipe | Hardwood sawdust, supplemented logs | Baby et al. (2015) |
| | *Trametes versicolor* (Turkey tail) | Multicolored zonate cap; thin leathery brackets | Sawdust, logs, agricultural waste | K et al. (2024) |

(continued)

**Table 2.3** (continued)

| Order | Species | Morphological features | Substrate in cultivation | References |
|---|---|---|---|---|
| Pezizales (Ascomycota) | *Morchella* spp. (Morels, used medicinally in some traditions) | Honeycomb ridges and pits; hollow ascocarps | Forest soils, spring fruiting | Richard et al. (2015), Sanz-Rocha et al. (2023), and Roux and Eyssartier (2024) |
| Russulales | *Grifola frondosa* (Maitake/Hen of the woods) | Rosette of gray-brown fronds; clustered growth | Sterilized hardwood sawdust blocks | Kodama et al. (2003), Chen et al. (2010a), and Hobbs (2020) |
| Auriculariales | *Auricularia auricula-judae* (Wood ear) | Ear-shaped gelatinous brown fruit body | Hardwood logs, sawdust bags | Looney (2013) and Wu et al. (2021) |
| Hericiales | *Hericium erinaceus* (Lion's mane) | White globose fruiting body with pendant spines | Hardwood sawdust, supplemented blocks | Lai et al. (2013), Hallenberg et al. (2013), and Spelman et al. (2017) |
| Hypocreales (Ascomycota) | *Cordyceps militaris* | Orange clavate fruiting bodies; produces cordycepin | Sterilized rice, insect pupae substrates | Sun et al. (2019) and Das et al. (2021) |
| | *Ophiocordyceps sinensis* (Caterpillar fungus, cultivated forms) | Club-shaped stromata on insect hosts; rare in nature | Artificial cultivation on insect larvae substitutes, grain substrates | Li et al. (2009), Chen et al. (2010b), Jiraungkoorskul and Jiraungkoorskul (2016), Chellapandi and Saranya (2023), and Yao et al. (2024) |

(Zhang et al. 2022). Although still emerging, these digital tools enhance rapid field identification and biodiversity monitoring (Jeewon 2016; DemiRel and DemiRel 2023).

Collectively, morphological, chemical, microscopic, molecular, and AI-based approaches form an integrative framework that expands the precision and scope of fungal taxonomy. Each method has distinct strengths and limitations, yet their combined application ensures the most reliable taxonomic resolution. Table 2.4 summarizes the main identification strategies, outlining their advantages, limitations, and key references.

**Table 2.4** Comparative overview of mushroom identification methods

| Method | Type | Advantages | Limitations | References |
|---|---|---|---|---|
| Macroscopic features | Classical | Simple, rapid, field-based; requires no equipment | High variability due to environment/ age; cryptic species not resolved | Jeewon (2016), Phukhamsakda et al. (2022), and Ikechi-Nwogu and Egere (2024) |
| Microscopy (spores, hyphae) | Classical | Detailed structural insights (spores, cystidia, clamp connections) | Requires expertise, overlapping features between taxa | Jeewon (2016) and Laessøe and Petersen (2019) |
| Chemical reagents | Classical | Inexpensive, diagnostic color reactions (e.g., Melzer's, KOH) | Limited specificity; some reagents are hazardous; declining use | Hongsanan et al. (2025) |
| DNA barcoding (ITS) | Molecular | Universal marker for fungi; standardized protocols | Limited resolution in some groups; requires sequencing facilities | Schoch et al. (2012) and Zhang et al. (2021) |
| Multilocus phylogenetics | Molecular | Higher resolution; clarifies complex clades | Time- and cost-intensive; requires multilocus datasets | Lambert (2024) and Asif et al. (2025) |
| Phylogenomics | Molecular | Genome-wide resolution; resolves deep nodes in the fungal tree | High cost; requires advanced bioinformatics expertise | Nagy and Szöllősi (2017), Keuler et al. (2020), and Asif et al. (2025) |
| Chemotaxonomy (MALDI-TOF) | Molecular/ Chemical | High-throughput; species-specific metabolite fingerprints | Databases still incomplete; requires specialized equipment | Patel (2019) and Horisawa and Iwamoto (2022) |
| AI & deep learning | Computational | Rapid, automated image-based identification; scalable with large datasets | Dependent on quality/size of training data; limited to macroscopic features | DemİRel and DemİRel (2023) |

## 2.3 Fungal Ecology

Fungi exhibit remarkable diversity in their trophic strategies, which determine their ecological roles and evolutionary success. The three dominant modes—mycorrhizal symbiosis, saprotrophic decomposition, and parasitism—illustrate how fungi mediate nutrient flows, regulate host populations, and maintain biodiversity across ecosystems.

### 2.3.1 Mycorrhizal Fungi

Mycorrhizal fungi comprise three major functional groups: arbuscular, ectomycorrhizal, and ericoid fungi. Arbuscular mycorrhizal (or endomycorrhizal) fungi, belonging to the Glomeromycota, form intracellular arbuscules and vesicles within plant roots, enhancing phosphorus and micronutrient uptake in exchange for photosynthates (Rillig et al. 2010; Chandrasekaran 2022). Dominant in grasslands and croplands, they produce glomalin-related soil proteins that improve aggregation and stabilize organic carbon, contributing to soil fertility and carbon sequestration. Ectomycorrhizal fungi, mainly Basidiomycota and Ascomycota, develop a root sheath and a Hartig net between cortical cells, facilitating access to organic nitrogen and phosphorus for dominant forest trees in boreal and temperate ecosystems (Mahmood et al. 2024). Ectomycorrhizal fungi exhibit diverse exploration strategies: long-distance cord-formers such as *Cortinarius* spp. mobilize nutrients over meters, while short-distance foragers like *Russula* spp. exploit nutrient-rich microsites (Van Der Linde et al. 2018). Ericoid mycorrhizal fungi, symbiotic with Ericaceae, colonize fine root hairs in acidic, nutrient-poor soils, producing enzymes that release nitrogen from recalcitrant organic matter (Grelet et al. 2016).

Mycorrhizal networks interconnect plants, mediating transfers of nutrients, water, and signals that influence growth and defense (Simard 2021; Luo et al. 2023). Such networks may regulate forest regeneration and competition, although the magnitude of these interactions remains actively debated.

### 2.3.2 Saprotrophic Fungi

Saprotrophs drive ecosystem decomposition and nutrient cycling. White-rot fungi, including *Trametes versicolor* and *Pleurotus ostreatus*, deploy oxidative enzymes such as lignin peroxidases and laccases, breaking down lignin and allowing complete mineralization of plant cell walls (Liu et al. 2021). Brown-rot fungi, conversely, depolymerize cellulose and hemicellulose while leaving modified lignin residues, significantly influencing soil organic matter chemistry. By recycling nutrients, saprotrophs underpin soil fertility and carbon turnover (Crowther et al. 2012).They also play critical roles in post-disturbance ecology. Following wildfires or windthrows, saprotrophic fungi rapidly colonize dead wood and litter, releasing immobilized nutrients and facilitating the successional recovery of microbial and plant communities (Espinosa et al. 2023).

### 2.3.3 Parasitic Fungi

Although often considered destructive, parasitic fungi are fundamental regulators of biodiversity. Tree pathogens such as *Armillaria* spp. decompose root systems and accelerate turnover of weakened hosts, shaping forest dynamics. Foliar parasites, such as rust fungi (*Puccinia* spp.), regulate plant populations and prevent competitive exclusion (Hadziabdic et al. 2022). In the animal kingdom, entomopathogenic fungi such as *Cordyceps militaris* suppress insect outbreaks, maintaining balance in arthropod communities. Collectively, parasites act as selective pressures that maintain genetic diversity and ecosystem stability.

## 2.4 Ecological Roles of Mushrooms in Ecosystems

Mushrooms, the macroscopic fruiting bodies of Basidiomycota and some Ascomycota, are a complement to the ecological functions of their underlying mycelia. Their ecological contributions include nutrient and carbon cycling, ecosystem stability, disturbance recovery, and bioremediation.

In nutrient and carbon cycling, fungal decomposition regulates the biogeochemical turnover of nitrogen, phosphorus, and carbon (Liu et al. 2021). Saprotrophic fungi release inorganic nutrients from organic matter, while mycorrhizal symbionts channel nutrients directly to plant roots. The balance between decomposition and symbiosis governs soil fertility and productivity (Crowther et al. 2012). Fungi also modulate carbon dynamics. Ectomycorrhizal fungi in boreal forests contribute to carbon sequestration by slowing litter decay and stabilizing organic matter (Clemmensen et al. 2015), whereas saprotrophs accelerate mineralization, maintaining global carbon equilibrium.

Fungi also function as ecosystem engineers, enhancing stability and recovery after disturbance. Their hyphae bind soil particles, improving aggregation, porosity, and water retention, thus preventing erosion (Rillig et al. 2010). After disturbances such as wildfires, persistent mycelia facilitate rapid recolonization and reestablishment of symbioses with regenerating vegetation (Brace et al. 2024). Decomposition of flammable litter reduces fuel loads, mitigating subsequent wildfire severity (Espinosa et al. 2023).

Fungi can act as key agents for bioremediation and environmental services, as they can degrade recalcitrant pollutants (mycoremediation). White-rot fungi degrade polycyclic aromatic hydrocarbons, pesticides, and dyes via oxidative enzymes, detoxifying polluted substrates (Ikehata 2015). Metal-tolerant species immobilize toxic elements, improving soil quality in contaminated environments. Collectively, these processes position fungi as key agents in ecosystem resilience and climate change mitigation.

Overall, mushrooms and their mycelial networks act as ecological axes, supporting nutrient fluxes, soil health, vegetation dynamics, and resilience against environmental stressors (Fig. 2.4).

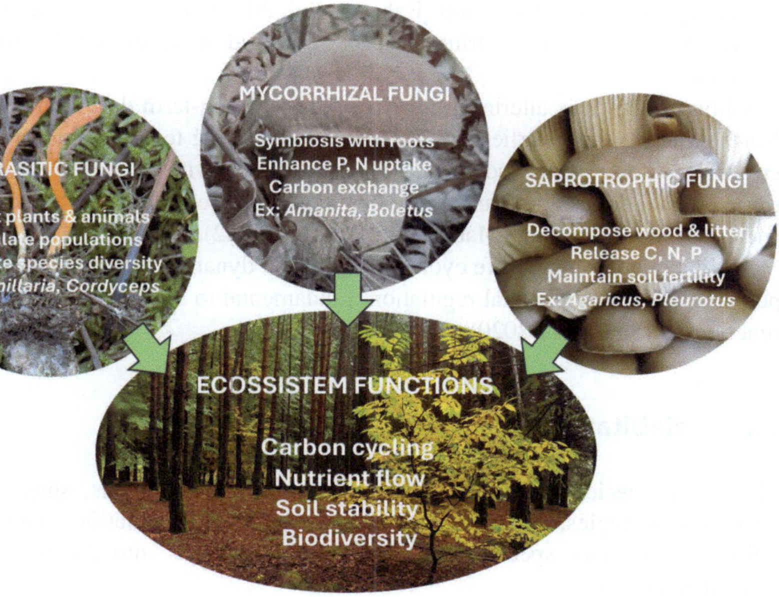


**Fig. 2.4** Trophic modes of fungi and their ecological contributions. (Original work and photos by José Costa)

## 2.5 Life Cycle and Environmental Factors Influencing Growth

The life cycle of fungi is a finely tuned adaptation to environmental constraints, ensuring their survival and dispersal. It includes several steps, such as spore dispersal, germination, and fruiting body development.

The cycle begins with spore dispersal, which occurs predominantly by wind, but also via animals, insects, or water currents. Spores germinate into monokaryotic hyphae, which fuse via plasmogamy to form dikaryotic mycelia, capable of persisting for decades (Kües 2015). These mycelia colonize substrates, competing with other microbes and storing resources until environmental conditions favor fruiting.

Fruiting body development is governed by external stimuli such as temperature, humidity, photoperiod, and nutrient availability. Rainfall frequently triggers fruiting in *Boletus*, *Amanita*, and *Cantharellus* spp., while *Morchella* spp. emerge after fire events, exploiting post-disturbance nutrient pulses (Buscardo et al. 2022; Mridu and Kumar 2024). Soil pH and texture further influence species distribution, with calciphilous taxa confined to alkaline soils and acidophilous fungi to podzolic substrates (Van Der Linde et al. 2018).

Biotic interactions also shape reproductive output. Host plant carbon allocation modulates mycorrhizal fruiting, whereas antagonistic microbes and soil fauna affect mycelial growth and sporulation (Bahram et al. 2020). Invertebrate grazers, such as collembolans, may reduce fruiting success but aid spore dispersal through fecal transport.

Climate change is altering fungal phenology. Long-term data from Europe and North America show earlier fruiting linked to warming temperatures and extended growing seasons (Kauserud et al. 2008). Combined with habitat degradation, soil compaction, deforestation, and pollution, these pressures disrupt reproductive cycles and community structure (Maurya and Pachauri 2022).

Ultimately, the fungal life cycle exemplifies a dynamic balance between genetic potential and environmental regulation, fundamental to biodiversity and ecosystem functioning (Sheldrake 2020).

## 2.6 Habitat-Specific Distribution

Mushroom species exhibit highly habitat-specific distributions, shaped by their ecological strategies, substrate affinities, and symbiotic associations. These patterns reflect evolutionary specialization and provide insights into the functioning of ecosystems worldwide.

A primary determinant of distribution is substrate availability. Saprotrophic fungi dominate organic-rich soils, decaying wood, and litter layers, where their enzymatic capacity enables efficient nutrient turnover (Liu et al. 2021). Wood-decaying basidiomycetes such as *Ganoderma* and *Trametes* colonize lignocellulosic substrates, whereas coprophilous species thrive on dung and compost. Symbiotic fungi, particularly ectomycorrhizal taxa, depend on host plants and are restricted to habitats dominated by compatible tree lineages such as Pinaceae and Fagaceae (Van Der Linde et al. 2018). Arbuscular mycorrhizal fungi are more generalist but still respond to soil fertility and disturbance gradients, shaping their distribution in grasslands and croplands (Chandrasekaran 2022).

Microclimatic and edaphic factors further influence species ranges. Temperature, soil moisture, and pH often act as filters that determine local dominance of acidophilous species such as *Russula emetica* or calciphilous taxa like *Tricholoma columbetta* (Suz et al. 2015). Disturbances, including fire and logging, also restructure fungal communities, favoring pioneering or fire-adapted taxa such as *Morchella* and *Pyronema* (Espinosa et al. 2023).

Consequently, mushroom communities exhibit clear biogeographic patterns. Temperate forests are dominated by ectomycorrhizal fungi forming extensive underground networks. Tropical rainforests host diverse saprotrophic and arbuscular mycorrhizal assemblages, while alpine zones support cold-adapted, short-season taxa. Grasslands and agricultural fields sustain arbuscular and dung-associated fungi adapted to frequent disturbance. These contrasts underscore the importance of habitat-based analyses for understanding fungal diversity at a global scale.

### 2.6.1 Temperate Forests

Temperate forests represent some of the richest ecosystems for mushroom diversity, with seasonality and heterogeneous vegetation creating niches for a broad range of fungal taxa. The complex canopy structure of mixed deciduous and coniferous forests supports diverse assemblages of mushrooms that contribute simultaneously to nutrient cycling, soil formation, and forest resilience (Baldrian 2016; Corrales et al. 2022). These communities are structured by interactions with trees and shrubs, substrate availability, and disturbance regimes such as storms, logging, and fire.

Many of the most recognizable temperate forest mushrooms form ectomycorrhizal associations with dominant tree species. Porcini (*Boletus edulis*), chanterelles (*Cantharellus cibarius*), and fly agarics (*Amanita muscaria*) establish mutualistic relationships with oaks, beeches, birches, and conifers, exchanging carbon for nitrogen and phosphorus (Suz et al. 2014; Rakić et al. 2022). These partnerships influence tree growth, competitive dynamics, and resource redistribution among neighboring trees through extensive underground mycorrhizal networks, enhancing forest stability under environmental stress (Simard et al. 2012).

Alongside these symbionts, temperate forests host abundant decomposers that recycle organic matter. Leaf-litter fungi such as *Marasmius rotula* and *Gymnopus dryophilus* accelerate decomposition, while wood-decaying basidiomycetes, including *Trametes versicolor* and *Fomes fomentarius,* degrade lignin and cellulose, releasing stored carbon and facilitating nutrient turnover (Hoppe et al. 2016). Their metabolic activity underpins soil fertility and long-term carbon balance, processes increasingly critical under climate change (Baldrian 2016).

Not all fungi are benign partners or decomposers. Pathogenic species, notably *Armillaria* species, form extensive subterranean networks that can cause widespread tree mortality. Yet, by creating canopy gaps and promoting regeneration, they also enhance structural diversity and successional dynamics (Oliva et al. 2014). This dual role underscores the balance between destructive and regenerative processes mediated by fungi.

Conservation concerns are mounting for species dependent on old-growth habitats. The iconic *Hericium erinaceus* ("lion's mane"), restricted to senescent hardwoods, has declined due to logging and loss of ancient trees (Mueller et al. 2022). *Amanita caesarea,* once common in Mediterranean oak forests, is threatened by habitat degradation and overharvesting, while *Boletus regius* and *Clavariadelphus pistillaris* have suffered regional declines (Dahlberg and Mueller 2011).

In sum, temperate forest mushrooms exemplify the ecological versatility of fungi, ranging from mutualism to decomposition and pathogenicity, while also representing vital cultural and economic resources. Their conservation demands recognition of fungi as integral components of forest ecosystems, essential to maintaining biodiversity and ecological balance.

### 2.6.2 Tropical Rainforests

Tropical rainforests host the planet's highest fungal diversity, driven by exceptional plant richness, climatic stability, and structural complexity. Unlike temperate regions with pronounced seasonality, tropical forests offer continuously warm and humid conditions, supporting mushroom fruiting year-round (Tedersoo et al. 2020). The diversity of tropical fungi spans symbionts, decomposers, and pathogens, each adapted to the extraordinary heterogeneity of rainforest ecosystems.

Ectomycorrhizal associations are particularly prominent in Southeast Asia, Central Africa, and the Neotropics, where dominant tree families such as Dipterocarpaceae, Fabaceae, and Myrtaceae sustain extensive fungal networks (Corrales et al. 2018, 2022). Genera such as *Russula* and *Lactarius* form dense ectomycorrhizal communities with dipterocarps, while *Cantharellus congolensis* is widely harvested across Central Africa (Buyck 2014; Yorou et al. 2024). These symbioses enhance host nutrient uptake in nutrient-poor tropical soils, shaping forest productivity and carbon dynamics (Peay et al. 2016).

Saprotrophic fungi dominate decomposition processes, processing immense quantities of litter and woody debris. Wood-decaying taxa like *Ganoderma applanatum* and *Trametes elegans* colonize fallen logs, whereas *Marasmius* species form intricate "leaf-tethering" networks that accelerate nutrient cycling (Kerekes et al. 2013). Bioluminescent mushrooms such as *Mycena chlorophos* in Asia and *Neonothopanus gardneri* in Brazil exemplify the unique evolutionary traits fostered by tropical conditions, possibly linked to nocturnal spore dispersal by insects (Oliveira et al. 2015).

Pathogenic fungi also exert strong ecological control by regulating host populations and maintaining tree diversity. *Armillaria* spp. and *Rigidoporus microporus* cause root diseases in plantations, while *Moniliophthora roreri*, the agent of cacao frosty pod rot, illustrates the risk of native fungi becoming major agricultural pathogens under monoculture expansion (Burgess and Wingfield 2004; Gomdola et al. 2022). These dynamics highlight the dual roles of rainforest fungi as biodiversity sustainers and potential threats to food security.

Deforestation, habitat fragmentation, and overharvesting are increasingly threatening tropical mushrooms. The termite-associated *Termitomyces titanicus*, the world's largest edible fungus, is declining across sub-Saharan Africa due to forest loss (Paloi et al. 2023). Similarly, *Lentinus squarrosulus*, a wood-decaying edible species, is diminishing in logged regions. Their conservation depends on safeguarding host trees, substrates, and symbiotic partners.

In sum, tropical rainforest mushrooms epitomize the intimate coupling between fungal diversity and ecosystem function. They sustain nutrient turnover, bolster forest productivity, and provide vital cultural and nutritional resources. Yet, ongoing deforestation and climatic shifts endanger many species, from *Termitomyces* to bioluminescent *Mycena*, underscoring the urgency of integrating fungi into tropical conservation frameworks as keystone elements of global biodiversity.

### 2.6.3 Alpine and Subalpine Environments

Alpine and subalpine ecosystems represent some of the most challenging environments for fungal life, characterized by low temperatures, intense ultraviolet (UV) radiation, short growing seasons, and nutrient-poor soils. Despite these constraints, they sustain distinctive mushroom communities with remarkable physiological and ecological adaptations. Given the sparse vegetation and limited productivity, fungal interactions are critical for maintaining ecosystem stability and nutrient cycling (Geml et al. 2015).

Mycorrhizal fungi dominate these high-elevation habitats, forming symbioses with dwarf shrubs, grasses, and conifers. Ectomycorrhizal associations with *Pinus cembra* in the Alps or *Abies* spp. in the Rocky Mountains are essential for tree survival under cold, nutrient-limited conditions (Xie et al. 2020). Iconic species such as *Suillus luteus*, *Cortinarius caperatus*, and *Inocybe lacera* enhance nutrient uptake in soils where mineralization is slow. Ericoid mycorrhizas associated with *Vaccinium* and *Rhododendron* support shrub growth in acidic, organic-rich soils, contributing significantly to primary production (Tedersoo et al. 2020).

Saprotrophic fungi, though limited by cold temperatures, play key roles in decomposition and soil formation. *Mycena galopus* and *Gymnopus androsaceus* degrade litter from dwarf shrubs and grasses, while *Clitocybe albirhiza* and *Arrhenia* spp. specialize in moss and lichen decomposition (Ryberg et al. 2009). Their activity contributes to the gradual accumulation of organic matter and the formation of fragile alpine soils, and many display psychrotolerant metabolism, enabling enzymatic activity at low temperatures and extending decomposition throughout brief alpine summers.

Parasitic fungi add further ecological complexity. *Hypomyces chrysospermus*, the bolete mold, parasitizes *Boletus* fruiting bodies, reducing reproduction yet enriching trophic diversity. Plant pathogens such as *Chrysomyxa rhododendri*, the alpine rhododendron rust, regulate host populations and maintain species balance in limited plant communities (Kohler et al. 2015). These parasitic interactions, while often detrimental to individual hosts, help maintain diversity and check dominance within limited plant communities.

Several alpine fungi are rare or vulnerable due to their narrow ecological ranges and climate sensitivity. Distinct alpine ecotypes of *Boletus edulis* may decline as treelines shift upward, while *Sarcodon leucopus* and *Albatrellus caeruleoporus*, restricted to subalpine conifer forests, are red-listed in parts of Europe (Kauserud et al. 2008, 2012). Habitat degradation from warming, grazing, and tourism further exacerbates their decline.

The ecological significance of alpine mushrooms extends beyond nutrient turnover: by facilitating shrub and tree establishment at the treeline, they stabilize slopes, influence hydrology, and sustain wildlife forage. Their persistence illustrates extraordinary adaptation to extreme climates, while their vulnerability highlights the urgent need to integrate fungi into mountain biodiversity and climate adaptation strategies (Tedersoo et al. 2020).

### 2.6.4 Grasslands and Agricultural Fields

Grasslands and agricultural fields, though often regarded as simplified ecosystems, sustain remarkable fungal diversity that underpins soil fertility, plant productivity, and ecosystem resilience. These habitats are shaped by frequent disturbances such as grazing, mowing, plowing, and fertilization, which continuously restructure fungal communities (Romdhane et al. 2022). Despite these pressures, mushrooms in grass-dominated landscapes perform vital ecological functions including decomposition, nutrient cycling, symbiosis, and biological control.

Ectomycorrhizal fungi, though less dominant than in forests, occur in association with scattered trees or shrubs embedded in open landscapes. *Agaricus campestris*, the cosmopolitan "field mushroom," thrives in organically enriched meadows and pastures (Kauserud et al. 2008, 2012). Species of *Inocybe* and *Hebeloma* form mycorrhizas with isolated woody plants, extending fungal influence into grasslands (Suz et al. 2014, 2015). These symbioses enhance the uptake of nitrogen and phosphorus, nutrients typically limiting in grassland soils, and stabilize transitional ecotones between forests and open areas.

Saprotrophic fungi are highly diverse and abundant. Dung-inhabiting mushrooms such as *Panaeolus semiovatus* and *Coprinellus micaceus* recycle nutrients in grazed systems (Ainsworth and Bisby 2011), while fairy-ring species like *Marasmius oreades* modify soil chemistry and vegetation patterns (Zotti et al. 2025). In agricultural fields, straw- and stubble-decomposers such as *Volvariella volvacea* and *Coprinopsis cinerea* accelerate residue breakdown, improving organic matter turnover and preparing soils for subsequent cultivation.

Parasitic and pathogenic fungi also shape grassland and agroecosystem dynamics. Rusts (*Puccinia* spp.) and smuts (*Ustilago maydis*) exemplify plant pathogens of major agronomic importance, while entomopathogenic species such as *Cordyceps militaris* regulate insect populations and contribute to natural pest control (Sheldrake 2020; Venkatesh et al. 2022). Together, these diverse trophic modes demonstrate the balance fungi maintain between promoting and constraining productivity in managed and natural systems.

Several grassland mushrooms are of high conservation value. European "waxcap grasslands" host distinctive, colorful species such as *Hygrocybe conica*, *H. punicea*, and *Gliophorus psittacinus*, indicators of low-intensity, semi-natural pastures now threatened by intensification and nutrient enrichment (Griffith 2013). Comparable declines affect *Clavaria* and *Entoloma* species in North American prairies and South American pampas. Maintaining traditional grazing regimes and reducing chemical inputs are thus essential for conserving both plant and fungal biodiversity.

Fungi in grasslands and agricultural fields exemplify the duality of their ecological roles, as allies that sustain productivity and as adversaries that challenge cultivation. Their diversity and functional versatility emphasize that these landscapes, though heavily managed, remain dynamic ecosystems dependent on fungal activity for long-term sustainability.

### 2.6.5 Urban and Peri-urban Environments

Urban and peri-urban environments are among the most anthropogenically altered ecosystems, yet they sustain a surprising diversity of mushrooms adapted to fragmented habitats and artificial substrates. Found in parks, gardens, roadside verges, cemeteries, and abandoned lots, urban fungi form intricate relationships with ornamental vegetation and even built structures. Despite frequent neglect in biodiversity assessments, these organisms play crucial ecological and cultural roles, enhancing nutrient cycling, supporting vegetation health, and enriching human interaction with nature (Newbound et al. 2010; Candido and Alvim Carvalho 2025).

Ectomycorrhizal fungi commonly colonize urban trees such as oaks, birches, lindens, and introduced ornamentals including *Eucalyptus* and *Pinus*. Genera such as *Amanita*, *Russula*, and *Cortinarius* occur in city parks and suburban woodlands, maintaining tree vigor despite soil compaction and pollution (Kauserud et al. 2012; Baldrian et al. 2022). Species like *Amanita muscaria* and *Laccaria bicolor* are typical under birch and pine, while truffle-forming *Tuber aestivum* occasionally establishes in peri-urban green spaces, providing both ecological and economic value (Van Der Heijden et al. 2015).

Saprotrophic fungi are especially abundant due to plentiful organic debris and artificial substrates. *Coprinellus micaceus*, *Coprinopsis atramentaria*, and *Agrocybe pediades* thrive in lawns, compost heaps, and mulched flowerbeds (Van Der Heijden et al. 2015; Tedersoo et al. 2020). Wood-decaying species such as *Ganoderma applanatum* and *Trametes versicolor* recycle biomass but can also weaken urban trees, presenting management challenges. Some fungi even colonize human structures. *Serpula lacrymans*, the dry rot fungus, degrades timber in buildings, illustrating the interface between fungal ecology and urban pathology (Schmidt 2006).

Pathogenic and allergenic fungi are equally significant. *Armillaria mellea* spreads through green spaces, killing stressed ornamentals, while powdery mildews flourish in densely planted, humid cityscapes. Airborne spores of *Alternaria* and *Cladosporium* are major contributors to respiratory allergies in peri-urban populations (Grinn-Gofroń and Bosiacka 2015). Though not all produce visible fruiting bodies, these taxa underscore the medical and ecological relevance of the urban mycobiome.

Urban habitats can also serve as refugia for rare fungi. Semi-natural grasslands and old cemeteries host waxcaps (*Hygrocybe* spp.) and red-listed species such as *Entoloma bloxamii* (Griffith 2013). However, soil sealing, pollution, and habitat homogenization threaten this diversity, while climate change alters fruiting phenology and distributions (Kauserud et al. 2012).

Despite these threats, urban fungi play essential ecological roles, buffering against nutrient loss, supporting vegetation health, and fostering biodiversity in cities. Moreover, their presence offers opportunities for citizen science, environmental education, and recreational foraging, linking human societies more closely with fungal ecology in daily life.

## 2.7 Environmental Drivers of Diversity

Mushroom diversity is strongly shaped by environmental drivers that influence community assembly, species richness, and ecosystem functions. Among these drivers, climate and seasonality, soil type and pH, host plant associations, and natural or anthropogenic disturbances are dominant factors shaping fungal distribution and abundance (Fig. 2.5). Together, they create ecological filters that regulate both the presence of taxa and the ecosystem services they sustain across forests, grasslands, and agricultural systems worldwide.

*Climate and seasonality* strongly govern fungal phenology and spatial patterns. Temperature, precipitation, and humidity control fruiting and mycelial growth, producing marked interannual and seasonal variation. In temperate and boreal regions, warming advances fruiting onset, shifts community composition, and favors generalist taxa (Kauserud et al. 2008, 2012; Andrew et al. 2016). In tropical ecosystems, where rainfall seasonality rather than temperature is the main limiting factor, mushroom diversity peaks during wet seasons, with dramatic reductions in dry periods (Han et al. 2023). Experimental warming studies confirm that fungal biomass and enzyme activity respond rapidly to temperature changes, underscoring the tight coupling between fungal diversity and global change drivers (Treseder et al. 2016; Glassman et al. 2018).

*Soil type and pH* further define fungal community structure by regulating nutrient availability and microhabitat conditions. Acidic soils favor genera such as *Russula* and *Cortinarius*, while calcareous substrates host *Hebeloma* and *Tricholoma* (Suz et al. 2014). High organic content supports saprotrophs, whereas sandy or nutrient-poor soils select stress-tolerant mycorrhizal partners (Van Der Linde et al. 2018). Moreover, nitrogen deposition and acidification can suppress specialist fruiting while promoting nitrophilous taxa (Treseder et al. 2018). Evidence from large-

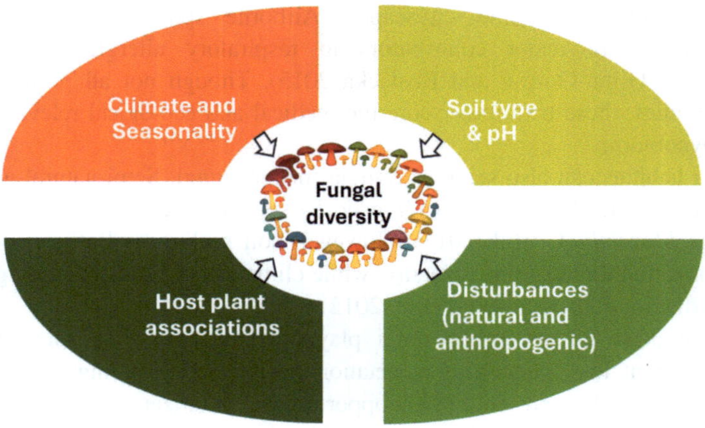


**Fig. 2.5** Environmental drivers of fungal diversity. (Original work by José Costa)

scale metabarcoding has confirmed soil pH as one of the most powerful global predictors of fungal community composition (Tedersoo et al. 2014; Bahram et al. 2018; Simard 2021).

*Host plant associations* represent a major axis of fungal diversity, especially among mycorrhizal symbionts. Different host plants act as ecological filters, structuring specialized fungal assemblages. *Pinus* and *Betula*, for instance, support distinct mycorrhizal communities, while mixed forests sustain greater fungal richness due to multiple compatible hosts (Tedersoo et al. 2018). In contrast, monocultures often reduce community complexity. Grasses and herbaceous plants also harbor endophytic and mycorrhizal fungi that enhance nutrient acquisition and drought tolerance (Bahram et al. 2018, 2020). Advances in genomic approaches reveal the coevolutionary signatures between hosts and fungi, highlighting that symbiosis is a major driver of diversification in both partners (Van Der Heijden et al. 2015; Martin et al. 2016).

Finally, *natural and anthropogenic disturbances* exert complex and sometimes contrasting effects. Wildfires, storms, and pest outbreaks create resource pulses that promote pyrophilous and saprotrophic species such as *Morchella* spp. (Dahlberg 2002; Hewitt et al. 2023). However, repeated or severe disturbances reduce community stability and functional redundancy. Human-driven pressures, like logging, agricultural intensification, urbanization, and pollution, have caused the decline of many specialists, especially in Europe and North America (Collado et al. 2019). Meanwhile, disturbance-tolerant taxa such as *Coprinellus micaceus* and *Agaricus bisporus* thrive in altered environments. Habitat fragmentation further disrupts dispersal and mycelial continuity (Looney et al. 2016). Emerging research suggests that land-use intensity, particularly agricultural expansion, is among the strongest anthropogenic pressures shaping fungal diversity worldwide (Větrovský et al. 2019).

Overall, these environmental drivers interact synergistically. Climate modulates soil chemistry, host distribution, and disturbance intensity, shaping the structure and resilience of fungal communities. Understanding these intertwined factors is essential for forecasting fungal responses to global change and guiding conservation strategies for the world's mycobiota.

## 2.8 Edible Mushrooms: Nutrition and Food Security

### 2.8.1 Nutritional Value and Dietary Benefits

Edible mushrooms have exceptional culinary attributes and organoleptic features, and for that reason, they have been consumed by humans since ancient times (Raut and Adhikari 2021). The Romans consumed mushrooms, which they considered to be food of the Gods, and the Chinese called them the elixir of life (Bashir et al. 2014). Reports of the consumption of mushrooms also date back to the Egyptian and the Greek civilizations, and similar evidence was found in some archaeological sites in the United Kingdom (O'Regan et al. 2016). Archaeologists in Italy found a food container with mushroom residue, confirming that humans in the Bronze Age supplemented their diets with wild-gathered mushrooms. Similarly, puffball remains have been identified at Paleolithic sites in Austria, Germany, and Switzerland, indicating early foraging practices. Remarkably, spores of Boletus and agaric mushrooms were found between the teeth of the "Red Lady of El Mirón," a Magdalenian-era skeleton from northern Spain dating back nearly 19,000 years (Wasson and McKenna 2015). Ancient peoples also practiced sun-drying to preserve mushrooms for consumption all year round (O'Regan et al. 2016).

Some mushrooms that can be safely consumed by humans have been highlighted for their valuable nutritive properties, in particular for their high protein content. Edible mushrooms have been recommended as a potential alternative to meat for human consumption, due to their meaty texture and flavor, allied to high protein content, including essential amino acids, polysaccharides, including β-glucan and chitin, vitamins, and dietary minerals. On the other hand, their caloric value is relatively low, conferring them the potential to provide high nutrition for relatively low energy (Khatun et al. 2012; Boro et al. 2025).

A review by Hu and Xu (2025) compiles the nutritional profile of mushrooms of Russula species, highlighting the high protein content (typically between 15% and 30%, although reporting higher values for *R. griseocarnosa*—32.2% *and R. brevipes*—31.6%, while some other species had lower protein content: *R. rosea*—12.2%, *R. lepida*—12.1%, *R. mairei*—11.0%, *R. virescens*—5.0%). The fat content varied between 0.2% for *R. mairei* and 7.2% for *R. griseocarnosa*. The fiber content ranged from 0.2% in *R. lepida* to 10.1% in *R. alatoreticula*. The mineral content, measured as ash, was found to be between 0.1% for *R. mairei* and 7.8% for *R. griseocarnosa*. Carbohydrates varied in the range from 40.8% for *R. brevipes* to 82.0% for *R. rosea*. The moisture content was relatively low, varying typically in the range from 5% to 10%, with *R. griseocarnosa* presenting a moisture content of 4.6% and *R. alatoreticula* having 9.1% water (Hu and Xu 2025).

Xu et al. (2019) present the chemical composition of *Lactarius deliciosus*, a mycorrhizal type of fungus, which presents 66.61 g/g dry mass (dm) carbohydrates, 17.19 g/100 gdm protein, 31.81 g/100 g dm fiber, 8.62 g/100 g dm of ash, 4.82 g/100 g dm fat, having 98% moisture content (wet basis). Su and Xu (2024) review the macronutrient composition of the same type of mushroom (*L. deliciosus*), showing the following contents: 60.30%–85.87% dry weight (dw) carbohydrates, 2.42%–6.47% dw fat, 6.95%–17.87% dw protein, 4.75%–14.28% dw ash, and 85.75%–90.05% fresh weight (fw) of moisture. The same work reports the energetic value of mushrooms between 371 and 399 Kcal/100 g dw.

Senila et al. (2024) evaluated the chemical composition of four wild edible mushroom species collected in Romania (*Agaricus campestris, Boletus edulis, Macrolepiota procera, and Russula cyanoxantha*). They found that protein content varied from 22.49 to 27.63 g/100 g dw, respectively for *B. edulis* and *M. procera*, total fat varied from 1.97 to 4.75 g/100 g dw, respectively for *A. campestris* and *R. cyanoxantha*, carbohydrates varied from 61.29 to 68.87 g/100 g dw, respectively for *R. cyanoxantha* and *B. edulis*, and ash content varied from 5.40 to 8.30 g/100 g dw, respectively for *B. edulis* and *A. campestris*.

Demirtas et al. (2024) evaluated the chemical composition and nutritional properties of six edible mushrooms collected in Turkey (*Hericium coralloides, Sparassis crispa, Suillus granulatus, Macrolepiota procera, Lepista nuda, and Suillus collinitus*). Moisture content varied from 65.22 to 93.3 g/100 g (respectively for *M. procera* and *S. granulatus*), proteins varied from 6.44 to 55.05 g/100 g dw (respectively for *H. corallaides* and *L. nuda*), carbohydrates varied from 32.51 to 80.08 g/100 g dw (respectively for *L. nuda* and *H. corallaides*), fat content varied from 2.09 to 8.87 g/100 g dw (respectively for *M. procera* and *S. crispa*), and ash varied from 4.53 to 9.82 g/100 g dw (respectively for *S. crispa* and *H. corallaides*). The caloric value ranged from 170 to 318 kcal/100 g dw, respectively for *H. corallaides* and *M. procera*.

Carbohydrates comprise sugars and fiber, including dietary soluble and insoluble fiber. Among the sugars present in *Lactarius deliciosus*, mannitol and trehalose were present in higher concentrations (1.50 and 0.24 g/100 g, respectively) when compared to ribose or sorbitol (0.14 and 0.16 g/100 g, respectively) (Su and Xu 2024). Regarding the dietary fiber content, Xu et al. (2019) present the contents in soluble and insoluble dietary fiber for the same type of mushrooms, 26.51 and 5.30 g/100 g dw, respectively. Soluble fiber dissolves in water, thus helping to regulate blood sugar and cholesterol levels. On the other hand, insoluble fiber, which does not dissolve, increases stool volume and promotes the regulation of the intestinal tract. Hence, consuming a diet rich in both types of fiber contributes to maintaining a healthy digestive system and reducing the risk of chronic diseases (Buttriss and Stokes 2008; Opperman et al. 2025).

Demirtas et al. (2024) reported the sugar and dietary fiber content in six edible mushrooms from Turkey (*Hericium coralloides, Sparassis crispa, Suillus granulatus, Macrolepiota procera, Lepista nuda, and Suillus collinitus*). Total dietary fiber ranged from 16.54 to 54.82 g/100 g dw (for *M. procera* and *H. coralloides*, respectively), the insoluble fraction varied from 15.79 to 50.60 g/100 g dw (for *M. procera* and *H. coralloides*, respectively), and the soluble fraction varied from 0.75 to 4.85 g/100 g dw (for *M. procera* and *S. granulatus*, respectively). Regarding the free sugars present, arabinose was present in higher concentrations in *H. coralloides* and *S. collinitus* (15.61 and 8.83 g/100 g dw), fructose was only detected in *L. nuda* (0.13 g/100 g dw), glucose was detected in the highest concentration in *S. collinitus* and *H. coralloides* (0.92 and 0.86 g/100 g dw, respectively), mannitol and trehalose were particularly high for *M. procera* (respectively 7.88 and 12.43 g/100 g dw).

Proteins in edible mushrooms contain essential amino acids vital for human health, as these cannot be synthesized by the body and must be obtained through diet. The nine essential amino acids—histidine, isoleucine, leucine, lysine, methionine, phenylalanine, threonine, tryptophan, and valine—are involved in many body functions, like growth, tissue repair, and the synthesis of hormones, enzymes, and neurotransmitters. They also support energy production, immune system, healthy skin, hair, and nails. Ingestion of specific amino acids in higher concentrations may improve gastrointestinal health, enhance immune responses, and influence behavior, while high levels of other amino acids can be associated with obesity (Dai et al.

2022; Lee et al. 2023). The most abundant amino acids in *Lactarius deliciosus* are arginine (6285 mg/100 g), glutamic acid (2631 mg/100 g), phenylalanine (1986 mg/100 g), aspartic acid (1863 mg/100 g), threonine (1274 mg/100 g), glycine (1264 mg/100 g), and lysine (1192 mg/100 g), with all others in concentrations below 1000 mg/100 g (Su and Xu 2024).

Fatty acids (FA) are crucial for human health, acting as major energy sources and key structural components of cell membranes. They regulate numerous biological processes, including cell signaling, inflammation, and chronic disease prevention. Essential fatty acids support membrane integrity, blood pressure regulation, coagulation, immune system, and absorption of fat-soluble vitamins. They also influence the synthesis of inflammatory mediators and protect cardiovascular health Among FA, the unsaturated fatty acids, especially the polyunsaturated fatty acids, are highly beneficial (Mititelu et al. 2025). The most abundant unsaturated fatty acids reported in *Lactarius deliciosus* mushrooms were oleic acid, a monounsaturated fatty acid (MUFA) (48.37%) and linoleic acid, a polyunsaturated fatty acid (PUFA) (29.49%). Stearic acid, which is a saturated fatty acid (SFA), was also found in high concentrations (44.38%) (Su and Xu 2024).

Fatty acid composition was evaluated by Senila et al. (2024) in four wild edible mushroom species (*Agaricus campestris, Boletus edulis, Macrolepiota procera, and Russula cyanoxantha*). They found values of SFA in the range from 19.65% to 32.18% (respectively for *B. edulis* and *R. cyanoxantha*), unsaturated fatty acids (UFA) varying in the range from 67.75% to 80.11% (respectively for *R. cyanoxantha* and *B. edulis*), of which the contents in MUFA were between 17.19% and 40.05% (respectively for *R. cyanoxantha* and *B. edulis*), and the PUFA was between 40.06% and 59.74% (respectively for *B. edulis* and *A. campestris*). The most abundant UFA were oleic and linoleic acids, and the most abundant SFA were palmitic and stearic acids.

The FA contents were evaluated in six edible mushrooms (*Hericium coralloides, Sparassis crispa, Suillus granulatus, Macrolepiota procera, Lepista nuda, and Suillus collinitus*) by Demirtas et al. (2024). The UFA oleic and linoleic acids were present in high concentrations (respectively in the ranges 15.51%–57.64% and 21.07%–64.61%). The SFA stearic and palmitic acids were also present in relative abundance (respectively in the ranges 1.58%–9.84% and 9.48%–29.40%).

Studies on mushroom vitamins are limited and mainly focus on vitamin D, as mushrooms are the only non-animal source of this nutrient (Boro et al. 2025). They primarily contain vitamin $D_2$, with smaller amounts of $D_3$ and $D_4$. Vitamin D promotes calcium absorption, reducing the risk of rickets in children and osteomalacia in adults. It also supports muscle function and may protect against diabetes, cardiovascular and respiratory diseases, certain cancers, and neurodegenerative diseases (Cardwell et al. 2018). A review by Cardwell et al. (2018) focused on the vitamin D contents in mushrooms, reporting contents in g/100 g fw for some wild mushrooms: *Craterellus tubaeformis* (3–30), *Cantharellus cibarius* (10.7), *Boletus edulis* (58.7), *Agaricus* species (1.5). They also reported that consistently, the cultivated mushrooms contain much lower amounts due to a lack of exposure to UV radiation in comparison to wild mushrooms. A few studies identified other vitamins present in mushrooms. Mattila et al. (2001) evaluated the vitamin contents in some cultivated mushrooms (*Agaricus bisporus*/white, *Agaricus bisporus*/brown, *Lentinula edodes*, and *Pleurotus ostreatus*). These were reported as being a source of vitamin $B_2$ (riboflavin) (1.2–5.1 g/100 g dw), niacin (31–65 g/100 g dw), and folates (0.30–0.64 g/100 g dw). Furlani and Godoy (2008) evaluated the contents of vitamins $B_1$ (thiamine) and $B_2$ in cultivated mushrooms: *Agaricus bisporus* (white button mushroom and portobello), *Lentinula edodes* (shiitake) and *Pleorotus* spp. (shimeji and oyster mushroom). The results showed thiamine in concentrations 0.004–0.08 mg/100 g and riboflavin 0.04–0.3 mg/100 g.

In comparison with green plants, mushrooms tend to present higher mineral content because they possess an active mechanism for collecting metals through the ecosystem (Ouzouni et al. 2007). Hence, wild edible mushrooms are considered particularly adequate to meet dietary mineral needs. The mineral content of *Lactarius deliciosus* was reported by Hu and Xu (Su and Xu 2024), highlighting the presence of potassium, sodium, calcium, phosphorus, magnesium, iron, copper, zinc, and manganese (26, 310, 390, 4, 960, 216, 16, 123, 23 g/kg dw). However, some heavy metals were also detected, namely chromium, arsenic, cadmium, lead, cobalt, and nickel (4.02, 0.75, 1.91, 0.85, 0.12, 1.58 mg/kg dw). Heavy metals can pose significant health risks if they are present in the human body at toxic levels. Their bioaccumulation may lead to diverse toxic effects affecting different body tissues and organ systems. Exposure to heavy metals can trigger several health problems, including damage to the nervous system, kidneys, liver, and cardiovascular system. Also, reproduction could be at risk, as well as certain types of cancer (Witkowska et al. 2021; Jomova et al. 2025).

Senila et al. (2024) evaluated the mineral elements in four edible mushrooms (*Agaricus campestris, Boletus edulis, Macrolepiota procera, and Russula cyanoxantha*). The minerals with nutritional interest were sodium (849 mg–1460 mg/kg dw), magnesium (410–659 mg/kg dw), potassium (19400–27,300 mg/kg dw), calcium (405–959 mg/kg dw), phosphorus (4600–6460 mg/kg dw), sulfur (1300–2540 mg/kg dw), iron (69.3–148 mg/kg dw), copper (36.1–76.9 mg/kg dw), zinc (67.4–157 mg/kg dw), manganese (9.21–15.1 mg/kg dw), boron (9.6–22.6 mg/kg dw), and strontium (2.13–2.60 mg/kg dw). On the other hand, some toxic heavy metals were also detected, like mercury (0.60–2.27 mg/kg dw), chromium (1.79–2.68 mg/kg dw), nickel (0.87–1.39 mg/kg dw), cadmium (0.17–0.23 mg/kg dw), lead (0.27–0.45 mg/kg dw), and arsenic (0.24–0.37 mg/kg dw).

Ouzouni et al. (2007) determined the metal content in eight wild edible mushrooms collected in Greece (*Cantharellus cibarius, Hydnum repandum, Lactarius salmonicolor, Xerocomus, chrysenteron, Agaricus cupreobrunneus, Amanita franchetii, Hygrophorus eburneus, and Hygrophorus chrysodon*). The dietary minerals' contents, expressed in μg/gdw, were: 739–1165 for magnesium, 0.41–13.1 for chromium, 11.3–100 for manganese, 46.3–317 for iron, 0.05–3.34 for cobalt, 0.28–6.53 for nickel, 3.80–27.4 for copper, 35.9–69.9 for zinc. Also, some heavy metals were analyzed, namely lead (0.05 μg–1.37 μg/g dw) and cadmium (0.09–0.41 μg/g dw).

Besides having macro and micronutrients, edible mushrooms also have other phytochemicals, like phenolic compounds, that have bioactive properties, conferring them health properties when ingested as dietary elements. These bioactive and health properties will be discussed further down in this chapter.

### 2.8.2 Potential for Sustainable Agriculture and Food Systems

Edible mushrooms are increasingly recognized as vital components of sustainable agriculture and global food security. Their unique ability to grow on lignocellulosic agricultural residues and organic waste positions them as a biotechnological solution for circular food systems (Royse et al. 2017). Unlike most crops, mushrooms do not require arable land, can be cultivated in controlled environments, and achieve high yields with minimal water and energy inputs, aligning closely with sustainable intensification principles (Carrasco et al. 2018).

A major contribution of mushroom cultivation lies in *waste valorization*. Species such as *Pleurotus ostreatus*, *Lentinula edodes*, and *Agaricus bisporus* can be grown on substrates derived from wheat straw, rice husks, corn stover, and other agricultural by-products (Sánchez 2009; Huang et al. 2024). This bioconversion process simultaneously mitigates waste accumulation and produces nutrient-dense food. After harvest, the residual spent mushroom substrate serves as organic fertilizer, animal feed supplement, or soil conditioner, effectively closing nutrient loops in agroecosystems (Baptista et al. 2023).

Mushroom farming is also among the *most efficient biological systems for protein production*. Protein yields per unit area can match or exceed those of traditional crops while using far less land (Chang and Wasser 2017). Furthermore, mushrooms provide high-quality protein, essential amino acids, dietary fiber, and micronutrients, making them crucial in regions facing food and nutrient insecurity (Valverde et al. 2015). Importantly, mushroom farming is viable in diverse climatic regions, from tropical to temperate zones, making it accessible for smallholders and commercial producers alike (Gateri et al. 2009; Khan et al. 2024).

From a socioeconomic perspective, mushroom cultivation supports *income diversification and rural resilience*. As a low-entry agribusiness, it can range from household-scale setups using local substrates to industrialized operations (Phan and Sabaratnam 2012; Royse et al. 2017). The relatively low capital requirements and rapid crop cycles empower farmers to enhance food security while generating income. Increasing adoption in sub-Saharan Africa and Southeast Asia reflects its growing role in rural development and poverty reduction initiatives (Chand and Singh 2022; Jayaraman et al. 2024).

Environmentally, edible mushrooms offer a *low-impact alternative to animal protein*. Life-cycle assessments reveal that mushroom production emits only a fraction of the greenhouse gases and uses far less water than livestock farming (Goglio et al. 2024). Expanding mushroom cultivation can therefore advance several Sustainable Development Goals, particularly SDG 2 (Zero Hunger) and SDG 12 (Responsible Consumption and Production).

Beyond food provision, mushrooms deliver broader **ecosystem services**. By recycling organic waste and enhancing soil fertility, they contribute to biodiversity conservation and the long-term resilience of agroecosystems. As global food systems confront escalating pressures from climate change, population growth, and land degradation, edible mushrooms exemplify a sustainable agricultural model that unites productivity, nutritional value, and environmental stewardship.

### 2.8.3 Toxicity Risks and Species Misidentification

Despite the substantial nutritional and economic potential of edible mushrooms, the risks associated with toxic species and misidentification remain a critical global concern. Many wild mushrooms closely resemble edible species in their morphology, leading to accidental poisonings that can cause severe illness or death (Govorushko et al. 2019; He et al. 2022a; Roux and Eyssartier 2024; Tuğcan and Akpınar 2025). As wild mushroom foraging remains a cultural tradition in rural areas and a growing gastronomic trend in urban centers, toxicological awareness and reliable identification methods are essential for ensuring food safety.

The most notorious toxic species belong to the genus *Amanita*, particularly *Amanita phalloides,* also known as death cap, which is probably the deadliest mushroom in the world, both because of its high toxicity and its frequency and abundance globally (Fig. 2.6). This species contains potent amatoxins that inhibit RNA polymerase II, leading to hepatic necrosis and multiorgan failure (Enjalbert

**Fig. 2.6** *Amanita phalloides.* (Original photo by José Costa)

et al. 2002). *Amanita virosa*, the "destroying angel," possesses similar toxins and causes equally severe poisoning, though less frequently. Misidentification with edible species such as *Agaricus* or *Volvariella volvacea* is a common cause of accidental ingestion across Europe, North America, and Asia (Wennig et al. 2020; Roux and Eyssartier 2024). Other genera like *Galerina* and *Lepiota*, also produce lethal amatoxins, further complicating field identification and emphasizing the limits of morphological diagnostics alone.

Toxicity risks also extend to species traditionally regarded as edible. *Gyromitra esculenta*, consumed in parts of Eastern Europe, contains gyromitrin, a volatile toxin causing hemolysis and hepatotoxicity; improper preparation can result in poisoning even after parboiling (Horowitz et al. 2025). Although parboiling reduces toxicity, the margin of safety is narrow, and cumulative consumption increases health risks. Likewise, *Coprinopsis atramentaria*, though edible when cooked, induces disulfiram-like reactions with alcohol due to the presence of coprine (Roux and Eyssartier 2024). These examples highlight the complex interplay between cultural practices and toxicological safety.

Misidentification remains a critical issue in regions where wild mushrooms contribute substantially to diets and livelihoods. In sub-Saharan Africa, South Asia, and Eastern Europe, foraging is common, yet formal training in taxonomy and toxicology is limited (Li et al. 2025). Poisonings are frequently underreported or misdiagnosed due to a lack of clinical expertise in mycetism, complicating epidemiological assessments (Wu et al. 2025). Although molecular tools such as ITS-region DNA barcoding have significantly improved species identification (Stockinger et al. 2010), access to such technologies remains confined to research or diagnostic settings.

Public health efforts increasingly focus on education and prevention. Campaigns in Europe and East Asia emphasize recognition of toxic look-alikes and advocate expert verification before consumption (Kaaronen 2020). Emerging mobile applications and digital identification platforms integrating morphological and molecular data offer promise for safer foraging (Yin et al. 2022; Hodgson et al. 2023), though their reliability and user overconfidence remain challenges.

Regulatory frameworks also vary widely. Some European nations require certification by trained mycologists for the sale of wild mushrooms, markedly reducing poisoning incidents (Kaaronen 2020). In contrast, informal markets in developing regions often lack oversight, increasing public health risks. Strengthening regulations and promoting collaboration between mycologists, health authorities, and local communities are crucial for balancing safety with cultural and economic values.

Ultimately, the close coexistence of edible and toxic mushrooms in natural habitats underscores the need for rigorous identification, consumer education, and coordinated policy action. Ensuring safe mushroom use is not merely a public health imperative, it is essential to sustaining the ecological, cultural, and nutritional benefits that mushrooms offer to societies worldwide.

## 2.9 Medicinal Mushrooms: Bioactive Compounds and Applications

### 2.9.1 Uses in Traditional and Modern Medicine

Since ancient times, mushrooms have been used as medicine. The Greek physician Hippocrates (~ 450 BC), classified *Fomes fomentarius* as a potent anti-inflammatory agent used for cauterizing wounds. Ötzi, the Ice Man, who lived nearly 5300 years ago in the Northern Italian Alps, carried amadou derived from *Fomes fomentarius* and the birch polypore (*Piptoporus betulinus*) for survival. In fifth-century AD China, the alchemist Tao Hongjing described several medicinal mushrooms, including *Ganoderma lucidum* and *Dendropolyporus umbellatus*, already in use for centuries. Similarly, the First Peoples of North America employed puffball mushrooms (*Calvatia* spp.) to heal wounds (Stamets and Zwickey 2014).

Although medicinal mushrooms have been used globally since antiquity, modern science has only recently rediscovered and confirmed their pharmacological potential, validating traditional knowledge (Stamets and Zwickey 2014). Numerous pharmaceutical substances with strong and unique properties were recently extracted from mushrooms and are now spread around the world as supplements. The Ganodermataceae family includes nearly forty species with hard basidiocarp (Rai et al. 2005).

Medicinal mushrooms, used as extracts, powders, or dietary supplements, are employed to prevent, alleviate, or treat diverse diseases. Their use in treating infections, lung diseases, and cancer has a long history, particularly in Asia. In China and Japan, mushroom-derived products have been officially approved as adjuvant therapies for cancer treatment for over three decades. These supplements are widely integrated into complementary and integrative medicine, especially within oncology, where they enhance treatment efficacy, reduce adverse effects, and improve patients' quality of life (Venturella et al. 2021).

Contemporary research provides strong evidence supporting these applications. A recent review focuses on the use of medicinal mushrooms for the treatment of multiple sclerosis (Pourmohammadi and Moghimi 2025). Patel and Goyal reviewed the use of mushrooms as anticancer therapeutics (Patel and Goyal 2012). Yong et al. presented a review of the use of mushrooms as anti-angiogenic agents (Yong et al. 2025). Porcaro et al. reviewed the role of mushrooms in the management of HPV (human papillomavirus) (Porcaro et al. 2025). Bell et al. reviewed how mushroom consumption can enhance neurologic health (Bell et al. 2025). The relevance of medicinal mushrooms as therapeutic agents against insulin resistance has been the scope of the review presented by Németh et al. (Németh et al. 2025). Sharika et al. reviewed the use of mushroom bioactive compounds against aging and lifestyle-related diseases (Sharika et al. 2024). These examples, among many more, are elucidative of the great amount of scientific evidence supporting the utilization of mushrooms as therapeutic agents, offering promising avenues for preventive healthcare and integrative medicine in the twenty-first century.

### 2.9.2 Bioactive Constituents and Pharmacological Properties

Medicinal mushrooms exhibit a broad spectrum of pharmacological activities, such as anti-allergic, antibacterial, anti-depressive, antidiabetic, antifungal, anti-hyperlipidemic, anti-inflammatory, antimicrobial, anti-obesity, antioxidative, antitumor, antiviral, cardiovascular-protective, cytotoxic, digestive, hepatoprotective, hypotensive, immunomodulatory, nephroprotective, neuroprotective, osteoprotective activities, and various other health-promoting properties (Venturella et al. 2021; Boro et al. 2025).

A range of medicinal properties are attributed to mushrooms due to their bioactive compounds. These bioactive compounds include the polyphenols (including phenolic acids and flavonoids), polysaccharides, terpenes, proteins, fatty acids, nucleoside antagonists, terpenoids, sesquiterpenes, lanostanoids, sterols, and laccases (copper-containing oxidases) (Popović et al. 2012; Elkhateeb 2020).

Polysaccharides are the structuring components of cell walls and have the capacity to carry biological information. These components demonstrated antidiabetic, anti-inflammatory, antimicrobial, antioxidative, anti-tumor, and immunomodulatory, activities. However, the effects are dependent on the specific structural characteristics of the molecules involved (namely, the type of monosaccharides that constitute the molecule, the degree of branching, among others). Alpha and beta-glucans are the most abundant, but also heteroglycans, peptidoglycans, and polysaccharide–protein complexes have been attributed biological activity. Their immunomodulatory effects derive from their ability to bind to specific cell wall receptors and stimulate particular immune responses (Zhao et al. 2020; Venturella et al. 2021).

The terpenes are a class of compounds with relevant bioactivity. These molecules are composed of units of five-carbon isoprene atoms, to which are added functional groups to produce the terpenoids. Their immunomodulatory activity is due to the stimulation of gene expression for proteins involved in the immune response. Other terpenes' activities include anti-inflammatory, antioxidant, and anti-tumor effects (Enshasy and Hatti-Kaul 2013; Venturella et al. 2021).

Some proteins found in mushrooms have cytotoxic and anticancer properties, with a strong immunomodulatory effect. Proteins also include lectins, which bind reversibly to highly specific mono- and oligosaccharides. These will recognize and interact with several carbohydrates and proteoglycans, being involved in many biological activities, such as innate immunity, immunomodulatory, antitumor, and antiproliferative properties (Enshasy and Hatti-Kaul 2013; Venturella et al. 2021).

Many different types of phenolic compounds are present in mushrooms, conferring bioactive properties. Phenolic compounds are responsible for antioxidant activities in mushroom extracts through acting as decomposers of peroxidase, inactivators of metals, oxygen scavengers, or inhibitors of free radicals. Phenolic compounds include phenolic acids, oxidized polyphenols, hydroxybenzoic acids, flavonoids, tannins, hydroxycinnamic acids, stilbenes, and lignans (Elkhateeb et al.

2019; Elkhateeb 2020). Phenolic compounds have attracted much attention owing to their outstanding properties as antioxidant, anti-inflammatory or antitumor agents, among other activities (Palacios et al. 2011).

### 2.9.3 Key Medicinal Species

Many species of mushrooms can be used in therapy applications. Some of the most used can be found in Table 2.5 Some of the most relevant medicinal species of mushrooms and their applications(Moretti 2024; Real Mushrooms 2025). These examples highlight the vast domain of applicability of mushrooms in preventing and treating diseases.

## 2.10 Regional Case Studies

Regional case studies provide crucial insights into the ecology, taxonomy, and socioeconomic importance of edible and medicinal mushrooms across the globe. Because fungi exhibit highly habitat-specific distributions, their uses are deeply shaped by cultural traditions, ecological conditions, and the availability of resources. Exploring regional diversity not only enhances scientific understanding but also preserves traditional knowledge and fosters sustainable management practices relevant to both biodiversity conservation and food security.

### 2.10.1 Examples of Diversity and Use Across Different World Regions

The distribution and utilization of edible and medicinal mushrooms are shaped by ecological conditions, biogeography, and cultural traditions. Regional patterns reveal both the taxonomic diversity of fungi and their varied roles in diets, economies, and healthcare systems.

In *Europe*, ectomycorrhizal species such as *Boletus edulis* (porcini), *Cantharellus cibarius* (chanterelle), *Hydnum repandum* (hedgehog mushroom), and *Tricholoma matsutake* are culturally emblematic of temperate and boreal forests (Pérez-Moreno 2021; Pérez-Moreno et al. 2021; Roux and Eyssartier 2024). These mushrooms hold central roles in regional cuisines, like porcini in Italian risottos or chanterelles in Scandinavian dishes, and foraging remains a strong rural tradition, especially in Eastern Europe, where it contributes to food and income security (Chauvin-Payan 2024). In addition, medicinal fungi such as *Inonotus obliquus* (chaga) have long been used in Northern and Eastern Europe, and modern studies confirm their antioxidant and immunomodulatory activities (Lu et al. 2021; Szychowski et al. 2021; Windsor et al. 2025; Wang et al. 2025).

In East Asia, mushroom diversity and utilization are even more pronounced, reflecting millennia of domestication and ethnomedicinal traditions. *Lentinula*

**Table 2.5** Some of the most relevant medicinal species of mushrooms and their applications

| Species | Therapeutic applications | References |
|---|---|---|
| *Agaricus bisporus* | Improve memory<br>Improve motor function (including balance)<br>Reduce cancer<br>Support cardiovascular health<br>Improve gut health<br>Modulate prostate-specific antigen levels<br>Antioxidant activity | Twardowski et al. (2015), Rahi and Malik (2016), and Hess et al. (2018) |
| *Agaricus subrufescens (sin. Agaricus blazei)* | Has strong immune-enhancing properties<br>Anti-tumor activity | Patel and Goyal (2012) and Rahi and Malik (2016) |
| *Cordyceps militaris*<br>*Ophiocordyceps sinensis* | Treat asthma<br>Alleviate chronic obstructive pulmonary disease<br>Support kidney transplant patients<br>Offer kidney protection<br>Improve energy<br>Improve physical performance<br>Has hormone-regulatory properties<br>Antitumor activity | Yi et al. (2004), Li et al. (2009), Chen et al. (2010b), Patel and Goyal (2012), Jiraungkoorskul and Jiraungkoorskul (2016), Sun et al. (2019), and Elkhateeb et al. (2019) |
| *Ganoderma lucidum* (sin. *G. lingzhi*) | Reduce fatigue<br>Improve memory and cognitive functions<br>Support the immune system<br>Improve sleep<br>Promote longevity<br>Improve quality of life<br>Improve lung function<br>Decrease symptoms of urination difficulty in men<br>Maintain healthy blood pressure<br>Reduce precancerous adenomas<br>Antitumor activity<br>Anti-diabetic activity | Noguchi et al. (2008), Patel and Goyal (2012), Phan et al. (2015), Lam et al. (2020), and Khursheed et al. (2020) |
| *Grifola frondosa* | Antitumor activity<br>Increase natural killer cell response<br>Increase macrophage and cytotoxic T-cell activity<br>Improve antibody response<br>Activate immune complement pathways<br>Improve memory and cognitive functions | Inoue et al. (2002), Kodama et al. (2003), Chen et al. (2010a), Patel and Goyal (2012), and Phan et al. (2015) |

(continued)

**Table 2.5** (continued)

| Species | Therapeutic applications | References |
|---|---|---|
| *Hericium erinaceus* | Enhance brain health and memory<br>Improve memory in Alzheimer's patients<br>Reduce irritability<br>Reduce symptoms of depression<br>Increase nerve growth<br>Reduce inflammation<br>Antitumor activity<br>Anti-diabetic activity | Patel and Goyal (2012), Lai et al. (2013), Phan et al. (2015), Spelman et al. (2017), and Khursheed et al. (2020) |
| *Inonotus obliquus* | Reduce allergies<br>Reduce fatigue<br>Reduce diabetes by improving glucose and insulin control<br>Reduce cholesterol<br>Improve memory<br>Support cancer treatment<br>Reduce HIV viral load<br>Improve skin conditions<br>Slow aging<br>Reduce inflammation<br>Boost digestion<br>Protect the skin<br>Antitumor activity<br>Immunomodulator | Patel and Goyal (2012), Rahi and Malik (2016), and Khursheed et al. (2020) |
| *Lentinula edodes* | Treat gastric cancer (through lentinan that boosts the body's response to cancer cells)<br>Support cardiovascular health<br>Control cholesterol<br>Modulate blood sugar levels<br>Enhance immunity<br>Decrease inflammation | Bisen et al. (2010) and Patel and Goyal (2012) |
| *Morchella esculenta* | Reduce cancer cells' growth<br>Reduce blood glycaemia<br>Protect the liver<br>Cure pneumonia<br>Cure cold, fever, and cough<br>Treats respiratory ailments | Rahi and Malik (2016) |
| *Phellinus Linteus* | Immunomodulatory benefits for autoimmune diseases<br>Modifies biological response<br>Antitumor activity<br>Anti-inflammatory activity<br>Antihistamine<br>Antioxidant effect<br>Anti-diabetic activity | Patel and Goyal (2012), Rahi and Malik (2016), and Khursheed et al. (2020) |

(continued)

**Table 2.5** (continued)

| Species | Therapeutic applications | References |
|---|---|---|
| *Pleurotus ostreatus* | Support heart functions<br>Improve metabolic health<br>Improve the immune system<br>Maintain healthy blood cholesterol levels<br>Antitumor activity | Schneider et al. (2011), Patel and Goyal (2012), Anandhi et al. (2013), Jesenak et al. (2014, 2017), and Minov et al. (2025) |
| *Trametes versicolor* (sin. *Coriolus versicolor*) | Improve immune system<br>Reduce HPV infections<br>Support a healthy gut and immunity (through polysaccharide-K extract)<br>Improve microbiome (ex. reduce E.coli)<br>Protect the body (by increasing antioxidant activity)<br>Antitumor activity | Wong et al. (2004), Patel and Goyal (2012), and Lam et al. (2020) |
| *Tremella fuciformis* | Enhance immunity<br>Skin anti-age effect<br>Has neurological benefits<br>Has antitumor activity<br>Improve memory<br>Improve gut function<br>Has antioxidant activity | Wu et al. (2016, 2019) and Shen et al. (2017) |

*edodes* (shiitake), *Auricularia auricula-judae* (wood ear), *Hericium erinaceus* (lion's mane), and *Ganoderma lucidum* (reishi) dominate cultivation systems, while wild species such as *Ophiocordyceps sinensis* remain among the world's most expensive medicinal fungi (Wei et al. 2021; Chellapandi and Saranya 2023). China produces over 75% of global mushrooms and also leads the advances in substrate recycling using agricultural residues (Royse et al. 2017). Japan and Korea also maintain large-scale cultivation of *Flammulina velutipes* (enoki) and *Pholiota nameko* (Chang and Wasser 2017). Beyond their economic role, mushrooms also hold symbolic value; *Ganoderma lucidum* (lingzhi) is revered as the "mushroom of immortality" in Chinese culture.

In *Africa*, wild edible mushrooms are crucial for food security and rural livelihoods, especially during rainy seasons. *Termitomyces* species, symbiotic with termites, are highly prized for their size, flavor, and nutritional value (Koné et al. 2018). These fungi support local markets in Uganda, Zambia, and Nigeria, where seasonal trade contributes significantly to household income (Paloi et al. 2023). Other wild mushrooms, such as *Lactarius* spp. and *Russula* spp. are also consumed, although local knowledge is crucial to avoid misidentification. Despite their importance, african mycobiota remains underdocumented, posing challenges to conservation and sustainable exploitation (Chelela 2015).

In the *Americas*, mushroom use reflects a blend of indigenous traditions and contemporary gastronomy. In North America, wild-foraged species such as *Morchella esculenta* (morel), *Cantharellus formosus* (Pacific chanterelle), and *Laetiporus sulphureus* (chicken of the woods) support significant seasonal industries, with morels alone forming a multimillion-dollar market (Pilz et al. 2007). Indigenous groups continue to use *Fomitopsis betulina* (birch polypore) medicinally, particularly for wound treatment (Arora and Shepard 2008). In Latin America, *Mexico* is a center of fungal diversity and ethnomycological heritage. Edible species such as *Amanita caesarea* var. *heibapha* (Caesar's mushroom) and *Lactarius indigo* (blue milk cap) are esteemed in local cuisines, while hallucinogenic *Psilocybe* species retain ceremonial roles among Mazatec and Mixtec peoples (López-García et al. 2025; Toxqui-Munguía et al. 2025). Farther south, in the Amazon basin, fungi form part of indigenous diets and cosmologies, though much traditional knowledge remains undocumented.

Beyond their cultural and nutritional roles, mushrooms are increasingly targeted by *conservation initiatives*, as many emblematic taxa face threats from climate change, habitat loss, and overharvesting. In Europe, the IUCN Red List of Fungi identifies species such as *Hericium erinaceus* and *Lyophyllum fumosum* as threatened, prompting habitat protection and monitoring efforts (Dahlberg and Mueller 2011). In Asia, overexploitation of *Ophiocordyceps sinensis* has led to community-based conservation in Tibet and Nepal, involving quotas, rotational harvesting zones, and cultivation trials (Shrestha and Bawa 2013; Chandra et al. 2016; Wei et al. 2021; Chellapandi and Saranya 2023). In North America, organizations such as the North American Mycological Association collaborate with forestry agencies to safeguard old-growth specialists like *Bondarzewia mesenterica* (Kunca et al. 2022). In Africa, community-driven programs in Zambia and Tanzania employ mapping and seasonal restrictions to sustain *Termitomyces* populations (Chelela 2015).

These regional examples underscore the intertwined ecological, cultural, and economic significance of fungi. Mushrooms act as biodiversity indicators, nutritional resources, medicinal agents, and drivers of local economies. Recognizing this diversity also highlights urgent challenges, like unsustainable harvesting, habitat degradation, and climate change threaten species ranging from *Ophiocordyceps sinensis* in Tibet to *Boletus edulis* in Europe (Wei et al. 2021; Pérez-Moreno et al. 2021). Addressing these threats requires interdisciplinary strategies that integrate conservation biology, ethnomycology, and sustainable development approaches.

### 2.10.2 Traditional Knowledge and Ethnomycological Insights

The cultural significance of mushrooms extends far beyond their nutritional and medicinal properties. Across human history, wild fungi have occupied a central place in mythology, religion, healing, and subsistence. Ethnomycology—the study of human-fungus relationships—reveals how societies have perceived, classified, and utilized mushrooms, intertwining practical experience with symbolic meaning (Yamin-Pasternak and Pasternak 2018; Toxqui-Munguía et al. 2025).

In ancient civilizations, mushrooms frequently appeared as sacred organisms. Mesoamerican cultures such as the Aztecs and Maya revered hallucinogenic *Psilocybe* species, known as *teonanácatl* or "flesh of the gods," employing them in rituals to communicate with deities and ancestors (Moreno Fuentes 2014; Lazcano-Flores and Torres-Montúfar 2024). Spanish chroniclers of the sixteenth century documented these practices, though filtered through colonial suppression. Similarly, in Siberia, *Amanita muscaria* (fly agaric), with its red cap and white spots, played a role in shamanic traditions, facilitating trance states and spiritual journeys. Its imagery later permeated European folklore and may have influenced Christmas iconography (Znamenski 2003; Chauvin-Payan 2024).

Traditional knowledge also encodes fine distinctions between edible and toxic fungi. In Eastern Europe, folk taxonomies and oral transmission sustain detailed recognition systems that distinguish edible *Boletus* and *Cantharellus* from deadly *Amanita* species (Chauvin-Payan 2024). In Africa, indigenous knowledge identifies seasonal edible species such as *Termitomyces*, associated with termite mounds and integral to communal diets and rituals (Oyetayo 2012; Koné et al. 2018). Ethnographic research highlights that misidentification is rare within traditional frameworks, though globalization and migration sometimes disrupt knowledge transmission, increasing the risk of poisoning (Antonelli et al. 2020).

Medicinal uses are equally prominent. In Traditional Chinese Medicine, *Ganoderma lucidum* (lingzhi) and *Cordyceps sinensis* have been used for millennia to promote vitality, longevity, and balance (Chandra et al. 2016; Wei et al. 2021). In Ayurveda and Japanese Kampo medicine, *Lentinula edodes* (shiitake) and *Hericium erinaceus* (lion's mane) are prescribed for immune and digestive health. These traditional applications anticipated modern pharmacological findings identifying β-glucans and triterpenoids as immunomodulatory and therapeutic compounds (Wasser 2017).

The fascination with psychoactive fungi bridges ritual, science, and counterculture. In the 1950s, R. Gordon Wasson's ethnomycological research publicized Mazatec rituals in Oaxaca, Mexico, where *Psilocybe mexicana* was used ceremonially under the guidance of Maria Sabina. This discovery inspired early psychiatric research and the psychedelic movement, leading to contemporary clinical trials testing psilocybin for depression and post-traumatic stress disorder (Carhart-Harris and Goodwin 2017). The path from indigenous ritual to modern biomedicine illustrates the enduring relevance of traditional mushroom knowledge.

Mushrooms have also inspired myth and caution. In European folklore, fairy rings, a circular growth patterns of fungi such as those of *Marasmius oreades*, were believed to mark the dance sites of fairies or witches, symbolizing enchantment and danger (Dugan 2008). Similarly, *Amanita phalloides* (death cap), the cause of most fatal mushroom poisonings worldwide, has long represented both fear and respect, shaping European attitudes toward wild foraging (Enjalbert et al. 2002; White et al. 2019). The dual symbolism of mushrooms as sacred and deadly underscores their complex role in human imagination.

Beyond food and medicine, fungi have supported traditional technologies. In Europe, *Fomes fomentarius* (tinder fungus) served as a fire starter and was carried by Ötzi the Iceman 5300 years ago alongside *Piptoporus betulinus*, likely used medicinally or for fire-making (Pérez-Moreno et al. 2021). Indigenous peoples of North America used polypores as wound dressings and as materials for tools or decoration, reflecting fungi's integration into material culture.

Collectively, ethnomycological research reveals mushrooms as organisms at the intersection of ecology, culture, and spirituality. They function as foods, medicines, symbols, and teachers—agents of healing and wonder as well as danger. Preserving this intertwined biological and cultural heritage is essential, as globalization, environmental degradation, and cultural change threaten both fungal biodiversity and the traditional knowledge systems that sustain it.

## 2.11 Future Perspectives and Challenges

The study and utilization of edible and medicinal mushrooms are entering a transformative phase, shaped by rapid advances in biotechnology, changing land-use practices, and an urgent need for biodiversity conservation. While traditional uses remain central to cultural heritage and food security, modern research highlights mushrooms as key players in global sustainability agendas. To fully realize this potential, it is necessary to integrate cutting-edge molecular tools, sustainable cultivation strategies, and conservation frameworks, while also addressing the persistent gaps in knowledge and infrastructure that limit broader applications.

### 2.11.1 Advances in Fungal Biotechnology and Genomics

Recent breakthroughs in fungal genomics and biotechnology have greatly expanded understanding of mushroom biology. Genome sequencing of *Lentinula edodes*, *Ganoderma lucidum*, and *Pleurotus ostreatus* have revealed the genetic basis for lignocellulose degradation, secondary metabolite biosynthesis, and stress tolerance, enabling targeted strain improvement (Nagy et al. 2016; Tarafder et al. 2024; Kijpornyongpan et al. 2025). CRISPR-Cas9–based genome editing, initially applied to model fungi, is now being adapted to mushroom-forming species to enhance yields, nutritional quality, and bioactive compound production (Zhang et al. 2023). In industrial biotechnology, mushrooms are recognized as biofactories for enzymes, polysaccharides, and novel metabolites, supporting applications in pharmaceuticals and biodegradable materials (Llanaj et al. 2023). Synthetic biology further enables customized biosynthetic pathways for compounds such as terpenoids and β-glucans. Collectively, these innovations position mushrooms as key resources for sustainable biotechnological and bioeconomic development.

### 2.11.2 Integration into Agroforestry and Ecosystem Restoration

Mushrooms are vital to ecosystem functioning, and their integration into agroforestry and restoration systems is gaining prominence. Edible ectomycorrhizal fungi, such as *Tuber* and *Boletus* species, can be cultivated in association with tree crops, producing food while enhancing soil fertility through nutrient cycling (Yamada 2022). Saprotrophic species like *Pleurotus* and *Volvariella* are increasingly used in agroecological models to recycle agricultural residues, contributing to circular food systems (Royse et al. 2017). Fungal inoculation is also applied in degraded or burned landscapes, where mycorrhizal symbioses accelerate soil regeneration, plant establishment, and biodiversity recovery (Větrovský et al. 2019). These practices provide income opportunities for rural communities and support global ecosystem restoration goals. However, scaling up requires improved inoculation protocols, host compatibility studies, and viable economic frameworks to make mushroom-based restoration sustainable at larger scales.

### 2.11.3 Research Gaps and Priority Areas

Despite notable progress, critical gaps persist in mushroom research. Taxonomic knowledge remains incomplete, particularly in tropical regions, where many edible and medicinal species are still undescribed or poorly documented (Hawksworth and Lücking 2017). Standardized methods for the cultivation and domestication of wild edible fungi are lacking, limiting their sustainable use and market expansion. Furthermore, the ecological roles of many mushrooms in soil carbon dynamics and climate feedback are underexplored, requiring integration of fungal data into Earth system models (Treseder and Lennon 2015). Another priority is the development of conservation strategies for threatened fungal species, as most global biodiversity policies still overlook fungi compared to plants and animals (Lofgren and Stajich 2021). Advances in ethnomycological documentation are also urgent, given the erosion of traditional knowledge due to cultural shifts and habitat loss. Addressing these research gaps requires interdisciplinary collaboration, capacity building in mycology, and policy frameworks that explicitly include fungi in biodiversity and sustainability agendas.

## 2.12 Conclusion

Mushrooms represent a remarkable intersection of ecology, culture, and sustainability. Their diversity across ecosystems reflects evolutionary innovation and the intricate ecological networks that sustain soil fertility, plant health, and forest resilience. Far from marginal organisms, edible and medicinal fungi are structural elements of the biosphere, regulating carbon and nutrient cycles and supporting ecosystem recovery after disturbance. Their ecological functions highlight the profound interdependence between fungal biodiversity and human well-being.

Simultaneously, mushrooms bridge tradition and innovation. For millennia, they have nourished and healed communities, shaping rituals, myths, and medical practices. Today, they drive progress in biotechnology, genomics, pharmaceuticals, and sustainable agriculture. While wild species retain cultural and economic importance, cultivated mushrooms exemplify how food production can align with ecological stewardship. This duality between heritage and modern science underscores their value as both symbols and instruments of sustainability.

However, the future of fungal diversity faces mounting threats from habitat loss, climate change, and the erosion of traditional knowledge. Conservation, equitable benefit-sharing, and stronger recognition of fungi within biodiversity policies are essential.

Ultimately, mushrooms remind us that ecological integrity, food security, and cultural identity are inseparable. By valuing their diversity and fostering responsible innovation, we open pathways toward more resilient ecosystems and societies, redefining humanity's place within the living fabric of the planet.

**Acknowledgments** This work was supported by the FCT-Foundation for Science and Technology, I.P. Furthermore, we would like to thank the Research Centre CERNAS throughout project UIDB/00681, as well as the Polytechnic University of Viseu, for their support.

**Declaration** Authors do not have any competing interests.

## References

Adamčík S, Cai L, Chakraborty D et al (2015) Fungal biodiversity profiles 1–10. Cryptogam Mycol 36:121–166. https://doi.org/10.7872/crym/v36.iss2.2015.121

Aerts R (2003) The role of various types of mycorrhizal fungi in nutrient cycling and plant competition. In: Van Der Heijden MGA, Sanders IR (eds) Mycorrhizal ecology. Springer, Berlin/Heidelberg, pp 117–133

Agrawal DC, Dhanasekaran M (eds) (2019) Medicinal mushrooms: recent progress in research and development. Springer, Singapore

Ainsworth GC, Bisby GR (2011) Ainsworth & Bisby's dictionary of the fungi, 10th edn. CABI, Wallingford

Anandhi R, Annadurai T, Anitha TS et al (2013) Antihypercholesterolemic and antioxidative effects of an extract of the oyster mushroom, Pleurotus ostreatus, and its major constituent, chrysin, in Triton WR-1339-induced hypercholesterolemic rats. J Physiol Biochem 69:313–323. https://doi.org/10.1007/s13105-012-0215-6

Andrew C, Heegaard E, Halvorsen R et al (2016) Climate impacts on fungal community and trait dynamics. Fungal Ecol 22:17–25. https://doi.org/10.1016/j.funeco.2016.03.005

Antonelli A, Fry C, Smith RJ et al (2020) State of the world's plants and fungi 2020. Royal Botanic Gardens, Kew

Arora D, Shepard GH (2008) Mushrooms and economic botany1. Econ Bot 62:207–212. https://doi.org/10.1007/s12231-008-9046-3

Asif M, Saba M, Maula F et al (2025) Multilocus phylogeny contributes to fungal taxonomy with three novel species of *Conocybe* (Bolbitiaceae) from southern Punjab, Pakistan. Mycologia 117: 463–479. https://doi.org/10.1080/00275514.2025.2463279

Baby S, Johnson AJ, Govindan B (2015) Secondary metabolites from Ganoderma. Phytochemistry 114:66–101. https://doi.org/10.1016/j.phytochem.2015.03.010

Bahram M, Hildebrand F, Forslund SK et al (2018) Structure and function of the global topsoil microbiome. Nature 560:233–237. https://doi.org/10.1038/s41586-018-0386-6

Bahram M, Netherway T, Hildebrand F et al (2020) Plant nutrient-acquisition strategies drive topsoil microbiome structure and function. New Phytol 227:1189–1199. https://doi.org/10.1111/nph.16598

Baldrian P (2016) Forest microbiome: diversity, complexity and dynamics. FEMS Microbiol Rev: fuw040. https://doi.org/10.1093/femsre/fuw040

Baldrian P, Bell-Dereske L, Lepinay C et al (2022) Fungal communities in soils under global change. Stud Mycol 103:1–24. https://doi.org/10.3114/sim.2022.103.01

Balestrini R, Mello A (2015) Truffle research in the post-genomics era. Food Anal Methods 8: 1815–1823. https://doi.org/10.1007/s12161-014-0064-8

Baptista F, Almeida M, Paié-Ribeiro J et al (2023) Unlocking the potential of Spent Mushroom Substrate (SMS) for enhanced agricultural sustainability: from environmental benefits to poultry nutrition. Life 13:1948. https://doi.org/10.3390/life13101948

Bashir A, Vaida N, Dar MA (2014) Medicinal importance of mushrooms—a review. Int J Adv Res 2:1–4

Bashir H, Chen J, Jabeen S et al (2021) An overview of Agaricus section Hondenses and Agaricus section Xanthodermatei with description of eight new species from Pakistan. Sci Rep 11:12905. https://doi.org/10.1038/s41598-021-92261-5

Bell V, Dimitrov P, Fernandes T (2025) Supporting neurologic health with mushroom nutrition. Nutrients 17:1568. https://doi.org/10.3390/nu17091568

Binder M, Hibbett DS (2006) Molecular systematics and biological diversification of Boletales. Mycologia 98:971–981. https://doi.org/10.1080/15572536.2006.11832626

Bisen PS, Baghel RK, Sanodiya BS et al (2010) Lentinus edodes: a macrofungus with pharmacological activities. Curr Med Chem 17:2419–2430. https://doi.org/10.2174/092986710791698495

Boa ER (2004) Wild edible fungi: a global overview of their use and importance to people. Food and Agriculture Organization of the United Nations, Rome

Boro S, Kambhampati V, Das S, Saikia D (2025) Edible mushrooms as meat analogues: a comprehensive review of nutritional, therapeutic, and market potential. Food Res Int 214: 116632. https://doi.org/10.1016/j.foodres.2025.116632

Brace AJ, Ruthrof KX, Miller BP et al (2024) Short-term soil fungal community dynamics following fire in mediterranean climate-type banksia woodlands. Soil Biol Biochem 199: 109579. https://doi.org/10.1016/j.soilbio.2024.109579

Büntgen U, Kauserud H, Egli S (2012) Linking climate variability to mushroom productivity and phenology. Front Ecol Environ 10:14–19. https://doi.org/10.1890/110064

Burgess T, Wingfield MJ (2004) Impact of fungal pathogens in natural forest ecosystems: a focus on eucalypts. In: Sivasithamparama K, Dixon KW, Barrett RL (eds) Microorganisms in plant conservation and biodiversity. Kluwer Academic Publishers, Dordrecht, pp 285–306

Buscardo E, Geml J, Schmidt SK et al (2022) Nitrogen pulses increase fungal pathogens in Amazonian lowland tropical rain forests. J Ecol 110:1775–1789. https://doi.org/10.1111/1365-2745.13904

Buttriss JL, Stokes CS (2008) Dietary fibre and health: an overview. Nutr Bull 33:186–200. https://doi.org/10.1111/j.1467-3010.2008.00705.x

Buyck B (2014) Exploring the diversity of "smooth chanterelles" (*Cantharellus*, Cantharellales). Cryptogam Mycol 35:23–40. https://doi.org/10.7872/crym.v35.iss1.2014.23

Candido HMN, Alvim Carvalho F (2025) Global trends in urban forests' ecosystem services: a bibliometric analysis. Arboric J 47:1–19. https://doi.org/10.1080/03071375.2025.2533653

Cao B, He M-Q, Ling Z-L et al (2021) A revision of *Agaricus* section *Arvenses* with nine new species from China. Mycologia 113:191–211. https://doi.org/10.1080/00275514.2020.1830247

Cardwell G, Bornman JF, James AP, Black LJ (2018) A review of Mushrooms as a potential source of dietary vitamin D. Nutrients 10:1498. https://doi.org/10.3390/nu10101498

Carhart-Harris RL, Goodwin GM (2017) The therapeutic potential of psychedelic drugs: past, present, and future. Neuropsychopharmacology 42:2105–2113. https://doi.org/10.1038/npp.2017.84

Carrasco J, Zied DC, Pardo JE et al (2018) Supplementation in mushroom crops and its impact on yield and quality. AMB Express 8:146. https://doi.org/10.1186/s13568-018-0678-0
Chan GC-F, Chan WK, Sze DM-Y (2009) The effects of β-glucan on human immune and cancer cells. J Hematol Oncol 2:25. https://doi.org/10.1186/1756-8722-2-25
Chand S, Singh B (2022) Mushroom cultivation for increasing income and sustainable development of small and marginal farmers. AJAHR:11–16. https://doi.org/10.9734/ajahr/2022/v9i430148
Chandra SN, Mukesh P, Paras J, Sachin B (2016) Conserving the caterpillar fungus [Ophiocordyceps sinensis (Berk.) G.H. Sung et al.]: a case study of habitat ecology and sustainability in district Pithoragarh, Western Himalaya. India Int J Biodivers Conserv 8:196–205. https://doi.org/10.5897/IJBC2014.0742
Chandrasekaran M (2022) Arbuscular Mycorrhizal fungi mediated enhanced biomass, root morphological traits and nutrient uptake under drought stress: a meta-analysis. JoF 8:660. https://doi.org/10.3390/jof8070660
Chang ST, Wasser SP (2017) The cultivation and environmental impact of mushrooms. In: Oxford research encyclopedia of environmental science. Oxford University Press
Chauvin-Payan C (2024) Wild mushrooms: uses, consumption and popular beliefs from Europe to Eastern Siberia. ethnoecologie 25. https://doi.org/10.4000/12a68
Chelela B (2015) Wild mushrooms from Tanzania: characterization and their importance to the rural communities. CREAM 5:307–321. https://doi.org/10.5943/cream/5/4/2
Chellapandi P, Saranya S (2023) *Ophiocordyceps sinensis:* a potential Caterpillar fungus for the production of bioactive compounds. Explor Res Hypothesis Med 000:000–000. https://doi.org/10.14218/ERHM.2023.00040
Chen J-T, Tominaga K, Sato Y et al (2010a) Maitake mushroom (Grifola frondosa) extract induces ovulation in patients with polycystic ovary syndrome: a possible monotherapy and a combination therapy after failure with first-line clomiphene citrate. J Altern Complement Med 16:1295–1299. https://doi.org/10.1089/acm.2009.0696
Chen S, Li Z, Krochmal R et al (2010b) Effect of Cs-4® (Cordyceps sinensis) on exercise performance in healthy older subjects: a double-blind, placebo-controlled trial. J Altern Complement Med 16:585–590. https://doi.org/10.1089/acm.2009.0226
Chen W, Tan H, Liu Q et al (2019) A review: the bioactivities and pharmacological applications of Phellinus linteus. Molecules 24:1888. https://doi.org/10.3390/molecules24101888
Clemmensen KE, Finlay RD, Dahlberg A et al (2015) Carbon sequestration is related to mycorrhizal fungal community shifts during long-term succession in boreal forests. New Phytol 205: 1525–1536. https://doi.org/10.1111/nph.13208
Collado E, Bonet JA, Camarero JJ et al (2019) Mushroom productivity trends in relation to tree growth and climate across different European forest biomes. Sci Total Environ 689:602–615. https://doi.org/10.1016/j.scitotenv.2019.06.471
Corrales A, Henkel TW, Smith ME (2018) Ectomycorrhizal associations in the tropics—biogeography, diversity patterns and ecosystem roles. New Phytol 220:1076–1091. https://doi.org/10.1111/nph.15151
Corrales A, Koch RA, Vasco-Palacios AM et al (2022) Diversity and distribution of tropical ectomycorrhizal fungi. Mycologia 114:919–933. https://doi.org/10.1080/00275514.2022.2115284
Crowther TW, Boddy L, Hefin Jones T (2012) Functional and ecological consequences of saprotrophic fungus–grazer interactions. ISME J 6:1992–2001. https://doi.org/10.1038/ismej.2012.53
Dahlberg A (2002) Effects of fire on ectomycorrhizal fungi in Fennoscandian boreal forests. Silva Fenn 36. https://doi.org/10.14214/sf.551
Dahlberg A, Mueller GM (2011) Applying IUCN red-listing criteria for assessing and reporting on the conservation status of fungal species. Fungal Ecol 4:147–162. https://doi.org/10.1016/j.funeco.2010.11.001
Dai Z, Zheng W, Locasale JW (2022) Amino acid variability, tradeoffs and optimality in human diet. Nat Commun 13:6683. https://doi.org/10.1038/s41467-022-34486-0

Das G, Shin H-S, Leyva-Gómez G et al (2021) Cordyceps spp.: a review on its immune-stimulatory and other biological potentials. Front Pharmacol 11:602364. https://doi.org/10.3389/fphar.2020.602364

Dasgupta D, Basu P, Paul A et al (2025) *Schizophyllum commune*, an underrated edible and medicinal mushroom: farm to industry. Stud Fungi 10:0–0. https://doi.org/10.48130/sif-0025-0003

DemiRel Y, DemiRel G (2023) Deep learning based approach for classification of mushrooms. GU J Sci Part A: Eng Innov 10:487–498. https://doi.org/10.54287/gujsa.1355751

Demirtas N, Sengul GF, Dizeci N, Yildirim O (2024) Exploring the chemical composition and nutritional properties of six edible mushroom species from Turkey. J Food Compos Anal 133: 106477. https://doi.org/10.1016/j.jfca.2024.106477

Dugan F (2008) Fungi, folkways and fairy tales: mushrooms & mildews in stories, remedies & rituals, from Oberon to the internet. North Am Fungi:23–72. https://doi.org/10.2509/naf2008.003.0074

Elkhateeb WA (2020) What medicinal mushroom can do? Chem Res J 5:106–118

Elkhateeb WA, Daba GM, Thomas PW, Wen T-C (2019) Medicinal mushrooms as a new source of natural therapeutic bioactive compounds. Egypt Pharm J 18:88–101. https://doi.org/10.4103/epj.epj_17_19

Enjalbert F, Rapior S, Nouguier-Soulé J et al (2002) Treatment of Amatoxin poisoning: 20-year retrospective analysis. J Toxicol Clin Toxicol 40:715–757. https://doi.org/10.1081/CLT-120014646

Enshasy HAE, Hatti-Kaul R (2013) Mushroom immunomodulators: unique molecules with unlimited applications. Trends Biotechnol 31:668–677. https://doi.org/10.1016/j.tibtech.2013.09.003

Espinosa J, Dejene T, Guijarro M et al (2023) Fungal diversity and community composition responses to the reintroduction of fire in a non-managed Mediterranean shrubland ecosystem. For Ecosyst 10:100110. https://doi.org/10.1016/j.fecs.2023.100110

Fang H, Yang S, Yang T (2025) Ganoderma lucidum polysaccharides: a comprehensive overview of pharmacological effects and future perspectives. Food Biosci 64:105990. https://doi.org/10.1016/j.fbio.2025.105990

Frank JL, Siegel N, Schwarz CF et al (2020) Xerocomellus (Boletaceae) in western North America. Fung Syst Evol 6:265–288. https://doi.org/10.3114/fuse.2020.06.13

Fufa BK, Tadesse BA, Tulu MM (2021) Cultivation of Pleurotus ostreatus on agricultural wastes and their combination. Int J Agron 2021:1–6. https://doi.org/10.1155/2021/1465597

Furlani RPZ, Godoy HT (2008) Vitamins B1 and B2 contents in cultivated mushrooms. Food Chem 106:816–819. https://doi.org/10.1016/j.foodchem.2007.06.007

Gateri MW, Muriuki AW, Waiganjo MW, Ngeli P (2009) Cultivation and commercialization of edible mushrooms in Kenya: a review of prospects and challenges for smallholder production. Acta Hortic:473–480. https://doi.org/10.17660/ActaHortic.2009.806.59

Geml J, Morgado LN, Semenova TA et al (2015) Long-term warming alters richness and composition of taxonomic and functional groups of arctic fungi. FEMS Microbiol Ecol 91:fiv095. https://doi.org/10.1093/femsec/fiv095

Geng Y, Wang Y, Li H et al (2024) Optimization of manure-based substrate preparation to reduce nutrients losses and improve quality for growth of Agaricus bisporus. Agriculture 14:1833. https://doi.org/10.3390/agriculture14101833

Glassman SI, Weihe C, Li J et al (2018) Decomposition responses to climate depend on microbial community composition. Proc Natl Acad Sci USA 115:11994–11999. https://doi.org/10.1073/pnas.1811269115

Goglio P, Ponsioen T, Carrasco J et al (2024) An environmental assessment of Agaricus bisporus ((J.E.Lange) Imbach) mushroom production systems across Europe. Eur J Agron 155:127108. https://doi.org/10.1016/j.eja.2024.127108

Gomdola D, Bhunjun C, Hyde K et al (2022) Ten important forest fungal pathogens: a review on their emergence and biology. Mycosphere 13:612–671. https://doi.org/10.5943/mycosphere/13/1/6

Govorushko S, Rezaee R, Dumanov J, Tsatsakis A (2019) Poisoning associated with the use of mushrooms: a review of the global pattern and main characteristics. Food Chem Toxicol 128: 267–279. https://doi.org/10.1016/j.fct.2019.04.016
Grelet G, Martino E, Dickie IA et al (2016) Ecology of ericoid mycorrhizal fungi: what insight have we gained with molecular tools and what's missing? In: Martin F (ed) Molecular mycorrhizal symbiosis, 1st edn. Wiley, pp 405–419
Grienke U, Zöll M, Peintner U, Rollinger JM (2014) European medicinal polypores—a modern view on traditional uses. J Ethnopharmacol 154:564–583. https://doi.org/10.1016/j.jep.2014.04.030
Griffith GW (2013) The international conservation importance of Welsh 'waxcap' grasslands. Mycosphere 4:969–984. https://doi.org/10.5943/mycosphere/4/5/10
Grimm D, Wösten HAB (2018) Mushroom cultivation in the circular economy. Appl Microbiol Biotechnol 102:7795–7803. https://doi.org/10.1007/s00253-018-9226-8
Grinn-Gofroń A, Bosiacka B (2015) Effects of meteorological factors on the composition of selected fungal spores in the air. Aerobiologia 31:63–72. https://doi.org/10.1007/s10453-014-9347-1
Hadziabdic D, Onufrak A, Gazis R (2022) Fungal pathogens in forested ecosystems. In: Mycoagroecology, 1st edn. CRC Press, Boca Raton, pp 175–195
Hallenberg N, Nilsson RH, Robledo G (2013) Species complexes in Hericium (Russulales, Agaricomycota) and a new species – Hericium rajchenbergii – from southern South America. Mycol Progress 12:413–420. https://doi.org/10.1007/s11557-012-0848-4
Han X, Liu D, Zhang M et al (2023) Macrofungal diversity and distribution patterns in the primary forests of the Shaluli Mountains. J Fungi 9:491. https://doi.org/10.3390/jof9040491
Hawksworth DL, Lücking R (2017) Fungal diversity revisited: 2.2 to 3.8 million species. Microbiol Spectr 5(4). https://doi.org/10.1128/microbiolspec.FUNK-0052-2016
He M-Q, Wang M-Q, Chen Z-H et al (2022a) Potential benefits and harms: a review of poisonous mushrooms in the world. Fungal Biol Rev 42:56–68. https://doi.org/10.1016/j.fbr.2022.06.002
He Z, Lin J, He Y, Liu S (2022b) Polysaccharide-peptide from Trametes versicolor: the potential medicine for colorectal cancer treatment. Biomedicine 10:2841. https://doi.org/10.3390/biomedicines10112841
Heleno SA, Barros L, Sousa MJ et al (2010) Tocopherols composition of Portuguese wild mushrooms with antioxidant capacity. Food Chem 119:1443–1450. https://doi.org/10.1016/j.foodchem.2009.09.025
Heng G (2016) Mycosphere essays 4. Mycorrhizal-associated nutrient dynamics in key ecosystems and their response to a changing environment. Mycosphere 7:190–203. https://doi.org/10.5943/mycosphere/7/2/8
Hess J, Wang Q, Gould T, Slavin J (2018) Impact of Agaricus bisporus mushroom consumption on gut health markers in healthy adults. Nutrients 10:1402. https://doi.org/10.3390/nu10101402
Hewitt RE, Day NJ, DeVan MR, Taylor DL (2023) Wildfire impacts on root-associated fungi and predicted plant–soil feedbacks in the boreal forest: research progress and recommendations. Funct Ecol 37 2110–2125. https://doi.org/10.1111/1365-2435.14205
Hobbs C (2020) Christopher Hobbs's medicinal mushrooms: the essential guide: boost immunity, improve memory, fight cancer, stop infection, and expand your consciousness. Storey Publishing, North Adams
Hodgson SE, McKenzie C, May TW, Greene SL (2023) A comparison of the accuracy of mushroom identification applications using digital photographs. Clin Toxicol 61:166–172. https://doi.org/10.1080/15563650.2022.2162917
Hongsanan S, Khuna S, Manawasinghe I et al (2025) Mycosphere notes 521–571: a special edition of fungal biodiversity to celebrate Kevin D. Hyde's 70th birthday and his exceptional contributions to mycology. Mycosphere 16:1–178. https://doi.org/10.5943/mycosphere/16/2/1
Hoppe B, Purahong W, Wubet T et al (2016) Linking molecular deadwood-inhabiting fungal diversity and community dynamics to ecosystem functions and processes in central European forests. Fungal Divers 77:367–379. https://doi.org/10.1007/s13225-015-0341-x

Horisawa S, Iwamoto K (2022) Identification and typing of strains of wood-rotting basidiomycetes by protein profiling using MALDI-TOF MS. Biotech 11:30. https://doi.org/10.3390/biotech11030030

Horowitz KM, Kong EL, Regina AC, Horowitz BZ (2025) Gyromitra mushroom toxicity. In: StatPearls. StatPearls publishing, Treasure Island

Hu X, Xu B (2025) Chemical compositions and health promoting effects of edible mushrooms from genus *Russula*. Phytomedicine Plus 5:100677. https://doi.org/10.1016/j.phyplu.2024.100677

Huang J, Wang J, Liu S (2024) Advances in the production of fungi-derived lignocellulolytic enzymes using agricultural wastes. Mycology 15:523–537. https://doi.org/10.1080/21501203.2023.2253827

Hyde KD, Xu J, Rapior S et al (2019) The amazing potential of fungi: 50 ways we can exploit fungi industrially. Fungal Divers 97:1–136. https://doi.org/10.1007/s13225-019-00430-9

Ikechi-Nwogu CG, Egere MU (2024) Macroscopic and molecular approaches for identification of fungi associated with Azanza garckeana. Asian J Biol Sci 17:324–330. https://doi.org/10.3923/ajbs.2024.324.330

Ikehata K (2015) Use of fungal laccases and peroxidases for enzymatic treatment of wastewater containing synthetic dyes. In: Sharma SK (ed) Green chemistry for dyes removal from wastewater, 1st edn. Wiley, pp 203–260

Inoue A, Kodama N, Nanba H (2002) Effect of Maitake (Grifola frondosa) D-fraction on the control of the T lymph node Th-1/Th-2 proportion. Biol Pharm Bull 25:536–540. https://doi.org/10.1248/bpb.25.536

Iryna B, Omoanghe S. I, Alina K, et al (2022) Mushroom fruiting body yield and morphological characteristics from different strains of Pleurotus eryngii. J Appl Biol Biotech https://doi.org/10.7324/JABB.2021.100101

Jayachandran M, Xiao J, Xu B (2017) A critical review on health promoting benefits of edible Mushrooms through gut microbiota. IJMS 18:1934. https://doi.org/10.3390/ijms18091934

Jayaraman S, Yadav B, Dalal RC et al (2024) Mushroom farming: a review focusing on soil health, nutritional security and environmental sustainability. Farming System 2:100098. https://doi.org/10.1016/j.farsys.2024.100098

Jeewanthi LAMN, Ratnayake K, Rajapakse P (2017) Growth and yield of Reishi mushroom [Ganoderma lucidum (Curtis) P. Karst] in different sawdust substrates. J Food Agric 10:8. https://doi.org/10.4038/jfa.v10i1-2.5208

Jeewon R (2016) Establishing species boundaries and new taxa among fungi: recommendations to resolve taxonomic ambiguities. Mycosphere 7:1669–1677. https://doi.org/10.5943/mycosphere/7/11/4

Jesenak M, Hrubisko M, Majtan J et al (2014) Anti-allergic effect of Pleuran (β-glucan from Pleurotus ostreatus) in children with recurrent respiratory tract infections. Phytother Res 28:471–474. https://doi.org/10.1002/ptr.5020

Jesenak M, Urbancikova I, Banovcin P (2017) Respiratory tract infections and the role of biologically active polysaccharides in their management and prevention. Nutrients 9:779. https://doi.org/10.3390/nu9070779

Jiraungkoorskul K, Jiraungkoorskul W (2016) Review of naturopathy of medical mushroom, Ophiocordyceps sinensis, in sexual dysfunction. Pharmacogn Rev 10:1–5. https://doi.org/10.4103/0973-7847.176566

Jomova K, Alomar SY, Nepovimova E et al (2025) Heavy metals: toxicity and human health effects. Arch Toxicol 99:153–209. https://doi.org/10.1007/s00204-024-03903-2

Jorjadze A, Kupradze I, Batsatsashvili K, Bussmann RW (2024) Sparassis crispa (Wulfen) Fr. Sparassidaceae. In: Bussmann RW (ed) Ethnobotany of the Caucasus. Springer, Cham, pp 1–3

K D, S P S, B V et al (2024) Medicinal potential of Turkey tail mushroom (Trametes versicolor): a comprehensive review. S Afr J Bot 172:254–266. https://doi.org/10.1016/j.sajb.2024.07.013

Kaaronen RO (2020) Mycological rationality: heuristics, perception and decision-making in mushroom foraging. Judgm Decis Mak 15:630–647. https://doi.org/10.1017/S1930297500007841

Kalač P (2016) Edible mushrooms: chemical composition and nutritional value. Academic, Amsterdam/Boston/Heidelberg/London/New York/Oxford/Paris/San Diego/San Francisco/ Singapore/Sydney/Tokyo

Kauserud H, Stige LC, Vik JO et al (2008) Mushroom fruiting and climate change. Proc Natl Acad Sci USA 105:3811–3814. https://doi.org/10.1073/pnas.0709037105

Kauserud H, Heegaard E, Büntgen U et al (2012) Warming-induced shift in European mushroom fruiting phenology. Proc Natl Acad Sci USA 109:14488–14493. https://doi.org/10.1073/pnas.1200789109

Kerekes J, Kaspari M, Stevenson B et al (2013) Nutrient enrichment increased species richness of leaf litter fungal assemblages in a tropical forest. Mol Ecol 22:2827–2838. https://doi.org/10.1111/mec.12259

Keuler R, Garretson A, Saunders T et al (2020) Genome-scale data reveal the role of hybridization in lichen-forming fungi. Sci Rep 10:1497. https://doi.org/10.1038/s41598-020-58279-x

Khan A, Murad W, Salahuddin et al (2024) Contribution of mushroom farming to mitigating food scarcity: current status, challenges and potential future prospects in Pakistan. Heliyon 10: e40362. https://doi.org/10.1016/j.heliyon.2024.e40362

Khatun S, Islam A, Cakilcioglu U, Chatterjee NC (2012) Research on mushroom as a potential source of nutraceuticals: a review on Indian perspective. J Exp Agric Int:47–73. https://doi.org/10.9734/AJEA/2012/492

Khursheed R, Singh SK, Wadhwa S et al (2020) Therapeutic potential of mushrooms in diabetes mellitus: role of polysaccharides. Int J Biol Macromol 164:1194–1205. https://doi.org/10.1016/j.ijbiomac.2020.07.145

Kijpornyongpan T, Kuatsjah E, Komp E et al (2025) Cross-kingdom comparative genomics reveal the metabolic potential of fungi for lignin turnover in deadwood. Nat Ecol Evol 9:1599–1613. https://doi.org/10.1038/s41559-025-02785-6

Kodama N, Komuta K, Nanba H (2003) Effect of Maitake (Grifola frondosa) D-fraction on the activation of NK cells in cancer patients. J Med Food 6:371–377. https://doi.org/10.1089/109662003772519949

Kohler A, Kuo A, Nagy LG et al (2015) Convergent losses of decay mechanisms and rapid turnover of symbiosis genes in mycorrhizal mutualists. Nat Genet 47:410–415. https://doi.org/10.1038/ng.3223

Koné NA, Soro B, Vanié-Léabo LPL et al (2018) Diversity, phenology and distribution of Termitomyces species in Côte d'Ivoire. Mycology 9:307–315. https://doi.org/10.1080/21501203.2018.1500498

Kües U (2015) From two to many: multiple mating types in basidiomycetes. Fungal Biol Rev 29: 126–166. https //doi.org/10.1016/j.fbr.2015.11.001

Kunca V, Peiger M, Tomka P, Vampola P (2022) Old-growth forest fungi—new localities and habitat and host preferences in Slovakia (I). Czech Mycol 74:33–55. https://doi.org/10.33585/cmy.74103

Laessøe T, Petersen JH (2019) Fungi of temperate Europe. Princeton University Press, Princeton Oxford

Lai P-L, Naidu M, Sabaratnam V et al (2013) Neurotrophic properties of the lion's mane medicinal mushroom, *Hericium erinaceus* (Higher basidiomycetes) from Malaysia. IJM 15. https://doi.org/10.1615/IntJMedMushr.v15.i6.30

Lam CS, Cheng LP, Zhou LM et al (2020) Herb–drug interactions between the medicinal mushrooms Lingzhi and Yunzhi and cytotoxic anticancer drugs: a systematic review. Chin Med 15:75. https://doi.org/10.1186/s13020-020-00356-4

Lambert H (2024) Fungal systematics and evolution: FUSE 10. Sydowia 76:297–365. https://doi.org/10.12905/0380.sydowia76-2024-0297

Lazcano-Flores E, Torres-Montúfar A (2024) The edible mushrooms from small community Tierras Coloradas in Hidalgo, Mexico. GSC Biol Pharm Sci 28:251–261. https://doi.org/10.30574/gscbps.2024.28.1.0282

Lee H, Wissitrassameewong K, Park MS et al (2019) Taxonomic revision of the genus Lactarius (Russulales, Basidiomycota) in Korea. Fungal Divers 95:275–335. https://doi.org/10.1007/s13225-019-00425-6

Lee JT, Rochell SJ, Kriseldi R et al (2023) Functional properties of amino acids: improve health status and sustainability. Poult Sci 102:102288. https://doi.org/10.1016/j.psj.2022.102288

Leonardi M, Iotti M, Pacioni G et al (2021) Truffles: biodiversity, ecological significances, and biotechnological applications. In: Abdel-Azeem AM, Yadav AN, Yadav N, Usmani Z (eds) Industrially important fungi for sustainable development. Springer, Cham, pp 107–146

Li Y, Xue W-J, Tian P-X et al (2009) Clinical application of *Cordyceps sinensis* on immunosuppressive therapy in renal transplantation. Transplant Proc 41:1565–1569. https://doi.org/10.1016/j.transproceed.2009.02.085

Li H, Zhang Y, Zhang H et al (2025) Mushroom poisoning outbreaks — China, 2024. China CDC Wkly 7:645–649. https://doi.org/10.46234/ccdcw2025.106

Liu R, He Y, Zhou G et al (2021) Mycorrhizal effects on decomposition and soil $CO_2$ flux depend on changes in nitrogen availability during forest succession. J Ecol 109:3929–3943. https://doi.org/10.1111/1365-2745.13770

Llanaj X, Törős G, Hajdú P et al (2023) Biotechnological applications of mushrooms under the water-energy-food nexus: crucial aspects and prospects from farm to pharmacy. Foods 12:2671. https://doi.org/10.3390/foods12142671

Lofgren LA, Stajich JE (2021) Fungal biodiversity and conservation mycology in light of new technology, big data, and changing attitudes. Curr Biol 31:R1312–R1325. https://doi.org/10.1016/j.cub.2021.06.083

Looney B (2013) Systematics of the genus Auricularia with an emphasis on species from the southeastern United States. N Am Fungi. https://doi.org/10.2509/naf2013.008.006

Looney BP, Ryberg M, Hampe F et al (2016) Into and out of the tropics: global diversification patterns in a hyperdiverse clade of ectomycorrhizal fungi. Mol Ecol 25:630–647. https://doi.org/10.1111/mec.13506

López-García A, Gómez-Hernández M, Gándara E (2025) Folk taxonomy of wild mushrooms in communities of the indigenous groups Chatino, Chontal, and Chinantec in Oaxaca, Mexico. J Ethnobiol Ethnomed 21:30. https://doi.org/10.1186/s13002-025-00779-z

López-Hortas L, Flórez-Fernández N, Torres MD, Domínguez H (2022) Update on potential of edible mushrooms: high-value compounds, extraction strategies and bioactive properties. Int J of Food Sci Tech 57:1378–1385. https://doi.org/10.1111/ijfs.15544

Lu Y, Jia Y, Xue Z et al (2021) Recent developments in Inonotus obliquus (Chaga mushroom) polysaccharides: isolation, structural characteristics, biological activities and application. Polymers 13:1441. https://doi.org/10.3390/polym13091441

Luo X, Liu Y, Li S, He X (2023) Interplant carbon and nitrogen transfers mediated by common arbuscular mycorrhizal networks: beneficial pathways for system functionality. Front Plant Sci 14:1169310. https://doi.org/10.3389/fpls.2023.1169310

Lupala AS, Kim D, Cho Y et al (2025) Species diversity of the amanita section Caesareae in South Korea. Mycol Prog 24:29. https://doi.org/10.1007/s11557-025-02032-z

Ma G, Yang W, Zhao L et al (2018) A critical review on the health promoting effects of mushrooms nutraceuticals. Food Sci Hum Wellness 7:125–133. https://doi.org/10.1016/j.fshw.2018.05.002

Mahmood S, Fahad Z, Bolou-Bi EB et al (2024) Ectomycorrhizal fungi integrate nitrogen mobilisation and mineral weathering in boreal forest soil. New Phytol 242:1545–1560. https://doi.org/10.1111/nph.19260

Martin F, Kohler A, Murat C et al (2016) Unearthing the roots of ectomycorrhizal symbioses. Nat Rev Microbiol 14:760–773. https://doi.org/10.1038/nrmicro.2016.149

Mattila P, Könkö K, Eurola M et al (2001) Contents of vitamins, mineral elements, and some phenolic compounds in cultivated Mushrooms. J Agric Food Chem 49:2343–2348. https://doi.org/10.1021/jf001525d

Maurya GK, Pachauri S (2022) Fungi: the indicators of pollution. In: Freshwater mycology. Elsevier, pp 277–296

Minov J, Bislimovska-Karadzhinska J, Petrova T et al (2025) Effects of Pleuran (B-Glucan from Pleurotus Ostreatus) supplementation on incidence and duration of COPD exacerbations. Open Access Maced J Med Sci 5:893–898. https://doi.org/10.3889/oamjms.2017.198

Mishra A, Shankar S (2025) Edible mushrooms for improved human health, food security and environmental sustainability: a critical review. Sci Total Environ 995:180093. https://doi.org/10.1016/j.scitotenv.2025.180093

Mititelu M, Lupuliasa D, Neacșu SM et al (2025) Polyunsaturated fatty acids and human health: a key to modern nutritional balance in association with polyphenolic compounds from food sources. Foods 14:46. https://doi.org/10.3390/foods14010046

Moreno Fuentes Á (2014) Un recurso alimentario de los grupos originarios y mestizos de México: los hongos silvestres. Anales de Antropología 48:241–272. https://doi.org/10.1016/S0185-1225(14)70496-5

Moretti H (2024) Twelve medicinal mushrooms and their unique benefits | The Healthy RD. https://thehealthyrd.com/12-medicinal-mushrooms-you-will-want-to-try/. Accessed 16 Aug 2025

Mridu AN, Kumar J (2024) Impact of climatic variations on fungi, mycorrhiza and lichens –an appraisal. CREAM 14:89–107. https://doi.org/10.5943/cream/14/1/5

Mueller GM, Cunha KM, May TW et al (2022) What do the first 597 global fungal red list assessments tell us about the threat status of fungi? Diversity 14:736. https://doi.org/10.3390/d14090736

Nagy LG, Szöllősi G (2017) Fungal phylogeny in the age of genomics: insights into phylogenetic inference from genome-scale datasets. In: Advances in Genetics. Elsevier, pp 49–72

Nagy LG, Riley R, Tritt A et al (2016) Comparative genomics of early-diverging mushroom-forming fungi provides insights into the origins of lignocellulose decay capabilities. Mol Biol Evol 33:959–970. https://doi.org/10.1093/molbev/msv337

Németh Z, Paulinné Bukovics M, Sümegi LD et al (2025) The importance of edible medicinal mushrooms and their potential use as therapeutic agents against insulin resistance. Int J Mol Sci 26:827. https://doi.org/10.3390/ijms26020827

Newbound M, Mccarthy MA, Lebel T (2010) Fungi and the urban environment: a review. Landsc Urban Plan 96:138–145. https://doi.org/10.1016/j.landurbplan.2010.04.005

Nilsson RH, Larsson K-H, Taylor AFS et al (2019) The UNITE database for molecular identification of fungi: handling dark taxa and parallel taxonomic classifications. Nucleic Acids Res 47: D259–D264. https://doi.org/10.1093/nar/gky1022

Noguchi M, Kakuma T, Tomiyasu K et al (2008) Randomized clinical trial of an ethanol extract of *Ganoderma lucidum* in men with lower urinary tract symptoms. Asian J Androl 10:777–785. https://doi.org/10.1111/j.1745-7262.2008.00361.x

Nuhn ME, Binder M, Taylor AFS et al (2013) Phylogenetic overview of the Boletineae. Fungal Biol 117:479–511. https://doi.org/10.1016/j.funbio.2013.04.008

O'Regan HJ, Lamb AL, Wilkinson DM (2016) The missing mushrooms: searching for fungi in ancient human dietary analysis. J Archaeol Sci 75:139–143. https://doi.org/10.1016/j.jas.2016.09.009

Olariaga I, Moreno G, Manjón JL et al (2017) Cantharellus (Cantharellales, Basidiomycota) revisited in Europe through a multigene phylogeny. Fungal Divers 83:263–292. https://doi.org/10.1007/s13225-016-0376-7

Oliva J, Stenlid J, Martínez-Vilalta J (2014) The effect of fungal pathogens on the water and carbon economy of trees: implications for drought-induced mortality. New Phytol 203:1028–1035. https://doi.org/10.1111/nph.12857

Oliveira AG, Stevani CV, Waldenmaier HE et al (2015) Circadian control sheds light on fungal bioluminescence. Curr Biol 25:964–968. https://doi.org/10.1016/j.cub.2015.02.021

Opperman C, Majzoobi M, Farahnaky A et al (2025) Beyond soluble and insoluble: a comprehensive framework for classifying dietary fibre's health effects. Food Res Int 206:115843. https://doi.org/10.1016/j.foodres.2025.115843

Ouzouni PK, Veltsistas PG, Paleologos EK, Riganakos KA (2007) Determination of metal content in wild edible mushroom species from regions of Greece. J Food Compos Anal 20:480–486. https://doi.org/10.1016/j.jfca.2007.02.008

Oyetayo VO (2012) Wild termitomyces species collected from Ondo and Ekiti states are more related to African species as revealed by ITS region of rDNA. ScientificWorldJournal 2012: 689296. https://doi.org/10.1100/2012/689296

Palacios I, Lozano M, Moro C et al (2011) Antioxidant properties of phenolic compounds occurring in edible mushrooms. Food Chem 128:674–678. https://doi.org/10.1016/j.foodchem.2011.03.085

Paloi S, Kumla J, Paloi BP et al (2023) Termite mushrooms (Termitomyces), a potential source of nutrients and bioactive compounds exhibiting human health benefits: a review. J Fungi 9:112. https://doi.org/10.3390/jof9010112

Patel R (2019) A moldy application of MALDI: MALDI-ToF mass spectrometry for fungal identification. J Fungi 5:4. https://doi.org/10.3390/jof5010004

Patel S, Goyal A (2012) Recent developments in mushrooms as anti-cancer therapeutics: a review. 3 Biotech 2:1–15. https://doi.org/10.1007/s13205-011-0036-2

Pawlik A, Malinowska A, Siwulski M et al (2015) Determination of biodiversity of Coprinus comatus using genotyping and metabolic profiling tools. Acta Biochim Pol 62:683–689. https://doi.org/10.18388/abp.2015_1102

Peay KG, Kennedy PG, Talbot JM (2016) Dimensions of biodiversity in the earth mycobiome. Nat Rev Microbiol 14:434–447. https://doi.org/10.1038/nrmicro.2016.59

Pérez-Moreno J (2021) Global perspectives on the ecological, cultural and socioeconomic relevance of wild edible fungi. SIF 6:408–424. https://doi.org/10.5943/sif/6/1/31

Pérez-Moreno J, Guerin-Laguette A, Rinaldi AC et al (2021) Edible mycorrhizal fungi of the world: what is their role in forest sustainability, food security, biocultural conservation and climate change? Plants People Planet 3:471–490. https://doi.org/10.1002/ppp3.10199

Pettifor BJ, Denman S, McDonald JE (2022) Using a systematic approach to synthesize existing knowledge on *Gymnopus fusipes* (syn. *Collybia fusipes*), the cause of *Collybia* root rot. Forensic Pathol 52:e12766. https://doi.org/10.1111/efp.12766

Phan C-W, Sabaratnam V (2012) Potential uses of spent mushroom substrate and its associated lignocellulosic enzymes. Appl Microbiol Biotechnol 96:863–873. https://doi.org/10.1007/s00253-012-4446-9

Phan C-W, David P, Naidu M et al (2015) Therapeutic potential of culinary-medicinal mushrooms for the management of neurodegenerative diseases: diversity, metabolite, and mechanism. Crit Rev Biotechnol 35:355–368. https://doi.org/10.3109/07388551.2014.887649

Phukhamsakda C, Nilsson RH, Bhunjun CS et al (2022) The numbers of fungi: contributions from traditional taxonomic studies and challenges of metabarcoding. Fungal Divers 114:327–386. https://doi.org/10.1007/s13225-022-00502-3

Pilz D, McLain R, Alexander S et al (2007) Ecology and management of morels harvested from the forests of western North America. U.S. Department of Agriculture, Forest Service, Pacific Northwest Research Station, Portland

Popović BM, Štajner D, Slavko K, Sandra B (2012) Antioxidant capacity of cornelian cherry (Cornus mas L.)—comparison between permanganate reducing antioxidant capacity and other antioxidant methods. Food Chem 134:734–741. https://doi.org/10.1016/j.foodchem.2012.02.170

Porcaro G, Calcagno M, Tinelli A (2025) Medicinal mushrooms, probiotics and combination of natural compounds in the management of HPV: a comparative look at viral clearance and lesion resolution. Viruses 17:942. https://doi.org/10.3390/v17070942

Pourmohammadi G, Moghimi H (2025) New approach of using medicinal mushrooms for managing and curing multiple sclerosis. Fitoterapia 185:106684. https://doi.org/10.1016/j.fitote.2025.106684

Rahi DK, Malik D (2016) Diversity of mushrooms and their metabolites of nutraceutical and therapeutic significance. J Mycol 2016:7654123. https://doi.org/10.1155/2016/7654123

Rai M, Tidke G, Wasser SP (2005) Therapeutic potential of mushrooms. Nat Prod Radiance 4:246–257

Rakić M, Marković M, Galić Z et al (2022) Diversity and distribution of macrofungi in protected mountain forest habitats in Serbia and its relation to abiotic factors. J Fungi 8:1074. https://doi.org/10.3390/jof8101074

Rathore H, Prasad S, Sharma S (2017) Mushroom nutraceuticals for improved nutrition and better human health: a review. Pharma Nutr 5:35–46. https://doi.org/10.1016/j.phanu.2017.02.001

Raut JK, Adhikari MK (2021) Mushroom: a true super food. In: Comprehensive insights in vegetables of Nepal. Nepal Academy of Science and Technology, Lalitpur, pp 201–234

Real Mushrooms (2025) Medicinal mushrooms: 7 kinds and their unique health benefits. In: Real mushrooms. https://www.realmushrooms.com/7-medicinal-mushroom-benefits-for-health/. Accessed 16 Aug 2025

Richard F, Bellanger J-M, Clowez P et al (2015) True morels (Morchella, Pezizales) of Europe and North America: evolutionary relationships inferred from multilocus data and a unified taxonomy. Mycologia 107:359–382. https://doi.org/10.3852/14-166

Rillig MC, Mardatin NF, Leifheit EF, Antunes PM (2010) Mycelium of arbuscular mycorrhizal fungi increases soil water repellency and is sufficient to maintain water-stable soil aggregates. Soil Biol Biochem 42:1189–1191. https://doi.org/10.1016/j.soilbio.2010.03.027

Romdhane S, Spor A, Banerjee S et al (2022) Land-use intensification differentially affects bacterial, fungal and protist communities and decreases microbiome network complexity. Environ Microbiome 17:1. https://doi.org/10.1186/s40793-021-00396-9

Roncero-Ramos I, Delgado-Andrade C (2017) The beneficial role of edible mushrooms in human health. Curr Opin Food Sci 14:122–128. https://doi.org/10.1016/j.cofs.2017.04.002

Roux P, Eyssartier G (2024) Guide Des Champignons: France Et Europe, 5th edn. BELIN

Royse DJ, Baars J, Tan Q (2017) Current overview of mushroom production in the world. In: Diego CZ, Pardo-Giménez A (eds) Edible and medicinal mushrooms, 1st edn. Wiley, pp 5–13

Ryberg M, Larsson E, Molau U (2009) Ectomycorrhizal diversity on Dryas octopetala and Salix reticulata in an alpine cliff ecosystem. Arct Antarct Alp Res 41:506–514. https://doi.org/10.1657/1938-4246-41.4.506

Saldanha A, Gomes LC, Pinela J et al (2023) Sustainable cultivation of edible mushrooms: preserving biodiversity and ensuring product quality. In: IECAG 2023. MDPI, p 6

Sánchez C (2009) Lignocellulosic residues: biodegradation and bioconversion by fungi. Biotechnol Adv 27:185–194. https://doi.org/10.1016/j.biotechadv.2008.11.001

Sánchez-Hernández D, Larsen J, Reverchon F (2025) Diversity of culturable endophytic fungi associated with avocado orchards under organic and conventional management. Curr Microbiol 82:466

Sanz-Rocha M, Gerding M, Quezada T et al (2023) Molecular and cultural characterization of Morchella spp. from disturbed environments of Central-Southern Chile. Fungal Biol 127:938–948. https://doi.org/10.1016/j.funbio.2023.01.009

Schmidt O (2006) Wood and tree fungi – biology, damage, protection, and use, Dr. Dieter Czeschlik. Springer, Heidelberg

Schneider I, Kressel G, Meyer A et al (2011) Lipid lowering effects of oyster mushroom (*Pleurotus ostreatus*) in humans. J Funct Foods 3:17–24. https://doi.org/10.1016/j.jff.2010.11.004

Schoch CL, Seifert KA, Huhndorf S et al (2012) Nuclear ribosomal internal transcribed spacer (ITS) region as a universal DNA barcode marker for *fungi*. Proc Natl Acad Sci USA 109:6241–6246. https://doi.org/10.1073/pnas.1117018109

Senila M, Senila L, Resz M-A (2024) Chemical composition and nutritional characteristics of popular wild edible mushroom species collected from North-Western Romania. J Food Compos Anal 134:106504. https://doi.org/10.1016/j.jfca.2024.106504

Shanmugaraj C, Biswas MK (2024) Evaluation of different substrates in composting for cultivation of Agaricus bisporus under agro-ecological conditions of West Bengal. J Adv Biol Biotechnol 27:199–207. https://doi.org/10.9734/jabb/2024/v27i91290

Sharika R, Mongkolpobsin K, Rangsinth P et al (2024) Experimental models in unraveling the biological mechanisms of mushroom-derived bioactives against aging- and lifestyle-related diseases: a review. Nutrients 16:2682. https://doi.org/10.3390/nu16162682

Sheldrake M (2020) Entangled life: how fungi make our worlds, change our minds and shape our futures. The Bodley Head, London

Shen T, Duan C, Chen B et al (2017) Tremella fuciformis polysaccharide suppresses hydrogen peroxide-triggered injury of human skin fibroblasts via upregulation of SIRT1. Mol Med Rep 16:1340–1346. https://doi.org/10.3892/mmr.2017.6754

Shen H, Yang B, Wang H et al (2022) Changes in soil ectomycorrhizal fungi community in oak forests along the urban–rural gradient. Forests 13:675. https://doi.org/10.3390/f13050675

Shrestha UB, Bawa KS (2013) Trade, harvest, and conservation of caterpillar fungus (Ophiocordyceps sinensis) in the Himalayas. Biol Conserv 159:514–520. https://doi.org/10.1016/j.biocon.2012.10.032

Simard SW (2021) Finding the mother tree: discovering the wisdom of the forest. Alfred A. Knopf, New York

Simard SW, Beiler KJ, Bingham MA et al (2012) Mycorrhizal networks: mechanisms, ecology and modelling. Fungal Biol Rev 26:39–60. https://doi.org/10.1016/j.fbr.2012.01.001

Smith ME, Henkel TW, Catherine Aime M et al (2011) Ectomycorrhizal fungal diversity and community structure on three co-occurring leguminous canopy tree species in a Neotropical rainforest. New Phytol 192:699–712. https://doi.org/10.1111/j.1469-8137.2011.03844.x

Song J, Sun Y-F, Ji X et al (2018) Phylogeny and taxonomy of Laetiporus (Basidiomycota, Polyporales) with descriptions of two new species from western China. MC 37:57–71. https://doi.org/10.3897/mycokeys.37.26016

Song X, Shang X, Zhang M et al (2025) Cultivation methods and biology of Lentinula edodes. Appl Microbiol Biotechnol 109:63. https://doi.org/10.1007/s00253-024-13387-w

Spelman K, Sutherland E, Bagade A (2017) Neurological activity of lion's mane (Hericium erinaceus). J Restor Med 6:19. https://doi.org/10.14200/jrm.2017.6.0108

Stamets P, Zwickey H (2014) Medicinal Mushrooms: ancient remedies meet modern science. Integr Med (Encinitas) 13:46–47

Stockinger H, Krüger M, Schüßler A (2010) DNA barcoding of arbuscular mycorrhizal fungi. New Phytol 187:461–474. https://doi.org/10.1111/j.1469-8137.2010.03262.x

Su Z, Xu B (2024) Chemical compositions and health promoting effects of wild edible mushroom milk-cap (*Lactarius deliciosus*): a review. Food Biosci 62:105118. https://doi.org/10.1016/j.fbio.2024.105118

Sun T, Dong W, Jiang G et al (2019) Cordyceps militaris improves chronic kidney disease by affecting TLR4/NF-κB redox signaling pathway. Oxidative Med Cell Longev 2019:7850863. https://doi.org/10.1155/2019/7850863

Šutara J (2008) *Xerocomus* s. l. in the light of the present state of knowledge. Czech Mycol 60:29–62. https://doi.org/10.33585/cmy.60104

Suz LM, Barsoum N, Benham S et al (2014) Environmental drivers of ectomycorrhizal communities in Europe's temperate oak forests. Mol Ecol 23:5628–5644. https://doi.org/10.1111/mec.12947

Suz LM, Barsoum N, Benham S et al (2015) Monitoring ectomycorrhizal fungi at large scales for science, forest management, fungal conservation and environmental policy. Ann For Sci 72:877–885. https://doi.org/10.1007/s13595-014-0447-4

Szychowski KA, Skóra B, Pomianek T, Gmiński J (2021) Inonotus obliquus – from folk medicine to clinical use. J Tradit Complement Med 11:293–302. https://doi.org/10.1016/j.jtcme.2020.08.003

Tarafder E, Nizamani MM, Karunarathna SC et al (2024) Advancements in genetic studies of mushrooms: a comprehensive review. World J Microbiol Biotechnol 40:275. https://doi.org/10.1007/s11274-024-04079-8

Tedersoo L, Bahram M, Põlme S et al (2014) Global diversity and geography of soil fungi. Science 346:1256688. https://doi.org/10.1126/science.1256688

Tedersoo L, Sánchez-Ramírez S, Kõljalg U et al (2018) High-level classification of the fungi and a tool for evolutionary ecological analyses. Fungal Divers 90:135–159. https://doi.org/10.1007/s13225-018-0401-0

Tedersoo L, Bahram M, Zobel M (2020) How mycorrhizal associations drive plant population and community biology. Science 367:eaba1223. https://doi.org/10.1126/science.aba1223

Toxqui-Munguía M, Romero-Natale A, Cid-Pérez TS et al (2025) Ethnomycology as a biocultural preservation strategy for wild mushrooms in San Miguel Canoa, Puebla, México. Sustainability 17:5852. https://doi.org/10.3390/su17135852

Treseder KK, Lennon JT (2015) Fungal traits that drive ecosystem dynamics on land. Microbiol Mol Biol Rev 79:243–262. https://doi.org/10.1128/MMBR.00001-15

Treseder KK, Marusenko Y, Romero-Olivares AL, Maltz MR (2016) Experimental warming alters potential function of the fungal community in boreal forest. Glob Chang Biol 22:3395–3404. https://doi.org/10.1111/gcb.13238

Treseder KK, Berlemont R, Allison SD, Martiny AC (2018) Nitrogen enrichment shifts functional genes related to nitrogen and carbon acquisition in the fungal community. Soil Biol Biochem 123:87–96. https://doi.org/10.1016/j.soilbio.2018.05.014

Tuğcan MO, Akpınar AA (2025) Mushroom poisoning: an updated review. Turk J Emerg Med 25: 10–16. https://doi.org/10.4103/tjem.tjem_129_24

Twardowski P, Kanaya N, Frankel P et al (2015) A phase I trial of mushroom powder in patients with biochemically recurrent prostate cancer: roles of cytokines and myeloid-derived suppressor cells for Agaricus bisporus–induced prostate-specific antigen responses. Cancer 121:2942–2950. https://doi.org/10.1002/cncr.29421

Vaario L-M, Yang X, Yamada A (2017) Biogeography of the Japanese gourmet fungus, Tricholoma matsutake: a review of the distribution and functional ecology of matsutake. In: Tedersoo L (ed) Biogeography of mycorrhizal symbiosis. Springer, Cham, pp 319–344

Valverde ME, Hernández-Pérez T, Paredes-López O (2015) Edible Mushrooms: improving human health and promoting quality life. Int J Microbiol 2015:1–14. https://doi.org/10.1155/2015/376387

Van Der Heijden MGA, Martin FM, Selosse M, Sanders IR (2015) Mycorrhizal ecology and evolution: the past, the present, and the future. New Phytol 205:1406–1423. https://doi.org/10.1111/nph.13288

Van Der Linde S, Suz LM, Orme CDL et al (2018) Environment and host as large-scale controls of ectomycorrhizal fungi. Nature 558:243–248. https://doi.org/10.1038/s41586-018-0189-9

Vellinga EC (2004) Ecology and distribution of Lepiotaceous fungi (Agaricaceae) A Review. Nova Hedwigia 78:273–299. https://doi.org/10.1127/0029-5035/2004/0078-0273

Venkatesh G, Sakthi Priya P, Anithaa V et al (2022) Chapter 19 role of entomopathogenic fungi in biocontrol of insect pests. In: Soni R, Suyal DC, Goel R (eds) Plant protection. De Gruyter, pp 505–548

Venturella G, Ferraro V, Cirlincione F, Gargano ML (2021) Medicinal mushrooms: bioactive compounds, use, and clinical trials. IJMS 22:634. https://doi.org/10.3390/ijms22020634

Větrovský T, Kohout P, Kopecký M et al (2019) A meta-analysis of global fungal distribution reveals climate-driven patterns. Nat Commun 10:5142. https://doi.org/10.1038/s41467-019-13164-8

Vieira FR, Di Tomassi I, O'Connor E et al (2023) Manipulating *Agaricus bisporus* developmental patterns by passaging microbial communities in complex substrates. Microbiol Spectr 11: e01978–23. https://doi.org/10.1128/spectrum.01978-23

Wang P, Zhang Y, Mi F et al (2015) Recent advances in population genetics of ectomycorrhizal mushrooms *Russula* spp. Mycology 6:110–120. https://doi.org/10.1080/21501203.2015.1062810

Wang Y, Gu J, Wu J et al (2025) Natural products and health care functions of Inonotus obliquus. Curr Issues Mol Biol 47:269. https://doi.org/10.3390/cimb47040269

Wasser SP (2017) Medicinal mushrooms in human clinical studies. Part I. Anticancer, oncoimmunological, and immunomodulatory activities: a review. Int J Med Mushrooms 19: 279–317. https://doi.org/10.1615/IntJMedMushrooms.v19.i4.10

Wasson RG, McKenna T (2015) Mushrooms in history—the Greeks and Egyptians. In: Mushroom hunting and more.. http://blog.mushroomanna.com/mushrooms-history-greeks-egyptians/. Accessed 15 Aug 2025

Wei Y, Zhang L, Wang J et al (2021) Chinese caterpillar fungus (Ophiocordyceps sinensis) in China: current distribution, trading, and futures under climate change and overexploitation. Sci Total Environ 755:142548. https://doi.org/10.1016/j.scitotenv.2020.142548

Wennig R, Eyer F, Schaper A et al (2020) Mushroom poisoning. Dtsch Arztebl Int. https://doi.org/10.3238/arztebl.2020.0701

White J, Weinstein SA, De Haro L et al (2019) Mushroom poisoning: a proposed new clinical classification. Toxicon 157:53–65. https://doi.org/10.1016/j.toxicon.2018.11.007

Windsor C, Kreynes AE, Chilton JS et al (2025) Comparative study of Chaga (Inonotus obliquus) dietary supplements using complementary analytical techniques. Int J Mol Sci 26:2970. https://doi.org/10.3390/ijms26072970

Witkowska D, Słowik J, Chilicka K (2021) Heavy metals and human health: possible exposure pathways and the competition for protein binding sites. Molecules 26:6060. https://doi.org/10.3390/molecules26196060

Wong CK, Tse PS, Wong ELY et al (2004) Immunomodulatory effects of Yun Zhi and Danshen capsules in health subjects—a randomized, double-blind, placebo-controlled, crossover study. Int Immunopharmacol 4:201–211. https://doi.org/10.1016/j.intimp.2003.12.003

Wu Y, Choi M-H, Li J et al (2016) Mushroom cosmetics: the present and future. Cosmet 3:22. https://doi.org/10.3390/cosmetics3030022

Wu Y, Wei Z, Zhang F et al (2019) Structure, bioactivities and applications of the polysaccharides from *Tremella fuciformis* mushroom: a review. Int J Biol Macromol 121:1005–1010. https://doi.org/10.1016/j.ijbiomac.2018.10.117

Wu F, Tohtirjap A, Fan L-F et al (2021) Global diversity and updated phylogeny of Auricularia (Auriculariales, Basidiomycota). J Fungi 7:933. https://doi.org/10.3390/jof7110933

Wu S, Wang Q, Duan H et al (2025) Mushroom poisoning deaths and prevention practices in hunan, 2014–2023. PLoS One 20:e0326107. https://doi.org/10.1371/journal.pone.0326107

Xie F, Ma A, Zhou H et al (2020) Revealing fungal communities in alpine wetlands through species diversity, functional diversity and ecological network diversity. Microorganisms 8:632. https://doi.org/10.3390/microorganisms8050632

Xu T, B. Beelman R, D. Lambert J (2012) The cancer preventive effects of edible mushrooms. Anti-Cancer Agents Med Chem 12:1255–1263. https://doi.org/10.2174/187152012803833017

Xu Z, Fu L, Feng S et al (2019) Chemical composition, antioxidant and Antihyperglycemic activities of the wild Lactarius deliciosus from China. Molecules 24:1357. https://doi.org/10.3390/molecules24071357

Yamada A (2022) Cultivation studies of edible ectomycorrhizal mushrooms: successful establishment of ectomycorrhizal associations in vitro and efficient production of fruiting bodies. Mycoscience 63:235–246. https://doi.org/10.47371/mycosci.2022.08.004

Yamin-Pasternak S, Pasternak I (2018) Ethnomycology. In: Callan H (ed) The international encyclopedia of anthropology, 1st edn. Wiley, pp 1–2

Yao L, Zhu L, Chen C et al (2024) A systematic review on polysaccharides from fermented Cordyceps sinensis: advances in the preparation, structural characterization, bioactivities, structure-activity relationships. Int J Biol Macromol 282:137275. https://doi.org/10.1016/j.ijbiomac.2024.137275

Yi X, Xi-zhen H, Jia-shi Z (2004) Randomized double-blind placebo-controlled clinical trial and assessment of fermentation product of Cordyceps sinensis (Cs-4) in enhancing aerobic capacity and respiratory function of the healthy elderly volunteers. Chin J Integr Med 10:187–192. https://doi.org/10.1007/BF02836405

Yin H, Yi W, Hu D (2022) Computer vision and machine learning applied in the mushroom industry: a critical review. Comput Electron Agric 198:107015. https://doi.org/10.1016/j.compag.2022.107015

Yong GY, Lau BF, Kue CS (2025) A comprehensive review of the anti-angiogenic potential of edible and medicinal mushrooms: bioactive compounds, mechanisms of action and clinical status. Food Biosci 71:107309. https://doi.org/10.1016/j.fbio.2025.107309

Yorou NS, Dramani R, Minter DW (2024) Cantharellus congolensis. Descriptions of fungi and bacteria. https://doi.org/10.1079/DFB/20242382357

Zhang Y, Mo M, Yang L et al (2021) Exploring the species diversity of edible mushrooms in Yunnan, Southwestern China, by DNA barcoding. J Fungi 7:310. https://doi.org/10.3390/jof7040310

Zhang L, Wang Y, Zhang R et al (2022) Effects of three probiotics and their interactions on the growth performance of and nutrient absorption in broilers. PeerJ 10:e13308. https://doi.org/10.7717/peerj.13308

Zhang Y, Chen S, Yang L, Zhang Q (2023) Application progress of CRISPR/Cas9 genome-editing technology in edible fungi. Front Microbiol 14:1169884. https://doi.org/10.3389/fmicb.2023.1169884

Zhang J, Yang Z, Zhao Z, Zhang N (2024) Structural and pharmacological insights into cordycepin for neoplasms and metabolic disorders. Front Pharmacol 15:1367820. https://doi.org/10.3389/fphar.2024.1367820

Zhao S, Gao Q, Rong C et al (2020) Immunomodulatory effects of edible and medicinal mushrooms and their bioactive Immunoregulatory products. J Fungi 6:269. https://doi.org/10.3390/jof6040269

Znamenski AA (2003) Shamanism in Siberia, 1st edn. Springer, Dordrecht

Zotti M, Bonanomi G, Mazzoleni S (2025) Fungal fairy rings: history, ecology, dynamics and engineering functions. imafungus 16:e138320. https://doi.org/10.3897/imafungus.16.138320

Zou C-X, Dong S-H, Hou Z-L et al (2020) Modified lanostane-type triterpenoids with neuroprotective effects from the fungus Inonotus obliquus. Bioorg Chem 105:104438. https://doi.org/10.1016/j.bioorg.2020.104438

Zsigmond AR, Kántor I, May Z et al (2019) Data on elemental composition of Russula cyanoxantha along an urbanization gradient in Cluj-Napoca (Romania). Data Brief 27:104572. https://doi.org/10.1016/j.dib.2019.104572

# 3 Conservation, Cultivation, and Sustainability of Edible and Medicinal Mushrooms

Iwan Saskiawan, Rini Riffiani, Rizki Rabeca Elfirta, Ryan Haryo Setyawan, Atik Retnowati, and Evan Febriansyah

**Abstract**

Indonesia, a tropical country with high biodiversity, is home to a vast variety of edible and medicinal mushrooms, many of which have been used for generations by indigenous communities. The conservation of fungal diversity is very important, even though it faces several challenges such as deforestation, climate change, and research activity. Cultivation presents a promising solution to these challenges. While only a fraction of the country's mushroom species is currently cultivated, there is significant potential for commercial development. For example, the cultivation of species like oyster mushrooms is increasing, driven by high demand for both food and medicinal products. However, there are still obstacles for farmers, including a lack of knowledge in advanced cultivation techniques, limited access to quality seeds, and issues with capital. Cultivation, particularly of wild species, also offers an opportunity to reduce pressure on wild populations and provide a sustainable source of income for local communities. This chapter reveals the basic biology of mushrooms, the biodiversity of wild edible and medicinal mushrooms as well as the mushroom cultivation technology in Indonesia.

I. Saskiawan (✉) · R. Riffiani · R. R. Elfirta · R. H. Setyawan
Research Center for Applied Microbiology, National Research and Innovation Agency, The Republic of Indonesia, Bogor, Indonesia
e-mail: iwan007@brin.go.id

A. Retnowati
Research Center for Biosystematics and Evolution, National Research and Innovation Agency, The Republic of Indonesia, Bogor, Indonesia

E. Febriansyah
Department of Biology, Faculty of Mathematics and Natural Sciences, Universitas Indonesia, Depok, Indonesia

 
B. Giri et al. (eds.), *Advances in Mushroom Research*,
https://doi.org/10.1007/978-981-95-7692-0_3

**Keywords**

Edible mushrooms · Medicinal mushroom · Mushroom cultivation

## 3.1 Introduction

Mushrooms are the fruiting bodies of macrofungi, which have been part of the human diet since ancient times. Mushrooms are a very nutritious food and well known as a functional food that will give a benefit to human health. Furthermore, several active compounds found in the mushroom can also function as medicine for several diseases. Mushrooms can be generated by the bioconversion process of agricultural waste such as sawdust, paddy straw, cotton waste, and palm sugar waste that mainly contains lignocellulosic compounds. Mushroom cultivation technology is friendly to the environment and could solve the problem of the handling of agricultural wastes. Furthermore, the spent substrate left after harvesting the mushrooms could be biochemically modified by the mushroom enzymes into simpler and more readily digestible compounds. It is more palatable to livestock when used as a livestock feed supplement. On the other hand, this spent mushroom substrate could also be utilized as organic fertilizer, which is good for the soil.

Mushrooms are "macrofungi" whose fruiting bodies are large enough to be seen and picked, whether they grow on the surface or underground (Miles and Chang 1997). Taxonomically, mushrooms belong to the Kingdom of Fungi, a group very distinct from Plantae, Animalia, Protista, and Monera in the classification of Whittaker (Hagen 2012). Most mushroom species are under the Basidiomycota and Ascomycota, the two phyla of the Kingdom of Fungi. In 1991, Hawksworth reported that the number of fungi species in the world was about 1,500,000, which comprised micro and macrofungi. About 25 years later, the fungal diversity has been revisited: 2.2–3.8 million species with about 120,000 currently described species or about 8% of the total fungal diversity (Hawksworth and Lücking 2017). The increasing number of those species is caused by the increasing use of molecular approaches in many areas of fungal studies. The species number in tropical and temperate countries is still questionable; however, a tropical country like Indonesia contains a rich diversity of mushrooms as compared to temperate countries. Retnowati et al. (2020) identified 2273 species documented across seven major biogeographical regions of Indonesia, particularly from Java, Sumatra, Kalimantan, Sulawesi, the Indonesian part of Papua New Guinea, Moluccas, and Lesser Sunda Island. However, this figure represents only a part of the country's biodiversity, as thousands of species remain unexplored and undescribed in Indonesia.

Most fungi have plant-like cells, but they do not have chlorophyll. Consequently, they depend on the environment for their food since they cannot produce nutrients by photosynthesis. Nutritionally, the fungi are described as saprophytic when they obtain nutrients from nonliving organic matter and as parasitic or symbiotic if they obtain nutrients from living organisms. Furthermore, fungi are different from other heterotrophic organisms in the way they take in food to the cell. They do not ingest their food but obtain it by absorbing it into their cells.

A wide variety of lignocelluloses, including agricultural and agro-industrial waste such as paddy straw, corncobs, cotton waste, sugar palm waste, sawdust, oil palm empty fruit bunches, banana leaves, water hyacinth, coconut fiber, and other agricultural wastes, can be degraded efficiently by the extracellular enzymes secreted by fungi from the Basidiomycota phylum. The extracellular lignocellulolytic enzymes are produced by fungi to assist in hydrolyzing complex compounds into simpler ones so that they can be more easily absorbed by the fungal cell. The extracellular enzymes that play an important role in the degradation of lignocelluloses are cellulases, hemicellulases, MnP (manganese peroxidase), LiP (lignin peroxidase), and Lac (laccase) (Lechner and Papinutti 2006; Pandey et al. 2014; Kabel et al. 2017; Kantharaj et al. 2017).

Mushrooms demonstrate diverse nutritional strategies to obtain the organic compounds they need for survival and growth. These strategies broadly fall into the categories of saprophytes, parasites, and symbiosis (specifically mycorrhiza associations). Saprophytic nutrition is the most common nutritional mode for many mushrooms. Saprophytic fungi obtain their nutrients from dead or decomposing organic matter. The lignocellulolytic enzymes break down complex organic polymers (like cellulose, lignin, proteins, starches) into simpler, soluble molecules (sugars, amino acids, fatty acids). The fungal hyphae then absorb these smaller, soluble nutrient molecules directly through their cell walls and membranes. The extensive network of hyphae (mycelium) provides a large surface area for efficient absorption. Saprophytic mushrooms are crucial decomposers, recycling nutrients back into the ecosystem, making them available for other organisms, particularly plants. Many of the most popular and widely cultivated edible mushrooms are saprophytic. They are shiitake (*Lentinula edodes*), oyster mushrooms (*Pleurotus* species), button mushrooms (*Agaricus bisporus*), maitake (*Grifola frondosa*), lion's mane (*Hericium erinaceus*), etc. All of them are cultivable in Indonesia.

Parasitic mushrooms obtain nutrients from living host organisms, often causing harm or disease to the host. Like saprophytes, they also secrete enzymes, but these enzymes are often designed to break down living host tissues, facilitating nutrient extraction and sometimes causing cell death or disease. The white rot fungus, particularly species like *Ganoderma boninense*, is a facultative parasite, meaning it can live as a saprophyte on decaying organic matter but becomes a destructive parasite when it infects a living host, especially the oil palm tree. Although *Ganoderma* is a destructive pathogen to oil palm trees, it is also highly revered as a medicinal mushroom, particularly the species *Ganoderma lucidum*, which is widely known as Reishi (in Japan) or Lingzhi (in China). For thousands of years, it has been a foundation of traditional Asian medicine, prized for its recognized ability to promote longevity, vitality, and well-being.

Mycorrhiza is a mutualistic symbiotic relationship between a fungus and the roots of a plant. Both organisms benefit from this association. The fungus forms an extensive network of hyphae. It is known as ectomycorrhiza if around the plant roots or endomycorrhiza/arbuscular mycorrhiza if in the plant roots. These hyphae extend far into the soil, acting as an extension of the plant's root system. This significantly increases the surface area for nutrient and water absorption, especially for less mobile nutrients like phosphorus, nitrogen, and micronutrients. The fungus can also release enzymes and organic acids that help solubilize tightly bound nutrients in the soil, making them accessible to the plant. In return for these essential nutrients and water, the plant provides the fungus with carbohydrates (sugars) and sometimes lipids, which are produced by the plant through photosynthesis. Fungi are heterotrophic and cannot photosynthesize, so they rely on the plant for these organic carbon sources. The most famous and culturally significant mycorrhizal mushroom is matsutake (*Tricholoma matsutake*).

Furthermore, the relationship between mushrooms of the genus *Termitomyces* and termites (specifically those in the subfamily *Macrotermitinae*, often called fungus-growing termites) is a classic and fascinating example of an obligate mutualistic symbiosis in terms of nutrition. Both organisms are entirely dependent on each other for their survival and nutrient acquisition. *Termitomyces* fungi are "farmed" by the termites. The termites collect tough plant material (like wood, leaves, grasses), which they partially chew and digest. This predigested plant material, mixed with termite feces, is then used to construct specialized structures within the termite nest called fungus combs or "fungus gardens." The fungus then secretes powerful extracellular enzymes that are highly efficient at breaking down the complex, recalcitrant compounds in the plant material, particularly lignin and cellulose, which termites themselves cannot fully digest. In essence, the termites act as farmers, providing the crop (*Termitomyces*) with raw materials and ideal growing conditions, while the mushroom acts as a highly efficient digestive system, transforming indigestible plant matter into a nutritious food source for the termites. This complex organization allows both organisms to thrive in environments where nutrient acquisition would otherwise be extremely challenging. In Indonesia, people pick up this muhsroom and cook it to be a delicious food.

The structure of fungi, except yeast, is filaments with a tubular structure called hyphae. Hyphae come from the germination of a spore in proper environmental conditions. Hyphae branch out, and in some species employing fusion hyphae (anastomosis), penetrate substrates and absorb nutrients. Fungal hyphae secrete enzymes that break down nutrients into smaller molecules before being absorbed. When hyphae come together and fuse, they form a mycelium. The mycelium derived from a single spore is called primary mycelium (monokaryotic hyphae). Following fusion of the two compatible primary mycelia, a new structure of mycelium is formed, which is called the secondary mycelium (dikaryotic hyphae). This secondary mycelium, however, is not diploid since the two compatible nuclei do not fuse (Fig. 3.1). The hyphae of the secondary mycelium preserve this dikaryotic condition by the process of clamp connection formation (Fig. 3.2). The formation of a fruiting body is normally derived from the secondary mycelium (Hood 2006).

The fungi are varied in matters of reproduction, which may be accomplished through sexuality or asexuality. Sexual reproduction is characterized by the union of two nuclei and consists of three important stages. The first is plasmogamy, which is

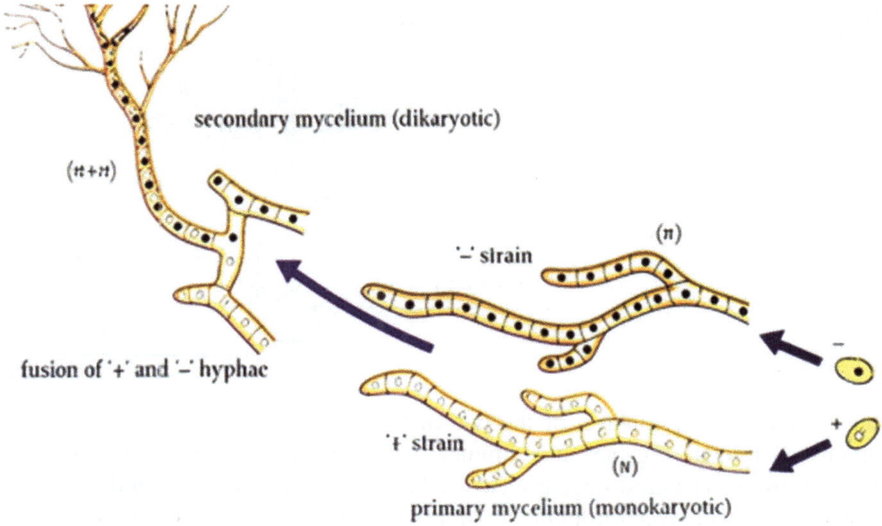


**Fig. 3.1** The formation of secondary mycelium (https://o.quizlet.com/ozH9uPapXUfm8yIGUaD2AA.png)

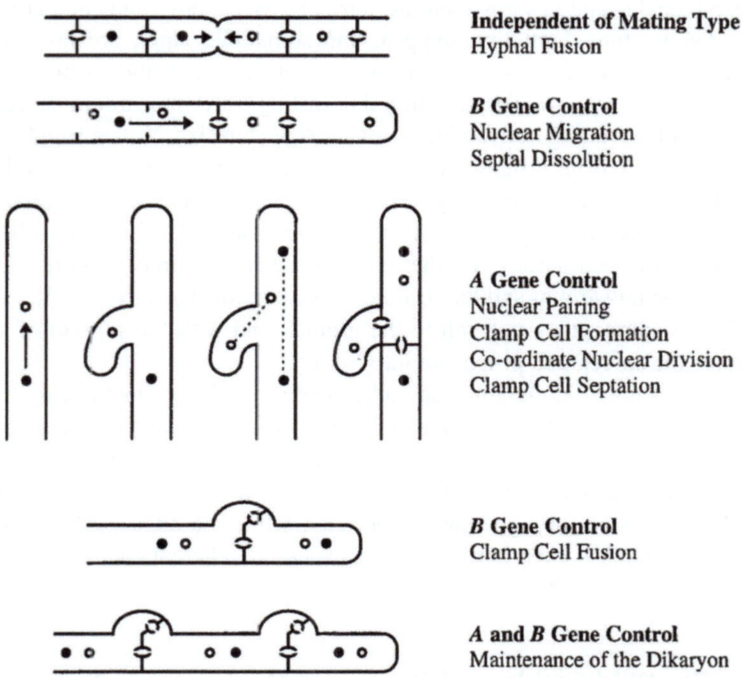


**Fig. 3.2** The role of clamp connections in maintaining a dikaryon (Casselton and Olesnicky 1998)

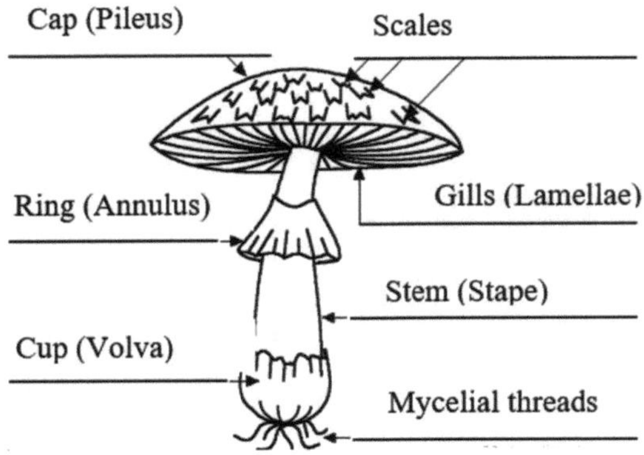


**Fig. 3.3** The morphology of the fruit body

the fusion of the cytoplasm of the two mating individuals. The nuclei from two strains are brought together in a common cytoplasm. The second is karyogamy or nuclear fusion. The third stage is meiosis, the nuclear division in which the chromosome number is reduced from the diploid to the haploid number (Chiu 1996).

Mushrooms have two stages based on their life cycle, vegetative and reproductive growth. Vegetative growth shows the linear growth of fungal mycelia. It produces the extracellular lignocellulosic enzymes that hydrolyze complex lignocellulose into simpler molecules and absorbs them as nutrients. When environmental conditions, such as temperature, humidity, oxygen, and sometimes light, are favorable, the mycelia halt vegetative growth and begin to produce fruit bodies, which are commonly called mushrooms. The structure that distinguishes mushroom species from other fungi is the fruiting body, which is macroscopic in size. Young fruit bodies are called pins and will be differentiated into cap and stem-forming fruit bodies that release spores to produce the next generation. Under the cap, spores are produced in the gills. If a section of the gills is cut and examined under the microscope, the spores will be found on their surface. Furthermore, as the cap expands fully, it signals the maturity of the mushroom and the readiness for spore dispersal (Fig. 3.3). After the dispersion of spores, they will fall to the ground, and with the support of suitable conditions, the spores will propagate into mycelia.

Mushroom cultivation requires an understanding of the optimal environment for the growth of each mushroom species as well as how to make a favorable environment for both the vegetative and the reproductive growth of mushrooms. Therefore, people can say that mushroom cultivation is a combination of science and art. The science is developed through research activity, while the art is perfected through curiosity and practical experience. Every place has an inherent character that affects mushroom cultivation technology (Chang and Wasser 2017).

## 3.2 The Wild Edible and Medicinal Mushroom in Indonesia

Indonesia is the world's largest archipelago with almost 17,000 islands spreading between the Pacific and the Indian Ocean. It has geographical characteristics that play a profound role in shaping mushroom biodiversity. As a vast archipelago straddling the equator, Indonesia offers a very specific range of habitats, climates, and ecological niches, all of which directly influence the distribution and diversity of fungal species. Under the tropical climate, Indonesia is characterized by constantly high humidity and warm temperatures during the year, which provide ideal conditions for fungal growth. Most fungi thrive in moist environments, and the abundant rainfall in many parts of Indonesia supports lush vegetation and decaying organic matter, which are essential for mushroom development. The extensive tropical rainforests across islands like Sumatra, Kalimantan (Borneo), Sulawesi, and Papua are primary hotspots for mushroom diversity. These forests offer huge amount of substrates (dead wood, leaf litter, animal dung, living trees) and microclimates that support a wide range of saprophytic, mycorrhizal, and parasitic fungi. Fungi, especially saprophytic mushrooms, are crucial decomposers in these forest ecosystems. Their ability to break down complex organic matter contributes significantly to nutrient cycling, which is vital for the health and productivity of the rainforests.

On the other hand, as altitude increases, temperature generally decreases, and humidity patterns can vary. This creates distinct climate gradients that support different fungal communities. For instance, some mushrooms might thrive in cooler, higher-altitude cloud forests, while others prefer warmer lowland areas. Research at the Gunung Gede Pangrango National Park (GGPNP), West Java Province, noted that dry conditions and low rainfall during a research period resulted in lower mushroom diversity, emphasizing the influence of moisture on fungal growth (Putra et al. 2019). They observed that the impact of dry conditions on mushroom diversity in GGPNP is consistent with broader concerns about climate change. Warmer and drier conditions, predicted for many regions, including parts of Southeast Asia, are expected to reduce mushroom productivity due to decreased soil water availability. Conversely, increased moisture in other areas due to heavier rainfall can also have complex effects, with some species thriving and others being inhibited by excessive wetness.

Ramadhani et al. (2019) reported a total of 14 macrofungi isolates belonging to the members of the Basidiomycota and Ascomycota phyla. However, only eight of the 14 isolates were molecularly approved. These macrofungi are classified into two phyla, two classes, four genera, and seven families. A total of three isolates were identified by molecular techniques up to the species level, namely, *Xylaria schweinitzii*, *Agaricus flocculosipes*, and *Fomitopsis feei*. The other five isolates could only be identified up to the genus level, namely, *Ganoderma* sp., *Pleurotus* sp., *Rigidoporus* sp., *Gymnopus* sp., and *Agaricus* sp. The obtained isolates can be screened to obtain potential isolates and can be utilized for bioprospecting.

With rich biodiversity and traditional knowledge, these mushrooms play a significant role in food security, traditional medicine, and economic development for local communities. Indigenous communities like the Bunggu tribe in West Sulawesi and the Kaili tribe in Central Sulawesi possess extensive traditional knowledge regarding edible and medicinal wild mushrooms. Documenting and preserving this ethnomycological knowledge is vital for identifying species that require conservation and informing future cultivation efforts. Various tribes, such as Bunggu, Baduy, and Pamona, have identified numerous species used for food, medicine (e.g., *Dacryopinax spathularia* for headaches, certain species for fatigue), and even other purposes like lighting. Table 3.1 shows the species of mushrooms that can be found in the Pasangkayu District, West Sulawesi Province, Indonesia (Yusran et al. 2024).

Nion et al. (2024) reported 18 genera of wild mushrooms and 44 species from the Tanjung Puting National Park and the Sebangau National Park of Indonesia. The Dayak tribe in Central Kalimantan generally use wild edible mushrooms as food, are *Auricularia* sp., *Campanella* sp., *Cookeina* sp., *Favolaschia* sp., *Hygrocybe conica*, *Hygrocybe chloropana*, *Lentinus* sp., *Oudemansiella* sp., *Pleurotus* sp., *Schizophillum commune*, and *Tremella fuciformis*. Furthermore, Kusuma et al. (2020) explored potential wild medicinal mushrooms from the Pocut Meurah Intan forest park, Nangroe Aceh Darusalam Province in Sumatra Island, Indonesia. They reported 15 species of mushrooms with potential as both food and medicine, namely, *Cookeina tricholoma*, *Sarcoscyphacoccinea*, *Tremella fuciformis*, *Auricularia auricula-judae*, *Flammulina velutipes*, *Schizophyllum commune*, *Scleroderma citrinum*, *Boletus edulis*, *Trametes versicolor*, *Lentinus tigrinus*, *Lentinus squarosulus*, *Ganoderma applanatum*, *Ganoderma resinaceum*, *Thelephora ganbajun*, and *Lycoperdon echinatum*, and indicated that the Pocut Meurah Intan forest park has great resources that can be utilized for the bioprospecting of mushrooms.

Despite the traditional knowledge of local communities, wild mushroom poisoning is a serious issue in Indonesia. In a 10-year period (2010–2020), there were 76 reported cases of wild mushroom poisoning, affecting over 500 people and resulting in nine deaths. This highlights the critical importance of caution. The primary danger lies in misidentification. Many poisonous mushrooms closely resemble edible ones. Popular myths or unreliable “rules of thumb” for identification are often misleading and can be fatal. Therefore, the conservation of wild mushrooms in Indonesia is a complex issue that requires a multi-pronged approach. By combining scientific research, traditional knowledge, community involvement, and proactive policy, Indonesia can protect its invaluable fungal heritage for future generations.

## 3.3 Edible and Medicinal Mushroom Cultivation in Indonesia

In nature, macrofungi function as decomposers, which are divided into primary, secondary, and tertiary decomposers (Rahi et al. 2009). This function of macrofungi in nature become a basic knowledge to develop a cultivation method. Primary decomposers such as the oyster mushrooms (*Pleurotus* spp.), ear mushroom (*Auricularia auricula*), shiitake (*Lentinula edodes*), and lingzhi (*Ganoderma lucidum*) degrade (hemi)cellulose, lignin, and other components of plant material. They do not depend on other organisms and their metabolites, which happens in secondary and tertiary decomposers. Furthermore, secondary decomposers such as button mushrooms (*Agaricus* spp.) typically colonize composted materials, while tertiary decomposers such as *Agrocybe* spp. are generally found in soils.

**Table 3.1** Species of mushrooms utilized by the indigenous people of the Bunggu tribe in Pasangkayu Regency, West Sulawesi, Indonesia (Yusran et al. 2024)

| No. | Vernacular name | Species | Family | Substrate | Mushrooms parts used | Utilization and processing | RFC | FL |
|---|---|---|---|---|---|---|---|---|
| 1 | Tanggidi/Tanggojo | *Schizophyllum commune* | *Schizophyllaceae* | Deadwood | Fruits bodies | As a food ingredient, the method is mixed with coconut milk and smoked fish or sauteed with vegetables | 1 | 1 |
| 2 | Lungo/Lungo Ombe/ Tambata Talinga | *Auricularia auricular-judae* | *Auriculariaceae* | Deadwood | Fruits bodies | As a food source, the method is to make fresh fruit bodies directly into soup or dry them and then store them so that they can be processed again at any time | 1 | 1 |
| 3 | Lungo Anggo/Lungo Ibo | *Auricularia polythrica* | *Auriculariaceae* | Deadwood | Fruits bodies | As a food source, the method is to make fresh fruit bodies directly into soup or dry them and then store them so that they can be processed again at any time | 1 | 1 |
| 4 | Ntolu Ntana/Ntolu Uja/NtoluAnggo/Limo | *Trichaleurina javanica* (Rehm) M. Carbone | *Pyronemataceae* | Soils | The fluids and fruits bodies | The liquid squeezed from the fruit body is used as a hair wash and shampoo hair tonic | 0.53 | 0.52 |
| 5 | Tombe/Palindo/ Tambata Sanjo | *Cookeina* spp. (Red color white orange) Cookeina trizholoma (Mont.) Kuntze (hairy) and *Cookeina speciosa* (Fr.) Dennis (Hairless) | *Sarcoscyphaceae* | Deadwood | Cups | As a medicine, the method is to fill the hood with water and then drink it. Besides that, rub it on the back/waist when it hurts or because of | 1 | 0.93 |

(continued)

**Table 3.1** (continued)

| No. | Vernacular name | Species | Family | Substrate | Mushrooms parts used | Utilization and processing | RFC | FL |
|---|---|---|---|---|---|---|---|---|
| | | | | | | fatigue. They also believe that pouring water from the hood on the babies makes it easy for the babies to sleep soundly, and it is also used as eye drops | | |
| 6 | Rava Rava Lalari Rava Ntonji (if it grows in clusters) dan Pusa (if it grows individually) | *Termytomyces* spp. | *Lyophyllaceae* | Soils | All parts of fruits bodies | As a food source, it is stir-fried and cooked in clear water mixed with fish. It appears 14 days after heavy rain accompanied by thunder and lightning, and lots of moths flying. According to local customs, it cannot be consumed by orphans (taboo) | 1 | 1 |
| 7 | Tubikire/Ntondo/Tambata/Guni | *Dacryopinax spathularia* | *Dacrymycetaceae* | Deadwood | All parts of fruits bodies | In medicine, the fruit body is rubbed over the eyebrows/on the forehead to relieve headaches (migraines) | 0.88 | 1 |
| 8 | Mariada/Rada/Tambata/Kalipopo | *Mycena* spp. | *Mycenaceae* | Deadwood | The fruit's body and the wood where it grows | Used as a lighting tool when walking in the garden/forest at night | 0.38 | 0.37 |

| | | | | | | | | |
|---|---|---|---|---|---|---|---|---|
| 9 | Tatingge | *Ganoderma* sp. | *Ganodermatceae* | Trees/ Deadwood | All parts of fruits bodies | As a medicine for headaches, the method is to rub it on the forehead | 0.52 | 0.52 |
| 10 | Tambata Sawit/Rava/ Loka | *Volvariella volvaceae* | *Pluteaceae* | Empty palm oil bunches/ Around banana trees | All parts of fruits bodies | As a food sources it is stir-fried and mixed with vegetables and spices then boiled in water | 1 | 1 |
| 11 | Tambata duyu/ Tambata/Jou | *Lentinus* sp. | *Polyporaceae* | Deadwood | All parts of fruits bodies | As a food source, it is mixed with spices and then stir-fried or boiled in water. Harvest them when they are young because when they are old, the taste is chewy like rubber, and they are difficult to chew | 1 | 1 |
| 12 | Tambata Neira | *Favolus* sp. | *Polyporaceae* | Deadwood | All parts of fruits bodies | All parts of the fruit body are dried, and the powder is used as a seasoning to flavor food | 0.45 | 0.64 |
| 13 | Tambata puti/Betue | *Pleurotus* sp. | *Pleurotaceae* | Deadwood | All parts of fruits bodies | Mixed with various kinds of vegetables, added spices, then stir-fried or boiled in water | 1 | 1 |
| 14 | Tambata tai | *Psilocybe* sp. | *Hymenogastraceae* | Animal dung | Fruits bodies | Roasted and mixed with drinks such as coffee, it is intoxicating and causes hallucinatory effects | 0.65 | 0.58 |

A B

**Fig. 3.4** Paddy straw mushroom (*Volvariella volvacea*) (**a**) and button mushroom (*Agaricus* sp.) (**b**), two edible mushrooms cultivated in Indonesia using paddy straw as the main substrate

Indonesia is a tropical country that extends out horizontally with the equator running through it, giving most of the country a tropical climate and making it well suited for agricultural activity. With this advantage, Indonesia emphasizes national food security programs through agricultural activities. As a result, the production of these agricultural activities produces large amounts of agricultural waste in the form of biomass every year. The huge amount of agricultural waste has the potential to be the substrate for edible and medicinal mushroom cultivation. Based on the materials for the substrates of mushroom cultivation, there are two categories of mushroom production in Indonesia. The first is using rice straw as the main growth substrate. Mushrooms cultivated in rice straw as the substrate are paddy straw mushroom (*Volvariella volvacea*) and champignon (*Agaricus* sp.) (Fig. 3.4). The second is using sawdust, sugar palm waste, banana leaf, coconut fiber, or corn cob. Mushrooms cultivated in those substrates are oyster mushroom (*Pleurotus* sp.), ear mushroom (*Auricularia* sp.), shiitake (*Lentinus* sp.), lion's mane (*Herecium* sp.), milky (*Callocybe indica*), Lingzhi (*Ganoderma* sp.), and maitake (*Grifola fondosa*) (Fig. 3.5).

Organic material as plant waste, is the basic substrate utilized for the cultivation of edible mushrooms. After mushroom cultivation, the byproduct, which is the partially degraded organic material, has been termed spent mushroom substrate (SMS). Management of SMS by piling up, released after mushroom crop harvesting, may cause various environmental problems, including groundwater contamination and nuisance (Ahlawat and Sagar 2007). However, some studies have reported utilizing SMS for bio-fertilizer and substrate for crop production. Priadi et al. (2016) reported that growing mediums containing spent paddy straw mushroom compost was the best treatment to improve the growth of local tomato seedlings (*Lycopersicon esculentum* Miller) in general. On the other hand, Ishihara et al.

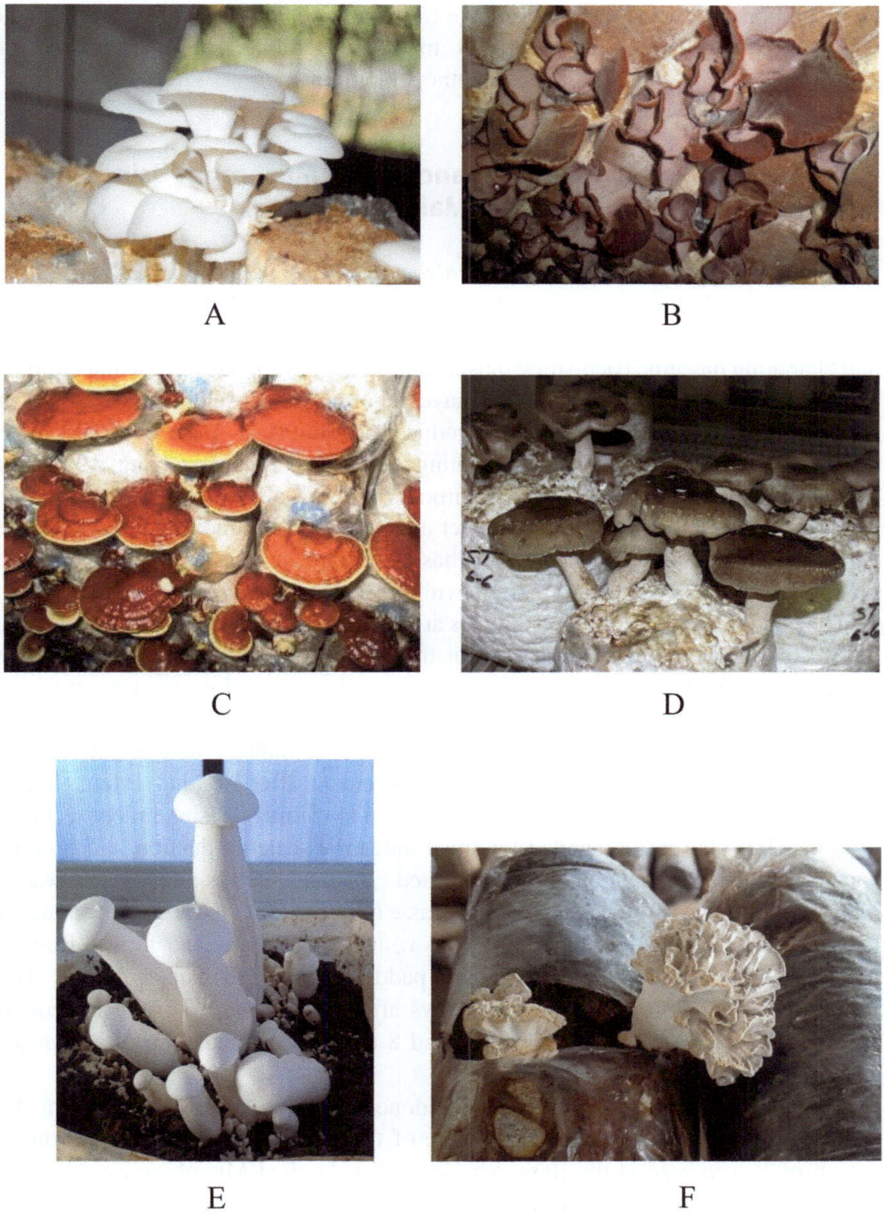


**Fig. 3.5** Oyster mushroom (*Pleurotus* sp.) (**a**), ear mushroom (*Auricularia* sp) (**b**), Lingzhi (*Ganoderma* sp.) (**c**), shitake (*Lentinus* sp.) (**d**), milky (*Callocybe indica*) (**e**), and maitake (*Grifola frondosa*) (**f**), the cultivated mushrooms in Indonesia using the sawdust as the main substrate

(2018) reported that the SMS of shiitake (*Lentinula edodes*) contains an antifungal compound. The extract of the SMS markedly inhibited the germination of *Pyricularia oryzae* conidia, the phytopathogenic fungi.

### 3.3.1 Cultivation of Medicinal and Edible Mushroom Using Paddy Straw Mushroom as the Main Substrate

Paddy straw mushroom (*Volvariella volvacea*) is one of the well-known edible mushrooms in Indonesia. It is called "jamur merang" in Indonesia, "choku" in China, "fukurotake" in Japan, "hed-fang" in Thailand, "kauk-yo-hmo" in Myanmar, and "kabuting dayami" (rice-straw mushroom) in the Philippines. These mushrooms originally are from China, and they have reportedly been cultivated for almost 300 years. Furthermore, they were introduced to Southeast Asian countries more than a hundred years later by migrating overseas Chinese (Chang 1977). In Indonesia, naturally, paddy straw mushrooms grow in a heap of rice straw in the rice field after harvesting. People collect and sell it in the traditional market. The cultivation of paddy straw mushrooms has been popular since around 1960 in the outdoors. Now, it is cultivated indoors with the semi-intensive method.

The paddy straw mushroom cultivates at a relatively high temperature. Therefore, it is also called the warm mushroom or the mushroom from a tropical area. The mycelia grow well at 33–36 °C while their fruiting bodies are formed at around 3–4° lower. It is a fast-growing mushroom and fruits 10–14 days after spawning. Cultivation of this mushroom employs easy cultivation technology and is suitable for farmers in rural areas where rice straw is abundant after rice harvesting. It can provide an additional income and nutritious food for farmers. Using indoor technology, which is also recommended for small industrial-scale production, mushrooms can grow with better yield on other composted substrates such as oil palm tree waste, corn cobs, cotton waste, and sugarcane bagasse (Quimio 2002). The study conducted by Zikriyani et al. (2018) reported that *V. volvacea* can grow and form a fruiting body in the substrate of cotton wastes, paddy straws, and banana leaves. The formation of the fruiting body took 7 days after inoculation of spawn for cotton wastes and paddy straws as a substrate and 8 days after inoculation of spawn for those of banana leaves.

Paddy straw mushroom growers in Indonesia usually obtain spawn from the spawn industry or research institutes. One of the research institutes that produces spawn of *V. volvacea* is the Research Center for Applied Microbiology, National Research and Innovation Agency, the Republic of Indonesia. There are two types of straw mushroom spawn: the mother spawn and the planting spawn. The mother spawn is grown on grain media such as sorghum, corn, and millet with the addition of 2% lime ($CaCO_3$). Then, the media are put into a bottle and sterilized at 121 °C, 1 atmosphere, for 20 minutes using an autoclave. After sterilization, each bottle is inoculated with an agar disk 6 mm in diameter containing mycelium of *V. volvacea* and then incubated at 30 °C in the absence of light (darkroom) for 18 days. The mother spawn containing sorghum as a substrate is used as a seed of the planting spawn. Media or substrate for planting spawn are chopped paddy straw (60%), cotton waste (15%), rice bran (15%), lime (5%), and horse manure (5%). The media are composed for 1 week, filled into a plastic bag (15 × 25 cm), and sterilized at 121 °C, 1 atm, for 20 minutes using an autoclave. The media or substrate is then inoculated with the mother spawn after cooling it down. It needs about 1 month for the mycelium to grow completely on the planting spawn media.

Cultivation of paddy straw mushroom using rice straw as the main substrate is conducted in a semi-intensive method. The mushroom is grown in a mushroom house that was constructed using bamboo frames covered with plastic sheeting, although contemporary growers often substitute Styrofoam for bamboo mats. The structure measures 6 m long, 4 m wide, and 4 m high, featuring a 1 × 0.5 m window on both the front and back. Inside, the total growing area was about 69 $m^2$, spread across two shelving units, each containing five layers. Bamboo is used in the construction of the mushroom house for the cultivation of paddy straw mushrooms in Indonesia because this material is cheap. Dark plastic sheets or bamboo mats are laid down and bound firmly on the roof to reduce light (Fig. 3.6). However, recently, paddy straw mushroom growers have used cold-formed steel instead of bamboo as a frame in mushroom house construction. It is expensive but long-lasting. They also utilize Styrofoam instead of bamboo mats (Fig. 3.7). The size of the mushroom house is about 7 m in length, 4 m in width, and 5 m in height. The house usually has two tiers of shelves, each of which has five layers. The total bed area is about 69 $m^2$.

Paddy straw (70%, w/w) and cotton waste (20%, w/w) are the raw substrates for bedding with a total weight of about 1000 kg. These materials are mixed with 8% rice bran and 2% lime and then composted for 6 days with two turnings of the substrates. The substrates are then filled into the shelves. The mushroom house is pasteurized using steam at a temperature of 65–75 °C for 7 hours. The inoculation of

**Fig. 3.6** The construction of the bamboo mushroom house for straw mushroom

**Fig. 3.7** The utilization of used Styrofoam to maintain the temperature, humidity, and reduce the light in the mushroom house of the straw mushroom

**Fig. 3.8** The different stages of the fruiting body of the paddy straw mushroom based on physical characteristics (1. Pinhead, 2. Button, 3. Broken button, and 4. Elongation)

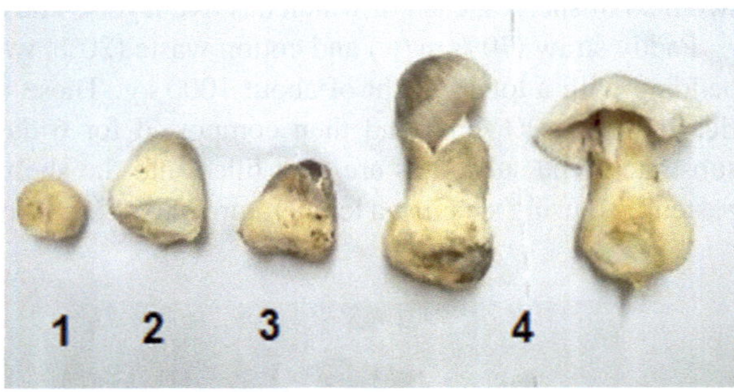


spawn is carried out when the temperature in the house is around 33–35 °C (usually the next day after pasteurization). Then, the mushroom house is closed tightly for 6 days. On day 7, proper ventilation is provided and a small amount of water is introduced onto the beds to maintain temperature and humidity in the house as well as in the substrate. The fruiting body usually forms 9–10 days after spawning. The fruiting body is divided into four different stages with its physical characteristics: 1. Pinhead, 2. Button, 3. Broken button, and 4. Elongation (Fig. 3.8). The period of fruiting body production could be achieved for 14–18 days. During this period, the maintenance of temperature and humidity is carried out by sprinkling clean water on the floor of the mushroom house. Harvesting is usually done in the morning by gently grasping, twisting, and pulling the fruiting bodies from the substrate to prevent damage.

The study by Saskiawan (2012) reported that the total production of fruiting bodies of paddy straw mushroom at the button stage for 20 days was 264.6 kg per 1000 kg of substrate/media by using paddy straw as the substrate. Thus, the Biological Efficiency Ratio (BER) of this study was 26.46%. This value of BER is higher than that of the cultivation of *V. volvacea* on banana leaves, with a BER value of 21.53% (Belewu and Belewu 2005). In Nigeria, *Volvariella* mushroom is cultivated on various lignocellulosic agricultural wastes such as palm fibers, rice husks, and sawdust (Ukoima et al. 2009). On the other hand, in India *Volvariella* mushroom is cultivated on rice straw with a slightly different method with a BER value of 22.35% (Ahlawat and Tewari 2007).

Once the mushrooms are harvested, they are placed in a plastic basket and covered with banana leaves. Collectors then visit the growers to pick up and sell the mushrooms, paying the growers the following day. Fresh mushrooms are primarily sold through supermarkets, with wet markets and grocery stores also serving as key distribution points.

### 3.3.2 Cultivation of Medicinal and Edible Mushrooms Using Sawdust as a Main Substrate

The medicinal and edible mushrooms cultivated using sawdust include oyster mushroom (*Pleurotus* sp.), ear mushroom (*Auricularia* sp.), shiitake (*Lentinus* sp.), lion's mane (*Hericium* sp.), milky (*Callocybe indica*), Lingzhi (*Ganoderma* sp.), and maitake (*Grifola fondosa*). The substrate preparation for the cultivation of the mushrooms described above is essentially similar. The only difference is that the spawn is inoculated with the culture according to the species to be cultivated. In this chapter, oyster mushrooms are used as a model for cultivation technology.

Oyster mushrooms (*Pleurotus* sp.), the big three of the largest commercially produced mushrooms in the world, are a prospective source of valuable food protein with the ability to effectively convert various lignocellulosic materials (Obodai et al. 2003; Mandeel et al. 2005; Sanchez 2010). Recently, in Indonesia, *Pleurotus* became the second most cultivated mushroom. The consumption of oyster mushrooms increases because of their taste, medicinal, and nutritional properties. The mushroom is rich in protein, fiber, carbohydrates, vitamins, and minerals with a low content of calories, fats, and sodium (Manzi et al. 1999; Cohen et al. 2002). Furthermore, Kües and Liu (2000) reviewed the medicinal properties of the fruiting bodies of basidiomycetes including *Pleurotus* spp. *Pleurotus ostreatus* is very popular in Indonesia and is generally cultivated on sawdust supplemented with rice bran (Saskiawan et al. 2016).

The cultivation of oyster mushrooms should pay attention into three main factors, (i) spawn, (ii) substrate, and (iii) environmental control of the mushroom house. The limiting factor and the first stage of the mushroom cultivation is spawn production. Therefore, the production of high-quality spawn with extensive mycelia in the substrates is very important. The function of spawn is to boost the mycelia to a state of vigor and penetrate the bulk substrates of mushrooms. For that reason, the substrate of spawn is not only a vehicle for evenly distributing the mycelia but also a source of nutrients. Cereal grains are a suitable carrier of spawn production because they have a hard shell with an aleurone layer containing protein and starch. Furthermore, the grains will attract water and swell. Spawn grains such as wheat, millet, and corn have been reported to affect fruiting body production (Elhami and Ansari 2008). Mushroom farmers in Indonesia usually use sawdust supplemented with cereal grains, such as wheat, sorghum, corn, rice husk, or millet, to increase the nutrition of the substrate. Almost similar to paddy straw mushroom farmers, the oyster mushroom farmers also obtain the spawn from research institutes or spawn factories.

The main substrate for the mother spawn is sorghum grains of good quality, which contain a few broken kernels, little debris, and have been recently harvested. The grain was soaked overnight, and the floating grains are discarded. The grains are drained and cooked in water for 10–15 minutes. The grains are drained again and added with 1% (w/w) of ground limestone ($CaCO_3$). Then, the substrates are put into a spawn bottle and sterilized at 121 °C for 30 minutes. After sterilization, each bottle is inoculated with 5 mm square mycelium of *P. ostreatus* growth on PDA in a petridish. It is then incubated at 30 °C in a dark room for 14 days.

The mother spawn is used as a source for inoculation into the final spawn. The main substrate of final spawn is sawdust (73%), rice husk (20%), corn flour (5%), limestone (1%), and gypsum ($CaSO_4$) (1%). All substrates are thoroughly mixed and clean water is added so that the substrate water content reaches 65%. It is then put into a polypropylene bag and covered with cotton. One day after sterilization, it is inoculated with 30 g of mother spawn and incubated at 30 °C in the dark room for 14 days (Fig. 3.9). Saskiawan et al. (2016) reported that the addition of 30%

A B

**Fig. 3.9** Mother spawn (**a**) and final spawn (**b**) of *Pleurotus* sp.

sorghum grain in the final spawn gave the best treatments in growth of mycelium and BER value of oyster mushroom.

The main content of substrate in *P. ostreatus* cultivation is sawdust, with the addition of 2% $CaCO_3$, 1% $CaSO_4$, 15% rice bran, and 4% corn flour. First, the sawdust and $CaCO_3$ are composted for 5 days. Then, the other materials are added and mixed properly. Tap water is added to make the water concentration of the substrate up to about 60%–70%. Polypropylene bags (15 × 35 cm) are filled with 1.0 kg of substrate, covered with cotton, sterilized at 100 °C for 2 hours, and cooled down to the desired inoculation temperature. The next day after sterilization, the bags are inoculated with the final spawn of 5% of the substrate weight and incubated in the incubation room at a temperature of 28 ± 5 °C at a relative humidity of 70%–75% ± 5%.

After the substrate is fully colonized with mycelium, the bags are replaced in a mushroom house and incubated at a temperature of 25 ± 5 °C at a relative humidity of 80%–85% ± 5%. Misting is carried out twice a day to maintain the temperature and humidity in the mushroom house (Fig. 3.10). In the mushroom house, the bags are opened. After the opening of the bags, 7–10 days later, the primordial is formed. The fruiting body is harvested by clasping the stipe and gently twisting and pulling the fruiting body from the substrate. The fruiting body of the oyster mushroom is then packaged in a plastic bag of 5 kg each. The market chain of oyster mushrooms is almost like that of paddy straw mushrooms. The collector will come to several oyster mushroom farmers to collect and market the mushrooms in a traditional or wet market.

**Fig. 3.10** Condition of the mushroom house. Misting is done twice a day to maintain temperature and humidity

## 3.4 Conclusion

Sustainable management of edible and medicinal mushrooms in Indonesia requires a multi-faceted approach. This includes the following:

- Ethnomycology: Documenting the traditional knowledge of local tribes, who possess valuable information on wild edible and medicinal mushrooms, is vital for conservation and identifying species for potential cultivation.
- Responsible Harvesting: Foraging must be done sustainably, with practices like leaving some mushrooms behind to ensure spore dispersal.
- Agricultural Waste Utilization: The use of agricultural waste, such as sawdust and rice straw, as a substrate for cultivation provides an eco-friendly way to manage waste while simultaneously producing a valuable commodity.
- Community Involvement: Engaging local communities and indigenous groups in both conservation and cultivation efforts is critical for success. This includes training and empowering them to develop their own mushroom enterprises.
- Technological Advancement: Research and innovation in mushroom cultivation, such as developing local seeds from wild species and applying green technology for humidity control, are essential for improving productivity and ensuring long-term sustainability.

Ultimately, a synergistic approach that combines traditional knowledge, modern science, and community empowerment is necessary to ensure the conservation and sustainable use of Indonesia's rich fungal resources for future generations.

**Acknowledgments** I acknowledge the use of AI (https://gemini.google.com) on September 2, 2025, to generate a list of potential arguments for my paper on "Conservation, Cultivation and Sustainability of Edible and Medicinal Mushrooms in Indonesia." The output was used to brainstorm ideas, and I incorporated some of these suggestions into my final draft.

## References

Ahlawat OP, Sagar MP (2007) Management of spent mushroom substrate. Technical bulletin. National Research Centre for Mushroom (ICAR), Chambaghat, Solan

Ahlawat OP, Tewari RP (2007) Cultivation technology of paddy straw mushroom (*Volvariella volvacea*). Technical bulletin. National Research Centre for Mushroom (ICAR), Chambaghat, Solan

Belewu MA, Belewu KY (2005) Cultivation of mushroom (*Volvariella volvacea*) on banana leaves. Afr J Biotechnol 4(12):1401–1403

Casselton LA, Olesnicky NS (1998) Molecular genetics of mating recognition in basidiomycete fungi. Microbiol Mol Biol Rev 62(1):55–70

Chang ST (1977) The origin and development of straw mushroom cultivation. Econ Bot 31:374–376

Chang ST, Wasser SP (2017) The cultivation and environmental impact of mushrooms. In: Oxford research encyclopedia, environmental science. Oxford University Press USA

Chiu SW (1996) Nuclear changes during fungal development. In: Chiu SW, Moore D (eds) Patterns in fungal development. Cambridge University Press, pp 105–125

Cohen R, Persky L, Hadar Y (2002) Biotechnological applications and potential of wood-degrading mushrooms of the genus *Pleurotus*. Appl Microbiol Biotechnol 58:582–594

Elhami B, Ansari NA (2008) Effect of substrates of spawn production on mycelium growth of oyster mushroom species. J Biol Sci 8:474–477

Hagen JB (2012) Five kingdom, more or less: Robert Whittaker and the broad classification of organisms. BioScience 64(1):72–74

Hawksworth D, Lücking R (2017) Fungal diversity revisited: 2.2 to 3.8 million species. Microbiol Spectr 5(4):FUNK-0052-2016. https://doi.org/10.1128/microbiolspec.FUNK-0052-2016

Hood IA (2006) The mycology of the Basidiomycetes. Heart rot and root rot in tropical Acacia plantations. In: Proceeding of a workshop held in Yogyakarta, Indonesia, 7–9 February 2006. Canberra, ACIAR Proceedings No. 12

Ishihara A, Goto N, Kikkawa M, Ube N, Ushijima S, Ueno M, Ueno K, Oka KO (2018) Identification of antifungal compouns in the spent mushroom substrate of *Lentinula edodes*. J Pestic Sci 43(2):1–16

Kabel MA, Jurak E, Makela MR, de Vries RP (2017) Occurrence and function of enzymes for lignocellulose degradation in commercial *Agaricus bisporus* cultivation. Appl Microbiol Biotechnol 101:4363–4369

Kantharaj P, Boobalan B, Sooriamuthu S, Mani R (2017) Lignocellulose degrading enzymes from fungi and their industrial applications. Int J Curr Res Rev 9(21):1–12

Kües U, Liu Y (2000) Fruiting body production in Basidiomycetes. Appl Microbiol Biotechnol 54: 141–152

Kusuma HI, Harnelly E, Thomy Z, Fitra MA, Samingan (2020) Exploration of potential wild medicinal mushrooms from Pocut Meurah Intan forest park, Aceh, Indonesia. J Nat 20(3): 66–73. https://doi.org/10.24815/jn.v20i3.16916

Lechner BE, Papinutti VL (2006) Production of lignocellulosic enzymes during growth and fruiting of the edible fungus *Lentinus trigrinus* on wheat straw. Process Biochem 41:594–598

Mandeel QA, Al-Laith AA, Mohamed SA (2005) Cultivation of oyster mushrooms (*Pleurotus* spp.) on various lignocellulosic wastes. World J Microbiol Biotechnol 21:601–607

Manzi P, Gambelli L, Marconi S, Vivanti V, Pizzoferrato L (1999) Nutrients in edible mushrooms: an inter-species comparative study. Food Chem 65:477–482

Miles PG, Chang ST (1997) Mushroom biology concise basics and current developments. World Scientific Publishing Co. Pte. Ltd, Singapore

Nion YA, Djaya AA, Kamillah, Agnestisia R, Mashabhi S, Nion KA (2024) Wild edible mushroom from Central Kalimantan and its potential for other benefits. IOP Conf Ser Earth Environ Sci 1421(2024):012011. https://doi.org/10.1088/1755-1315/1421/1/012011

Obodai M, Okine JC, Vowotor VA (2003) Comparative study on the growth and yield of *Pleurotus ostreatus* mushroom on different lignocellulosic by-products. J Ind Microbiol Biotechnol 30: 146–149

Pandey AK, Vishwakarma SK, Srivastava AK, Pandey VK, Agrawal, Singh MP (2014) Production of ligninolytic enzymes by White Rot Fungi on lignocellulosic wastes using novel retreatments. Cell Mol Biol 60(5):41–46

Priadi D, Arfani A, Saskiawan I, Mulyaningsih ES (2016) Use of grass and spent mushroom compost as growing medium of local tomato (*Lycopersicon esculentum* Miller) seedling in the nursery. Agrivita J Agric Sci 38(3):242–250

Putra IP, Nasrullah MA, Dinindaputri TA (2019) Study on diversity and potency of some macro mushroom at Gunung Gede Pangrango National Park. Bull Plasma Nutfah 25(2):1–14

Quimio TH (2002) Tropical mushroom cultivation: Published by Natural Books MC store. Mandaluyong, p 163

Rahi DK, Rahi S, Pandey AK, Rajak RC (2009) Enzymes from mushrooms and their industrial application. In: Rai M (ed) Advances in fungal biotechnology. I K International Publishing House Pvt, New Delhi, pp 136–184

Ramadhani I, Idris, Masrukhin, Nurcahyanto DA, Setiawan RH, AZN I, Elfirta. RR (2019) Isolation and identification of macrofungi from Gunung Halimun Salak National Park. J Mikol Indones 3(2):104–117

Retnowati A, Kusmawaty I, Apandi A, Haryadi (2020) Two newly recorded species of *Agaricales* (Basidiomycota) from Banggai Kepulauan, Sulawesi, Indonesia. Biodiversitas 21(12): 5615–5621. https://doi.org/10.13057/biodiv/d211217

Sanchez C (2010) Cultivation of *Pleurotus ostreatus* and other edible mushrooms. Appl Microbiol Biotechnol 85:1321–1337

Saskiawan I (2012) Bioconversion of rice straw to produce straw mushroom (Volvariella volvacea) in Indonesia. In: Proceeding of the 12th Sciences Council of Asia (SCA) conference and international symposium, 10–12 July 2012, Bogor, Indonesia, pp 405–413

Saskiawan I, Hasanah N, Shimomura N (2016) Cultivation of *Pleurotus ostreatus* using sorghum-supplemented spawn on various substrates. Mushroom Sci Biotechnol 23(4):179–182

Ukoima HN, Ogbonnaya LO, Arikpo GE, Ikpe FN (2009) Cultivation of mushroom (*Volvariella volvacea*) on various farm wastes in Obubra Local Government of Cross River State, Nigeria. Pak J Nutr 8(7):1059–1061

Yusran Y, Erniwati E, Rukmi R (2024) Ethnomycology of wild edible mushrooms by the Bunggu Tribe in West Sulawesi, Indonesia. Int J Des Nat Ecodyn 19:2097–2107

Zikriyani H, Saskiawan I, Mangunwardoyo W (2018) Utilization of agricultural waste for cultivation of paddy straw mushrooms (*Volvariella volvacea* (Bull.) Singer 1951). Int J Agric Technol 14(5):805–881

# Mushroom Cultivation as a Sustainable Bioconversion Strategy for Agrowaste Management

# 4

Hom-Singli Mayirnao, Gurpreet Kaur, and Surinder Kaur

**Abstract**

Food insecurity remains one of the most persistent global challenges, driven by rapid population growth, climate change, and constrained agricultural resources. Alongside, improper agrowaste management and the prevalent practice of crop residue burning contribute to environmental degradation and health risks. Promoting effective waste recycling and improving resource efficiency within agro-industrial sectors are therefore essential for the development of sustainable food systems. Within this context, mushroom cultivation emerges as a promising approach by serving as an efficient agrowaste management strategy and delivering socioeconomic benefits, which frames the basis for the synthesis of this chapter on sustainable bioconversion. It first examines the types and physico-chemical composition of common agrowastes. Representative agrowastes and their attributes are discussed with emphasis on their relevance to mushroom cultivation, followed by insights into substrate combinations that optimize fungal growth and productivity. Biological efficiency and yield patterns are analyzed to assess cultivation performance across different agro-substrate systems. Furthermore, the recycling potential of spent mushroom substrate is explored, with attention to its applications in soil amendment, animal feed, bioenergy production, bioremediation, seed protection and mulch, and second-generation cultiva-

H.-S. Mayirnao
Department of Botany, University of Delhi, Delhi, India

G. Kaur · S. Kaur (✉)
Department of Botany, SGTB Khalsa College, University of Delhi, Delhi, India
e-mail: surinder@sgtbkhalsa.du.ac.in



B. Giri et al. (eds.), *Advances in Mushroom Research*,
https://doi.org/10.1007/978-981-95-7692-0_4

tion of mushrooms. Beyond technical aspects, this chapter also addresses the socioeconomic dimensions of mushroom cultivation and its role in rural livelihoods, waste repurposing, and sustainable development.

**Keywords**

Agrowaste · Mushroom cultivation · Biological efficiency · Sustainable agriculture · Spent mushroom substrate · Circular economy

## 4.1 Introduction

With the increase in the production of agricultural products driven by population growth, the agricultural residues generated each year have increased considerably, amounting to millions of tons of waste being generated. Agro-industrial residues, collectively termed agrowastes, are rich in various types of nutrients, yet pose significant challenges in terms of their disposal. Mostly, they are disposed of by means of incineration, contributing to environmental pollution and resource wastage. Open-field burning of agricultural residues remains a widespread yet environmentally detrimental waste management practice across Asia, including India, where it significantly contributes to air pollution and global warming (Gupta et al. 2004; Ahmed and Ahmad 2013; Sarkar et al. 2018; Kaushal and Prashar 2021). This practice results in the emission of substantial quantities of greenhouse gases, atmospheric pollutants, and fine particulate matter, collectively deteriorating air quality (Gadde et al. 2009). As agricultural activities expand and modernize, agrowaste is being produced faster than it can be sustainably recycled. Therefore, sustainable strategies for its management are imperative to protect the environment and future food production. Among the available options, mushroom cultivation has emerged as a prospective agro-industrial activity due to its potential to produce high-quality food using low-cost and readily available raw materials (Kosre et al. 2021). Hence, it effectively tackles two major challenges: agricultural waste disposal and the need for affordable nutrition and income-generating opportunities.

Production of edible mushrooms represents exclusive utilization of microbial technology for bioconversion of the agrowastes into nutritious food (Gupta et al. 2018). Edible mushrooms are nutritionally dense macrofungi, distinguished by high-quality protein (19%–35% dry weight), providing all essential amino acids. They contain substantial B-complex vitamins (thiamine, riboflavin, niacin, folate, and biotin), as well as vitamin E, and trace amounts of B12 and D, making mushrooms the only significant vegan source of vitamin D. Their mineral profile boasts potassium, phosphorus, magnesium, calcium, copper, iron, and zinc. Mushrooms are hypolipidemic, with the low-fat content dominated by unsaturated fatty acids such as linoleic and oleic acids. Additionally, mushrooms house complex polysaccharides (β-glucans, chitin, and hemicellulose) and dietary fibers, which impart immunomodulatory, hypocholesterolemic, and prebiotic effects. Some of the most cultivated edible and medicinal mushrooms include *Agaricus bisporus* (button mushroom),

*Pleurotus ostreatus* (oyster mushroom), *Flammulina filiformis* (enoki), *Volvariella volvacea* (paddy straw mushroom), *Calocybe indica* (milky mushroom), *Lentinula edodes* (shiitake), and *Ganoderma lucidum* (reishi/lingzhi) (Royse et al. 2017).

Mushroom cultivation offers a sustainable pathway for agrowaste management while delivering socioeconomic benefits. Within this context, this chapter synthesizes current knowledge on mushroom-based bioconversion, highlighting the suitability of various agrowastes as substrates and the role of substrate combinations in optimizing yield and biological efficiency (BE). The recycling potential of spent mushroom substrate is examined, with applications in soil amendment, animal feed, bioenergy, bioremediation, mulching, seed protection, and second-generation mushroom cultivation. Beyond technical aspects, this chapter emphasizes the contribution of mushroom farming to rural livelihoods, waste repurposing, and sustainable development.

## 4.2 Sustainable Agrowaste Management Through Bioconversion

Globally, an estimated 998 million tonnes of agricultural waste is produced annually (Karić et al. 2022). In India alone, approximately 141 million tonnes of crop residues are generated each year, of which nearly 92 million tonnes are burned in fields, exacerbating air pollution and soil degradation (Gatkal et al. 2024). Conventional waste management strategies, such as open dumping and incineration, also squander resources that can be valorized through sustainable pathways.

Fungi, particularly basidiomycetes, represent the most efficient natural decomposers of lignocellulosic biomass owing to their highly evolved enzymatic machinery capable of mineralizing complex organic polymers. Edible and medicinal mushrooms encapsulate this potential by converting low-value agricultural residues into high-value biomass (Hyde et al. 2019). Mushroom cultivation has thus emerged as a bioconversion model system that addresses the issue of agrowaste disposal and contributes to food and nutritional security as well as sustainable agriculture.

The lignocellulosic residues commonly employed for mushroom cultivation include paddy straw, wheat straw, maize cobs, sugarcane bagasse, cotton stalks, sawdust, and coffee pulp. These substrates are abundant, underutilized, and otherwise pose disposal challenges. The fungal mycelium secretes an array of ligninolytic and hydrolytic enzymes—such as laccases, manganese peroxidases, lignin peroxidases, cellulases, and xylanases—which synergistically depolymerize cellulose, hemicellulose, and lignin (Kumla et al. 2020; Sahu et al. 2025). *Pleurotus ostreatus* has been reported to produce high laccase activity ($>3 \times 10^5$ U/L under solid-state fermentation), enabling efficient delignification of rice straw and subsequent carbohydrate utilization (Mazumder et al. 2009). The selective delignification capacity allows mushrooms to colonize substrates without complete mineralization, leaving behind partially degraded lignocellulosic material enriched in humified organic matter. This biological transformation reduces lignin by 30%–50% while enhancing crude protein content of the spent substrate by 15%–20% (Gupta et al. 2011; Kumla et al. 2020).

Globally, mushroom production exceeds 44 million tonnes annually (FAO 2023), with China contributing more than 75% of the total output. The valorization potential is substantial, i.e., every tonne of dry straw diverted into mushroom cultivation can generate 800–1000 kg of fresh mushrooms worth approximately USD 1000–1200, besides producing nearly 2 tonnes of spent mushroom substrate (SMS) as a secondary resource (Singh et al. 2021; Leong et al. 2022a; Ravlikovsky et al. 2024).

A vital factor influencing the success of mushroom production is the choice of substrate, which serves as the nutrient-rich medium for its growth. Different species require specific substrates that best support their mycelial growth and fruiting (Suwannarach et al. 2022; Lakdawala et al. 2025). Common substrates are cost-effective, encompassing crop residues (straw, husk, stalks, and leaves), fruit and vegetable wastes, animal manure, and agro-industrial by-products such as bagasse, oilseed cakes, and brewery residues. The choice of substrate affects several aspects of mushroom cultivation, including spawn run time, yield, BE, and the quality of the fruiting bodies (Mohamed et al. 2016; Desisa et al. 2023). Additionally, the physical properties (e.g., moisture content, porosity) and chemical composition (e.g., carbon-to-nitrogen ratio) of the substrate directly influence the growth rate and productivity of mushrooms (Fig. 4.1). Elucidating the physicochemical properties, nutrient composition, and biodegradability of different substrates is fundamental for advancing sustainable and resource-efficient mushroom cultivation systems.

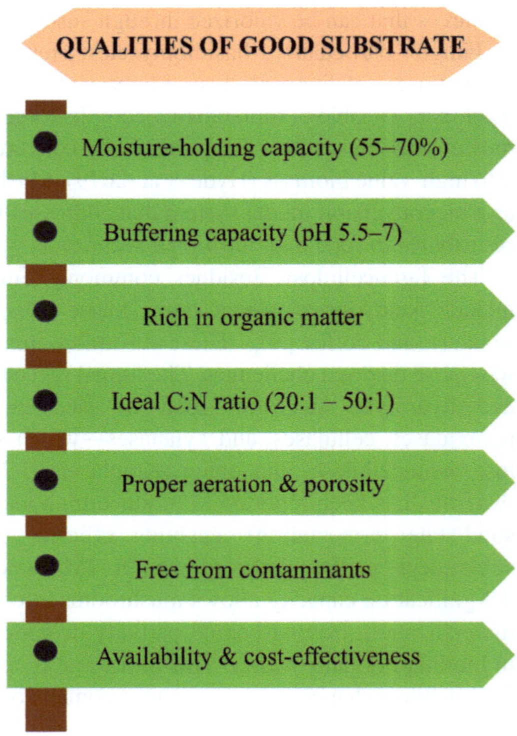


**Fig. 4.1** Qualities of a good substrate for mushroom cultivation

## 4.3 Agrowastes: Types and Physicochemical Composition

Agricultural and agro-industrial systems generate a wide spectrum of biomass residues, which differ in their origin, structural complexity, and biochemical composition. Depending on these factors, agrowastes can be broadly categorized into primary, secondary, and tertiary biomass resources (Shah et al. 2025):

I. *Primary biomass resources* (*crop residues*): These are the inedible plant parts left in the field after harvesting the main crop. They include straw, stover, stalks, stubble, leaves, branches, twigs, haulms, roots, prunings, and trimmings. Such residues represent the largest fraction of agricultural waste generated globally.
II. *Secondary biomass resources* (*agro-industrial residues*): These wastes are produced after crops are processed into primary products in agro-industries. Examples include husks, hulls, peels, sawdust, bagasse, pomace, corncobs, dust, and other by-products from food- and wood-processing sectors.
III. *Tertiary biomass resources* (*processing residues*): These arise from further processing of raw materials into refined products. They include rice hulls, wheat chaff, corn cobs, and soybean seed coats, which represent the by-products of milling, polishing, and seed-processing industries.

The physicochemical composition of agrowastes is highly variable and largely influenced by crop type, processing methods, and storage conditions, yet certain common characteristics are consistently reported. Physically, agro-residues exhibit high moisture content (10%–60%), low bulk density, and variable particle size, which influence porosity and aeration potential. Chemically, they are rich in organic matter (70%–95% dry weight) and typically exhibit pH values ranging from slightly acidic to alkaline (4.5–8.5) (Zhang et al. 2012a; Ishii and Furuichi 2014; Gatkal et al. 2024). A key determinant of their biodegradability and valorization potential is the carbon-to-nitrogen (C:N) ratio, which is often high (60–120:1) in lignocellulosic materials such as straw and husks, but comparatively lower (20–40:1) in fruit pomace, press mud, and oilseed residues (Pan et al. 2021; Yang et al. 2021). The lignocellulosic matrix generally comprises cellulose (25%–45%), hemicellulose (15%–35%), and lignin (5%–30%), together with minor extractives such as polyphenols and waxes. Proximate composition indicates that agrowastes are dominated by carbohydrates (40%–70%), followed by proteins (3%–15%), and lipids (1%–10%), with ash contents typically in the range of 2%–15%. The structural polysaccharides (cellulose, hemicellulose) provide fermentable sugars upon hydrolysis, while lignin contributes to recalcitrance and potential as a source of aromatic compounds. The mineral fraction is enriched in essential macronutrients (N, P, K, Ca, Mg, and S) and micronutrients (Fe, Zn, and Mn), while some residues, such as rice husk, are also notable for high silica content (Thomsen 2005; Ingle et al. 2025). Mineral profiles vary widely and influence both microbial processing and end-use applications. These characteristics determine their suitability for valorization into mushroom cultivation substrates and other potential applications in bioenergy production, composting, and other biorefinery applications.

## 4.4 Representative Agrowastes and Their Attributes with Relevance to Mushroom Cultivation

The global mushroom industry is heavily reliant on lignocellulosic agrowaste as substrates. These residues provide a physical matrix for fungal colonization in addition to the carbon source and nutrients required for mushroom growth. The repurposing of agrowaste in mushroom farming represents a sustainable approach to agricultural waste management. The most extensively used waste for edible mushroom production has been wheat and paddy straws, sawdust, hardwood chips, sugarcane bagasse, cottonseed hulls, banana leaves, peanut hull and corn leaves, mango fruits and seeds, corn cobs, rice, and wheat bran (Orts et al. 2008; Raman et al. 2021). Various agrowastes used for cultivation are listed in Table 4.1. The major categories of agrowaste are discussed with reference to their relevance in mushroom cultivation.

### 4.4.1 Cereal Straws

Cereal straws are among the most abundant agrowastes, with a global annual production exceeding 1 billion tonnes. Cereal straws are the most widely utilized agricultural residues in mushroom farming due to their abundance as well as balanced nutrient composition (Raman et al. 2021). They are primarily composed of cellulose (30%–45%), hemicellulose (20%–30%), and lignin (10%–20%), along with small amounts of ash, proteins, and soluble sugars. The high lignocellulosic fraction renders cereal straws intractable to enzymatic degradation. However, their abundance and favorable C:N ratios (60–80:1) make them widely used in mushroom cultivation. Their lignocellulosic framework provides the essential carbon skeletons for fungal metabolism. Therefore, they serve as primary substrates for the commercial cultivation of various commercially produced mushrooms.

Wheat straw is rich in cellulose and hemicellulose and has a moderate lignin content, which supports mycelial growth. Its physical structure allows for good aeration, and it requires only mild pasteurization to be used effectively. Wheat straw accounts for 35%–40% N, 10%–15% P, and 80%–90% K (Kamthan and Tiwari 2017; Ma et al. 2022). Due to its abundance and suitability as a substrate, most of the commercially available mushrooms (*Pleurotus* spp., *Lentinula edodes*, *Flammulina velutipes*, *Hericium erinaceus*, *Agrocybe cylindracea*, *Hypsizygus ulmarius*, among others) are cultivated on wheat straw, including medicinal mushrooms such as *Ganoderma lucidum* (Ćilerdžić et al. 2018; Aditya et al. 2024). Biological efficiency has been reported to increase when wheat straw is properly supplemented with nitrogen sources like wheat bran (Vieira and de Andrade 2016). Rice/paddy straw is traditionally used for *Volvariella volvacea* cultivation in tropical and subtropical regions, while also serving as a low-cost alternative for *Pleurotus* production. Despite slower degradation due to silica, it supports the rapid growth of heat-loving species under high moisture field conditions in *F. velutipes*. The BE of *Calocybe indica* was found to be the highest when using

**Table 4.1** Overview of agrowastes based on their type, physicochemical properties, benefits, and limitations for cultivation of edible mushrooms

| S. no. | Type of agrowaste | Physicochemical properties | Benefits | Limitations | Use for | References |
|---|---|---|---|---|---|---|
| 1. | Cereal straws (wheat, rice, maize) | High cellulose and hemicellulose, moderate lignin content | Readily available, easy to process, good yield | Requires pasteurization, seasonal availability | *Pleurotus* spp., *Volvariella volvacea*, *Agaricus bisporus*, *Lentinula edodes* | Girmay et al. (2016), Ćilerdžić et al. (2018), Aditya et al. (2024) |
| 2. | Sugarcane bagasse | Fibrous, high lignin content | Abundant in sugar-producing regions | Needs pretreatment (alkaline or steam) for lignin breakdown | *Pleurotus ostreatus*, *P. eryngii*, *Ganoderma lucidum* | Aguilar-Rivera and De Jesús-Merales (2010), Rashad et al. (2019), Okano et al. (2007), De et al. (2023) |
| 3. | Cotton waste and gin trash | High nitrogen and cellulose | Enhances protein content in fruiting bodies | Susceptible to contamination if not treated properly | *V. volvacea*, *A. bisporus*, *P. sajor-caju* | Akhtar et al. (2016), Ahlawat and Kaur (2019) |
| 4. | Sawdust and wood chips | Lignin-rich, slow decomposing | Long fruiting cycles, good flavor profiles | Requires supplementation with bran or cornmeal | *L. edodes*, *G. lucidum* | Telang et al. (2010), Shrikhandia et al. (2022), Wang et al. (2025) |
| 5. | Corncobs and stalks | High porosity, good aeration | Enhances water retention and aeration | Often mixed with straw to optimize nutrient profile | *P. pulmonarius*, *P. florida* | Janet (2012), Salama et al. (2016) |
| 6. | Coffee husk and tea waste | Rich in polyphenols and tannins | Potential in coffee/tea-producing regions | May inhibit growth if not properly composted | *Pleurotus* spp. | Chai et al. (2021), Fan et al. (2000), Peksen and Yakupoglu (2009) |
| 7.. | Banana leaves and pseudostems | High moisture, moderate lignocellulose | Biodegradable and locally abundant | Needs chopping and fermentation | *V. volvacea* in tropical regions | Patra and Pani (1995), Yue-Lian and Qing-Fang (2011), Richard et al. (2019), de Siqueira et al. (2011) |

rice straw as the substrate (Amin et al. 2010). Compost prepared from cereal straws has been reported to exhibit high BE% for cultivation of *Agaricus* spp., including A. *bisporus*, *A. campestris*, *A. subrufescens*, and *A. blazei* (Suwannarach et al. 2022). Other desirable cereal straw substrates for the cultivation of *Pleurotus* spp. include sorghum, barley, and oats (Dedousi et al. 2024; Jahedi et al. 2024). Collectively, cereal straws are the main reference agrowaste for mushroom production against which other agrowastes are evaluated.

### 4.4.2 Sugarcane Bagasse

Sugarcane bagasse, a fibrous by-product of the sugar industry, has been successfully utilized as a bulk substrate for mushroom cultivation because its loose structure allows good air flow that facilitates rapid mycelial colonization. For *Pleurotus* spp., bagasse supplemented with wheat straw supports BE of up to 60%, while for *A. bisporus*, composted bagasse can replace up to 40%–50% of cereal straw in conventional formulations (Hasan et al. 2015; Zakil et al. 2020; Karpagavalli et al. 2024). Partially composted sugarcane bagasse mixed with wheat straw or poultry manure was substituted for up to 40%–50% of wheat straw substrate for growing *A. bisporus*, without reducing yield (Noble et al. 2002; Kumar et al. 2021). Composting enhances microbial activity and substrate selectivity, which led to competitive colonization by *A. bisporus*. Multiple reports confirmed that *P. ostreatus*, *P. djamor*, *P. sajor-caju*, and *L. edodes* thrive on pasteurized bagasse, producing BE ranging from 70% to 100% (Hasan et al. 2015; Suwannarach et al. 2022). Supplementation with bran or gypsum further improves yield and reduces crop cycle duration. Bagasse has been successfully used as a substitute for rice straw for the cultivation of *V. volvacea* (Thiribhuvanamala et al. 2021).

### 4.4.3 Sawdust and Wood Chips

Sawdust and wood chips are classic substrates for the cultivation of several high-value mushrooms. Their slow decomposition rate, high porosity, and ability to sustain long cropping cycles make them especially suitable for species that thrive on lignin-rich substrates. Therefore, it stands out as the substrate of choice for wood-loving mushrooms such as *L. edodes*, *Auricularia polytricha*, *A. heimuer*, *Hericium erinaceus*, and *Macrocybe gigantea* (Shrikhandia et al. 2022; Wang et al. 2025). Hardwood sawdust blocks enriched with bran provide stable yields exceeding 100% BE. Wood chips, when incorporated into compost, improve aeration and nutrient release, supporting *A. bisporus* cultivation. For species of *Pleurotus*, sawdust substrates offer durability and higher fruiting body quality compared to straws, especially when blended with crop residues. Sawdust is also widely used for growing wood-rotting medicinal mushrooms, such as *Lenzites sterioides*, *Trametes versicolor*, and *G. lucidum*, particularly from hardwoods including *Alnus nepalensis*, *Swietenia mahagoni*, *Dipterocarpus turbinatus*, *Tectona grandis*, *Gmelina arborea*, and *Michelia champaca* (Deshpande and Arya 2022; Shrikhandia et al. 2022). Furthermore, the use of agroforest wastes, including vineyard pruning and olive pruning, has proven to be successful for the cultivation of *L. edodes* (Sassine et al. 2024). It has high lignin and low nitrogen content and must be supplemented with bran to increase yields. A study by Hassan et al. (2022) demonstrated that rubberwood sawdust gave the best yield for the cultivation of *Schizophyllum commune* due to the high cellulose content and mildly destructed structure of the substrate in the absence of silica (rice straw). Liquid spawn production of an edible mushroom, *Sparassis latifolia*, in submerged fermentation led to quick mycelial growth on sterile pine sawdust substrate compared to the traditional sawdust spawn (Ma et al. 2016). While cereal straws are commonly used, sawdust and wood chips provide longer-lasting substrates with higher structural integrity.

### 4.4.4 Cotton Waste and Gin Trash

Cotton residues, particularly gin trash (a mixture of stalks, leaves, burrs, and lint produced during cotton ginning) and textile mill wastes, have been successfully utilized as alternative substrates for mushroom cultivation. The residual lint fibers improve substrate aeration, supporting rapid colonization and fruiting. Their fibrous, lignocellulosic nature provides a suitable growth matrix, though nitrogen supplementation and proper pretreatment are typically required for optimal yields (Deora et al. 2021; Zou et al. 2020). It has been reported to be highly suitable for mushrooms such as *Pleurotus* spp. and *L. edodes*, among others. *Pleurotus* spp. colonize cotton waste effectively, with biological efficiencies up to 120% under optimized conditions. Composted gin trash blended with poultry manure has also been employed in *A. bisporus* production, substituting up to 30%–40% of wheat straw (Ahlawat and Kaur 2019; Aditya et al. 2024). Moreover, textile cotton waste mixed with sawdust tested for *L. edodes* and *Pleurotus* spp. offered promising results. Studies indicate that up to 30%–40% replacement of cereal straw with cotton waste in compost formulations maintains good yields while lowering substrate costs. Cotton waste from spinning mills also exhibited enhanced fruiting and shortened crop cycles (Oh et al. 2004; Ahmad et al. 2011; Nguyen et al. 2025).

### 4.4.5 Corncobs and Stalks

Corn residues represent a promising and underutilized lignocellulosic feedstock for mushroom cultivation. Corncobs, corn stovers, maize husks, and groundnut shells are often used either alone or in combination with other substrates to enhance yield. These agrowastes support sustainable mushroom production and promote efficient agricultural waste management in the interim (Nanje Gowda and Chennappa 2021). Owing to their porous structure and balanced physicochemical profile, corn residues are promising alternatives to cereal straws. *Pleurotus* spp., and *A. auricula*, achieve elevated BE on corncobs, with superior aeration and reduced compaction compared to wheat straw (Lu et al. 2022; Ma et al. 2011). When composted, corn stalks can substitute 30%–40% of straw in *A. bisporus* production without compromising yield. Various combinations of corncobs with sawdust are increasingly used in *L. edodes* and *Pleurotus* spp. cultivation due to their porosity and water retention (Yu et al. 2022). Corn-based substrates thus offer both agronomic and environmental benefits, especially in maize-dominant regions. Studies show that *Pleurotus* species such as *P. ostreatus*, *P. florida*, and *P. pulmonarius* grow vigorously on corncob-based substrates, with fruit bodies exhibiting high protein and polysaccharide content (Jin et al. 2018; Wu et al. 2019). Maize stalks and straw have gained prominence as partial replacements for wheat straw in composts, improving substrate aeration and reducing reliance on seasonal straw availability.

### 4.4.6 Coffee Husk and Tea Waste

Agro-industrial residues from beverage crops, such as coffee husk and spent tea waste, have gained attention as unconventional substrates for mushroom farming. Recent studies demonstrate positive effects on BE and nutritional quality when coffee husk and spent tea waste are used as partial replacements (typically ≤30%–50% wet basis) combined with straw or sawdust (Saratale et al. 2020; Ahmed et al. 2024; Torres-Martinez et al. 2024). Coffee husk and spent tea waste typically have lower C:N ratios and elevated levels of organic N, K, and trace minerals compared with pure cereal straws, which can accelerate mycelial colonization but may also shift fungal metabolism (protein synthesis, secondary metabolites) (Torres-Martinez et al. 2024). Among the most studied taxa, *P. ostreatus* demonstrates superior adaptability to spent coffee grounds and tea leaf waste-amended substrates. When coffee husk replaces 20%–30% of the basal wheat straw matrix, *P. ostreatus* achieves high BE, with reports of yields comparable to or surpassing conventional substrates. Moreover, the incorporation of tea waste has been shown to enrich the protein fraction and antioxidant potential of the fruiting bodies, suggesting that bioactive constituents of tea leaves may modulate fungal secondary metabolism (Ahmed et al. 2024; Torres-Martinez et al. 2024). Other oyster mushroom species, including *P. pulmonarius* and *P. sajor-caju*, have also been successfully cultivated on spent coffee grounds, with enhanced laccase induction and lignin biodegradation, although their tolerance ceiling is generally higher, permitting supplementation up to ~40%–50% under optimized pH and moisture conditions (Alsanad et al. 2021). Beyond *Pleurotus*, other high-value fungi also exhibit compatibility with coffee and tea residues. For instance, *L. edodes* has been cultivated on sawdust blocks partially substituted with spent coffee grounds (≤20%), which not only supported satisfactory colonization but also resulted in mineral enrichment, particularly potassium and magnesium, in the harvested mushrooms (Saratale et al. 2020). Similarly, *G. lucidum* mycelia colonize spent coffee grounds-supplemented substrates effectively, though fruiting data remain limited. *A. auricula-judae* (wood ear mushroom) also tolerates up to 25%–30% incorporation of tea leaf residues, with studies reporting an increase in polysaccharide yield and potential nutraceutical enhancement (Ahmed et al. 2024).

### 4.4.7 Banana Leaves and Pseudostems

Banana agro-residues, including leaves and pseudostems, are generated in large volumes in tropical and subtropical regions, where bananas are cultivated extensively. These materials are rich in cellulose (30%–40%), hemicellulose (20%–25%), and possess moderate lignin content (~10%–15%), which makes them suitable lignocellulosic matrices for fungal colonization. Additionally, their relatively high moisture-holding capacity and favorable porosity facilitate gas exchange, which is critical for mycelial respiration. The C:N ratio of banana pseudostems, often ranging between 40:1 and 60:1, is comparable to that of cereal straws, positioning them as effective basal substrates or co-substrates in mushroom production systems (Carvalho et al. 2012; Doroški et al. 2022). *Pleurotus* spp. exhibit strong adaptability to banana residues. Banana stalks exhibited more efficiency in the production of *P. sajor-caju* when utilized without the addition of other substrates, with a BE of 74.4%. However, when combined with other substrates or grasses, there was a drop in BE, irrespective of their N-content (de Siqueira et al. 2011). A similar result was obtained in a study by Singh et al. (2016) on *P. florida*. *V. volvacea* has also been reported to colonize banana pseudostem-based substrates, though BE is typically lower than in *Pleurotus* (Yue-Lian and Qing-Fang 2011). Overall, banana leaves and pseudostems act as bulking agents, improving aeration in denser substrates like sawdust.

### 4.4.8 Fruit Pulp Residues

Fruit processing industries generate large quantities of fruit pulp and pomace after juice extraction, representing an underutilized lignocellulosic and nutrient-rich residue (Kim et al. 2024; Leong and Chang 2022). Mushroom cultivation on fruit pulp residues exploits the biotransformation ability of fungi, especially genera such as *Pleurotus* and *Agaricus*, to degrade complex lignocellulosic material via extracellular enzymatic hydrolysis (by laccases, cellulases, and hemicellulases) (Durán-Aranguren et al. 2021). *P. pulmonarius* and *P. ostreatus* grown on grape pomace blends show lignin degradation, increased protein in spent substrate, and in some cases, improved antioxidant and phenolic profiles in fruiting bodies (Papadaki et al. 2019). Industrial-scale adoption is facilitated through solid substrate fermentation, with fruit pulps such as tomato and pineapple successfully employed in the cultivation of *Pleurotus* spp., and *T. versicolor* (Otieno et al. 2022; Iandolo et al. 2011). Consequently, cultivation of mushrooms using fruit pulp wastes valorizes into protein-rich food and advances green biorefineries.

## 4.5 Combination of Agricultural Substrates Used for the Cultivation of Mushrooms

Different types of agrowastes affect mushroom growth, yield, and nutritional content, making substrate selection vital. Different mushroom species require specific substrates that best support their mycelial growth and fruiting. Agrowastes such as wheat straw, rice husks, sugarcane bagasse, cotton stalks, maize cobs, wheat bran, sawdust, and coffee grounds constitute renewable biomass resources that can be strategically used either as individual substrates or in synergistic combinations for bioconversion processes (Orts et al. 2008; Jeznabadi et al. 2016; Gasecka et al. 2020; Kumla et al. 2020). These substrates are rich in proteins, sugars, tannins, lignin, cellulose, hemicellulose, polyphenols, polyols, pectin, and lipids. They provide carbon, nitrogen, and energy needed for mushroom growth, and also determine the composition of the spent mushroom substrate (Altieri et al. 2009; Patra and Saxena 2011). In addition, animal manure is often added to the substrate to supplement nitrogen sources, and lime is added to sterilize it. These materials are not only cost-effective but also promote the recycling of agro-industrial by-products, reducing environmental pollution. The choice of substrate affects several aspects of mushroom cultivation, including spawn run time, yield, BE, and quality of the fruiting bodies. Additionally, the physical and chemical properties of the substrate directly influence the growth rate and productivity of mushrooms. Through the optimization of substrate formulations, mushroom yield can be enhanced, production costs reduced, and cultivation practices adapted to locally available raw materials.

A variety of combinations of agricultural wastes, in addition to the supplements, are used to improve the yield, nutritional, and medicinal values of cultivated mushrooms (Carrasco et al. 2018; Zięba et al. 2021). Vegetable waste, when used in combination with paddy straw, resulted in a higher yield of *Pleurotus* sp. (Singh and Singh 2012). The substrate composition of 3:1 ratio of cotton stalks and cotton seed hulls exhibited significantly higher yield with BE (23.2% and 22.5%) and improved nutritional composition of *P. ostreatus* and *P. florida*, respectively (Sardar et al. 2020). On barley straw and sugar beet pulp substrate complemented with rice bran, the highest mushroom fresh weight and moisture content were achieved (Zadrail and Dube 1992). Cereal straw supplemented with cottonseed powder, yeast mud, groundnut cake, and rice bran led best response concerning the yield of *Pleurotus* spp. Rice husks with sawdust are reported as an optimal substrate for the cultivation of *P. ostreatus* (Singh and Singh 2012). For *Pleurotus sajor-caju*, a combination of soybean straw with wheat straw showed a significantly higher yield as compared to soybean straw and sawdust combination (Rani et al. 2008; Jatwa et al. 2016). Agricultural wastes are used alongside supplements, including gypsum, lime, and urea. Gypsum provides Ca and regulates the acidity levels during the expansion of the mushrooms. Due to the high water-holding capacity of gypsum, excess wetting of the substrate can be prevented. Lime is used to regulate pH. Various components such as urea, bran, and molasses are incorporated in substrate to fulfill the appropriate nitrogen content requirement for mushroom cultivation (Nunes et al. 2012).

### 4.5.1 Analysis of Biological Efficiency and Yield Patterns

The biological efficiency (BE) of a substrate—calculated as the ratio of fresh mushroom yield to the dry weight of the substrate (expressed as a percentage)—is the key metric for evaluating its performance. Substrate combinations are often

**Table 4.2** Effect of substrate alone or in combination on biological efficiency

| Substrate combination | Mushroom | Biological efficiency (BE%) | Reference |
|---|---|---|---|
| 100% wheat straw | *Pleurotus ostreatus* | 78.5 | Mandeel et al. (2005) |
| 75% wheat straw + 25% cotton waste | | 92.3 | |
| 100% rice straw | *Volvariella volvacea* | 18.4 | Baskaran et al. (2020) |
| Rice straw (70%) + wheat bran (25%) + gypsum (5%) | | 27.9 | |
| 100% sugarcane bagasse | *Pleurotus sajor-caju* | 68.2 | Pant et al. (2006) |
| Sugarcane bagasse (80%) + rice bran (18%) + $CaCO_3$ (2%) | | 87.6 | |
| 100% sunflower seed hulls | *Pleurotus ostreatus* | 42.1 | Oseni et al. (2012) |
| Sunflower hulls (50%) + wheat straw (50%) | | 71.8 | |
| 100% banana leaves | *Pleurotus eous* | 55.5 | Jonathan and Fasidi (2003) |
| Banana leaves (60%) + paddy straw (40%) | | 82.0 | |
| 100% sawdust (hardwood) | *Lentinula edodes* | 64.8 | Moonmoon et al. (2010) |
| Sawdust (78%) + wheat bran (20%) + gypsum (2%) | | 98.2 | |
| 100% coffee pulp | *Pleurotus pulmonarius* | 48.3 | Salmones et al. (2005) |
| Coffee pulp (50%) + corn cobs (50%) | | 75.9 | |
| 100% groundnut shells | *Pleurotus florida* | 59.7 | Alam et al. (2008) |
| Groundnut shells (70%) + wheat straw (30%) | | 84.4 | |

designed to optimize the C:N ratio, improve porosity for aeration, and provide a complete nutritional profile for the fungus (Han et al. 2024). The research findings on various substrates and their combinations, detailing the cultivated species and their resulting biological efficiencies is given in Table 4.2.

Research consistently demonstrates that substrate formulation significantly impacts BE in mushroom cultivation. Pure substrates typically exhibit moderate BE values due to nutritional imbalances and suboptimal physical properties. The enhancement observed with supplemented formulations underscores the importance of nutrient balancing in substrate preparation (Adenipekun and Omolaso 2015). Mandeel et al. (2005) documented a substantial increase in BE from 78.5% to 92.3% for *P. ostreatus* when cotton waste was incorporated into wheat straw substrate, attributed to the improved C/N ratio and nutrient availability. The dramatic improvement in BE for *V. volvacea* from 18.4% on pure rice straw to 27.9% on supplemented substrate highlights the critical role of nitrogen supplementation through wheat bran and the structural benefits provided by gypsum (Baskaran et al. 2020). This pattern is particularly evident in wood-decaying species such as *L. edodes*, where Moonmoon et al. (2010) reported that supplementation of sawdust with wheat bran and gypsum elevated BE from 64.8% to 98.2%, approaching the theoretical maximum for commercial production.

Combination substrates frequently outperform single-component substrates by mitigating individual limitations. Oseni et al. (2012) observed that a 50:50 blend of sunflower hulls and wheat straw yielded 71.8% BE for *P. ostreatus*, significantly higher than the 42.1% obtained with pure sunflower hulls, suggesting that the wheat straw counteracts phenolic inhibitors while improving substrate texture. Similarly, Salmones et al. (2005) found that combining coffee pulp with corn cobs enhanced BE for *P. pulmonarius* to 75.9%, as the corn cobs effectively diluted the acidity and caffeine content while providing additional structural support.

These findings collectively emphasize that strategic substrate formulation, tailored to species-specific requirements, is essential for maximizing BE. Optimal combinations must address nutritional balance, moisture management, and gas exchange capabilities to achieve economically viable yields in mushroom cultivation systems.

## 4.6 Recycling of Spent Mushroom Substrate

Mushroom cultivation offers a sustainable route for agrowaste utilization and, in turn, generates vast quantities of spent mushroom substrate (SMS) as a by-product. SMS is the lignocellulosic residue that remains after commercial basidiomycete cultivation, which constitutes a globally abundant agro-industrial stream. Every 1 kg of fresh mushroom generates 2.5 kg–5 kg of fresh SMS (Zisopoulos et al. 2016; Cunha Zied et al. 2020). As per the data available, 64 million tons of SMS were generated worldwide by the mushroom industry in 2018, which is likely to escalate to above 100 million tons by 2026 (Atallah et al. 2021).

SMS consists of partially degraded lignocellulosic residues enriched with fungal mycelium, microbial consortia, and residual nutrients. Traditionally, SMS has been considered a disposal burden, frequently landfilled or incinerated, leading to secondary environmental problems such as methane emissions, leachate formation, and air pollution. However, advances in waste valorization research have positioned SMS as a multifunctional resource. The enzymatic breakdown of lignocellulosic fibers during mushroom cultivation enhances substrate bioavailability, while the accumulation of fungal metabolites such as polysaccharides, phenolics, and extracellular enzymes improves the functional quality of the resulting biomass and spent substrate. Fungal metabolites in SMS impart biofunctional properties to SMS. Consequently, SMS can be valorized across multiple domains, as illustrated in Fig. 4.2.

This multifaceted utilization situates SMS at the crossroad of circular bioeconomy frameworks, where primary agrowaste substrates are first converted into edible mushrooms, and the post-harvest residues are further cascaded into new value chains. Thus, SMS valorization offers both environmental dividends and economic opportunities (Martín et al. 2023; Viriato et al. 2024). Current research emphasizes developing scalable, eco-efficient strategies to integrate SMS into agricultural, industrial, and environmental systems, thereby transforming mushroom production into a zero-waste, resource-efficient enterprise, some of which are discussed below.

**Fig. 4.2** Flowchart showing the use of agricultural wastes for mushroom cultivation, leading to spent mushroom substrate with potential applications

### 4.6.1 Soil Amendment and Biofertilizer

SMS is a composite of spent casing/compost components partially depolymerized by fungal enzymatic systems and enriched with fungal biomass, residual substrate-bound macro- and micronutrients (N, P, K, Ca, Mg, Fe, and Zn), and extracellular hydrolytic/oxidative enzymes (cellulases, xylanases, and laccases) (Beckers et al. 2019; Baptista et al. 2023a; Aiduang et al. 2025). These features confer three principal functionalities when SMS is applied to soil: (1) supply of labile and slow-release nutrients; (2) amendment of soil physical properties (aggregability, porosity, and water retention) via organic matter input and mycelial-derived polysaccharides; and (3) stimulation or restructuring of the soil microbial community through inoculation with saprotrophic consortia and provision of carbon sources for heterotrophs (Hu et al. 2021; Martín et al. 2023; Aiduang et al. 2025). Quantitatively, SMS increases soil organic carbon and microbial biomass carbon in a dose-dependent manner and enhances extracellular enzyme activities linked to C, N, and P cycling (β-glucosidase, urease, and phosphatase) (Pintarič et al. 2024).

Meta-analytical and field studies report consistent yield and soil health benefits across horticultural and arable systems when SMS is properly matured. Reported effects include increased crop biomass, improved seedling establishment in substrate blends, higher available P and K, and ameliorated bulk density and water-holding capacity in sandy soils (Paula et al. 2017; Yagüe et al. 2025). At the rhizosphere level, SMS amendments have been associated with greater microbial diversity, increases in plant-growth-promoting taxa, and suppression of certain soilborne pathogens (such as *Fusarium oxysporum* f. sp. *cucumerinum*, *Fusarium oxysporum* f. sp. *radicis-lycopersici*, and *Phytophthora nicotianae*) via competition, antibiotic production, or induced systemic resistance (against *Septoria lycopersici*) (Ntougias et al. 2008; Wang et al. 2020; Mwangi et al. 2024). However, effect magnitude is contingent on feedstock origin, mushroom species, initial salinity, and maturity of the SMS; poorly matured or salt-rich SMS may transiently inhibit seed germination or cause nutrient imbalances (Aamlid and Landschoot 2007). Overall, well-conditioned SMS consistently demonstrates potential as an eco-efficient biofertilizer and soil amendment.

### 4.6.2 Animal Feed

Spent mushroom substrate retains substantial quantities of organic matter, fungal mycelium, partially degraded lignocellulosic biomass, and microbial metabolites. This composition enriches it in crude protein, nonstarch polysaccharides, amino acids, vitamins (notably B-complex), and bioactive compounds (Cunha Zied et al. 2020). The enzymatic degradation of cellulose and hemicellulose during mushroom cultivation increases fiber digestibility and reduces antinutritional lignin content, enhancing the feed value relative to untreated crop residues (Klausen et al. 2025). SMS also harbors probiotic microbial consortia and secondary metabolites that may improve gut health in livestock (Nasehi et al. 2017). These physicochemical and biochemical traits make SMS a promising unconventional feedstuff for ruminants, poultry, swine, and aquaculture species, especially in regions where conventional feed resources are limited or costly.

Using SMS as animal feed across different livestock systems has demonstrated that partial replacement of conventional feed ingredients with SMS improves animal performance while reducing feed costs. In ruminants, SMS supplementation enhances dry matter intake, improves rumen fermentation efficiency, and increases milk yield and fat content. Poultry diets incorporating SMS show improved weight gain, feed conversion ratio, and reduced intestinal pathogen loads (Fazaeli and Masoodi 2006; Song et al. 2007; Baptista et al. 2023b). A similar trend was observed in aquaculture, wherein the inclusion of SMS showed positive results on growth performance, immunomodulation, and intestinal microbiota of Nile tilapia and improved disease resistance against *Streptococcus agalactiae* using SMS derived from *A. bisporus* and *Cordyceps militaris* (Lee et al. 2022; Da-oh et al. 2024). Additionally, SMS-based feed reduced enteric methane emissions in cattle by altering the fiber profile and adding bioactive compounds (Rangubhet et al. 2017; Palangi et al. 2022).

### 4.6.3 Bioenergy and Bioproducts

Spent mushroom substrate is considered an excellent feedstock for bioenergy production because its chemical composition contains partially degraded cellulose, hemicellulose, lignin, residual nutrients, and fungal biomass. The enzymatic pretreatment by mushrooms reduces structural recalcitrance, enhancing substrate accessibility for microbial or thermochemical processing. SMS exhibits a moderate calorific value (~ 14–18 MJ $kg^{-1}$) and favorable C:N ratios, making it suitable for anaerobic digestion, bioethanol production, thermochemical conversion (pyrolysis, gasification, combustion), and as a precursor for biochar (Pérez-Chávez et al. 2019; Gao et al. 2021; Leong et al. 2022b; Londoño et al. 2024). Beyond energy, SMS also supports the production of bioproducts such as organic acids, enzymes, polysaccharides, and nanocellulose, which extend its utility to biorefineries aligned with circular bioeconomy frameworks. Anaerobic digestion of SMS yields biogas enriched in methane, with reported potentials of 150–250 L $CH_4 \cdot kg^{-1}$ volatile solids, particularly when co-digested with manure or food waste to balance C:N ratios. SMS-derived bioethanol production has been demonstrated after saccharification, achieving yields comparable to other agricultural residues (Ravlikovsky et al. 2025; Wattanavichean et al. 2025). Thermochemical conversion (e.g., pyrolysis) produces bio-oil and biochar; SMS-biochar has gained attention for its dual role as an energy carrier and soil amendment with high adsorption capacity for heavy metals and organic pollutants (Jiang et al. 2014; Pérez-Chávez et al. 2019). Gasification of SMS generates syngas (mainly CO) suitable for heat and electricity production (Lee et al. 2024). Moreover, SMS byproduct pipelines have been explored: lignocellulose-derived nanofibers for packaging, enzyme recovery such as laccases, xylanases for industrial catalysis, and extraction of antioxidant compounds for nutraceuticals (Phan and Sabaratnam 2012; Lim et al. 2013; Baptista et al. 2023a). These studies collectively position SMS as a versatile renewable resource for integrated bioenergy and bioproduct generation.

### 4.6.4 Bioremediation

The adsorptive organic matter of SMS makes it a potent tool for environmental remediation. Residual ligninolytic enzymes such as laccases, manganese peroxidases, and versatile peroxidases are retained in SMS and have been documented to catalyze oxidative breakdown of xenobiotic compounds, including pesticides, dyes, polycyclic aromatic hydrocarbons (PAHs), and endocrine-disrupting chemicals. The porous structure of SMS helps absorb heavy metals (Pb, Cd, Cr, and Zn) and other organic pollutants, while the residual fungi contribute to breaking down and detoxifying these contaminants (Barh et al. 2019; Asemoloye et al. 2020; Cai et al. 2021; Ma et al. 2025). Moreover, humified fractions and biochar derived from SMS act as stable adsorbents, improving remediation potential.

Treating contaminated soils with SMS reduces the bioavailability of heavy metals while simultaneously enhancing soil fertility via organic matter input. In wastewater treatment, SMS-based biofilters and biochar effectively remove dyes, phenols, and antibiotic residues. The coupled microbial activity further degrades recalcitrant organic contaminants (Barh et al. 2019). SMS helps stimulate native microbial communities, thereby accelerating the biotransformation of hydrocarbons in oil-contaminated soils (Chang et al. 2018; Antón-Herrero et al. 2022). Moreover, in phytoremediation systems, SMS compost enhances plant growth and metal uptake (Asemoloye et al. 2020).

### 4.6.5 Seed Protection and Mulch

It helps conserve soil moisture, regulate temperature, and suppress weeds through physical cover and allelopathic effects. Moreover, SMS carries beneficial microbes and residual enzymes that can establish in the rhizosphere, potentially protecting seeds from soilborne pathogens.

The fibrous, porous, and moisture-rich matrix enriched with fungal polysaccharides and humic-like substances confers excellent mulching attributes to SMS (Ma et al. 2021). Other desirable mulching properties include moisture conservation, soil temperature regulation, and suppression of weed emergence through physical coverage. Additionally, SMS harbors beneficial fungal consortia that colonize the rhizosphere microbes and provide seed protection against soilborne pathogens (Tazuba et al. 2025). Its nutrient content, along with organic matter input aid in improving soil fertility for leafy vegetables (De Bonis et al. 2025). SMS from *A. subrufescens* improved maize germination better than synthetic fertilizers (Alves et al. 2022).

When applied as seed coats, nursery bed covers, or surface mulch, SMS acts as both a physical barrier and a biologically active layer supporting early plant establishment (Zhang et al. 2012b; Meng et al. 2018). Moreover, SMS mulch derived from *P. sajor-caju* decreased the incidence of damping-off and other seedling diseases by promoting beneficial microbial antagonists and inducing systemic resistance against *Pythium aphanidermatum* in ginger (Remya and Paul 2014). In orchards and vegetable systems, SMS mulch contributes to improved root-zone microclimate, better seedling establishment, and enriches several beneficial microorganism genera, including bacteria (*Saccharimonadales*, *Gaiella*, and *Bacillus*) and fungi (*Thermomyces*, *Kernia*, and *Mortierella*), compared to conventional mulches (Huang et al. 2023).

### 4.6.6 Second-Generation Cultivation

SMS can also be used as a substrate for the cultivation of edible mushrooms. Species of *Agaricus* can be produced on straw-based SMS from *Pleurotus* mushroom cultivation complemented with 20% vermicompost or sunflower seed hulls (González Matute et al. 2011). These production levels are similar when compared to standard compost. The SMS produced after cultivation of *L. edodes* can be used for the production of *P. sajor-caju* mushrooms (Royse 1992), but a 20% higher production is obtained by supplementing it with 10% wheat bran and 10% millet (Grimm and Wösten 2018). SMS retains significant quantities of lignocellulosic residues, partially decomposed by primary fungal enzymes. Although nutrient levels are reduced compared to fresh substrate, SMS still contains cellulose, hemicellulose, lignin fragments, and bioavailable minerals, which can support further fungal growth (Klausen et al. 2025; Solórzano et al. 2025). The enzymatic pretreatment by the first cultivation cycle reduces substrate recalcitrance, improving accessibility for secondary fungal colonizers. These characteristics make SMS a potential substrate for "second-generation" mushroom cultivation such as *A. bisporus*, either directly after conditioning or in combination with fresh lignocellulosic feedstocks to restore nutrient balance. SMS inclusion rates of 20%–50% in fresh formulations have been reported to sustain acceptable yields and mushroom quality (Ma et al. 2025; Cunha Zied et al. 2020). Reuse of SMS also reduces production costs and minimizes waste disposal. However, successive reuses without nutrient supplementation can lead to reduced yields, longer crop cycles, and higher susceptibility to contaminants due to nutrient depletion and altered microbial dynamics.

## 4.7 Socioeconomic Impact of Mushroom Cultivation

The production of mushrooms is primarily done by local people and even specialists around the world. Mushroom cultivation has become a low-cost, high-value agribusiness that significantly supports rural livelihoods and socioeconomic development worldwide. By using agricultural wastes unlike capital-intensive cash crops, mushroom cultivation mainly depends on cheap, locally available lignocellulosic residues that would otherwise go to waste.

Mushroom cultivation has emerged as a low-cost, high-value agribusiness that plays an important role in rural livelihoods and socioeconomic development worldwide. Use of agrowastes lowers production costs and increases profitability. Unlike capital-intensive cash crops, mushroom production relies predominantly on locally available lignocellulosic raw materials which are otherwise discarded as agricultural residues (Royse et al. 2017). This minimizes input costs and, at the same time, promotes circular bioeconomy practices by recycling biomass into nutritionally dense and marketable commodities. The enterprise requires minimal land, thereby providing opportunities for the women and marginalized groups to engage in income-generating activities, such as substrate preparation, spawning, harvesting, and packaging (Mafe et al. 2025; Mayirnao et al. 2025; Sriram et al. 2019).

Economically, mushrooms serve as a source of rapid cash turnover due to their short cropping cycles and high market demand as functional foods, being rich in proteins, vitamins, minerals, and bioactive compounds (Valverde et al. 2015). The high BE of mushroom cultivation translates into a significant yield per unit of substrate. Thus, it ensures higher profitability compared to conventional crops (Sharma et al. 2024). Moreover, the generated SMS can be repurposed in animal feed, biofertilizers, and bioremediation processes, among other uses. Collectively, mushroom production strengthens food and nutritional security, diversifies rural incomes, generates employment, and promotes resource-efficient, sustainable agriculture.

## 4.8 Conclusion

The exploitation of agrowastes through mushroom cultivation offers a multidimensional solution to the global challenges of waste management and food security. Sustainable agrowaste management via bioconversion represents a circular approach in which low-value residues are transformed into nutritionally rich and commercially valuable products. Analyzing the physicochemical properties of agrowastes is key to selecting suitable substrates, with emphasis on their lignocellulosic content, which directly affects mushroom growth and yield. Representative agrowastes such as cereal straws, cotton stalks, sugarcane bagasse, and agro-industrial by-products demonstrate considerable potential in supporting mushroom growth, with their attributes tailored to fungal enzymatic systems. Moreover, combinations of different agricultural substrates often result in improved nutrient balance and structural properties. The recycling of SMS further extends the sustainability dimension of the cultivation process. SMS can be valorized as a valuable resource for soil amendment, animal feed, biogas generation, and as a second-generation substrate for cultivation, thereby minimizing environmental burden. From a socioeconomic perspective, recent advancements in production techniques, processing, and value addition have opened growth opportunities in the mushroom industry. It emerges as an inclusive enterprise, generating income opportunities and contributing to dietary diversification and nutrition security. Future research should focus on policy frameworks and integration into circular economy models to enhance the scalability and sustainability of mushroom-based enterprises.

**Conflicts of Interest** The authors confirm that they have no conflicts of interest.

## References


Aamlid TS, Landschoot PJ (2007) Effect of spent mushroom substrate on seed germination of cool season turfgrasses. HortScience 42(1):161–167. https://doi.org/10.21273/HORTSCI.42.1.161

Adenipekun CO, Omolaso PO (2015) Comparative study on cultivation, yield performance and proximate composition of *Pleurotus pulmonarius* Fries. (Quelet) on rice straw and banana leaves. World J Agric Sci 11(3):151–158. https://doi.org/10.5829/idosi.wjas.2015.11.3.1852

Aditya N, Jarial RS, Jarial K, Bhatia JN (2024) Comprehensive review on oyster mushroom species (Agaricomycetes): morphology, nutrition, cultivation and future aspects. Heliyon 10(5):e26539. https://doi.org/10.1016/j.heliyon.2024.e26539

Aguilar-Rivera N, De Jesús-Merales J (2010) Edible mushroom *Pleurotus ostreatus* production on cellulosic biomass of sugar cane. Sugar Tech 12(2):176–178. https://doi.org/10.1007/s12355-010-0034-4

Ahlawat OP, Kaur H (2019) Re-use of *Agaricus bisporus* spent compost for commercial scale compost making of the succeeding crop of *A. bisporus*. Int J Curr Microbiol App Sci 8(11): 177–186. https://doi.org/10.20546/ijcmas.2019.811.022

Ahmad W, Iqbal J, Ahmad I, Sarwar MA, Shehzad MA, Rafiq MA (2011) Performance of oyster mushroom (*Pleurotus ostreatus*) on cotton waste amended with maize and banana leaves. Pak J Nutr 10(6):509–513. https://doi.org/10.3923/pjn.2011.509.513

Ahmed T, Ahmad B (2013) Why do farmers burn rice residue? Examining farmers' choices in Punjab, Pakistan. Land Use Policy 47:448–458. https://doi.org/10.1016/j.landusepol.2015.05.004

Ahmed R, Niloy MAHM, Islam MS, Reza MS, Yesmin S, Rasul SB, Khandakar J (2024) Optimizing tea waste as a sustainable substrate for oyster mushroom (*Pleurotus ostreatus*) cultivation: a comprehensive study on biological efficiency and nutritional aspect. Front Sustain Food Syst 7:1308053. https://doi.org/10.3389/fsufs.2023.1308053

Aiduang W, Jatuwong K, Kiatsiriroat T, Kamopas W, Tiyayon P, Jawana R et al (2025) Spent mushroom substrate-derived biochar and its applications in modern agricultural systems: an extensive overview. Life 15(2):317. https://doi.org/10.3390/life15020317

Akhtar MM, Jahangir MM, Zeeshan M, Abdullah M (2016) Evaluation of gin waste as a growing substrate, enriched with different volume percentage of the wheat bran for cultivation of oyster mushroom (*Pleurotus ostreatus*). J Mt Area Res 1:21. https://doi.org/10.53874/jmar.v1i0.9

Alam N, Amin R, Khan A, Ara I, Shim MJ, Lee MW, Lee TS (2008) Nutritional analysis of cultivated mushrooms in Bangladesh—*Pleurotus ostreatus*, *Pleurotus sajor-caju*, *Pleurotus florida* and *Calocybe indica*. Mycobiology 36(4):228–232. https://doi.org/10.4489/MYCO.2008.36.4.228

Alsanad A, Sassine M, El Sebaaly YN, Abou Fayssal S (2021) Spent coffee grounds influence on *Pleurotus ostreatus* production, composition, fatty acid profile, and lignocellulose biodegradation capacity. CyTA J Food 19(1):11–20

Altieri R, Esposito A, Parati F, Lobianco A, Pepi M (2009) Performance of olive mill solid waste as a constituent of the substrate in commercial cultivation of *Agaricus bisporus*. Int Biodeterior Biodegradation 63(8):993–997. https://doi.org/10.1016/j.ibiod.2009.06.008

Alves LDS, Caitano CEC, Ferrari S, Vieira Júnior WG, Heinrichs R, de Almeida Moreira BR et al (2022) Application of spent Sun mushroom substrate in substitution of synthetic fertilizers at maize topdressing. Agronomy 12(11):2884. https://doi.org/10.3390/agronomy12112884

Amin R, Khair A, Alam N, Lee TS (2010) Effect of different substrates and casing materials on the growth and yield of *Calocybe indica*. Mycobiology 38(2):97–101. https://doi.org/10.4489/MYCO.2010.38.2.097

Antón-Herrero R, García-Delgado C, Baena N, Mayans B, Delgado-Moreno L, Eymar E (2022) Assessment of different spent mushroom substrates to bioremediate soils contaminated with petroleum hydrocarbons. Appl Sci 12(15):7720. https://doi.org/10.3390/app12157720

Asemoloye MD, Chukwuka KS, Jonathan SG (2020) Spent mushroom compost enhances plant response and phytoremediation of heavy metal polluted soil. J Plant Nutr Soil Sci 183(4):492–499. https://doi.org/10.1002/jpln.202000044

Atallah E, Zeaiter J, Ahmad MN, Leahy JJ, Kwapinski W (2021) Hydrothermal carbonization of spent mushroom compost waste compared against torrefaction and pyrolysis. Fuel Process Technol 216:106795. https://doi.org/10.1016/j.fuproc.2021.106795

Baptista F, Campos J, Costa-Silva V, Pinto AR, Saavedra MJ, Ferreira LM et al (2023a) Nutraceutical potential of *Lentinula edodes*' spent mushroom substrate: a comprehensive study on phenolic composition, antioxidant activity, and antibacterial effects. J Fungi 9(12):1200. https://doi.org/10.3390/jof9121200

Baptista F, Almeida M, Paié-Ribeiro J, Barros AN, Rodrigues M (2023b) Unlocking the potential of spent mushroom substrate (SMS) for enhanced agricultural sustainability: from environmental benefits to poultry nutrition. Life 13(10):1948. https://doi.org/10.3390/life13101948

Barh A, Kumari B, Sharma S, Annepu SK, Kumar A, Kamal S, Sharma VP (2019) Mushroom mycoremediation: kinetics and mechanism. In: Smart bioremediation technologies. Academic Press, pp 1–22. https://doi.org/10.1016/B978-0-12-818307-6.00001-9

Baskaran R, Vijayakumar S, Pazhanimuthu A, Shankar R, Velmurugan S (2020) Influence of substrate supplementation with nitrogen sources on the growth and yield of paddy straw mushroom (*Volvariella volvacea*). J Environ Biol 41(5):1059–1065. https://doi.org/10.22438/jeb/41/5/MRN-1217

Beckers SJ, Dallo IA, Del Campo I, Rosenauer C, Klein K, Wurm FR (2019) From compost to colloids—valorization of spent mushroom substrate. ACS Sustain Chem Eng 7(7):6991–6998. https://doi.org/10.1021/acssuschemeng.8b06710

Cai B, Chen Y, Du L, Liu Z, He L (2021) Spent mushroom compost and calcium carbonate modification enhances phytoremediation potential of *Macleaya cordata* to lead-zinc mine tailings. J Environ Manag 294:113029. https://doi.org/10.1016/j.jenvman.2021.113029

Carrasco J, Zied DC, Pardo JE, Preston GM, Pardo-Giménez A (2018) Supplementation in mushroom crops and its impact on yield and quality. AMB Express 8(1):146. https://doi.org/10.1186/s13568-018-0678-0

Carvalho CSMD, Aguiar LVBD, Sales-Campos C, Minhoni MTDA, Andrade MCND (2012) Applicability of the use of waste from different banana cultivars for the cultivation of the oyster mushroom. Braz J Microbiol 43:819–826. https://doi.org/10.1590/S1517-83822012000200048

Chai WY, Krishnan UG, Sabaratnam V, Tan JBL (2021) Assessment of coffee waste in formulation of substrate for oyster mushrooms *Pleurotus pulmonarius* and *Pleurotus floridanus*. Future Foods 4:100075. https://doi.org/10.1016/j.fufo.2021.100075

Chang BV, Fan SN, Tsai YC, Chung YL, Tu PX, Yang CW (2018) Removal of emerging contaminants using spent mushroom compost. Sci Total Environ 634:922–933. https://doi.org/10.1016/j.scitotenv.2018.03.366

Ćilerdžić JL, Vukojević JB, Klaus AS, Ivanović ŽS, Blagojević JD, Stajić MM (2018) Wheat straw–a promising substrate for *Ganoderma lucidum* cultivation. Acta Sci Pol Hortorum Cultus 17(1):11–25. https://doi.org/10.24326/asphc.2018.1.2

Cunha Zied D, Sánchez JE, Noble R, Pardo-Giménez A (2020) Use of spent mushroom substrate in new mushroom crops to promote the transition towards a circular economy. Agronomy 10(9): 1239. https://doi.org/10.3390/agronomy10091239

Da-oh H, Kiriratnikom S, Choosong K, Nitipan S (2024) Evaluation of spent mushroom substrate extracts from *cordyceps militaris*: impact on growth performance, immunomodulation, and intestinal microbiota of Nile tilapia. Egypt J Aquat Res 50(4):577–584. https://doi.org/10.1016/j.ejar.2024.11.001

De Bonis M, Sambo P, Zanin G, Cardarelli M, Nicoletto C (2025) Spent *Pleurotus* substrate as organic fertilizer to improve yield and soil fertility: the case of baby leaf lettuce production. J Sci Food Agric 105:5874–5886. https://doi.org/10.1002/jsfa.14303

de Siqueira FG, Martos ET, Silva RD, Dias ES (2011) Cultivation of *Pleurotus sajor-caju* on banana stalk and Bahia grass based substrates. Hortic Bras 29:199–204. https://doi.org/10.1590/S0102-05362011000200011

De A, Mridha D, Roychowdhury T, Bandyopadhyay B, Panja AS (2023) Substrate level optimization for better yield of oyster mushroom (*Pleurotus ostreatus*) production, using different ratio of rice straw and sugarcane bagasse. World J Microbiol Biotechnol 39(10):270. https://doi.org/10.1007/s11274-023-03714-0

Dedousi M, Melanouri EM, Diamantis I, Papanikolaou S, Diamantopoulou P (2024) Biochemical, functional and antioxidant potential of higher fungi cultivated on agro-industrial residues. Part II: cultures on mixtures of spent mushroom substrates and mushroom cropping by-products. Resour Chem Mater 3(3):175–187. https://doi.org/10.1016/j.recm.2024.04.001

Deora A, Sharma SS, Kumari P, Dahima V, Kumar S, Rohith M (2021) Cultivation of Kabul Dhingri (*Pleurotus eryngii*) mushroom by standardizing protocols in subtropical zones of world. Sci Rep 11(1):14692. https://doi.org/10.1038/s41598-021-94038-2

Deshpande AG, Arya A (2022) Mushroom biotechnology: developing cultivation protocol for four different mushrooms and accessing their potential in pollution management. In: Biology, cultivation and applications of mushrooms. Springer Singapore, Singapore, pp 457–485. https://doi.org/10.1007/978-981-16-6257-7_16

Desisa B, Muleta D, Dejene T, Jida M, Goshu A, Martin-Pinto P (2023) Substrate optimization for shiitake (*Lentinula edodes* (Berk.) Pegler) mushroom production in Ethiopia. J Fungi 9(8):811. https://doi.org/10.3390/jof9080811

Doroški A, Klaus A, Režek Jambrak A, Djekic I (2022) Food waste originated material as an alternative substrate used for the cultivation of oyster mushroom (*Pleurotus ostreatus*): a review. Sustainability 14(19):12509. https://doi.org/10.3390/su141912509

Durán-Aranguren DD, Meléndez-Melo JP, Covo-Ospina MC, Díaz-Rendón J, Reyes-Gutiérrez DN, Reina LC et al (2021) Biological pretreatment of fruit residues using the genus *Pleurotus*: a review. Bioresour Technol Rep 16:100849. https://doi.org/10.1016/j.biteb.2021.100849

Fan L, Pandey A, Mohan R, Soccol C (2000) Use of coffee industry residues for mushroom production. Braz Arch Biol Technol 43(4):441–446. https://doi.org/10.1590/S1516-89132000000400015

Food and Agriculture Organization of the United Nations (2023) FAOSTAT statistical database. [Rome]: FAO

Fazaeli H, Masoodi AR (2006) Spent wheat straw compost of *Agaricus bisporus* mushroom as ruminant feed. Asian Australas J Anim Sci 19(6):845–851. https://doi.org/10.5713/ajas.2006.845

Gadde B, Bonnet S, Menke C, Garivait S (2009) Air pollutant emissions from rice straw open field burning in India, Thailand and the Philippines. Environ Pollut 157(5):1554–1558. https://doi.org/10.1016/j.envpol.2009.01.004

Gao X, Tang X, Zhao K, Balan V, Zhu Q (2021) Biogas production from anaerobic co-digestion of spent mushroom substrate with different livestock manure. Energies 14(3):570. https://doi.org/10.3390/en14030570

Gasecka M, Magdziak Z, Siwulski M, Jasińska A, Budzyńska S, Rzymski P, Kalac P, Niedzielski P, Pankiewicz J, Mleczek M (2020) Effect of *Thymus vulgaris* post-extraction waste and spent coffee grounds on the quality of cultivated *Pleurotus eryngii*. J Food Process Preserv 44(9): e14648. https://doi.org/10.1111/jfpp.14648

Gatkal NR, Nalawade SM, Sahni RK, Walunj AA, Kadam PB, Bhanage GB, Datta R (2024) Present trends, sustainable strategies and energy potentials of crop residue management in India: a review. Heliyon 10(21):e39815. https://doi.org/10.1016/j.heliyon.2024.e39815

Girmay Z, Gorems W, Birhanu G, Zewdie S (2016) Growth and yield performance of *Pleurotus ostreatus* (Jacq. Fr.) Kumm (oyster mushroom) on different substrates. AMB Express 6(1):87. https://doi.org/10.1186/s13568-016-0253-9

González Matute R, Figlas D, Curvetto N (2011) *Agaricus blazei* production on non-composted substrates based on sunflower seed hulls and spent oyster mushroom substrate. World J Microbiol Biotechnol 27(6):1331–1339. https://doi.org/10.1007/s11274-010-0582-5

Grimm D, Wösten AB (2018) Mushroom cultivation in the circular economy. Appl Microbiol Biotechnol 102(17):7795–7803. https://doi.org/10.1007/s00253-018-9226-8

Gupta PK, Sahai S, Singh N, Dixit CK, Singh DP, Sharma C, Tiwari MK, Gupta RK, Garg SC (2004) Residue burning in rice–wheat cropping system: causes and implications. Curr Sci 87(12):1713–1717

Gupta R, Mehta G, Khasa YP, Kuhad RC (2011) Fungal delignification of lignocellulosic biomass improves the saccharification of cellulosics. Biodegradation 22(4):797–804. https://doi.org/10.1007/s10532-010-9404-6

Gupta S, Summuna B, Gupta M, Annepu SK (2018) Edible mushrooms: cultivation, bioactive molecules, and health benefits. In: Preedy VR (ed) Bioactive molecules in food. Springer, pp 1–33. https://doi.org/10.1007/978-3-319-54528-8_52-1

Han J, Sun R, Huang C, Xie H, Gao X, Yao Q et al (2024) Effects of different carbon and nitrogen ratios on yield, nutritional value, and amino acid contents of *Flammulina velutipes*. Life 14(5): 598. https://doi.org/10.3390/life14050598

Hasan MT, Khatun MHA, Sajib MAM, Rahman MM, Rahman MS, Roy M et al (2015) Effect of wheat bran supplement with sugarcane bagasse on growth, yield and proximate composition of pink oyster mushroom (*Pleurotus djamor*). Am J Food Sci Tech 3(6):150–157. https://doi.org/10.12691/ajfst-3-6-2

Hassan SAM, Ting SS, Zulkepli NN, Kasim FH (2022) Physicochemical characterization of rubberwood sawdust, mushroom spent medium, rice husk, and rice husk ash as potential growth substrates for wild *Schizophyllum commune* cultivation. Malays J Microsc 18(2):225–236

Hu Y, Mortimer PE, Hyde KD, Kakumyan P, Thongklang N (2021) Mushroom cultivation for soil amendment and bioremediation. Circ Agric Syst 1(1):1–14. https://doi.org/10.48130/CAS-2021-0011

Huang C, Han X, Luo Q, Nie Y, Kang M, Chen Y et al (2023) Agro-based spent mushroom compost substrates improve soil properties and microbial diversity in greenhouse tomatoes. Agronomy 13(9):2291. https://doi.org/10.3390/agronomy13092291

Hyde KD, Xu J, Rapior S, Jeewon R, Lumyong S, Niego AGT et al (2019) The amazing potential of fungi: 50 ways we can exploit fungi industrially. Fungal Divers 97(1):1–136. https://doi.org/10.1007/s13225-019-00430-9

Iandolo D, Piscitelli A, Sannia G, Faraco V (2011) Enzyme production by solid substrate fermentation of *Pleurotus ostreatus* and *Trametes versicolor* on tomato pomace. Appl Biochem Biotechnol 163(1):40–51. https://doi.org/10.1007/s12010-010-9014-0

Ingle AP, Saxena S, Moharil MP, Rivaldi JD, Ramos L, Chandel AK (2025) Production of biomaterials and biochemicals from lignocellulosic biomass through sustainable approaches: current scenario and future perspectives. Biotechnol Sustain Mater 2(1):3. https://doi.org/10.1186/s44316-025-00025-2

Ishii K, Furuichi T (2014) Influence of moisture content, particle size and forming temperature on productivity and quality of rice straw pellets. Waste Manag 34(12):2621–2626. https://doi.org/10.1016/j.wasman.2014.08.008

Jahedi A, Ahmadifar S, Mohammadigoltapeh E (2024) Revival of wild edible-medicinal mushroom (*Hericium erinaceus*) based on organic agro-industrial waste-achieving a commercial protocol with the highest yield; optimum reuse of organic waste. Sci Hortic 323:112510. https://doi.org/10.1016/j.scienta.2023.112510

Janet I (2012) Cultivation of mushroom (*Pleurotus ostreatus*) using corn cobs and saw dust as the major substrate. Glob J Agric Sci 11(1):47–54. https://doi.org/10.4314/gjass.v11i1.9

Jatwa TK, Apet KT, Wagh SS, Sayyed KS, Rudrappa KB, Sornapriya SP (2016) Evaluation of various agro-wastes for production of *Pleurotus* spp. (*P. florida*, *P. sajor-caju* and *P. eous*). J Pure Appl Microbiol 10(4):2783–2792. https://doi.org/10.22207/JPAM.10.4.37

Jeznabadi EK, Jafarpour M, Eghbalsaied S (2016) King oyster mushroom production using various sources of agricultural wastes in Iran. Int J Recycl Org Waste Agric 5(1):17–24. https://doi.org/10.1007/s40093-015-0113-3

Jiang H, Cheng Z, Zhao T, Liu M, Zhang M, Li J et al (2014) Pyrolysis kinetics of spent lark mushroom substrate and characterization of bio-oil obtained from the substrate. Energy Convers Manag 88:259–266. https://doi.org/10.1016/j.enconman.2014.08.006

Jin Z, Li Y, Ren J, Qin N (2018) Yield, nutritional content, and antioxidant activity of *Pleurotus ostreatus* on corncobs supplemented with herb residues. Mycobiology 46(1):24–32. https://doi.org/10.1080/12298093.2018.1454014

Jonathan SG, Fasidi IO (2003) Utilization of agricultural wastes for the cultivation of *Pleurotus tuber-regium* (Fr.) Singer in Nigeria. J Food Technol Afr 8(4):129–131. https://doi.org/10.4314/jfta.v8i4.19225

Kamthan R, Tiwari I (2017) Agricultural wastes-potential substrates for mushroom cultivation. Eur J Exp Biol 7(5):31. https://doi.org/10.21767/2248-9215.100031

Karić N, Maia AS, Teodorović A, Atanasova N, Langergraber G, Crini G et al (2022) Bio-waste valorisation: agricultural wastes as biosorbents for removal of (in) organic pollutants in wastewater treatment. Chem Eng J Adv 9:100239. https://doi.org/10.1016/j.ceja.2021.100239

Karpagavalli S, Manisha R, Mageshwari S, Sowbharnika M (2024) Influence of growth substrates on bioactive compounds and yield of oyster mushroom (*Pleurotus florida*). Sci Hortic 329: 112959. https://doi.org/10.1016/j.scienta.2024.112959

Kaushal LA, Prashar A (2021) Agricultural crop residue burning and its environmental impacts and potential causes—case of Northwest India. J Environ Plan Manag 64(3):464–484. https://doi.org/10.1080/09640568.2020.1714490

Kim IJ, Park S, Kyoung H, Song M, Kim SR (2024) Microbial valorization of fruit processing waste: opportunities, challenges, and strategies. Curr Opin Food Sci 56:101147. https://doi.org/10.1016/j.cofs.2024.101147

Klausen SJ, Romero-Soto LA, Liaqat A, Dehghanmanshadi Z, Strætkvern KO, Xiong S, Martín C (2025) Biorefining spent substrates of shiitake (Lentinula edodes) and oyster mushroom (*Pleurotus ostreatus*): enzymatic saccharification of cellulose and xylan, with lignin recovery from residues. Bioengineered 16(1):2536443. https://doi.org/10.1080/21655979.2025.2536443

Kosre A, Koreti D, Mahish PK, Chandrawanshi NK (2021) Current perspective of sustainable utilization of agro waste and biotransformation of energy in mushroom. In: Energy: crises, challenges and solutions. Wiley, pp 274–302. https://doi.org/10.1002/9781119741503.ch15

Kumar P, Kumar V, Goala M, Singh J, Kumar P (2021) Integrated use of treated dairy wastewater and agro-residue for *Agaricus bisporus* mushroom cultivation: experimental and kinetics studies. Biocatal Agric Biotechnol 32:101940. https://doi.org/10.1016/j.bcab.2021.101940

Kumla J, Suwannarach N, Sujarit K, Penkhrue W, Kakumyan P, Jatuwong K et al (2020) Cultivation of mushrooms and their lignocellulolytic enzyme production through the utilization of agro-industrial waste. Molecules 25(12):2811. https://doi.org/10.3390/molecules25122811

Lakdawala H, Desai P, Pandya HA, Solanki H (2025) Mushroom substrate selection: factors influencing productivity and sustainability. Int J Sci Res Sci Technol 12(2):352–361. https://doi.org/10.32628/IJSRST25122236

Lee PT, Wu YS, Tseng CC, Lu JY, Lee MC (2022) Dietary *Agaricus blazei* spent substrate improves disease resistance of Nile tilapia (*Oreochromis niloticus*) against *Streptococcus agalactiae* in vivo. J Mar Sci Eng 10(1):100. https://doi.org/10.3390/jmse10010100

Lee T, Choi D, Park J, Tsang YF, Lin KYA, Jung S, Kwon EE (2024) Valorizing spent mushroom substrate into syngas by the thermo-chemical process. Bioresour Technol 391:130007. https://doi.org/10.1016/j.biortech.2023.130007

Leong YK, Chang JS (2022) Valorization of fruit wastes for circular bioeconomy: current advances, challenges, and opportunities. Bioresour Technol 359:127459. https://doi.org/10.1016/j.biortech.2022.127459

Leong YK, Ma TW, Chang JS, Yang FC (2022a) Recent advances and future directions on the valorization of spent mushroom substrate (SMS): a review. Bioresour Technol 344:126157. https://doi.org/10.1016/j.biortech.2021.126157

Leong YK, Varjani S, Lee DJ, Chang JS (2022b) Valorization of spent mushroom substrate for low-carbon biofuel production: recent advances and developments. Bioresour Technol 363:128012. https://doi.org/10.1016/j.biortech.2022.128012

Lim SH, Lee YH, Kang HW (2013) Efficient recovery of lignocellulolytic enzymes of spent mushroom compost from oyster mushrooms, *Pleurotus* spp., and potential use in dye decolorization. Mycobiology 41(4):214–220. https://doi.org/10.5941/MYCO.2013.41.4.214

Londoño CR, Gil AG, Moreno A, Alvarado PN (2024) Valorization of spent mushroom compost through a cascading use approach. Energies 17(21):5458. https://doi.org/10.3390/en17215458

Lu X, Li F, Zhou X, Hu J, Liu P (2022) Biomass, lignocellulolytic enzyme production and lignocellulose degradation patterns by *Auricularia auricula* during solid state fermentation of corn stalk residues under different pretreatments. Food Chem 384:132622. https://doi.org/10.1016/j.foodchem.2022.132622

Ma F, Wang J, Zeng Y, Yu H, Yang Y, Zhang X (2011) Influence of the co-fungal treatment with two white rot fungi on the lignocellulosic degradation and thermogravimetry of corn stover. Process Biochem 46(9):1767–1773. https://doi.org/10.1016/j.procbio.2011.05.020

Ma L, Lin YQ, Yang C, Ying ZH, Jiang XL (2016) Production of liquid spawn of an edible mushroom, *Sparassis latifolia* by submerged fermentation and mycelial growth on pine wood sawdust. Sci Hortic 209:22–30. https://doi.org/10.1016/j.scienta.2016.06.001

Ma Z, Zhang YQ, Wang LJ, Hu GL, Gong XQ, Bai Q et al (2021) Short-term effects of spent mushroom substrate mulching thickness on the soil environment, weed suppression, leaf nutrients, and nut characteristics in a hazelnut orchard. Agronomy 11(6):1122. https://doi.org/10.3390/agronomy11061122

Ma H, Fu P, Zhao J, Lin X, Wu W, Yu Z et al (2022) Pretreatment of wheat straw lignocelluloses by deep eutectic solvent for lignin extraction. Molecules 27(22):7955. https://doi.org/10.3390/molecules27227955

Ma X, Yan S, Wang M (2025) Spent mushroom substrate: a review on present and future of green applications. J Environ Manag 373:123970. https://doi.org/10.1016/j.jenvman.2024.123970

Mafe AN, Otieno CA, Edo GI, Akpoghelie PO, Yousif E, Isoje EF et al (2025) Domestication and market potential of indigenous mushrooms in Nigeria and Kenya through cultivation techniques, nutritional enhancement, and consumer preference analysis. Discov Food 5(1):1–48. https://doi.org/10.1007/s44187-025-00452-0

Mandeel QA, Al-Laith AA, Mohamed SA (2005) Cultivation of oyster mushrooms (*Pleurotus* spp.) on various lignocellulosic wastes. World J Microbiol Biotechnol 21(4):601–607. https://doi.org/10.1007/s11274-004-3494-4

Martín C, Zervakis GI, Xiong S, Koutrotsios G, Strætkvern KO (2023) Spent substrate from mushroom cultivation: exploitation potential toward various applications and value-added products. Bioengineered 14(1):2252138. https://doi.org/10.1080/21655979.2023.2252138

Mayirnao HS, Sharma K, Jangir P, Kaur S, Kapoor R (2025) Mushroom-derived nutraceuticals in the 21st century: an appraisal and future perspectives. J Future Foods 5(4):342–360. https://doi.org/10.1016/j.jfutfo.2024.07.013

Mazumder S, Basu SK, Mukherjee M (2009) Laccase production in solid-state and submerged fermentation by *Pleurotus ostreatus*. Eng Life Sci 9(1):45–52. https://doi.org/10.1002/elsc.200700039

Meng X, Dai J, Zhang Y, Wang X, Zhu W, Yuan X et al (2018) Composted biogas residue and spent mushroom substrate as a growth medium for tomato and pepper seedlings. J Environ Manag 216:62–69. https://doi.org/10.1016/j.jenvman.2017.09.056

Mohamed MF, Refaei EF, Abdalla MM, Abdelgalil SH (2016) Fruiting bodies yield of oyster mushroom (*Pleurotus columbinus*) as affected by different portions of compost in the substrate. Int J Recycl Org Waste Agric 5(4):281–288. https://doi.org/10.1007/s40093-016-0138-2

Moonmoon M, Shelly NJ, Khan MA, Uddin MN, Hossain K, Tania M, Ahmed S (2010) Cultivation of different strains of king oyster mushroom (*Pleurotus eryngii*) on sawdust and rice straw in Bangladesh. Saudi J Biol Sci 17(4):341–345. https://doi.org/10.1016/j.sjbs.2010.05.004

Mwangi RW, Mustafa M, Kappel N, Csambalik L, Szabó A (2024) Practical applications of spent mushroom compost in cultivation and disease control of selected vegetables species. J Mater Cycles Waste Manag 26(4):1918–1933. https://doi.org/10.1007/s10163-024-01969-9

Nanje Gowda NE, Chennappa G (2021) Mushroom cultivation: a sustainable solution for the management of agriculture crop residues. In: Recent advances in mushroom cultivation technology and its application. Bright Sky, pp 15–26

Nasehi M, Torbatinejad NM, Zerehdaran S, Safaie AR (2017) Effect of solid-state fermentation by oyster mushroom (*Pleurotus florida*) on nutritive value of some agro by-products. J Appl Anim Res 45(1):221–226. https://doi.org/10.1080/09712119.2016.1150850

Nguyen BTT, Le VV, Nguyen HN, Nguyen HTT, Nguyen LT, Ngo NX (2025) Cotton waste as an optimal substrate for cultivation of the pink oyster mushroom *Pleurotus djamor*. J Appl Biol Biotechnol 13:184–191. https://doi.org/10.7324/JABB.2024.179930

Noble R, Hobbs PJ, Mead A, Dobrovin-Pennington A (2002) Influence of straw types and nitrogen sources on mushroom composting emissions and compost productivity. J Ind Microbiol Biotechnol 29(3):99–110. https://doi.org/10.1038/sj.jim.7000292

Ntougias S, Papadopoulou KK, Zervakis GI, Kavroulakis N, Ehaliotis C (2008) Suppression of soil-borne pathogens of tomato by composts derived from agro-industrial wastes abundant in Mediterranean regions. Biol Fertil Soils 44(8):1081–1090. https://doi.org/10.1007/s00374-008-0295-1

Nunes MD, Da Luz JMR, Paes SA, Ribeiro JJO, da Silva MDCS, Kasuya MCM (2012) Nitrogen supplementation on the productivity and the chemical composition of oyster mushroom. J Food Res 1(2):113. https://doi.org/10.5539/jfr.v1n2p113

Oh SJ, Park JS, Shin PG, Yoo YB, Jhune CS (2004) An improved compost using cotton waste and fermented sawdust substrate for cultivation of oyster mushroom. Mycobiology 32(3):115–118. https://doi.org/10.4489/MYCO.2004.32.3.115

Okano K, Fukui S, Kitao R, Usagawa T (2007) Effects of culture length of *Pleurotus eryngii* grown on sugarcane bagasse on in vitro digestibility and chemical composition. Anim Feed Sci Technol 136(3-4):240–247. https://doi.org/10.1016/j.anifeedsci.2006.08.024

Orts WJ, Holtman KM, Seiber JN (2008) Agricultural chemistry and bioenergy. J Agric Food Chem 56(11):3892–3899

Oseni TO, Dlamini SO, Earnshaw DM, Masarirambi MT (2012) Effect of wheat bran supplement on growth and yield of oyster mushroom (*Pleurotus ostreatus*) on fermented pineapple waste. Asian J Agric Sci 4(5):350–354

Otieno OD, Mulaa FJ, Obiero G, Midiwo J (2022) Utilization of fruit waste substrates in mushroom production and manipulation of chemical composition. Biocatal Agric Biotechnol 39:102250. https://doi.org/10.1016/j.bcab.2021.102250

Palangi V, Kaya A, Kaya A, Giannenas I (2022) Ecofriendly usability of mushroom cultivation substrate as a ruminant feed: anaerobic digestion using gas production techniques. Animals 12(12):1583. https://doi.org/10.3390/ani12121583

Pan SY, Tsai CY, Liu CW, Wang SW, Kim H, Fan C (2021) Anaerobic co-digestion of agricultural wastes toward circular bioeconomy. Iscience 24(7):102704. https://doi.org/10.1016/j.isci.2021.102704

Pant D, Reddy UG, Adholeya A (2006) Cultivation of oyster mushrooms on wheat straw and bagasse substrate amended with distillery effluent. World J Microbiol Biotechnol 22(3): 267–275. https://doi.org/10.1007/s11274-005-9028-9

Papadaki A, Kachrimanidou V, Papanikolaou S, Philippoussis A, Diamantopoulou P (2019) Upgrading grape pomace through *Pleurotus* spp. cultivation for the production of enzymes and fruiting bodies. Microorganisms 7(7):207. https://doi.org/10.3390/microorganisms7070207

Patra AK, Pani BK (1995) Evaluation of banana leaf as a new alternate substrate to paddy straw for oyster mushroom cultivation. J Phytol Res 8(2):145–148

Patra AK, Saxena J (2011) Exploitation of dietary tannins to improve rumen metabolism and ruminant nutrition. J Sci Food Agric 91(1):24–37. https://doi.org/10.1002/jsfa.4152

Paula FS, Tatti E, Abram F, Wilson J, O'Flaherty V (2017) Stabilisation of spent mushroom substrate for application as a plant growth-promoting organic amendment. J Environ Manag 196:476–486. https://doi.org/10.1016/j.jenvman.2017.03.038

Peksen A, Yakupoglu G (2009) Tea waste as a supplement for the cultivation of *Ganoderma lucidum*. World J Microbiol Biotechnol 25(4):611–618. https://doi.org/10.1007/s11274-008-9931-z

Pérez-Chávez AM, Mayer L, Albertó E (2019) Mushroom cultivation and biogas production: a sustainable reuse of organic resources. Energy Sustain Dev 50:50–60

Phan CW, Sabaratnam V (2012) Potential uses of spent mushroom substrate and its associated lignocellulosic enzymes. Appl Microbiol Biotechnol 96(4):863–873. https://doi.org/10.1007/s00253-012-4446-9

Pintarič M, Štuhec A, Tratnik E, Langerholc T (2024) Spent mushroom substrate improves microbial quantities and enzymatic activity in soils of different farming systems. Microorganisms 12(8):1521. https://doi.org/10.3390/microorganisms12081521

Raman J, Jang KY, Oh YL, Oh M, Im JH, Lakshmanan H, Sabaratnam V (2021) Cultivation and nutritional value of prominent *Pleurotus* spp.: an overview. Mycobiology 49(1):1–14. https://doi.org/10.1080/12298093.2020.1835142

Rangubhet KT, Mangwe MC, Mlambo V, Fan YK, Chiang HI (2017) Enteric methane emissions and protozoa populations in Holstein steers fed spent mushroom (*Flammulina velutipes*) substrate silage-based diets. Anim Feed Sci Technol 234:78–87. https://doi.org/10.1016/j.anifeedsci.2017.06.005

Rani P, Kalyani N, Prathiba K (2008) Evaluation of lignocellulosic wastes for production of edible mushrooms. Appl Biochem Biotechnol 151(2–3):151–159. https://doi.org/10.1007/s12010-008-8153-2

Rashad FM, El Kattan MH, Fathy HM, Abd El-Fattah DA, El Tohamy M, Farahat AA (2019) Recycling of agro-wastes for *Ganoderma lucidum* mushroom production and *Ganoderma* post mushroom substrate as soil amendment. Waste Manag 88:147–159. https://doi.org/10.1016/j.wasman.2019.03.040

Ravlikovsky A, Pinheiro MNC, Dinca L, Crisan V, Symochko L (2024) Valorization of spent mushroom substrate: establishing the foundation for waste-free production. Recycling 9(3):44. https://doi.org/10.3390/recycling9030044

Ravlikovsky A, Pinheiro MNC, Kucheruk P, Symochko L (2025) Spent mushroom substrate as a renewable energy resource: evaluating its biogas production potential. Sustainability 17(5): 1800. https://doi.org/10.3390/su17051800

Remya JS, Paul TS (2014) Effect of spent mushroom substrate on growth parameters, yield and soft rot incidence of ginger. Mushroom Res 23(1):61–67

Richard EN, Hilonga A, Machunda RL, Njau KN (2019) A review on strategies to optimize metabolic stages of anaerobic digestion of municipal solid wastes towards enhanced resources recovery. Sustain Environ Res 29(1):36. https://doi.org/10.1186/s42834-019-0037-0

Royse DJ (1992) Recycling of spent shiitake substrate for production of the oyster mushroom, *Pleurotus sajor-caju*. Appl Microbiol Biotechnol 38(2):179–182

Royse DJ, Baars J, Tan Q (2017) Current overview of mushroom production in the world. In: Zied DC, Pardo-Giménez A (eds) Edible and medicinal mushrooms. Wiley, pp 5–13. https://doi.org/10.1002/9781119149446.ch2

Sahu M, Chinara N, Swain S, Priyadarshini A, Behera P, Sahoo BB (2025) Evaluation of lignocellulosic substrates on productivity of pearl oyster mushroom (*Pleurotus ostreatus*). Arch Curr Res Int 25(5):63–69. https://doi.org/10.9734/acri/2025/v25i51188

Salama ANA, Abdou AAK, Helaly AA, Salem EA, Zhanxi L (2016) Cultivation of *Pleurotus ostreatus* L on different substrates based on JuNCAO technology in Egypt. Nat Sci 14(4):59–66

Salmones D, Mata G, Waliszewski KN (2005) Comparative culturing of *Pleurotus* spp. on coffee pulp and wheat straw: biomass production and substrate biodegradation. Bioresour Technol 96(5):537–544. https://doi.org/10.1016/j.biortech.2004.06.019

Saratale GD, Bhosale R, Shobana S, Banu JR, Pugazhendhi A, Mahmoud E et al (2020) A review on valorization of spent coffee grounds (SCG) towards biopolymers and biocatalysts production. Bioresour Technol 314:123800. https://doi.org/10.1016/j.biortech.2020.123800

Sardar A, Satankar V, Jagajanantha P, Mageshwaran V (2020) Effect of substrates (cotton stalks and cotton seed hulls) on growth, yield and nutritional composition of two oyster mushrooms (*Pleurotus ostreatus* and *Pleurotus florida*). J Cotton Res Dev 34(1):135–145

Sarkar S, Singh RP, Chauhan A (2018) Crop residue burning in northern India: increasing threat to Greater India. J Geophys Res Atmos 123(13):6920–6934. https://doi.org/10.1029/2018JD028428

Sassine YN, Nabhan S, Rachkidy E, El Sebaaly Z (2024) Valorization of agro-forest wastes (oak acorns, vineyard pruning, and olive pruning) through the cultivation of shiitake (*Lentinula edodes*) mushrooms. Heliyon 10(12):e32562. https://doi.org/10.1016/j.heliyon.2024.e32562

Shah AM, Zhang H, Shahid M, Ghazal H, Shah AR, Niaz M et al (2025) The vital roles of agricultural crop residues and agro-industrial by-products to support sustainable livestock productivity in subtropical regions. Animals 15(8):1184. https://doi.org/10.3390/ani15081184

Sharma D, Ramniwas S, Mugabi R, Uddin J, Nayik GA (2024) Revolutionizing mushroom processing: innovative techniques and technologies. Food Chem: X 23:101774. https://doi.org/10.1016/j.fochx.2024.101774

Shrikhandia SP, Devi S, Sumbali G (2022) Lignocellulosic waste management through cultivation of certain commercially useful and medicinal mushrooms: recent scenario. In: Biology, cultivation and applications of mushrooms. Springer Singapore, Singapore, pp 497–534. https://doi.org/10.1007/978-981-16-6257-7_18

Singh MP, Singh VK (2012) Biodegradation of vegetable and agro wastes by *Pleurotus sapidus*: a novel strategy to produce mushroom with enhanced yield and nutrition. Cell Mol Biol 58(1):1–7

Singh AR, Dutta SK, Singh SB, Boopathi T, Lungmauna S, Saha S, Singh LH, Singh TB (2016) Utilization of wild banana leaves for cultivation of oyster mushroom (*Pleurotus florida*) in Northeastern Hill of India. In: 6th International conference on plant, pathogens and people, New Delhi, India

Singh M, Kamal S, Sharma VP (2021) Species and region-wise mushroom production in leading mushroom producing countries-China, Japan, USA, Canada and India. Mushroom Res 30(2): 99. https://doi.org/10.36036/MR.30.2.2021.119394

Solórzano R, Dionisio L, Burga L, Javier-Astete R, Quispe-Apaza C, Oscco P, Johnson L (2025) Ichu valorization by *Pleurotus* spp. cultivation and potential of the residual substrate as a biofertilizer. Sustainability 17(15):6695. https://doi.org/10.3390/su17156695

Song YM, Lee SD, Chowdappa R, Kim HY, Jin SK, Kim IS (2007) Effects of fermented oyster mushroom (*Pleurotus ostreatus*) by-product supplementation on growth performance, blood parameters and meat quality in finishing Berkshire pigs. Animal 1(2):301–307. https://doi.org/10.1017/S1751731107683785

Sriram V, Deering K, Wright P (2019) Good practices for integrating gender equality and women's empowerment in climate-smart agriculture programmes. Food and Agriculture Organization of the United Nations (FAO)/CARE International

Suwannarach N, Kumla J, Zhao Y, Kakumyan P (2022) Impact of cultivation substrate and microbial community on improving mushroom productivity: a review. Biology 11(4):569. https://doi.org/10.3390/biology11040569

Tazuba AF, Ocimati W, Ogwal G, Nyangwire B, Onyilo F, Blomme G (2025) Spent *Pleurotus ostreatus* substrate has potential for controlling the plant-parasitic nematode, *Radopholus similis* in bananas. Agronomy 15(5):1040. https://doi.org/10.3390/agronomy15051040

Telang SM, Patil SS, Baig MMV (2010) Biological efficiency and nutritional value of *Pleurotus sapidus* cultivated on different substrates. Food Sci Res J 1(2):127–129

Thiribhuvanamala G, Krishnamoorthy AS, Kavitha C, Kamal S, Kumar A, Sharma VP (2021) Strategic approaches for outdoor cultivation of paddy straw mushroom (*Volvariella volvacea*) as intercrop under different cropping systems. Madras Agric J 108:1–8

Thomsen MH (2005) Complex media from processing of agricultural crops for microbial fermentation. Appl Microbiol Biotechnol 68(5):598–606. https://doi.org/10.1007/s00253-005-0056-0

Torres-Martinez BDM, Vargas-Sánchez RD, Pérez-Alvarez JÁ, Fernández-López J, Viuda-Martos-M, Esqueda M et al (2024) Bio-valorization of spent coffee grounds and potato peel as substrates for *Pleurotus ostreatus* growth. Foods 13(23):3774. https://doi.org/10.3390/foods13233774

Valverde ME, Hernández-Pérez T, Paredes-López O (2015) Edible mushrooms: improving human health and promoting quality life. Int J Microbiol 2015(1):376387. https://doi.org/10.1155/2015/376387

Vieira FR, de Andrade MCN (2016) Optimization of substrate preparation for oyster mushroom (*Pleurotus ostreatus*) cultivation by studying different raw materials and substrate preparation conditions (composting: phases I and II). World J Microbiol Biotechnol 32(11):190. https://doi.org/10.1007/s11274-016-2152-y

Viriato V, Carvalho SADD, Santoro BDL, Bonfim FPG (2024) A business model for circular bioeconomy: edible mushroom production and its alignment with the sustainable development goals (SDGs). Recycling 9(4):68. https://doi.org/10.3390/recycling9040068

Wang HW, Xu M, Cai XY, Feng T, Xu WL (2020) Application of spent mushroom substrate suppresses Fusarium wilt in cucumber and alters the composition of the microbial community of the cucumber rhizosphere. Eur J Soil Biol 101:103245

Wang YH, Yan CS, Deng YJ, Zhu ZF, Sun HA, Li HP et al (2025) Feasibility of bamboo sawdust as sustainable alternative substrate for *Auricularia heimuer* cultivation. J Fungi 11(5):387. https://doi.org/10.3390/jof11050387

Wattanavichean N, Phanthuwongpakdee J, Koedrith P, Laoratanakul P, Thaithatgoon B, Somrithipol S et al (2025) Mycelium-based breakthroughs: exploring commercialization, research, and next-gen possibilities. Circ Econ Sustain 5:1–43. https://doi.org/10.1007/s43615-025-00539-x

Wu N, Tian F, Moodley O, Song B, Jia C, Ye J et al (2019) Optimization of agro-residues as substrates for *Pleurotus pulmonarius* production. AMB Express 9(1):184. https://doi.org/10.1186/s13568-019-0907-1

Yagüe MR, González-Pérez JA, Almendros G, Lobo MC (2025) Assessing the potential agronomic value of spent mushroom substrates: evaluating their suitability to contribute to soil carbon storage. Sustainability 17(16):7335. https://doi.org/10.3390/su17167335

Yang H, Zhang H, Qiu H, Anning DK, Li M, Wang Y, Zhang C (2021) Effects of C/N ratio on lignocellulose degradation and enzyme activities in aerobic composting. Horticulturae 7(11): 482. https://doi.org/10.3390/horticulturae7110482

Yu H, Zhang D, Zhang L, Li Q, Song C, Shang X et al (2022) Corncob as a substrate for the cultivation of *Lentinula edodes*. Waste Biomass Valorization 13(2):929–939. https://doi.org/10.1007/s12649-021-01575-y

Yue-Lian L, Qing-Fang L (2011) Identification and cultivation of a wild mushroom from banana pseudo-stem sheath. Sci Hortic 129(4):922–925. https://doi.org/10.1016/j.scienta.2011.06.00

Zadrail F, Dube HC (1992) The oyster mushroom—importance and prospects. Mushroom Res 1(1): 25–32

Zakil FA, Hassan KM, Sueb MM, Isha R (2020) Growth and yield of *Pleurotus ostreatus* using sugarcane bagasse as an alternative substrate in Malaysia. In: IOP conference series: materials science and engineering, vol 736, No. 2. IOP Publishing, p 022021. https://doi.org/10.1088/1757-899X/736/2/022021

Zhang Y, Ghaly AE, Li B (2012a) Availability and physical properties of residues from major agricultural crops for energy conversion through thermochemical processes. Am J Agric Biol Sci 7(3):312–321. https://doi.org/10.3844/ajabssp.2012.312.321

Zhang RH, Zeng-Qiang DUAN, Zhi-Guo LI (2012b) Use of spent mushroom substrate as growing media for tomato and cucumber seedlings. Pedosphere 22(3):333–342. https://doi.org/10.1016/S1002-0160(12)60020-4

Zięba P, Sękara A, Bernaś E, Krakowska A, Sułkowska-Ziaja K, Kunicki E et al (2021) Supplementation with magnesium salts—a strategy to increase nutraceutical value of *Pleurotus djamor* fruiting bodies. Molecules 26(11):3273

Zisopoulos FK, Ramírez HAB, van der Goot AJ, Boom RM (2016) A resource efficiency assessment of the industrial mushroom production chain: the influence of data variability. J Clean Prod 126:394–408. https://doi.org/10.1016/j.jclepro.2016.03.066

Zou Y, Du F, Hu Q, Yuan X, Dai D, Zhu M (2020) Integration of *Pleurotus tuoliensis* cultivation and biogas production for utilization of lignocellulosic biomass as well as its benefit evaluation. Bioresour Technol 317:124042. https://doi.org/10.1016/j.biortech.2020.124042

# *Pleurotus Species*: Significance in Entrepreneurship

**5**

Rajni Gupta, Manisha Sharma, and Yashika Singh

**Abstract**

The genus *Pleurotus*, commonly known as oyster mushrooms, has emerged as a valuable resource in both nutritional science and rural entrepreneurship. Mushrooms play an active role in bioremediation, thus contributing significantly to ecosystem sustainability. They serve as a viable source of food loaded with nutraceuticals and serve as functional food. The *Pleurotus* species are characterized by their ability to grow efficiently on lignocellulosic agricultural wastes, requiring minimal inputs while offering high biological yield. Their cultivation is environmentally sustainable and especially suitable for rural and peri-urban areas with limited capital and infrastructure. *Pleurotus* mushrooms are rich in proteins, dietary fibers, essential amino acids, vitamins, and minerals. Bioactive compounds such as β-glucans, lovastatin, and ergosterol confer various therapeutic properties, thereby positioning these mushrooms as functional foods with significant pharmacological relevance. *Pleurotus* cultivation offers a low-cost, high-return enterprise model that supports rural livelihoods, especially for women and youth. The integration of *Pleurotus* cultivation into local agroeconomic systems not only contributes to nutritional security but also fosters inclusive growth and environmental sustainability. *Pleurotus* species represents a strategic crop with significant potential for promoting health, economic empowerment, and circular bioeconomy in developing regions. Their cultivation is environmentally safe as it is growing on agricultural wastes, providing a sustainable solution to the stubble burning problem. This review aims to consolidate

R. Gupta (✉) · M. Sharma · Y. Singh
Department of Botany, Kirori Mal College, University of Delhi, New Delhi, India
e-mail: rgupta@kmc.du.ac.in

 
B. Giri et al. (eds.), *Advances in Mushroom Research*,
https://doi.org/10.1007/978-981-95-7692-0_5

current findings on the biological, nutritional, medicinal, and economic importance of *Pleurotus* species, highlighting their role in sustainable agricultural practices and small-scale enterprise development.

**Keywords**

Pleurotus · Antidiabetic · Cholesterol · Pharma companies · Functional food

## 5.1 Mushroom Cultivation: A Global Perspective

Mushrooms are the fruiting bodies of higher fungi, belonging to the phyla Basidiomycota and Ascomycota. They play a major role in global ecosystems. They serve as primary decomposers by assisting in the breakdown of organic materials, such as dead wood, agricultural residues, and leaf litter, facilitating nutrient recycling and maintaining soil health. Besides their ecological significance, mushrooms have long captivated human interest due to their roles in culinary traditions and medicinal systems. In the last few decades, the remarkable growth in the global mushroom industry has been driven by a growing awareness of the nutritional/health benefits of mushrooms, an increasing demand for sustainable food options, vegan food, skin care products, and the rising trend of plant-based diets.

Global mushroom production is experiencing significant growth, which was approximately 17.25 million tonnes in 2023, and is projected to increase to 32.04 million tonnes by 2032. Asia is the dominant producer of mushrooms, driving this expansion at a CAGR of approximately 7.18%. China is the world's leading mushroom producer, accounting for the vast majority of global output. The mushroom market size has grown strongly in recent years. It will grow from $66.9 billion in 2024 to $73.14 in 2025 at a compound annual growth rate (CAGR) of 9.3%. Governments of various countries are supporting the mushroom sector to produce new solutions to better serve the industry. In 2022, the Russian Government initiated a new program to provide grants for the development of agricultural products and small businesses. The Japanese Government provides subsidies for agriculture and mushroom production, benefits by trade pacts. In Japan, mushroom production has increased 14 times from 32.9 to 469 million kg. In the USA, about 98% of the mushroom production is accounted for by button mushrooms. In Canada, mushroom production has increased many times, but it is at a slow rate. The Australian mushroom industry is dominated by *Agaricus bisporus*, which makes up about 98 percent of domestic mushroom production. In India, mushroom production has nearly doubled in the last 5 years from 1.00 to 2.01 lakh MT, highlighting its rising role in sustainable agriculture and income generation, especially among smallholders (Kapil 2023). All these key factors have made mushroom cultivation, majorly of species like *Pleurotus*, an exciting area for agricultural diversification and economic progress. The growth can be due to increasing Government support, rapid urbanization, the rising demand for food products, the increasing vegetarian population, and health-conscious consumers. The major companies in mushroom production are Monaghan Mushroom Ltd., Monterey Mushrooms Inc., Giorgio Fresh Co., Fujian Food Co Ltd., Soko Mushrooms, African Mushroom Growers, and Eco Fresh (Mushroom Global Market Report 2025).

The genus *Pleurotus* is distinguished by its rapid growth, high biological efficiency, and low-input requirements, making it an ideal candidate for sustainable entrepreneurship in the edible fungus sector, which has positioned *Pleurotus as* a sustainable and profitable business venture, attracting the attention of both investors and consumers worldwide. *Pleurotus* cultivation has many advantages over other edible and medicinal mushrooms, i.e., (1) Grows fast under a wide range of temperatures and pH; (2) Secretes a wide range of enzymes that are capable of degrading lignocellulosic biomass of substrates; (3) Substrate preparation is easy and low maintenance; and (4) Has high-yielding and high nutritional value compounds (Ejigu et al. 2022).

Several *Pleurotus* species including *P. ostreatus*, *P. florida*, *P. sajor-caju*, *P. eryngii*, and *P. flabellatus* are cultivated commercially across the globe, not only for their rich nutritional content, including high-quality proteins, essential amino acids, vitamins (particularly B-complex and $D_2$), dietary fibers, and bioactive compounds such as β-glucans, phenolics, and lovastatin (Khan and Tania 2012; Gąsecka et al. 2016), but also for their broad pharmacological properties which comprises antioxidant, anticancer, hypoglycemic, hepatoprotective, and anti-inflammatory effects (Elkhateeb et al. 2019; Wasonga et al. 2008). These properties have positioned them as functional foods with potential therapeutic applications. Several reports are available that mushroom consumption is increasing rapidly worldwide due to their ironic source of bioactive compounds such as functional protein, glucans, laccase, proteoglycan, pleuran, and pleurostrin (Sanchez 2010). Recent advancements in biotechnology, such as CRISPR gene editing and omics-based analyses, have further enabled the exploration of genetic traits associated with yield, stress tolerance, and bioactive metabolite synthesis (Smiderle et al. 2008; Raman et al. 2020).

## 5.2 Structural Organization and Key Characteristics of Commercially Known *Pleurotus* Species

### 5.2.1 Structure of *Pleurotus* Species

*Pleurotus species* are commonly known as oyster mushrooms and are currently ranked as the second most important cultivated species of edible mushrooms globally (Gupta and Singh 2024). The general structure of the fruiting body of *Pleurotus* usually includes three different parts named the pileus, stipe, and gills. However, notable morphological differences still persist between species. The pileus is a spatula-shaped cup or a fleshy shell with an eccentric or lateral stipe. Its length is about 1.9–7.8 inches, while its diameter ranges from 5 to 25 cm. Depending upon the species, it can be white, cream, yellow, pink, grey, or brown in color. The stipe is a short or long lateral or central stalked structure which is off-center with a later

**Fig. 5.1** Various *Pleurotus species.* (Courtesy Kirori Mal College)

attachment to the substrate. Underneath the pileus, gills are present as long ridges and furrows. They extend from the edge of the cap, down to the stalk and bear smooth, cylindrical, white- to cream-colored spores. Within 48 hours, these spores can germinate on any type of mycological media. *Pleurotus* sp. have a characteristic whitish to lilac grey spore prints. They are best viewed under a dark background. Purely white colored mycelium can be seen in the genus *Pleurotus* (Fig. 5.1).

### 5.2.2 Morphological Characteristics of Different Types of *Pleurotus* Species

Typically, the structure of oyster mushrooms more often appears to be similar; still, some variations exist in their morphology, listed in Table 5.1. These variations can be potentially rewarding for entrepreneurs in the area of upscale restaurants or premium food stores where visual appeal and variety are highly valued. Mala et al. (2023) provided details of all five listed species, whereas Muzaffar et al. (2023) covered only three species from Kashmir, viz., *P. florida*, *P. sajor-caju*, and *P. djamor*.

**Table 5.1** Morphological characterization of different types of *Pleurotus* species

| Features | Different types of *Pleurotus species* | | | | | Citations |
|---|---|---|---|---|---|---|
| | *P. ostreatus* | *P. florida* | *P. sajor caju* | *P. djamor* | *P. citrinopileatus* | |
| Colour of fruiting body | Gray to blue black | White to cream/ gray | White-gray/brown | Cream- peach to reddish pink | Yellow white to golden yellow | Muzaffar et al. (2023) and Mala et al. (2023) |
| Pileus: size; shape | 4–6 cm; kidney-fan shaped | 4–23 cm; flat fan-shell shaped | 3–17 cm; leaf/ petal/fan shaped | 5–6.5 cm; flat fan-shell shaped | Fan-narrowly infundibuliform shaped | |
| Gills | Decurrent, closed | | | | | |
| Veil | – | Absent | Present | Absent | – | |
| Stipe | 2–3 cm long & 0.8–1.1 cm thick | 1–4.3 cm long & 0.8–2 cm thick | 1–2.9 cm long & 0.9–1.6 cm thick | 1–3.5 cm long & 0.6–1.3 cm thick | 3–5.5 cm length & 1.5–1.9 cm diameter | |
| Spores: size (μm) | 8–11.5 × 4–4.4; | 7–8.1 × 3.4–4.5 | 6–8.5 × 3–4.4 | 7–8.1 × 3–4.5 | 5.2–7.9 × 3.5–4.9 | |
| Spore print | White | White | Gray white | Light pink | White | Mala et al. (2023) |

Source: Muzaffar et al. (2023) and Mala et al. (2023)

## 5.3 Nutritional and Therapeutic Value of *Pleurotus*

### 5.3.1 Nutritional Significances

The striking nutritional value and profitable composition, which includes high protein content, carbohydrates, minerals, vitamins, and amino acids, of *Pleurotus* mushrooms are the main reasons for their widespread recognition as a functional food. Moreover, the protein content of these species can be enhanced by cultivating them on nitrogen-rich substrates supplemented with additional nitrogen sources. This increased protein content is particularly beneficial for vegetarian diets (Naraian et al. 2009; Akyuz et al. 2010) and those searching for plant-based protein substitutes, therefore making it a competitive substitute for animal proteins. The Food and Agriculture Organization of the United Nations (FAO) reported about 30.4% protein content (Raman et al. 2020). However, Khan and Tania (2012) and Galappaththi et al. (2021) recorded it between 17% and 42%. Table 5.2 highlights the carbohydrate, protein, ash, fat, lipid, and fiber content, which aids in the understanding of dietary advantages and market positioning. These species are considered a novel food source due to the high variability in their dietary fiber and are also an excellent source of fatty acids (linoleic acid and oleic acid) while containing low lipids.

The carbohydrate content of various *Pleurotus* species ranges between 24.95% and 75.88%. In *P. tuber-regium*, *P. flabellatus*, and *P. ostreatus*, carbohydrate levels are recorded as 63.03%, 57.4%, and 50.9%, respectively (Raman et al. 2020). The presence of ergothioneine, oligosaccharides, monosaccharides, and disaccharides as carbohydrate content aids in maintaining gut microbiomes as well as performing the functions of a prebiotic. A total of nine disaccharides were found in *P. ostreatus*, among them, sucrose was present in higher concentration (51.60 g/100 g) when compared to other disaccharides such as maltose, chitobiose, trehalose, and cellobiose, which were 29.21, 11.9, 7.37, and 0.0068 g per 100 g, respectively (Effiong et al. 2024). Ash, being an incombustible remains of a substance, not only contains various minerals but also plays a pivotal role in disease prevention, immune regulation, maintenance of homeostasis, and metabolic processes (Sahoo et al. 2022) (Table 5.2).

Besides, these mushrooms contain a range of Vitamin B, with Vitamin B2 (92.97 mg/kg) being the highest. Other vitamins like A, C and E were present in increasing concentration but remain lower than that of Vitamin B (Ahmed et al. 2016; Effiong et al. 2024). *Pleurotus ostreatus* contain high amount of vitamins B1, B2, B12, C, D, E, K and minerals such as iron, magnesium, copper, zinc and potassium (Sharifi-Rad et al. 2020) and is exceptionally high in folic acid (B9). While edible mushrooms contain Zn content ranges between 4.22 and 7.70 μg/g (Patil et al. 2010), the calcium content in *P. ostreatus* is too low to contribute to the recommended daily intake, therefore, not generally considered as a good source of calcium. However, due to the low amount of sodium these mushrooms are an eligible source of food for hypertensive patients (Dundar et al. 2008). Tables 5.3 and 5.4 provide details on micronutrient and vitamin content of oyster mushroom species.

**Table 5.2** Nutritional composition of *Pleurotus species* in dry weight

| *P. species* | Protein | Fat/lipid | Ash | Carbohydrates | Fiber | Citations |
|---|---|---|---|---|---|---|
| PO | 20.33–42 | 0.5–4.68 | 6.1–10.91 | 3–56.20 | 7.15–31 | Patil et al. (2010), Khan and Tania (2012), Ahmed et al. (2016), Tolera and Abera (2017), and Raman et al. (2020) |
| PE | 11–12 | 7–8 | 4.89 | 39–40 | 28.29 | Mehmet and Sevda (2010) and Khan and Tania (2012) |
| PSC | 24.63–25.50 | 4–4.41 | 7.23–8.28 | 38–39.82 | 22.87–25.20 | Alam et al. (2008) and Ahmed et al. (2016) |
| PGI | 17.7 | 4.3 | – | 78.0 | – | Phan et al. (2012) |
| PF | 15–23.2 | 1–5 | 6.35–9.02 | 7–57.80 | 7.40–27 | Alam et al. (2007), Ahmed et al. (2009), Khan and Tania (2012), and Ahmed et al. (2016) |
| PFL | 21–27.57 | 1–3.70 | 7.20–10.7 | 57.4–60 | 10–24.37 | Khan and Tania (2012), Ahmed et al. (2016), and Raman et al. (2020) |
| PC | 20–30 | 1.32–3.9 | 7.65 | 42.5 | 20.78 | Musieba et al. (2013), Khatun et al. (2015), Raman et al. (2020), and Gąsecka et al. (2016) |
| PP | 15–26.82 | 1.05–2.62 | 5.86–7.33 | 34 | 4–9 | Silva et al. (2002), Khatun et al. (2015), and Raman et al. (2020) |
| PEO | 24.10 | 4.73 | 9.84 | 45.59 | 15.91 | Kortei and Wiafe-Kwagyan (2015) |
| PTR | 13–22.10 | 0.2–2 | 2.97 | 53–63.03 | 10.86–16 | Ijeh et al. (2009) and Khan and Tania (2012) |
| PCY | 15.4–18 | 3.10–6 | 9.62 | 43–63.1 | 8.74–26 | Yang et al. (2001) and Khan and Tania (2012) |
| PS | 2.362 | 0.72 | 1.28 | 86.73 | 2.97 | Atri et al. (2016) |
| PGE | 19–31.80 | 3–3.64 | 11.20–26.3 | 31.60–46 | 5.7–27 | Khan et al. (2008), Khan and Tania (2012), and Ahmed et al. (2016) |

Source: Galappaththi et al. (2021)

*PO Pleurotus ostreatus, PE P. eryngii, PSC P. sajor-caju, PGI P. giganteus, PF P. florida, PFL P. flabellatus, PC P. citrinopileatus, PP P. pulmonarius, PEO P. eous, PTR P. tuber regium, PCY P. cystidiosus, PS P. sapidus, PGE P. geestaranus*

The amino acid profile of *Pleurotus* spp. revealed a marked variation among different taxa, highlighting their impressive nutritional potential. These mushrooms are rich in both essential and nonessential amino acids, which makes them a valuable dietary source of high-quality protein. Among the five essential amino acids

**Table 5.3** Micronutrient content (mg/1000 g) in different species of *Pleurotus* (dried mushrooms)

| Strains | Se | Zn | Fe | Mg | K | Ca | Na | Citations |
|---|---|---|---|---|---|---|---|---|
| *P. ostreatus* | 0.11 | 38 | 15.80 | 900 | 1950 | 260 | 270 | Patil et al. (2010) and Yang et al. (2001) |
| *P. flabellatus* | ND | 141 | 248 | 40 | 1537 | 120 | 686 | Mshandete and Cuff (2008) |
| *P. florida* | 0.013 | 2.3 | 12.38 | 92 | 920 | 305 | 90 | Ahmed et al. (2009) and Alam et al. (2008) |
| *P. sajor-caju* | 0.025 | 20.9 | 33.45 | 20.22 | 2146 | 22.15 | 220 | Alam et al. (2008) |
| *P. citrinopileatus* | – | – | 590 | 102 | 162 | 80 | 90 | Gosh and Chakravarty (1990) |
| *P. eryngii* | 4.5 | 33.5 | 620.7 | 14.8 | 15.3 | 0.51 | 100 | Gąsecka et al. (2016) and Mehmet and Sevda (2010) |
| *P. pulmonarius* | – | – | 75 | 2593 | 2709 | 38 | – | Silva et al. (2002) |
| *P. eous* | – | 1.9 | 2.3 | 7.6 | 121.5 | 3.7 | 60 | Kortei and Wiafe-Kwagyan (2015) and Bano et al. (1982) |
| *P. djamor var. roseus* | – | 1.21 | 17.32 | 167.37 | 2218.33 | 49.67 | 67.12 | Jegadeesh et al. (2018) |
| *P. tuber regium* | 0.0025 | 0.14 | 0.5 | 0.04 | 0.02 | 0.08 | 0.08 | Ijeh et al. (2009) |

Source: Raman et al. (2020)

**Table 5.4** Vitamin content (g/100 g) in different species of *Pleurotus* (dried mushrooms)

| Strains | Thiamin (B1) | Riboflavin (B2) | Niacin (B3) | Folic acid (B9) | Ascorbic acid | Citations |
|---|---|---|---|---|---|---|
| *P. ostreatus* | 0.32 | 0.58 | 8.72 | 0.052 | 12.52 | Bano and Srivastava (1962) and Patil et al. (2010) |
| *P. flabellatus* | 1.46 | 7.10 | 73.3 | 1.22 | 144 | Bano and Rajarathnam (1986) |
| *P. Florida* | 1.36 | 7.88 | 72.9 | 1.41 | 113 | Bano and Rajarathnam (1988) |
| *P. sajor-caju* | 1.75 | 6.66 | 60 | 1.23 | 111 | Bano and Rajarathnam (1988) |
| *P. citrinopileatus* | 0.16 | 0.94 | 22.20 | 0.10 | <1 | Musieba et al. (2013) |
| *P. pulmonarius* | 0.68 | 0.26 | 0.48 | – | 6.74 | Okwulehie and Nosike (2015) |
| *P. eous* | 2.23 | 8.97 | 66.6 | 1.35 | 92 | Bano and Rajarathnam (1986) |

Source: Raman et al. (2020)

(leucine, threonine, methionine, phenylalanine, and lysine) and eight nonessential amino acids (alanine, aspartic acid, proline, serine, asparagine, hydroxyproline, cysteine, and glutamine) recorded, aspartic acid (492.12 mg/100 g) and lysine (23.18 mg/100 g) were found in higher concentration, however cysteine (9.32 mg/100 mg) was in the lowest concentration (Effiong et al. 2024). GABA, a nonessential amino acid that is required for brain functioning and mental activity, was also found to be present in high amounts (Lavelli et al. 2018; Gupta et al. 2019). In the category of basic amino acids, lysine was the only one to be present in *P. ostreatus* (Chatterjee et al. 2021). Table 5.5 represents the amino acid analysis of different *Pleurotus* spp. Among the essential amino acids' leucine, lysine, isoleucine, valine, and threonine were consistently found in notable amounts across species (Adebayo and Oloke 2017; Atri et al. 2012). *P. pulmonarius* was particularly outstanding with remarkably high levels of glutamine (18.24 g/100 g), serine (7.12 g/100 g), and leucine (7.96 g/100 g), underscoring its potential as a protein-rich mushroom (Adebayo 2013). Likewise, *P. florida* showed elevated concentrations of lysine (9.90), leucine (7.500), and valine (6.90) per 100 g of crude protein, while *P. sajor-caju* offered a more balanced amino acid profile with moderate yet consistent levels of isoleucine, leucine, and lysine (Hashem et al. 2013; Oyetayo and Ariyo 2013). In terms of nonessential amino acids, aspartic acid and glutamine dominated across most

**Table 5.5** Amino acids (AA) analysis in different *Pleurotus species*

| AA | PO | PC | PP | PF | PSC | PCY | PFR | PS |
|---|---|---|---|---|---|---|---|---|
| Ala | 0.23–3.75 | 0.83 | 0.19–6.35 | ND | 0.12–1.80 | 0.18 | 0.17 | 0.11 |
| Arg | 0.64–6.30 | 1.01 | 0.19–4.82 | ND | 0.2–2.52 | 0.19 | 0.22 | 0.23 |
| Asp | 1.77–4.30 | 1.82 | 0.48–9.65 | ND | 0.44–2.01 | 0.42 | 0.38 | 0.37 |
| Gln | 1.53–10.2 | 3.07 | 18.24 | ND | 4.22 | – | – | – |
| Gly | 0.09–1.65 | 0.84 | 5.99 | ND | 0.96 | – | – | – |
| His | 0.56–1.10 | 0.51 | 2.17 | 2.80 | 0.52–2.20 | – | – | – |
| Ile | 0.29–1.26 | 0.61 | 4.41 | 5.20 | 0.75–4.40 | – | – | – |
| Leu | 0.98–2.31 | 1.07 | 7.96 | 7.50 | 1.30–7 | – | – | – |
| Lys | 0.64–1.5 | 0.81 | 4.54 | 9.90 | 0.78–5.7 | – | – | – |
| Met | 0.08–0.53 | 0.30 | 2.30 | 3.00 | 0.34–1.8 | – | – | – |
| Phe | 0.75 | 0.65 | 3.72 | 3.50 | 0.86–5 | – | – | – |
| Pro | 0.10 | 0.33 | 0.06–4.98 | ND | 0.33–0.05 | 0.02 | 0.08 | 0.09 |
| Thr | 1.16 | 0.83 | 6.57 | 6.10 | 0.98–5 | – | – | – |
| Trp | 0.11 | 0.28 | ND | 1.10 | 0.09–1.2 | – | – | – |
| Tyr | 0.18 | 0.58 | 0.17–3 | ND | 0.15–0.67 | 0.18 | 0.14 | 0.2 |
| Val | 0.24 | 0.85 | 5.93 | 6.90 | 0.98–5.3 | – | – | – |
| Cys | 0.06–0.53 | 0.20 | 2.25 | ND | ND | – | – | – |
| Ser | 0.12 | 1.03 | 7.12 | ND | 0.99 | – | – | – |
| Citations | $a_1$, $a_2$ | $a_3$ | $a_5$, $a_4$ | $a_6$ | $a_7$, $a_6$, $a_5$ | $a_5$ | $a_5$ | $a_5$ |

Source: Adebayo and Oloke (2017) and Atri et al. (2012)

$a_1$: Hashem et al. (2013) (mg/ g fresh weight), $a_2$: Oyetayo and Ariyo (2013) (g/100 g), $a_3$: Musieba et al. (2013) (%), $a_4$: Adebayo (2013) (g/100 g protein dry weight), $a_5$: Atri et al. (2012) (%), $a_6$: Crisan and Sands (1978) (aa/100 g of corrected crude protein), $a_7$: Chirinang and Intarapichet (2009) (mg/g fresh weight)

*PO Pleurotus ostreatus, PSC P. sajor-caju, PFR P. florida, PFL P. flabellatus, PC P. citrinopileatus, PP P. pulmonarius, PEO P. eous, PCY P. cystidiosus, PS P. sapidus*

species, often surpassing the levels of several amino acids (Adebayo 2013). On the other hand, amino acids like proline, cysteine, and tryptophan were generally present in only small amounts or absent in certain species (Musieba et al. 2013; Crisan and Sands 1978). This table helps the business owners to promote these mushrooms as a complete or additional source of protein, especially to consumers preferring vegan and vegetarian diets.

### 5.3.2 Bioactive Compounds and Therapeutic Properties

Besides nutritional significance, *Pleurotus* mushrooms are also a rich source of bioactive compounds such as β-glucan, lectins, polysaccharides, ubiquinone-9, ergosterol, lovastatin, and other structurally distinct proteins, having therapeutic benefits, making them appealing components in the growing pharmaceutical and nutraceutical sectors. Kamilya et al. (2008) worked on the effect of β-glucan on immunostimulants in farmed Indian Carp, produced from *P. florida. Pleurotus* mushrooms contain antiviral compounds, thus helping the body in combating against viruses such as the Herpes simplex virus (HSV) and HIV by boosting the immune system (Gupta and Singh 2024). They have been found to contain beta-glucan and other macromolecules that are effective against cancer cell lines without harming the normal cells that may help fight cancer cells (Mishra et al. 2021). They encourage cancer cells to undergo a process called apoptosis, which may slow down or stop the growth of tumors and aid in cancer prevention or treatment. Wisbeck et al. (2014) reported the antitumor activity of *P. ostreatus* due to the presence of polysaccharides, which helps in controlling the Ehrlich tumor and sarcoma. *Pleurotus* mushrooms exhibit antimutagenic properties by shielding DNA from damage that could lead to cancer. Through their ability to reduce mutations in cells, they lower the risk of genetic alterations that can trigger carcinogenesis in the human body (Patel et al. 2012; Hu et al. 2006).

### 5.3.3 Antioxidant and Anti-Inflammatory Effects

The mushrooms are rich in antioxidants, which are substances that protect the body from harmful free radicals. Free radicals are unstable molecules that can cause cellular damage, potentially leading to diseases such as cancer and cardiovascular disorders, as well as contributing to the aging process. *Pleurotus* helps in neutralizing free radicals, thereby assisting in protecting the body from oxidative stress and damage (Hu et al. 2006; Elkhateeb et al. 2019). Jayakumar et al. (2011) reported the diminishing of age-related disorders from the extract of P. ostreatus by an increase in catalase gene expression. Besides, *Pleurotus* mushrooms demonstrate the capacity to reduce inflammation in the body. This property renders them beneficial in managing conditions where inflammation plays a significant role, such as in arthritic joint pain or other inflammatory diseases. Lectins produced by *P. eryngii* can be considered as a potential therapeutic agent in the development of strategies for cancer and other inflammatory diseases (Seliman et al. 2024).

### 5.3.4 Liver Protection, Blood Pressure, and Cholesterol Control

*Pleurotus* mushrooms exhibit hepatoprotective properties, shielding the liver from toxin-induced damage. The compounds present in these mushrooms contribute to reducing oxidative stress and improving hepatic health. Consequently, they may play a crucial role in preventing liver diseases or aiding in the recovery of a damaged liver (Patel et al. 2012). Hypertension and hypercholesterolemia are major risk factors for cardiovascular disease. *Pleurotus* mushrooms demonstrate the ability to lower both blood pressure and cholesterol levels. Therefore, they can contribute to maintaining cardiovascular health and potentially lowering the risk of myocardial infarction or cerebrovascular events (Elkhateeb et al. 2019).

### 5.3.5 Antidiabetic and Antiaging Effects

*Pleurotus* mushrooms may also prove beneficial for individuals with diabetes. They exhibit the capacity to lower blood glucose levels and improve insulin sensitivity. This makes them potentially advantageous for managing diabetes or preventing its onset in at-risk individuals (Wasonga et al. 2008; Lesa et al. 2022). As the aging process progresses, the human body's capacity to combat oxidative stress and damage diminishes. *Pleurotus* mushrooms can enhance the body's antioxidant defenses, potentially mitigating the signs of aging. This may contribute to the prevention of age-related neurodegenerative disorders such as Alzheimer's disease, Parkinson's disease, and certain types of cancer (Smiderle et al. 2008).

Table 5.6 summarizes the pharmacologically active compounds isolated from oyster species and their high therapeutic potential for application in therapeutic mediation and functional food development.

## 5.4 Entrepreneurial Potential and Market Trend of *Pleurotus* Mushrooms

The production of *Pleurotus species* presents a significant entrepreneurial opportunity, especially in rural and peri-urban areas. Such regions have limited formal sector employment as well as access to capital, but are abundant in agricultural residues, which are primary substrates for mushroom cultivation. Low capital investment, compact cultivation period, and increasing demand for health foods by consumers, these factors make *Pleurotus* increasingly important as a sustainable agribusiness. Additionally, the growing linkages of value chains and internet-based marketplaces have opened up this industry to new entrants, particularly small farmers, youth, and women.

### 5.4.1 Scalability and Low-Cost Cultivation

The main advantage of *Pleurotus* cultivation is its cost-effectiveness and scalability. Farmers can make use of locally abundant agrowastes like paddy straw, wheat husk, sugarcane bagasse, and cotton wastes as substrates, thus transforming waste into wealth (ICAR-DMR 2023). The process of production needs very little infrastructure: a sunny place, polybags, and watering at regular intervals are enough in most cases. A normal crop cycle of *Pleurotus* finishes within 25–35 days, with several harvests per year. This high turnover greatly enhances return on investment and increases scalability potential for smallholders.

One key challenge in mushroom entrepreneurship is the dependency on external sources for spawn production, especially F0–F2 generation seeds. A community-based intervention in Indonesia demonstrated that training entrepreneurs in in-house spawn preparation using low-cost tools such as autoclaves and simple laminar air flow systems significantly increased self-reliance and reduced costs. Spent bag log

**Table 5.6** Bioactive compounds and therapeutic properties of some oyster mushrooms (*Pleurotus* spp.)

| Name of the species | Active compounds/ extract | Properties | Citations |
|---|---|---|---|
| *P. ostreatus* | β glucan, methanol, lectins, ribonuclease, mevinolin polysaccharide | Antitumor, immunomodulatory, antioxidant, antihypercholesterolemic, anti HIV, Anti mutagenic, anti diabetic | Facchini et al. (2014), Nomura et al. (1994), Wang et al. (2000), Akanni et al. (2010), Yang et al. (2002), Hossain et al. (2003), Lakshmi et al. (2004), Stastny et al. (2022), and Varghese et al. (2019) |
| *P. pulmonarius* | Polysaccharides, α&β glucans | Anticancer, HIV-1 reverse transcriptase inhibition | Lavi et al. (2012), Wang and Ng (2007), Akanni et al. (2010), and Patel et al. (2012) |
| *P. citrinopileatus* | Lectins, PCP-3A protein | Antileukemia, antihyperglycemic | Chen et al. (2009) and Hu et al. (2006) |
| *P. florida* | Glucans, lectins | Immunostimulator, toxin modulator | Maiti et al. (2008), Rana et al. (2012), and Jose et al. (2004) |
| *P. sajor caju* | Ribonuclease, hot water ext. | Antitumor, anti microbial, anti HIV | Facchini et al. (2014), Ngai and Ng (2004), and Patel et al. (2012) |
| *P. eryngii* | Ubiquinone-9, pleurone, eryngin | Anticancer, anti aging, antioxidant | Bae et al. (2009), Lee et al. (2011), and Chen et al. (2012) |
| *P. tuberregium* | Sulfated β glucans | Anticancer, anti HIV | Zhang et al. (2001, 2003, 2004) |
| *P. cornucopiae* | Laccase, acetone | Antileukemia/human hepatoma cell | Wong et al. (2010) |
| *P. ferulae* | Ethanol | Hypocholesterolemic | Alam et al. (2011) |
| *P. geesteranus* | Polysaccharide | Cytotoxic to breast cancer cells (MCF-7) | Alam et al. (2011) |
| *P. saca* | Lovastatin | Hypocholesterolemic | Gunde Cimerman and Plemenitaš (1995) |
| *P. nebrodensis* | Hemolysin | Anti cancer/HIV-1 | Lv et al. (2009) |
| *P. cystidiosus* | Methanol | Antioxidant | Yang et al. (2002) |

(continued)

**Table 5.6** (continued)

| Name of the species | Active compounds/ extract | Properties | Citations |
|---|---|---|---|
| *p. salmoneostramineus* | Mushroom powder | Lipid/cholesterol reduction | Yoon et al. (2012) |
| *P. eous* | Methanol, lectin | Antioxidant | Ramkumar et al. (2010) |
| *P. abalonus* | Polysaccharide peptide | Anti aging | Li et al. (2007) |
| *P. djamor* | Polysaccharide precipitate | Anti tumor | Borges et al. (2013) |
| *P. floriandus* | Ethanol ext. | Antibacterial | Das et al. (2012) |

Sources: Paul et al. (2017), Stastny et al. (2022), Varghese et al. (2019), and Abidin et al. (2016)

waste was successfully composted and applied to vegetable cultivation, improving plant growth while reducing environmental pollution. This model showcases how small-scale mushroom enterprises can become more sustainable and resource-efficient through skill development and waste valorization (Hasanah et al. 2023).

The choice of substrate and mushroom species plays a critical role in maximizing yield and profitability in mushroom-based enterprises. Table 5.7 highlights the biological efficiency (BE) of various *Pleurotus* species grown on different agrowaste substrates. Selecting the right species–substrate combination enables cultivators to maximize profit margins and enhance biological efficiency, making informed decisions—a key entrepreneurial advantage.

Apart from biological efficiency, morphological traits such as the cap size and the number of fruiting bodies significantly influence commercial value. Table 5.8 compares how different agro-substrates affect the physical characteristics of *P. ostreatus*, which are critical for determining the yield, visual appeal, and market acceptability.

### 5.4.2 Rural Entrepreneurship in Practice (Case Study)

Practical applications validate the potential of oyster mushrooms. In Bihar, the State Horticulture Department has launched support schemes for women-owned self-help groups (SHGs), which have enabled them to transform agricultural residues into earnings through the production of oyster mushrooms. One SHG can produce some 200–250 kg/month, earning a net profit of between ₹15,000 and ₹20,000 (NHB 2023). Likewise, in Odisha and Jharkhand tribal belts, mushroom farming is a sustainable alternative livelihood option, and it is aided by the Krishi Vigyan Kendras (KVKs) and RKVY schemes (Aditya and Bhatia 2020). These instances reflect the congruence between people-centered government policy and grassroots entrepreneurship (Fig. 5.2).

**Table 5.7** Influence of various substrates on the biological efficiency of oyster mushroom species

| Mushroom species | Top substrates | Maximum BE (%) | Citations |
|---|---|---|---|
| *P. sajor-caju* | Paddy straw, spent compost | 137.40 | Bhatia (2007) |
| *P. eous* | Soyabean straw, pearl millet stalks | 96.36 | Telang et al. (2010) |
| *P. sapidus* | Wheat straw, sunflower stalk | 72.90 | Telang et al. (2010) |
| *P. platypus* | Wheat straw | 50.10 | Hoa et al. (2015) |
| *P. cystidiosus* | Saw dust | 62.50 | Atila (2017) |
| *P. djamor* | Oak, safflower hay/head residue | 78.20 | Paswal et al. (2017) |
| *P. flabellatus* | Shiris, wheat straw | 53.87 | Mishra et al. (2015) and Otunla and Idowu (2012) |
| *P. florida* | Wheat straw, popular + paddy straw | 96.38 | Kulshreshtha et al. (2010) |
| *P. eryngii* | Olive leaves + two phase olive meal waste | 87.50 | Zervakis et al. (2013) |
| *P. citrinopileatus* | Sunflower head residue, pea pod + paddy straw | 94.30 | Shevale and Deshmukh (2016) |
| *P. fossulatus* | Wheat straw | 42.15 | Pal et al. (2017) |
| *P. pulmonarius* | Wheat straw | 99.70 | Zervakis et al. (2013) |
| *P. ostreatus* | Saw dust + faba bean leaves, sugarcane bagasse, wheat straw | 114.90 | Rivera and Merales (2010) and Koutrotsios et al. (2018) |
| *P. cornucopiae* | Paddy straw, wheat straw, saw dust | 82.67 | Jarial et al. (2020) and Stanley (2010) |
| *P. tuber-regium* | Maize cob, water hyacinth | 99.65 | Apetorgber et al. (2013) |
| *P. nebrodensis* | Olive leaves + two phase olive meal waste | 97.50 | Mishra et al. (2015) |

Source: Aditya et al. (2024)
Here, *BE* biological efficiency

### 5.4.3 Involvement of Youth and Women in the Mushroom Economy

Perhaps the most significant feature of *Pleurotus* cultivation is its accessibility. Women make up almost 60% of micro-entrepreneurs in the industry (Aditya and Bhatia 2020). The ease of cultivation, presence of training, and adaptability to home-based farming make mushroom cultivation a compelling opportunity for women and rural youth seeking income-generating activities. Involvement in mushroom businesses has been linked with better household income, increased food security, and greater social empowerment for women.

**Table 5.8** Proximate size and number of fruiting bodies of *P. ostreatus* based on cultivation with common lignocellulosic substrates

| Substrate | Cap diameter (mm) | Stripe length & thickness (mm) | No. of effective fruiting bodies | Citations |
|---|---|---|---|---|
| Corn cob | 86.7 | 35.3; 11.1 | 7.9 | Hoa et al. (2015) and Philippoussis (2009) |
| Sawdust (Soft wood) | 17.6 | – | 29.6 | Ogundele et al. (2017) and Philippoussis (2009) |
| Saw dust (Hardwood) | 24.0 | – | 14.6 | Hoa et al. (2015), Ogundele et al. (2017), and Philippoussis (2009) |
| Sawdust (General) | 70.6 | 38.2; 8.5 | 11.5 | Girmay et al. (2016) and Hoa et al. (2015) |
| Sugarcane bagasse | 84.8 | 35.7; 10.2 | 8.1 | Hoa et al. (2015) and Philippoussis (2009) |
| Wheat straw | 86–89.3 | 28.1; 10.3 | 18.3 | Fanadzo et al. (2010), Girmay et al. (2016), Masevhe et al. (2016), Salama et al. (2016), and Upadhyay et al. (2002) |

Source: Karmani et al. (2022)

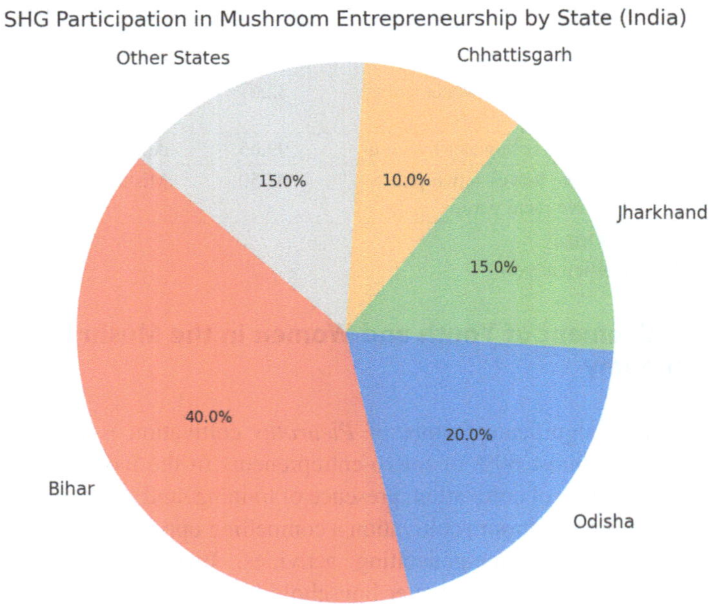


**Fig. 5.2** SHG participation in mushroom-based rural entrepreneurship across selected Indian states. (Data derived from NABARD (2021) and Odisha Livelihoods Mission (2022))

### 5.4.4 Profitableness and Job Creation

From a financial point of view, *Pleurotus* cultivation is one of the most lucrative low-cost agri-business ventures. The farmer invests approximately ₹2000 for every 100 kg of mushrooms grown utilizing straw-based substrates and receives a profit of ₹8000–₹10,000, with net gains of ₹6000–₹8000. Entrepreneurs working on a larger scale (approximately 1000 kg/month) can earn more than ₹70,000 per month (Thakur and Sharma 2017; Directorate of Mushroom Research 2021).

Recent studies have shown that using mixed agricultural residues as substrates significantly improves the yield and biological efficiency, thereby increasing profitability and resource utilization (Gebru et al. 2024).

In addition, the industry generates indirect employment in spawn production, processing, logistics, and retail. This labor-intensive nature of cultivation also provides sustainable livelihood opportunities, particularly in rural areas (Gebru et al. 2024).

#### 5.4.4.1 Real-World Feasibility

Building on the financial insights above, field-based feasibility studies offer concrete evidence of the profitability and sustainability of *Pleurotus* cultivation. In Pandowoharjo village of Indonesia, Yogyakarta, a study by Sumbodo et al. (2023) demonstrated that oyster mushroom farmed under a collaborative model with Jejamuran restaurant and academic institutions achieved a net present value (NPV) of Rp 6.3 million, an internal rate of return (IRR) of 26.86%, and net and gross B/C ratios of 1.54 and 1.10, respectively. These results reveal that *Pleurotus* cultivation is not only economically viable but also scalable and replicable in the rural sector. The study also emphasized the importance of triple-helix collaboration between local government, industry, and educational institutions for empowering farmers, improving product quality, and enhancing profitability (Fig. 5.3).

Oyster mushroom cultivation is emerging as a viable rural enterprise across various agri-climatic zones of India. In Chhattisgarh, Krishna and Rupsingh (2024) found that while large farms generated the highest gross (₹119617.41) and net income (₹79218.65), small farms were more cost-efficient, with lower production costs (₹13160.05) and the highest input-output ratio (1:4.11). Their lower break-even point (29.74 kg vs. 101.71 kg for large farms) also indicated quicker returns and reduced financial risk. Similarly, in Bhagalpur, Bihar, Kumari et al. (2022) reported that although large farms had lower production costs (₹75.76/kg) and a higher benefit-cost ratio (1.40:1), small farms still achieved a respectable 1.19:1 return. Together, these studies highlight that while profitability improves with scale, small-scale oyster mushroom farming remains a low-risk, high-potential livelihood option, especially for women and marginal farmers in rural areas.

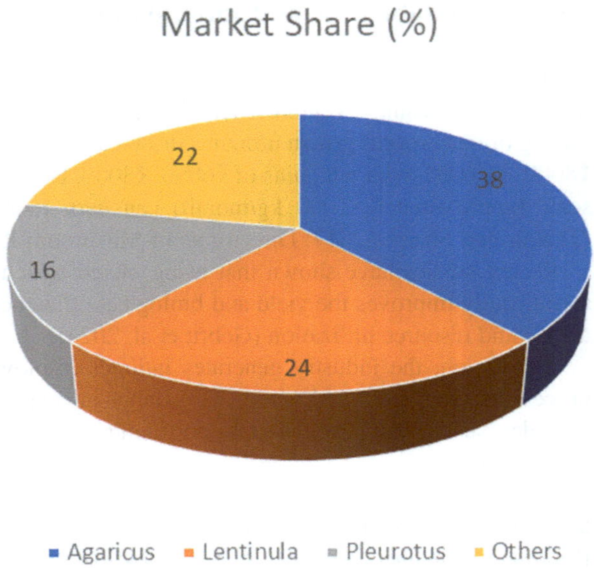


**Fig. 5.3** Global mushroom market share. (Adapted from IMARC Group (2024))

### 5.4.5 Market Linkage and Value Addition

With increasing awareness about health and sustainability, the demand for protein-rich, low-fat food products is slowly on the rise. Entrepreneurs are capitalizing on this need by creating a variety of value-added mushroom products, such as dehydrated mushroom slices, mushroom powder and soup mixes, pickles, curry cubes, cookies, and high-protein mushroom flour. These commodities fetch high prices and are well-suited for both domestic and export markets. Digital platforms, Farmer Producer Companies (FPCs), and cooperative societies also facilitate direct-to-consumer models, eliminating intermediaries and raising profitability (Fig. 5.3).

### 5.4.6 Export Potential and Future Growth

Internationally, the demand for powdered and dried mushrooms is increasing because of their nutrient concentration and long shelf life. While India's share in world exports is still limited, efforts under the One District One Product (ODOP) and PMFME schemes are fueling capacity creation and market growth. Technological advancements such as climate-controlled growing chambers and traceability based on digital platforms are likely to further improve the export scenario.

## 5.5 Conclusion

The cultivation and use of *Pleurotus* species have proven to be a practical and sustainable solution for promoting inclusive entrepreneurship, improving nutrition, and advancing health-based innovations. Due to their ability to grow on agricultural waste with minimal inputs, along with the generation of high biological efficiency, these mushrooms are especially well-suited for rural and peri-urban areas. Rich in nutrients and therapeutic compounds, *Pleurotus* mushrooms not only support plant-based diets but also offer great potential in developing functional foods and medicinal products. In India and beyond, *Pleurotus* cultivation is making a real difference, offering affordable and scalable business opportunities that empower women and youth. With growing interest in health-conscious food choices, supportive government programs, and easier access to online markets, the mushroom sector is gaining strong momentum. Overall, *Pleurotus* mushrooms are not just a promising food crop; they are a pathway toward healthier communities, stronger rural economies, and a more sustainable future.

**Declaration** The authors have no competing interests to declare that are relevant to the content of this review article.

## References


Abidin MHZ, Abdullah N, Abidin NZ (2016) Therapeutic properties of *Pleurotus species* (oyster mushrooms) for atherosclerosis: A review. Int J Food Prop 20:1251–1261

Adebayo EA (2013) Production of new strains of *Pleurotus pulmonarius* through cross breeding and mutagenesis. An unpublished thesis submitted to the Department of Pure and Applied Biology, Ladoke Akintola University of Technology, Ogbomoso

Adebayo EA, Oloke JK (2017) Oyster mushroom (*Pleurotus species*): a natural functional food. J Microbiol Biotechnol Food Sci 7:254–264

Aditya A, Bhatia JN (2020) Economic empowerment of rural women through mushroom farming. Int J Curr Microbiol Appl Sci 9:1081–1089

Aditya N, Jarial RS, Jarial K, Bhatia JN (2024) Comprehensive review on oyster mushroom species (Agaricomycetes): morphology, nutrition, cultivation and future aspects. Heliyon 10:e26539

Ahmed SA, Kadam JA, Mane VP et al (2009) Biological efficiency and nutritional contents of *Pleurotus florida* (Mont.) Singer cultivated on different agro-wastes. Nat Sci 7:44–48

Ahmed M, Abdullah N, Nuruddin MM (2016) Yield and nutritional composition of oyster mushrooms: an alternative nutritional source of rural people. Sains Malays 45:1609–1615

Akanni EO, Oloke JK, Adebayo EA, Ola IO (2010) Antitumour activities of *Pleurotus pulmonarius* and *Pleurotus ostreatus* metabolites on N-Nitroso-N-ethylurea induced solid tumour bearing Wister rats. Glob Res J 1:13–18

Akyuz M, Onganer AN, Erecevit P, Kirbag S (2010) Nutritive value of edible wild and cultured mushrooms. Turk J Biol 23:125–130

Alam N, Khan-Hossain MS, Amin SMR, Kahn LA (2007) Nutritional analysis of dietary mushrooms: *Pleurotus florida* Eger and *Pleurotus sajor-caju* (Fr.) Singer. Bangladesh J Mushroom 1:1–7

Alam N, Amin R, Khan A et al (2008) Nutritional analysis of cultivated mushrooms in Bangladesh: *Pleurotus ostreatus*, *Pleurotus sajor-caju*, *Pleurotus florida* and *Calocybe indica*. Microbiology 36:228–232

Alam N, Yoon KN, Lee TS (2011) Antihyperlipidemic activities of Pleurotus ferulae on biochemical and histological function in hypercholesterolemic rats. J Res Med Sci 16:776–786
Apetorgber AK, Dzomeku M, Apetorgber MM (2013) Growth factors and cultivation of Pleurotus-tuber regium on selected plant wastes. Int Food Res J 20:3387–3393
Atila F (2017) Evaluation of suitability of various agro-wastes for productivity of *Pleurotus djamor*, *Pleurotus citrinopileatus* and *Pleurotus eryngii* mushrooms. J Exp Agric Int 17:1–11
Atri NS, Sharma SK, Joshi R, Gulati A, Gulati A (2012) Amino acid composition of five wild Pleurotus species chosen from Northern West India. Eur J Biol Sci 4:31–34
Atri NS, Kumari B, Kumar S, Upadhyay RC, Gulati A, Lata, Gulati A (2016) FUNGI: applications and management strategies. Taylor & Francis Group LLC, pp 373–395
Bae JS, Park JW, Park SH, Park JB, Rho YH, Ryu YB, Lee KS, Park KH, Bae YS (2009) Apoptotic cell death of human leukemia U937 cells by ubiquinone-9 purified from *Pleurotus eryngii*. Nat Prod Res 23:1112–1119
Bano Z, Rajarathnam V (1986) Value of Pleurotus mushrooms. Plant Food Hum Nutr 36:11–15
Bano Z, Rajarathnam S (1988) *Pleurotus* mushrooms part II: chemical composition, nutritional value, post-harvest physiology, preservation, and role as human food. CRC Cr Rev Food Sci 27: 87–158
Bano Z, Srivastava HC (1962) Studies on cultivation of *Pleurotus sp.* on paddy straw. Food Sci 11: 363–365
Bano ZS, Bhagya S, Srinivaran FS (1982) Essential amino acid composition and proximate analysis of the mushrooms *Pleurotus eous* and *Pleurotus florida*. Mushroom Newslett Trop 3:6–10
Bhatia JN (2007) Evaluation of substrates for sporophore production of *Pleurotus sajor-caju* (Fr.) Singer in Haryana. Annu Rev Plant Physiol 21:281–284
Borges GM, De Barba FFM, Schiebbelbein AP, Pereira BP, Chaves MB, Silveira MLL (2013) Extracellular polysaccharide production by a strain of *Pleurotus djamor* isolated in the south of Brazil and antitumor activity on Sarcoma 180. Braz J Microbiol 44:1059–1065
Chatterjee D, Halder D, Das S (2021) Varieties of mushrooms and their nutraceutical importance: a systematic review. J Clin Diagn Res 15
Chen JN, Wang YT, Wu JSB (2009) A glycoprotein extracted from golden oyster mushroom *Pleurotus citrinopileatus* exhibiting growth inhibitory effect against U937 leukemia cells. J Agric Food Chem 57:6706–6711
Chen J, Mao D, Yong Y, Li J, Wei H, Lu L (2012) Hepatoprotective and hypolipidemic effects of water-soluble polysaccharidic extract of *Pleurotus eryngii*. Food Chem 130:687–694
Chirinang P, Intarapichet KO (2009) Amino acids and antioxidant properties of the oyster mushrooms, *Pleurotus ostreatus* and *Pleurotus sajor-caju*. Sci Asia 35:326–331
Crisan EV, Sands A 1978 Nutritional value. Academic Press, New York, pp. 137–165
Das N, Pasman B, Mishra S, Bhattacharya B, Sengupta C (2012) Comparative studies of antibacterial properties of three *Pleurotus species* (oyster mushroom). Nat Sci 10:178–183
Directorate of Mushroom Research (2021) Training manual on mushroom cultivation. ICAR-DMR, Solan
Dundar A, Acay H, Yildiz A (2008) Yield performances and nutritional contents of three oyster mushroom species cultivated on wheat stalk. Afr J Biotechnol 7:3497–3501
Effiong ME, Umeokwochi CP, Afolabi IS, Chinedu SN (2024) Assessing the nutritional quality of *Pleurotus ostreatus* (oyster mushroom). Front Nutr 10:1279208
Ejigu N, Sitotaw B, Girmay S, Assaye H (2022) Evaluation of oyster mushroom (Pleurotus ostreatus) production using water hyacinth (*Eichhornia crassipes*) biomass supplemented with agricultural wastes. Int J Food Sci 2022:9289043
Elkhateeb WA, Daba GM, Elnahas MO, Thomas PW (2019) Anticoagulant capacities of some medicinal mushrooms. ARC J Pharm Sci (AJPS) 5:1–9
Facchini JM, Alves EP, Aguilera C, Gern RM, Silveira MLL, Wisbeck E, Furlan SA (2014) Antitumor activity of *Pleurotus ostreatus* polysaccharide fractions on Ehrlich tumor and Sarcoma 180. Int J Biol Macromol 68:72–77

Fanadzo M, Zireva DT, Dube E et al (2010) Evaluation of various substrates and supplements for biological efficiency of *Pleurotus sajor-caju* and *Pleurotus ostreatus*. Afr J Biotechnol 9:2756–2761

Galappaththi MCA, Dauner L, Madawala S, Karunarathna SC (2021) Nutritional and medicinal benefits of Oyster (*Pleurotus*) mushrooms: a review. Fungal Biotech 1:65–87

Gąsecka M, Mleczek M, Siwulski M, Niedzielski P (2016) Phenolic composition and antioxidant properties of *Pleurotus ostreatus* and *Pleurotus eryngii* enriched with selenium and zinc. Eur Food Res Technol 242:723–732

Gebru H, Belete T, Faye G (2024) Growth and yield performance of *Pleurotus ostreatus* cultivated on agricultural residues. Mycobiology 52:388–397

Girmay Z, Gorems W, Birhanu G et al (2016) Growth and yield performance of *Pleurotus ostreatus* (Jacq. Fr.) Kumm (oyster mushroom) on different substrates. AMB Express 6:1–7

Gosh N, Chakravarty DK (1990) Predictive analysis of the protein quality of *Pleurotus citrinopileatus*. J Food Sci Technol 27:236–238

Gunde Cimerman N, Plemenitaš A (1995) Hypocholesterolemic activity of the genus *Pleurotus* (Jacq.: Fr.) P Kumm (Agaricales s. l., Basidiomycetes). Int J Med Mushrooms 3:395–397

Gupta R, Singh A (2024) Textbook of mushroom cultivation. Daya Publishing House, ISBN: 9789354617195

Gupta S, Summuna B, Gupta M, Annepu SK (2019) Edible mushrooms: cultivation, bioactive molecules, and health benefits. In: Mérillon JM, Ramawat KG (eds) Bioactive molecules in food. Springer Nature, Berlin, pp 1815–1847

Hasanah Y, Rahmawati N, Sembiring MB, Nursa'adah (2023) Increasing the independence of oyster mushroom (*Pleurotus ostreatus*) entrepreneurs in seedling preparation. J Saintech Transfer 6:10–16

Hashem A, Algarawi AA, Al-Huqail AA, Abd Allah EF (2013) Biochemical composition of *Pleurotus ostreatus* (jacq.) P Kumm Grown on sawdust of Leucaena leucocephala (lam.) de wit. Pak J Bot 45:1197–1201

Hoa HT, Wang CL, Wang CH (2015) The effects of different substrates on the growth, yield, and nutritional composition of two oyster mushrooms (*Pleurotus ostreatus* and *Pleurotus cystidiosus*). Mycobiology 43:423–434

Hossain S, Hashimoto M, Choudhury EK, Alam N, Hussain S, Hasan M et al (2003) Dietary mushroom (Pleurotus ostreatus) ameliorates atherogenic lipid in hypercholesterolaemic rats. Clin Exp Pharmacol Physiol 30:470–475

Hu SH, Wang JC, Lien JL, Liaw ET, Lee MY (2006) Antihyperglycemic effect of polysaccharide from fermented broth of Pleurotus citrinopileatus. Appl Microbiol Biotechnol 70:107–113

ICAR-DMR (2023) Annual report 2022–23. Indian Council of Agricultural Research Directorate of Mushroom Research

Ijeh II, Okwujiako IA, Nwosu PC, Nnodim HI (2009) Phytochemical composition of *Pleurotus tuber regium* and effect of its dietary incorporation on body/organ weights and serum triacylglycerols in albino mice. J Med Plants Res 3:939–943

IMARC Group (2024) India mushroom market: industry trends, share, size, growth, opportunity and forecast 2024–2033

Jarial RS, Sharma AK, Jarial K, Jandaik S (2020) Evaluation of different grain substrates for the spawn production of *Pleurotus cornucopiae*. Int J Curr Microbiol Appl Sci 7:1689–1700

Jayakumar T, Thomas PA, Sheu JR, Geraldine P (2011) In vitro and in vivo antioxidant effects of the oyster mushroom *Pleurotus ostreatus*. Food Res Int 44:851–861

Jegadeesh R, Lakshmanan H, Kab-Yeul J et al (2018) Cultivation of pink oyster mushroom Pleurotus djamor var. roseus on various agro-residues by low cost technique. J Mycopathol Res 56:213–220

Jose N, Ajith TA, Janandhanan KK (2004) Methanol extract of the oyster mushroom, *Pleurotus florida*, inhibits inflammation and platelet aggregation. Phytother Res 18:43–46

Kamilya D, Joardar SN, Mal BC, Maiti TK (2008) Effects of a glucan from the edible mushroom (*Pleurotus florida*) as an immunostimulant in farmed Indian major carp (Catla catla). Isr J Aquacult Bamidgeh 60:37–45

Kapil S (2023) Evolution and growth of mushroom cultivation in India. Int Educ Res J (IERJ) 9:91–92

Karmani M, Subramaniam G, Sivasamugham L, Cheng W, Wong L (2022) Effects of different substrates on the growth and nutritional composition of *Pleurotus ostreatus:* a review. J Exp Biol Agric Sci 10:481–486

Khan MA, Tania M (2012) Nutritional and medicinal importance of *Pleurotus* mushrooms: an overview. Food Rev Int 28:313–329

Khan MA, Amin S, Uddin MN, Tania M, Alam N (2008) Comparative study of the nutritional composition of oyster mushrooms cultivated in Bangladesh. Bangladesh J Mushroom 2:9–14

Khatun S, Islam A, Cakilcioglu U, Guler P, Chatterjee NC (2015) Nutritional qualities and antioxidant activity of three edible oyster mushrooms (*Pleurotus spp*.). NJAS Wageningen J Life Sci 72–73:1–5

Kortei NK, Wiafe-Kwagyan M (2015) Comparative appraisal of the total phenolic content, flavonoids, free radical scavenging activity and nutritional qualities of *Pleurotus ostreatus* (EM-1) and *Pleurotus eous* (P-31) cultivated on rice (*Oryzae sativa*) straw in Ghana. J Adv Biol Biotechnol 3:153–164

Koutrotsios G, Kalogeropoulos N, Kaliora AC, Zervakis GI (2018) Toward an increased functionality in oyster (*Pleurotus*) mushrooms produced on grape marc or olive mill wastes serving as sources of bioactive compounds. J Agric Food Chem 66:5971–5983

Krishna K, Rupsinh DJ (2024) Profitability and cost efficiency of oyster mushroom production in Bilaspur district of Chhattisgarh. Int J Adv Biochem Res 8:480–484

Kulshreshtha S, Mathur N, Bhatnagar P, Jain BL (2010) Bioremediation of industrial waste through mushroom cultivation. J Environ Biol 33:441–444

Kumari N, Anand A, Kumar S, Panda CK, Kumari M (2022) Economics of oyster mushrooms (*Pleurotus spp.*) production in Bhagalpur District of Bihar. Econ Aff 67:275–281

Lakshmi B, Jose N, Ajith TA, Janandhanan KK (2004) Antimutagenic activity of methanolic extract of culinary-medicinal oyster mushroom, *Pleurotus ostreatus* (Jacq.: Fr.) Kumm. (strain floric Eger nom. Nud.) and its protective effect against benzo(á) pyrene-induced hepatic damages. Int J Med Mushroom 6:139–149

Lavelli V, Proserpio C, Gallotti F et al (2018) Circular reuse of bio-resources: the role of *Pleurotus spp*. in the development of functional foods. Food Funct 9:1353–1372

Lavi I, Nimri L, Levinson D, Peri I, Hadar Y, Schwartz B (2012) Glucans from the edible mushroom Pleurotus pulmonarius inhibit colitis-associated colon carcinogenesis in mice. J Gastroenterol 47:504–518

Lee IS, Ryoo IJ, Kwon KY, Ahn JS, Yoo ID (2011) Pleurone, a novel human neutrophil elastase inhibitor from the fruiting bodies of the mushroom *Pleurotus eryngii* var. ferulae. J Antibiot 64: 587–589

Lesa KN, Khandaker MU, Rashed Iqbal FM, Sharma R, Islam F, Mitra S, Emran TB (2022) Nutritional value, medicinal importance and health promoting effects of dietary mushroom (*Pleurotus ostreatus*). J Food Qual:1–9

Li L, Ng TB, Song M, Yuan F, Liu ZK, Wang CL (2007) A polysaccharide–peptide complex from abalone mushroom (*Pleurotus abalonus*) fruiting bodies increases activities and gene expression of antioxidant enzymes and reduces lipid peroxidation in senescence-accelerated mice. Appl Microbiol Biotechnol 75:863–869

Lv H, Kong Y, Yao Q, Zhang B, Leng FW, Bian HJ et al (2009) Nebrodeolysin, a novel haemolytic protein from mushroom *Pleurotus nebrodensis* with apoptosis inducing and anti HIV1 effects. Phytomedicine 16:198–205

Maiti S, Bhutia SK, Mallick SK, Kumar A, Khadgi N, Maiti TK (2008) Antiproliferative and immunostimulatory protein fraction from edible mushrooms. Environ Toxicol Pharmacol 26: 187–191

Mala A, Thakur M, Kamal S, Attri BL, Jandaik S (2023) Morphological characterization of some *Pleurotus species*. TIJER- Int Res J 10:184–188

Masevhe MR, Taylor NJ, Soundy P (2016) Alternative substrates for cultivating oyster mushrooms (*Pleurotus ostreatus*). S Afr J Plant Soil 33:97–103

Mehmet A, Sevda K (2010) Effect of various agro-residues on nutritive value of *Pleurotus eryngii* (DC. ex Fr.) Quel. var. ferulae Lanzi. J Agric Sci 16:83–88

Mishra RP, Shahid M, Pandey S, Pandy M, Deepshika Singh M (2015) Characterization of *Pleurotus spp.* of mushroom based on phenotypic, biochemical and yield parameters. Afr J Microbiol Res 9:934–937

Mishra V, Tomar S, Yadav P, Singh MP (2021) Promising anticancer activity of polysaccahrides and other macromolecules derived from Oyster mushroom *Pleurotus* sp; an updated review. Indian J Macromol 182:1628–1637

Mshandete AM, Cuff J (2008) Cultivation of three types of indigenous wild edible mushrooms: Coprinus cinereus, *Pleurotus flabellatus* and *Volvariella volvocea* on composted sisal decortications residue in Tanzania. Afr J Biotechnol 7:4551–4562

Mushroom Global Market Report (2025) Published in 2025, p 200

Musieba F, Okoth S, Mibey RK et al (2013) Proximate composition, amino acids and vitamins profile of Pleurotus citrinopileatus Singer: an indigenous mushroom in Kenya. Am J Food Technol 8:200–206

Muzaffar M, Rasool F, Sofia TA, Shikari AB, Mir SA, Lone GM (2023) Morphological characterization of different *Pleurotus species* under Kashmir conditions. Pharm Innov J 12:3767–3770

NABARD (2021) Opportunities in mushroom cultivation for entrepreneurship development. National Bank for Agriculture and Rural Development

Naraian R, Sahu RK, Kumar S et al (2009) Influence of different nitrogen rich supplements during cultivation of *Pleurotus florida* on corn cob substrate. Environmentalist 29:1–7

Ngai PHK, Ng TB (2004) A ribonuclease with antimicrobial, antimitogenic and antiproliferative activities from the edible mushroom *Pleurotus sajor-caju*. Peptides 25:11–17

NHB (2023) Mushroom cultivation DPR guidelines. National Horticulture Board

Nomura H, Inokuchi N, Kobayashi H, Koyama T, Iwama M, Ohgi K, Irie M (1994) Purification and primary structure of a guanylic acid–specific ribonuclease from *Pleurotus ostreatus*. J Biochem 116:26–33

Odisha Livelihoods Mission (2022) SHG enterprise models: Mushroom value chain. Department of Mission Shakti Government of Odisha

Ogundele GF, Salawu SW, Abdulraheem IA, Bamidele OP (2017) Nutritional composition of oyster mushroom (*Pleurotus ostreatus*) grown on softwood (*Daniella oliveri*) sawdust and hardwood (*Anogeissus leiocarpus*) sawdust. Curr J Appl Sci Techol 20:1–7

Okwulehie IC, Nosike EN (2015) Phytochemicals and vitamin compositions of *Pleurotus pulmonarius* cultivated on barks of some indigenous fruit trees supplemented with agro-wastes. Asian J Plant Sci 5:1–7

Otunla CA, Idowu OO (2012) Comparative study of the growth and yield of Pleurotus florida (oyster mushroom) on some tropical trees. Food 6:86–89

Oyetayo VO, Ariyo OO (2013) Antimicrobial and antioxidant properties of Pleurotus ostreatus (Jacq: Fries) cultivated on different tropical woody substrates. J Waste Conver Bioprod Biotechnol 1:28–32

Pal J, Sharma R, Lal M, C Suman B (2017) Effect of different spawn supplementation on yield of Indian oyster mushroom, *Pleurotus pulmonarius* (Fr.) Quel. J Appl Nat Sci 9:1406–1410

Paswal S, Mughal N, Bharti V, Mahajan S, Majeed M (2017) Evaluation of locally available substrates for sporocarps production of Pink *Pleurotus* [*Pleurotus djamor* (Rumph.ex.Fr.) boedijn] mushroom. Int J Curr Microbiol Appl Sci 6:1677–1684

Patel Y, Naraian R, Singh VK (2012) Medicinal properties of *Pleurotus sp*: the review. World J Fungal Plant Biol 3:1–12

Patil SS, Ahmed SA, Telang SM et al (2010) The nutritional value of *Pleurotus ostreatus*(jacq.:fr.) Kumm cultivated on different lignocellulosic agrowastes. Innov Rom Food Biotechnol 7:66–76

Paul C, Roy T, Das N (2017) Potentiality of oyster mushroom (*Pleurotus spp.*) in medicine: a review. Ann Food Process Preserv 2:1014

Phan CW, Wong WL, David P, Naidu M, Sabaratnam V (2012) *Pleurotus giganteus* (Berk.) Karunarathna & K.D. Hyde: nutritional value and in vitro neurite outgrowth activity in rat pheochromocytoma cells. BMC Complement Altern Med 12:102

Philippoussis AN (2009) Chapter 9: Production of mushrooms using agro-industrial residues as substrates. In: Singh nee' Nigam P, Pandey A (eds) Biotechnology for agro-industrial residues utilisation. Springer, Dordrecht, pp 163–196

Raman J, Jang KY, Oh YL, Oh M, Im JH, Lakshmanan H, Sabaratnam V (2020) Cultivation and nutritional value of prominent *Pleurotus spp.*: an overview. Mycobiology 49:1–14

Ramkumar V, Ramanathan T, Thirunavukkarasu V, Arivuselvan N (2010) Antioxidant and radical scavenging activity of nine edible mushroom extracts. Int Pharmacol 6:950–953

Rana T, Bhattacharyya A, Dey S, Biswas D, Pal D, Bandyopadhyay S (2012) *Pleurotus florida* lectin normalizes duration dependent hepatic oxidative stress responses caused by arsenic in rat. Exp Toxicol Pathol 64:665–671

Rivera NA, Merales JD (2010) Edible mushroom *Pleurotus ostreatus* production on cellulosic biomass of sugarcane. Sugar Technol 12:176–178

Sahoo S, Gayakwad T, Shahi S (2022) Medicinal value of edible mushrooms. Int J Health Sci 6: 8760–8767

Salama ANA, Abdou AAK, Helaly AA et al (2016) Cultivation of *Pleurotus ostreatus* L. on different substrates based on JuNCAO technology in Egypt. Nat Sci 14:59–66

Sanchez C (2010) Cultivation of *Pleurotus ostreatus* and other edible mushrooms. Appl Microbiol Biotechnol 85:1321–1337

Seliman TMH, Ulbegi GA, Ozsoy N, Kavlo HA, Sagrili PA (2024) Anti inflammatory activity of a novel lectin isolated from Pleurotus eryngiivar. Ferulae mushroom. Istanbul J Pharm 54(2): 195–204. https://doi.org/10.26650/IstanbulJPharm.2024.1322171

Sharifi-Rad J, Butnariu M, Ezzat et al (2020) Mushrooms rich preparations on wound healing: from nutritional to medicinal attributes. Front Pharmacol 11:567518

Shevale SB, Deshmukh HV (2016) Yield performance and nutritional analysis of *Pleurotus species* on different agro-wastes and vegetable wastes. Int J Plant Protect 9:162–167

Silva SO, da Costa SMG, Clemente E (2002) Chemical composition of *Pleurotus pulmonarius* (Fr.) Quel., substrates and residue after cultivation. Braz Arch Biol Technol 45:531–535

Smiderle FR, Olsen LM, Carbonero ER, Marcon R, Baggio CH, Freitas CS, Santos ARS, Torri G, Gorin PAJ, Iacomini M (2008) A 3-O-methylated mannogalactan from *Pleurotus pulmonarius*: structure and antinociceptive effect. Phytochemistry 69:2731–2736

Stanley HO (2010) Effect of cultural substrates of spawn production on mycelial growth of oyster mushroom species. Agric Biol J N Am 1:817–820

Stastny J, Marsik P, Tauchen J, Bozik M, Mascellani A, Havlik J, Landa P, Jablonsky I, Treml J, Herczogova P, Bleha R, Synytsya A, Kloucek P (2022) Antioxidant and anti-inflammatory activity of five medicinal mushrooms of the genus *Pleurotus*. Antioxidants 11:1569

Sumbodo BT, Mujamil T, Ducati D, Ika SR (2023) Business feasibility and development strategy of oyster mushroom farming in Pandowoharjo Village, Yogyakarta. In: Setiawan B (ed) Proceedings of the international conference on management and business (ICoMB). Atlantis Press, pp 21–34

Telang SM, Patil SS, Baig MMV (2010) Comparative study on yield and nutritional aspect of *Pleurotus eous* mushroom cultivated on different substrates. Food Sci Res J 1:60–63

Thakur MP, Sharma VP (2017) Oyster mushroom cultivation: a profitable enterprise. ICAR-Directorate of Mushroom Res, Solan

Tolera KD, Abera S (2017) Nutritional quality of Oyster Mushroom (*Pleurotus ostreatus*) as affected by osmotic pretreatments and drying methods. Food Sci Nutr 5:989–996

Upadhyay RC, Verma RN, Singh SK et al (2002) Effect of organic nitrogen supplementation in *Pleurotus species*. Mushroom Biol Mushroom Prod 105:225–232

Varghese R, Dalvi YB, Lamrood PY, Shinde BP, Nair CKK (2019) Historical and current perspectives on therapeutic potential of higher basidiomycetes: an overview. 3 Biotech 9:362

Wang HX, Ng TB (2007) A lectin with HIV-1 reverse transcriptase inhibitory activity from *Pleurotus pulmonarius*. Life Sci 80:2098–2102

Wang HX, Gao J, Ng TB (2000) A new lectin with highly potent anti-hepatoma and anti-sarcoma activities from the oyster mushroom Pleurotus ostreatus. Biochem Biophys Res Commun 275: 810–816

Wasonga CGO, Okoth SA, Mukuria JC, Omwandho COA (2008) Mushroom polysaccharide extracts delay progression of carcinogenesis in mice. J Exp Ther Oncol 7:113–122

Wisbeck E, Facchini JM, Alves EP, Silveira MLL, Gern RMM, Ninow JL, Furlan SA (2014) Antitumor activity of *Pleurotus ostreatus* polysaccharide fractions on Ehrlich tumor and Sarcoma 180. Food Chem 145:1111–1118

Wong JH, Ng TB, Jiang Y, Liu F, Sze SCW, Zhang KY (2010) Purification and characterization of a laccase with inhibitory activity toward HIV-1 reverse transcriptase and tumor cells from an edible mushroom (*Pleurotus cornucopiae*). Protein Pept Lett 17:1040–1047

Yang JH, Lin HC, Mau JL (2001) Non-volatile taste components of several commercial mushrooms. Food Chem 72:465–471

Yang JH, Lin HC, Mau JL (2002) Antioxidant properties of several commercial mushrooms. Food Chem 77:229–235

Yoon KN, Alam N, Shim MJ, Lee TS (2012) Hypolipidemic and antiatherogenic effect of culinary-medicinal pink oyster mushroom, *Pleurotus salmoneostramineus* L. Vass. (Higher Basidiomycetes) in hypercholesterolemic rats. Int J Med Mushrooms 14:27–36

Zervakis GI, Kotrotsios G, Katsaris P (2013) Composted versus raw olive mill wastes as substrates for the production of medicinal mushrooms: an assessment of selected cultivation and quality parameters. Biomed Res Int 2013:546830

Zhang M, Cheung PCK, Zhang L (2001) Evaluation of mushroom dietary fiber (nonstarch polysaccharides) from sclerotia of Pleurotus tuber-regium (Fries) Singer as a potential antitumor agent. J Agric Food Chem 49:5059–5062

Zhang M, Zhang L, Cheung PCK (2003) Chain conformation of sulfated derivatives of β-glucan from sclerotia of *Pleurotus tuber-regium*. Carbohydr Res 338:2863–2870

Zhang M, Zhang L, Cheung PCK, Ooi VEC (2004) Evaluation of sulfated fungal β-glucans from the sclerotium of *Pleurotus tuber-regium* as a potential water-soluble anti-viral agent. Carbohydr Res 339:2297–2301

# Advancing Mushroom Cultivation Through IoT-Enabled Smart Systems

6

Rekha Mehrotra, Samiksha Sharma, and Payal Mago

**Abstract**

The global mushroom industry is a high-value agribusiness experiencing rapid expansion and recognition worldwide. This growth is primarily driven by increasing consumer demands for sustainable alternatives, as well as the minimal resource and land requirements for production. Mushroom farming offers several advantages over conventional farming; however, manual dependency and inconsistent regulation restrict the yield and profitability. Addressing these challenges requires the integration of advanced technologies that can ensure precision, consistency, and scalability. This chapter examines the transformative role of Internet-of-things (IoT) technologies in advancing mushroom production on local as well as commercial scales. It discusses how real-time monitoring and automated control algorithms can regulate growth parameters—temperature, humidity, carbon dioxide concentrations, substrate moisture, and ambient light—for proper mycelial development, primordia initiation, and fruit body maturation. It also emphasizes how tools like low-cost sensor networks, wireless communication systems, and artificial intelligence (AI) platforms could collectively support data-driven decision-making in mushroom cultivation. Case studies and recent innovations demonstrate how IoT-enabled smart farms enhance productivity, improve resource efficiency, reduce contamination risks, and provide resilience against climate variability. IoT-enabled systems can help establish

R. Mehrotra (✉)
Department of Microbiology, Shaheed Rajguru College of Applied Sciences for Women, University of Delhi, Delhi, India

S. Sharma
Department of Microbiology, University of Delhi, South Campus, New Delhi, India

P. Mago
Shaheed Rajguru College of Applied Sciences for Women, University of Delhi, Delhi, India



B. Giri et al. (eds.), *Advances in Mushroom Research*,
https://doi.org/10.1007/978-981-95-7692-0_6

a new paradigm of precision farming by linking the biological aspect of mushroom cultivation with digital agriculture technologies, bridging the gap, and improving the efficiency. This chapter provides a comprehensive overview of the structure, components, and mechanisms of smart farming linked with applications, challenges, and future prospects, offering valuable insights for researchers, entrepreneurs, and policymakers seeking sustainable and technologically advanced solutions in mushroom agriculture.

**Keywords**

IoT · Data-driven cultivation · Mushroom farming · Precision farming · Sustainable agriculture

## 6.1 Introduction

Mushroom cultivation is an emerging and rapidly expanding agribusiness sector owing to its highly profitable nature with minimum resource requirement and great significance. The business offers a sustainable market by promoting eco-friendly cultivation practices, promoting natural substrates and resources, and agricultural waste valorization while meeting consumer needs with organic and nutritious food products. In addition to being a popular food, mushrooms are becoming increasingly coveted for their medicinal and nutraceutical uses as a sustainable source of protein and bioactive compounds. But conventional mushroom farming techniques, which mainly depend on manual labor, are frequently characterized by uneven environmental control, susceptibility to contamination, and less than ideal yield. These restrictions present serious obstacles to productivity, particularly considering the shifting climate and increasing demand for food security (Guragain et al. 2024).

The foundation and advancements of Internet-of-things (IoT) have revolutionized many business sectors, with precision farming and protected cultivation being the two most prominent agricultural fields. A promising way to turn traditional mushroom farming into a data-driven, effective, and climate-resilient system is through the integration of smart sensors, automated control systems, and real-time data analytics. Mushrooms offer a wide diversity and have characteristic requirements, and therefore, the successful cultivation of each species is highly dependent on several critical growth parameters, including temperature, humidity, carbon dioxide levels, substrate moisture, and ambient light, which can be monitored and controlled by smart mushroom farms made possible by the introduction of Internet-of-things (Chong et al. 2023).

This chapter examines how smart systems powered by IoT are influencing mushroom farming in the present and the future. It explores the structure and elements of intelligent farming systems, talks about important automation and data analytics tools, and showcases recent examples and case studies that highlight how the digital revolution has affected the mushroom industry. IoT integration in mushroom farming has the potential to improve yields, lower operating costs, and ensure consistent quality by bridging the gap between biology and technology. This would provide a significant boost and a scalable model for both commercial and small-scale growers, thereby escalating ecological and economic growth.

## 6.2 Overview of Mushroom Cultivation

Mushrooms are the fruiting bodies of certain fungi, belonging primarily to the phyla Basidiomycota and Ascomycota. The life cycle begins with the germination of spores, which develop into thread-like structures called the hyphae, which then interconnect to form a dense network called mycelium that colonizes the substrate. Under favorable environmental conditions, the mycelium differentiates and gives rise to primordia, which mature into full-grown mushrooms that can produce and release spores, thus completing the cycle (Sakamoto 2018).

Mushroom cultivation involves several discrete and intricate successive stages which are highly dependent on various external environmental factors for attaining a yield at its best. The first step is the process of spawning, which involves introducing the mycelium of mushrooms into a pasteurized or sterilized substrate like compost, sawdust, or straw. Then follows the incubation or spawn run phase, where the mycelium colonizes the substrate in a warm, humid, and dark environment. The fruiting stage, during which visible mushroom primordia (pins) grow and mature into full-grown fruiting bodies, is initiated by adjusting factors like temperature, humidity, carbon dioxide levels, and light after full colonization is accomplished. To ensure freshness and market quality, mushrooms are carefully selected at their ripest during the harvesting stage (Higgins et al. 2017). The environmental parameters required for various commonly cultivated mushroom species are summarized in the following table (Table 6.1).

Despite being widely used, traditional mushroom cultivation frequently encounters a few difficulties that restrict consistency and productivity. Every one of the cultivation phases needs careful attention and monitoring as fluctuations in environmental control could lead to reduced yield, contamination, or malformed fruiting bodies. Variability brought on by manual monitoring and adjustment often can have a detrimental effect on the produce and quality. Another significant worry is contamination by rival bacteria or fungi, which can swiftly take over the substrate and destroy entire batches, particularly during spawning and incubation. Furthermore, uneven substrate composition, bad strain selection, or insufficient control over growing conditions cause yield variability. The absence of automation and data-driven decision-making can exacerbate these constraints, rendering conventional approaches time-consuming, ineffective, and inconsistent.

**Table 6.1** Optimal environmental parameters for commonly cultivated mushroom species

| Common name | Scientific name | Optimal temperature | Relative humidity (%) | $CO_2$ level | Light requirement | Substrate type |
|---|---|---|---|---|---|---|
| Button mushroom | *Agaricus bisporus* | 23–25 °C (spawn), 16–18 °C (fruiting) | 85–95 | High (spawn), low (fruiting) | Minimal | Compost (manure + straw-based) |
| Cremini mushroom | *Agaricus bisporus* (brown strain) | 22–24 °C | 85–90 | Low | Minimal | Compost |
| Oyster mushroom | *Pleurotus* spp. | 20–30 °C | 85–95 | Moderate to low | 100–200 lux (diffused light) | Straw, sawdust, agro-waste |
| King oyster mushroom | *Pleurotus eryngii* | 20–24 °C (spawn), 12–18 °C (fruiting) | 85–95 | Low (≤800 ppm) | Low to moderate (100–300 lux) | Sawdust enriched with bran |
| Paddy straw mushroom | *Volvariella volvacea* | 28–35 °C | 80–90 | Moderate | Not essential (shade-loving) | Paddy straw, banana leaves, cotton waste |
| Shiitake mushroom | *Lentinula edodes* | 22–26 °C (spawn), 12–18 °C (fruiting) | 80–90 | Low | 500–1000 lux (indirect light) | Sawdust, hardwood logs |
| Milky mushroom | *Calocybe indica* | 25–35 °C | 80–90 | Low | Indirect light preferred | Straw, sugarcane bagasse |
| Reishi mushroom | *Ganoderma lucidum* | 25–30 °C | 85–90 | Low | Indirect light essential | Sawdust, hardwood chips |
| Enoki mushroom | *Flammulina velutipes* | 13–15 °C | 85–90 | Low | Very low (dark conditions) | Sawdust, corn cobs |
| Cordyceps | *Cordyceps militaris* | 20–22 °C | 70–80 | Low | Moderate light (12 h light/dark) | Rice, silkworm pupae, or synthetic media |

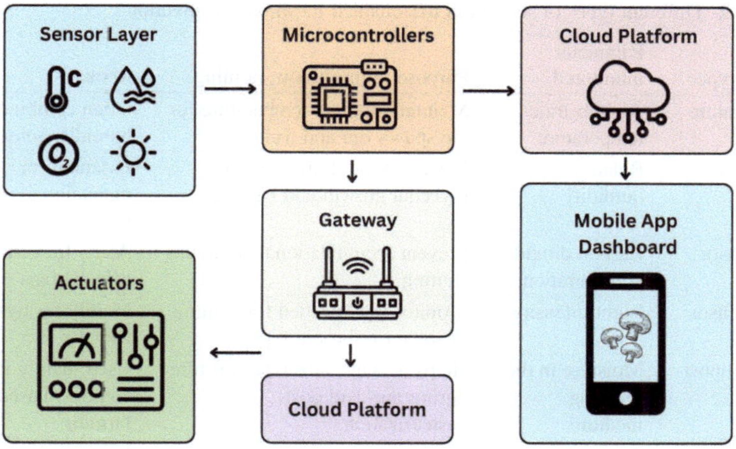


**Fig. 6.1** Architecture of an IoT-enabled smart mushroom cultivation system integrating sensors, microcontrollers, cloud platforms, and mobile dashboards for real-time monitoring and control. (Figure made using Canva)

## 6.3 IoT in Mushroom Farming: Concepts and Components

### 6.3.1 What Is IoT?

The Internet-of-things (IoT) refers to a network of physically connected devices mainly comprising sensors, software and connectivity built to enable real-time data collection, transmission, and action (Fig. 6.1). IoT technologies are revolutionizing traditional agricultural practices by enabling automated decision-making, resource optimization, and precision farming (Miller et al. 2025). Continuous monitoring and control of environmental parameters is made easier by IoT-driven systems, which ensure consistency, boost yield, reduce waste, and lessen dependency on labor. For mushroom cultivation, which necessitates exact control of microclimatic factors like humidity, temperature, $CO_2$ concentration, and light, IoT offers a very effective and scalable solution (Guragain et al. 2024).

### 6.3.2 Core Components of IoT Systems in Agriculture

#### 6.3.2.1 Sensors

These are the foundational element of any IoT system. In mushroom farming, commonly used sensors are mentioned in Table 6.2.

#### 6.3.2.2 Actuators

These receive signals from the system and take physical action to adjust conditions. Key examples include the following:

**Table 6.2** Different types of sensors in IoT-enabled mushroom cultivation

| Sensor type | Parameter monitored | Purpose in mushroom farming | Remarks |
|---|---|---|---|
| Temperature sensor | Air/substrate temperature | Maintain optimal temperature for the spawn run and fruiting | Often combined with humidity sensors |
| Humidity sensor | Relative humidity | Ensure proper humidity for mycelial growth and fruiting | Essential for maintaining >85% RH |
| $CO_2$ sensor | Carbon dioxide concentration | Prevent accumulation that inhibits fruiting | Alerts for activating exhaust fans |
| Light sensor (LDR) | Light intensity | Monitor light needed for fruiting in some mushroom types | Mainly for oyster, shiitake, etc. |
| Soil/compost moisture sensor | Moisture in the growing medium | Helps manage substrate moisture during pre- and post-pasteurization | Used mainly in compost-based farming |
| Air pressure sensor | Atmospheric pressure | Monitor microclimate pressure variations in sealed grow rooms | Useful in advanced farm setups |
| pH sensor | Substrate or casing layer pH | Maintain optimal pH for mycelial growth and contamination control | Used during composting or casing preparation |
| Gas sensors (optional) | Ammonia, Methane, etc. | Detect harmful gases from poor composting or contamination | For safety and contamination monitoring |

- *Heating and cooling systems*: Maintain ideal temperature ranges.
- *Humidifiers and dehumidifiers*: Regulate moisture content.
- *Fans and ventilation systems*: Control air circulation and $CO_2$ levels.

#### 6.3.2.3 Microcontrollers

These act as the brain of the IoT system, processing data from sensors and triggering actuators based on programmed logic or remote instructions.

#### 6.3.2.4 Communication Technologies

Stable and efficient communication is vital for real-time control and data transfer. Depending on the scale and location, common communication protocols include the following:

- *Wi-Fi*: Suitable for indoor setups with reliable internet.
- *LoRa (Long Range)*: Ideal for large farms requiring low power, long-distance data transmission.
- *ZigBee*: Enables mesh networking among devices.
- *Bluetooth*: Useful for short-range connectivity and mobile integration.

**Table 6.3** Key IoT components used in smart mushroom farming, their functions, and benefits

| IoT component | Function | Examples | Benefits |
|---|---|---|---|
| Sensors | Monitor environmental parameters | Humidity, temperature, $CO_2$, pH sensors | Precision monitoring and control |
| Actuators | Respond to sensor inputs | Misters, heaters, ventilators | Automated environmental adjustment |
| Microcontrollers | Process input and control output | Arduino, Raspberry Pi | Low-cost automation |
| Connectivity modules | Enable wireless communication | Wi-Fi, LoRa, Zigbee | Remote monitoring |
| Cloud platforms | Store and analyze data | AWS IoT, Azure IoT Hub, Google Cloud | Centralized data access and analytics |

#### 6.3.2.5 Cloud Platforms and Mobile Applications

The data collected from sensors is stored and processed in cloud platforms, enabling remote monitoring, historical analysis, and visualization through mobile apps or dashboards (Table 6.3). These interfaces allow users to:

- Monitor live environmental data
- Receive alerts or notifications
- Set thresholds and automation rules
- Access historical trends for informed decisions

### 6.3.3 A State-of-The-Art Technology

The integration of Internet-of-things (IoT) technologies into mushroom farming represents a state-of-the-art approach that is transforming traditional farming into a data-driven, automated, and intelligent system (Rukhiran et al. 2023). The use of a broad range of environmental sensors for real-time monitoring of vital parameters is one example of recent developments. Most sensors are usually connected to cloud-based platforms and microcontrollers, which allow for automated feedback control and accessing data remotely, coupled with continuous data collection (Fig. 6.1) (Irwanto et al. 2024). In even more advanced setups, machine learning (ML) algorithms are often employed to maximize yield and quality by predicting growth stages, identifying irregularities, and in turn, fine-tuning conditions to optimum (Moysiadis et al. 2023). Furthermore, wireless communication technologies like Wi-Fi/LoRa help ease connectivity between different units for large-scale cultivation. Many modern research prototypes and commercial systems now feature artificial intelligence (AI)-powered analytics dashboards, mobile app controls, and smart growth chambers (Chong et al. 2023). These technologies aid in reducing energy use, managing resources precisely, and waste valorization, thereby promoting sustainable agriculture. Altogether, this marks a big step forward in creating intelligent, scalable, and climate-resilient mushroom production systems (Irwanto et al. 2024).

### 6.3.4 Role of Data Analytics, AI, and Machine Learning

IoT-based advancements in agriculture have moved far beyond basic sensor automations to data integration, data analytics, artificial intelligence, and machine learning for sophisticated decision-making and maintaining optimum parameters (Miller et al. 2025). In mushroom farming, these technologies use both past records and real-time data to identify trends and improve growth conditions by understanding the crucial dependence of fruiting performance on various parameters. Predictive models help farmers forecast yields and detect early signs of stress/contamination, allowing them to act before problems accelerate. ML takes this a step further by learning from each consecutive cultivation cycle and thereby adjusting and adapting settings over time for enhanced growth patterns. Together, these intelligent systems reduce human error probability, boost productivity, and ensure consistent quality—making mushroom farming more efficient, scalable, resilient to environmental changes, and market-ready (Rukhiran et al. 2023).

### 6.3.5 IoT-Enabled Versus Traditional Mushroom Farming

IoT-enabled mushroom farming offers real-time monitoring and automated control of critical parameters like temperature, humidity, and $CO_2$, ensuring optimal growing conditions and consistent yields (Guragain et al. 2024). In contrast, traditional farming relies on manual monitoring and environmental adjustments, which are often less precise and labor-intensive. IoT systems also enable predictive analytics and remote access, improving efficiency and reducing crop loss. This technological shift enhances productivity and sustainability in mushroom cultivation. Comparison of key features and outcomes between IoT-enabled and traditional mushroom farming systems has been tabulated below (Table 6.4) (Gill and Chawla 2021; Geetha Rani et al. 2023).

## 6.4 Smart Mushroom Cultivation System Design

A Smart Mushroom Cultivation System is a systematic combination of precision farming, automation techniques, and sustainable methodologies to produce a highly effective and flexible farming environment. The critical requirement of optimum growth parameters like temperature, relative humidity, $CO_2$ levels, light intensity, and substrate moisture necessitates the use of sensors throughout the growing area, which provides real-time monitoring and information analysis to edge computing units, thereby enabling prompt responses by making intelligent decisions like turning on misting systems or modifying airflow (Nguyen et al. 2024). Predictive

**Table 6.4** Key differences between traditional and IoT-enabled mushroom farming

| Parameter | Traditional mushroom farming | IoT-enabled mushroom farming | Improvement |
|---|---|---|---|
| Environmental monitoring | Manual checks at intervals using thermometers, hygrometers, or no monitoring at all | Real-time, continuous monitoring using temperature, humidity, $CO_2$, and light sensors | Timely alerts prevent deviations, ensuring ideal conditions |
| Environmental control | Operated manually (fans, misters, heaters switched on/off by humans) | Automated systems are activated based on sensor feedback | Minimizes human error; stabilizes growing environment |
| Data logging & record keeping | Paper-based logs or verbal tracking | Cloud-based or local digital logging with visualization dashboards | Enables data-driven decision-making |
| Labor requirement | High labor dependence for monitoring and adjustments | Reduced labor for routine tasks; centralized control | 30%–50% reduction in labor costs |
| Yield consistency | Variable yields due to environmental fluctuations | More consistent yields due to better control | 10%–25% improvement in yield stability |
| Energy use | Often inefficient; systems run longer than needed | Smart automation optimizes usage (e.g., foggers only run when needed) | Up to 30% energy savings |
| Pest & disease detection | Late detection based on visual symptoms | Under research: early detection using cameras, gas sensors, or thermal imaging | Potential to revolutionize disease prevention |
| Scalability | Difficult; more area requires proportionally more labor | Easily scalable; add sensors and connect to existing network | Suitable for both small and commercial-scale growers |
| Cost of setup | Low initial investment; high operational cost | Higher upfront cost (sensors, controllers) but lower long-term costs | Break-even is typically in 6–12 months, depending on scale |
| Remote access | Not possible | Possible via mobile app/web dashboard | Enables 24/7 farm supervision |
| Integration with other systems | Standalone farming | Can integrate with weather APIs, power backups, and solar systems | Enhances automation and resilience |
| Sustainability | Often unsustainable due to overuse of inputs | Promotes sustainability through precision farming | Supports eco-certifications and organic branding |

maintenance, yield forecasting, and remote access through computers/smartphones through cloud computing platforms gather and store this data for future analysis; hence, this multi-layered system ensures continuous optimization with reduced human dependence (Karunathilake et al. 2023).

The physical layout of the system is centered on smart grow chambers and vertical farming modules which enable year-around cultivation even in constrained areas. These chambers are equipped with feedback-controlled heating, irrigation, and lighting systems that rely on automated adjustments as per crop-specific needs and real-time sensor data. The system can also be clubbed with more renewable energy sources like solar panels to power sensors and control units—these further lower dependency on traditional power grids and promote sustainability. Smart-learning and data-driven algorithms can be employed to determine the best harvesting time, spot early problems, and direct resource-saving techniques such as water/substrate supply. All things considered, a smart mushroom cultivation system supports scalable, environment-friendly farming that is appropriate for both commercial and small-scale producers (Chong et al. 2023; Singh et al. 2020).

## 6.5 Automation and Control Systems

### 6.5.1 Automated Irrigation and Misting Systems

Substrate and air moisture are one of the most critical parameters, and this delicate balance often relies on automated irrigation and misting systems. Real-time moisture and humidity sensors detect even minute changes in the growing environment, help maintain constant mycelial activity, and avoid desiccation by regulating precise amounts of water/mist application at ideal intervals as per the sensor input. Additionally, overwatering is avoided since it can encourage bacterial growth and contamination. Different mushroom species and different developmental stages require programming the system to increase misting during pinning and decrease it during harvesting. (Irwanto et al. 2024).

### 6.5.2 Intelligent Lighting Systems

While many mushroom species grow best in low light/dark conditions, some species like shiitake, enoki, and oyster mushrooms prefer a more controlled lighting system. Such systems often use programmable LEDs to imitate natural sunlight or any other specific light type (blue/red wavelengths) to trigger certain fungal physiological responses. These systems can be tailored to each species' photoreceptor responses and their stage-specific needs, such as pinning/cap expansion/harvesting. This reduces the need for constant manual adjustments, hence ensuring even exposure of light across all horizontal and vertical layers, particularly helpful in vertical farming setups (Irwanto et al. 2024).

### 6.5.3 Automated Climate Control (HVAC Integration)

Maintaining optimum environment parameters like temperature, humidity, $CO_2$ levels, and airflow is a necessity for good mushroom growth, as even small changes could deteriorate both quality and quantity significantly. HVAC systems include environmental sensors that provide necessary and efficient control over these parameters. These systems automatically manage fresh air intake, circulation, and exhaust. These ensure continuous removal of stale air and introduction of fresh oxygen at an appropriate and controlled rate for each growth stage. Humidifiers and dehumidifiers activate based on specific humidity levels. Temperature control units keep the environment within the ideal ranges for each mushroom species. This careful management helps prevent stress, lowers the risk of contamination and supports healthy fruiting body development (Santos et al. 2023).

### 6.5.4 AI-Based Decision Support Systems

AI adds a strong predictive and flexible layer to automation. AI decision support systems use historical data from past cycles and real-time sensor input to make optimum modifications about irrigation, lighting, temperature, and harvesting. For instance, if past patterns show that a specific strain tends to dry out in certain weather conditions, the AI can adjust misting schedules ahead of time. These systems can also suggest the best time for harvesting based on expected yield curves and spot setbacks, and offer ways to improve future cycles. As they learn continuously, these tools become more accurate and better suited to the farm's specific needs (Dhaked et al. 2025).

## 6.6 Data Management and Analytics

Data management and analytics play a key role in improving operations in smart mushroom farms. Real-time data collection from sensors that monitor temperature, humidity, $CO_2$ levels, substrate moisture, and light is centralized through integrated platforms. These data streams are visualized on user-friendly dashboards that can be accessed from computers or mobile devices. This allows growers to monitor conditions and get alerts right away when the parameters go outside of optimal ranges. This real-time visualization supports quick decision-making and improves control over the growing environment. In addition to basic monitoring, data is stored over time to create a rich historical dataset. This dataset serves as the basis for more detailed analysis, long-term planning, and system improvement (Fig. 6.2).

Beyond monitoring, analytics tools, especially predictive analytics and machine learning, provide smart decision-making support. ML algorithms learn to spot small changes, like gradual temperature shifts or unusual $CO_2$ levels, that might indicate fungal infections or substrate degradation, sending early alerts to help prevent crop loss. Some farms also use digital twins, which are virtual replicas of the physical

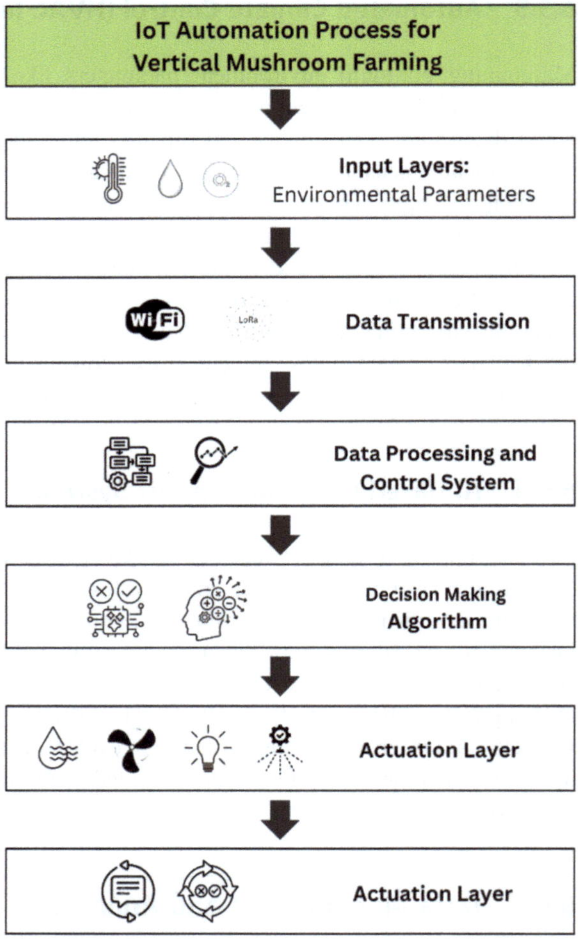


**Fig. 6.2** Process diagram of a vertical mushroom farming setup with IoT automation. (Figure made using Canva)

farm environment that simulate mushroom growth under different conditions. These digital models let growers experiment with various cultivation strategies without putting real crops at risk. Together, these tools turn data into useful insights, making mushroom farming more precise, flexible, and efficient (Kassim et al. 2019).

## 6.7 Case Studies and Applications

The integration of IoT in mushroom farming has moved beyond theoretical frameworks and pilot experiments to real-world deployment across diverse geographical and economic contexts. This section explores the practical implementations of smart-farming technologies in both global and Indian settings, highlighting how IoT-based systems—ranging from automated climate control to machine learning-based crop monitoring—have transformed cultivation practices.

**Table 6.5** International case studies on IoT-enabled mushroom farming

| Location | IoT system features | Key results | Reference |
|---|---|---|---|
| Universitas Gadjah Mada, Indonesia | IoT monitoring of temperature and humidity using DHT11 and Node MCU; web dashboard via Thinger.io | Enabled real-time monitoring; improved climate stability | Hendinata and Fikri (2023) |
| Universiti Malaysia Perlis, Malaysia | IoT + ML system with ESP8266 sensors; machine-learning classifiers predict environmental states | Fully automated control; high prediction accuracy | Omar et al. (2024) |
| E3S Web Conf. (Indonesia, 2025—Smart Kumbung) | Frugal IoT system with temperature/ humidity sensors, misting, photoelectric control, smartphone interface | Affordable automation for small and medium enterprises (SMEs); consistent yields over 2 months | Kawuryan et al. (2025) |
| Thailand (2022) King Mongkut's Institute | Low-cost smart wireless PLC-based IoT system for oyster mushroom production (temperature/ humidity sensors, automatic misting fans) | Higher crop yield compared to traditional farming methods | Thong-un and Wongsaroj (2022) |
| Bangladesh (2022)—Khwaja Yunus Ali University | ESP32 + sensors (temperature, humidity), farm automation, remote monitoring, ML ensemble model | 100% classification accuracy of edible vs. toxic mushrooms; SUS score 82% trust usability | Rahman et al. (2022) |
| Nepal—Centralized IoT Ecosystem for Oyster Mushrooms (2024) | Low-cost central IoT platform using DHT11, ESP32; automated misting & temperature regulation | Improved crop cycle efficiency and cost savings | Guragain et al. (2024) |

Through selected case studies, we illustrate the measurable benefits such as increased yield, reduced labor dependency, enhanced environmental control, and cost-effectiveness. These applications not only demonstrate the versatility of IoT solutions but also serve as models for scalable, data-driven agricultural innovations.

Table 6.5 summarizes notable international case studies on IoT-enabled mushroom farming across Indonesia, Malaysia, Thailand, Bangladesh, and Nepal. Each entry highlights the technological components used, such as sensors, PLCs, and cloud-based systems, along with reported outcomes like yield improvement and cost savings. The studies collectively illustrate how diverse innovations are enhancing mushroom cultivation efficiency worldwide.

Table 6.6 presents selected Indian case studies on IoT-integrated mushroom farming, focusing on low-cost automation, wireless monitoring, and climate control solutions. These initiatives have demonstrated measurable gains in productivity,

**Table 6.6** Indian case studies on IoT-enabled mushroom farming

| Location | IoT system features | Key results | Reference |
|---|---|---|---|
| Shri Vishwakarma Skill University, Haryana (2023) | Real-time automation using Arduino/ESP-control of ventilation, irrigation, and lighting | Improved control and monitoring; reduced labor and risk | Akhil et al. (2023) |
| R. L. Jalappa Institute of Technology, Karnataka (IJISAE) | ATmega328-based IoT monitoring of the gray oyster mushroom environment; automatic alerts to the farmer | Continuous control via IoT; better hygiene and environment stabilization | Kumar et al. (2023) |

reduced resource usage, and enhanced farm sustainability. The examples underscore India's progress in digitizing small-scale mushroom agriculture through scalable technologies.

Thus, the integration of IoT in mushroom cultivation has been successfully demonstrated through several real-world case studies across different geographies. These applications highlight improvements in productivity, resource efficiency, and disease control through sensor-based monitoring and automation. Such examples serve as proof-of-concept for the transformative potential of IoT in modern agriculture. Together, they lay the groundwork for wider adoption and continuous innovation in mushroom farming practices.

### 6.7.1 Getting Started with IoT in Mushroom Farming

For small-scale mushroom growers, starting with IoT does not require a massive investment—rather, one can often begin with affordable, easy-to-install sensors for temperature, humidity, and $CO_2$ levels inside the growing area. Using Wi-Fi or Bluetooth-enabled devices connected to a simple mobile app to monitor conditions in real time could provide a convenient way of control and management. One can start by automating one critical process, such as humidity control via a misting system linked to sensor readings, before expanding to other parameters. Data logs can be kept to identify trends and to adjust growing conditions for optimal yields. Partnering with local IoT vendors or agricultural extension services can also reduce setup complexity and ensure reliable system maintenance. By starting small and scaling gradually, growers can avoid overwhelming costs while steadily improving crop quality. With consistent monitoring and data-driven adjustments, IoT can turn even a modest mushroom farm into a highly efficient and productive operation (Fig. 6.3).

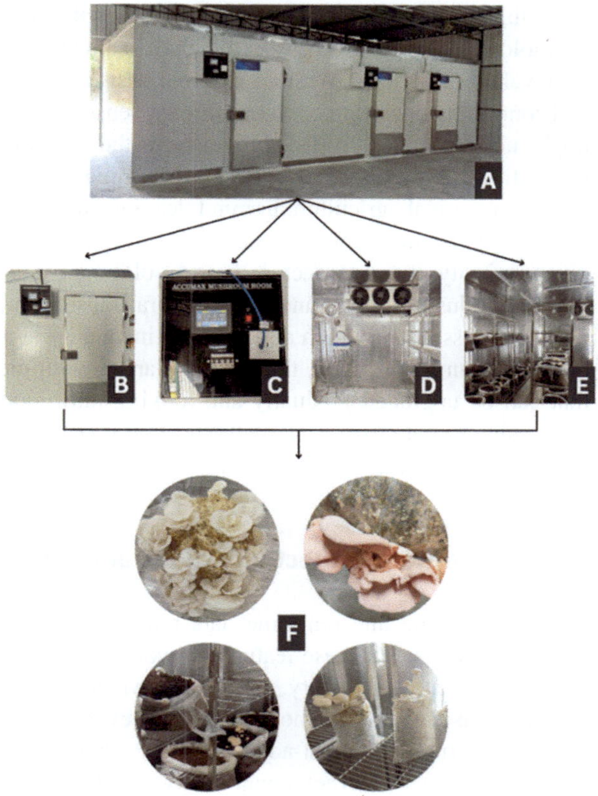


**Fig. 6.3** Demonstrates a realistic IoT set-up installed for mushroom cultivation. (**a**) Centralized IoT-enabled mushroom cultivation unit equipped with climate-controlled grow chambers. (**b**) Individual grow chamber with integrated control panel for real-time environment regulation. (**c**) Smart automation interface displaying environmental parameters and system controls. (**d**) Internal cooling and humidification system ensuring optimal growth conditions. (**e**) Cultivation chamber with substrate bags arranged under regulated conditions. (**f**) Different stages and varieties of mushroom growth are monitored under the smart-farming system. (Pictures taken from the IoT-based Mushroom Facility at Shaheed Rajguru College of Applied Sciences for Women, University of Delhi)

## 6.8 Challenges and Limitations

Despite the potential for increased yields, better resource efficiency, and increased sustainability, smart mushroom cultivation also presents certain challenges that might prevent its widespread use, especially among small and local growers. These issues are complex and include infrastructure, education, security, technological, and economic obstacles. It is therefore essential to recognize and address these constraints to fully realize the potential of smart farming for mushroom production (Irwanto et al. 2024).

### 6.8.1 Cost and Accessibility for Small-Scale Farmers

The high initial financial investment for equipment and installation presents one of the biggest obstacles to the adoption of smart mushroom cultivation systems. Real-time sensors, data loggers, computing units, AI-driven analytics platforms, and automated control systems are some of the technologies that frequently demand large financial outlays at the onset, coupled with appropriate maintenance. These technologies might be too costly for small-scale or subsistence growers, particularly in developing nations. Without external assistance or subsidies, the infrastructure and continuing operating costs (electricity, sensor replacement, and software license) might not be financially feasible, even in the case of inexpensive or open-source alternatives. Additionally, a lot of smart-farming technologies are created with large-scale commercial agriculture in mind, leaving smaller operations behind in terms of technology. Features and scales that are less important to small farms are frequently given priority by manufacturers and solution providers. The productivity gap between industrial farms and local or rural mushroom growers is widened by this lack of accessibility, which also sustains inequality in agricultural innovation. Such problems can be lessened, though, by starting with inexpensive, modular IoT kits that can be upgraded gradually and looking into grants, subsidies, or cooperative purchasing to cut costs. This methodical approach guarantees affordability without sacrificing scalability in the future.

### 6.8.2 Internet Connectivity in Rural Areas

Cloud computing and real-time data transmission are key components of smart-farming systems. For these features to work well, reliable and fast internet connectivity is frequently necessary, particularly when combining cloud-based dashboards, mobile alerts, and remote monitoring. However, many mushroom-growing regions, particularly rural or forest-adjacent areas where cultivation conditions are ideal, suffer from poor internet connectivity. Without a reliable internet connection, features like real-time monitoring, remote control, and cloud-based analytics cannot operate effectively. Offline edge computing solutions can sometimes work for a short time, but they make it harder for the grower to get to centralized data, get updates, or do a predictive analysis. Thus, bad infrastructure makes it difficult to use smart systems in areas that could benefit from them the most, slowing down the technological progress of rural farming. Limited internet access often makes it hard to monitor rural farms in real time. However, using devices that support offline data logging with delayed syncing or technologies like LoRa for stable rural coverage could make sure that data collection and decision-making happen on time, even in areas with poor connectivity.

### 6.8.3 Technical Literacy Among Growers

Even if the technology and infrastructure are available, the grower's level of technical expertise is what makes smart mushroom cultivation systems work. Many traditional mushroom growers, especially in rural or developing areas, may not have ever used sensors, data analytics platforms, or automated control systems before. To use mobile apps, calibration procedures, cloud dashboards, and AI tools, one needs to know at least the basics about digital technology and environmental engineering for a smooth workflow. Lack of training or user-friendly interfaces can lead to underutilization of the technology or errors in system operation. For example, improper sensor placement or incorrect interpretation of environmental data may lead to poor climate control, defeating the purpose of automation. Additionally, complex software interfaces without proper localization (language support, intuitive icons, etc.) may alienate users. Addressing this challenge requires a strong emphasis on capacity building, extension services, and training programs tailored to different literacy levels and local contexts. Offering visual manuals, easy-to-use apps, and community workshops focused on practical operation over technical theory can empower growers to manage systems confidently from day one.

### 6.8.4 Sensor Calibration, Maintenance, and Durability

Smart systems are only as reliable as their sensors and devices. Under the frequently humid and fluctuating conditions of mushroom farms, environmental sensors that measure humidity, temperature, $CO_2$, light, and substrate moisture must be very precise, well-calibrated, and long-lasting. Sensors may drift, break down, or deteriorate over time as a result of dust, condensation, microbial accumulation, or mechanical damage. Accurate readings must be ensured by routine calibration and maintenance, which calls for technical expertise, tool access, and frequently the replacement of costly components. Routine sensor calibration may be beyond the time, skill, or resources of many growers, particularly when the equipment is not readily replaceable or modular. Furthermore, environmental deterioration can reduce the lifespan of automated components and sensors, resulting in more frequent malfunctions, leading to interruptions and added costs.

Durability is particularly important in vertical farms or high-humidity environments where water exposure is constant. Failure to maintain sensor integrity can lead to false readings, triggering incorrect system responses—such as over-irrigation, poor ventilation, or delayed harvesting—that harm the crop and reduce profitability. However, investing in rugged, agriculture-grade sensors and following a regular calibration and cleaning schedule ensures long-term reliability and accuracy in environmental monitoring.

### 6.8.5 Data Privacy and Cybersecurity

Concerns about data privacy and cybersecurity are becoming more important as smart mushroom farms depend largely on cloud-based platforms, smartphone apps, and remote connectivity, which are prone to cyber threats. Environmental parameters, operating schedules, system configurations, and even proprietary cultivation methods are among the many sensitive data that these systems gather and send. Cyber thieves could access or alter such data if it is left unprotected. Cybersecurity threats—such as unauthorized access, data theft, ransomware attacks, or system hijacking—can have serious implications for growers. For instance, a compromised system could lead to intentional climate disruptions, crop loss, or misdirection of operational controls. Moreover, as smart farms become part of broader agricultural supply chains, a breach in one system could compromise connected stakeholders.

Unfortunately, many smart-farming platforms—especially those that are consumer-grade or entry-level ones—do not place a high priority on strong cybersecurity. Local growers also frequently lack the expertise necessary for best practices in network segmentation, secure login protocols, and encrypted communications. In remote locations, software updates that fix security flaws might also be unavailable or delayed. Technology companies, legislators, and users must work together to protect farm data and operations from cyberattacks. Sensitive farm data can be protected and system integrity preserved by using strong authentication, encrypted communication, secure cloud storage, and limited user access.

## 6.9 Future Perspectives

The future of intelligent mushroom farming is changing fast, with new technologies promising to make these systems more efficient, transparent, robust, and accessible. Perhaps the most revolutionary development on the horizon is integrating blockchain technology to provide end-to-end traceability and certification. With consumers and regulators now requiring food safety, transparency of origin, and responsible production methods in an era of heightened concern, blockchain provides an immutable, decentralized ledger that can track each step of mushroom development and supply. From the source of the spawn and the preparation of the substrate to growth conditions, dates of harvesting, packaging information, and routes of distribution, all may be tracked irretrievably on a blockchain. This traceability not only generates consumer confidence and meets certification norms (organic or fair-trade certifications) but also facilitates quick response in the event of contamination or product recalls. As the technology becomes more scalable and available, its use in mushroom cultivation will increasingly shift from pilot tests to widespread adoption, especially among export-focused and high-volume operations.

Another crucial force influencing the future of intelligent mushroom cultivation is the growing application of artificial intelligence (AI) and machine learning (ML) for automating advanced decision-making activities throughout the farm life cycle. Though existing systems provide rule-based automation (e.g., start misting when humidity falls below 85%), future systems can employ AI models with huge datasets to make predictive and adaptive decisions. These models could learn continuously from environmental information, yield history, and species behavior to make fine adjustments in irrigation, lighting, $CO_2$ levels, and harvest schedules with great accuracy. For example, an AI system may anticipate a disease outbreak as an early warning based on minute environmental changes or suggest changing light wavelengths for a species to speed up pinning. Aside from environmental management, AI might aid in autonomous farm management, where robotic systems carry out mundane tasks such as substrate loading, monitoring for contaminants, or even picking fragile fruiting bodies with limited human intervention. Drone surveillance or computer vision technologies may determine mushroom growth stages in real time, allowing for targeted interventions and quality inspection. With the use of IoT, AI can turn mushroom farms into autonomous systems that are optimized to perform under changing conditions, minimizing labor intensity and human mistakes and maximizing yield and quality.

Moreover, resilience and accessibility will also be important pillars of innovation. Mobile apps and intuitive interfaces need to be developed to make smart-farming technologies accessible to small-scale and rural mushroom farmers who are not formally technically trained. The applications need to be designed in simple terms, supporting multiple languages, and allowing offline capabilities to accommodate diverse literacy rates and infrastructural constraints. Functions like voice-assisted setup, real-time notifications, and graphical overviews will enable local farmers to engage with confidence in their farm infrastructure. Climate-resilient smart agriculture will also be a design and research focus. As climate change introduces capricious weather, increased heat, and escalating pest and disease pressure, mushroom-growing systems will need to be adaptive and resilient. Technologies like modular climate-controlled units with renewable power sources, fail-safe control loops, and weather response models based on AI will stabilize production in unstable environments. At a larger scale, coupling smart mushroom farms with digital marketplaces and supply chain systems will transform the process of mushrooms getting from farm to consumer. IoT-enabled packaging, live tracking of logistics, and dynamic pricing through demand forecasts can guarantee freshness, minimize postharvest loss, and increase farmer revenues. Such a system can connect local producers to urban markets, export points, or value-added processing facilities via smart supply-demand mapping. In the end, the future of intelligent mushroom farming does not just rest in technology but in the creation of inclusive, resilient, and transparent systems that empower both the producers and the consumers.

## 6.10 Conclusion

The adoption of Internet-of-things (IoT) technologies in mushroom culture systems is a revolutionary leap forward for agriculture. Through real-time monitoring, automatic environmental management, and data-based decision-making, IoT allows farmers to establish ideal conditions for every growth phase of mushrooms. Accurate control of factors like humidity, temperature, $CO_2$ levels, and substrate water content not only increases yield and product quality but also minimizes wastage of resources, labor dependence, and crop loss. Smart systems with sensors, cloud technology, and machine learning can automatically control intricate activities—from misting and lighting to climate control and harvest schedules—rendering mushroom farms efficient, scalable, and sustainable. Mobile-compatible dashboards and remote access capabilities also enable farmers to monitor their operations from nearly anywhere, adding to flexibility and responsiveness. The capacity to gather and compare large amounts of data longitudinally enables ongoing improvement in cultivation practices, with predictive analytics detecting trends in yield, disease incidence, or environmental variation. In addition, innovations such as blockchain traceability and digital twin simulations provide value additions on transparency, certification, and operational testing, particularly for high-value or export-oriented mushroom produce.

Despite these key advantages, it is imperative to facilitate wider use of IoT-based systems among mushroom farmers, especially small and medium-scale ones who have the most to benefit from higher productivity and minimized physical labor. Adoption, though, should go hand in hand with efforts toward making such technologies affordable, modular, and easy to use. Governments, agri-tech startups, and research centers must collaborate to offer training programs, localized assistance, and financial support that demystify and democratize smart-farming solutions. This involves developing mobile apps and interfaces that are user-friendly for users with minimal technical backgrounds, testing for functionality in offline or low-connectivity scenarios, and providing language support as required. Moreover, the creation of affordable sensor kits and open-source automation platforms can facilitate growers' adoption of smart systems in an incremental manner, depending on their size and capabilities. Even with the promising advancements, there are numerous research and implementation gaps. These encompass a requirement for low-maintenance and durable sensors to be used in mushroom farms to handle the environmental high-humidity environments, enhanced cybersecurity standards to safeguard farm data, and adaptation of AI models to address the unique mushroom species and farming methods. There are also insufficient long-term studies that assess the environmental, economic, and social implications of smart-farming technologies across various geographic locations. As the science evolves, filling these gaps through interdisciplinary research, field testing, and stakeholder outreach will be critical. Through creating strong, equitable, and scalable mushroom farming ecosystems, the future is vast with possibilities—not just for mushroom farming, but in establishing a model in sustainable technology-enabled agriculture globally.

## References

Akhil A, Kumar S, Kathuria S, Preeti P, Rathore SS (2023) Real time automation of mushroom farm using Arduino Akhil. J Adv Commun Syst 6(3):1–9

Chong JL, Chew KW, Peter AP, Ting HY, Show PL (2023) Internet of things (IoT)-based environmental monitoring and control system for home-based mushroom cultivation. Biosensors 13(1):98

Dhaked MK, Tomar DS, Kumar M, Sharma V (2025) Use of AI technique in mushroom farming. In: Smart farming with AI: innovations for sustainable agriculture, p 369

Geetha Rani E, Chalasani R, Anusha D, Dhanalakshmi M, Vadlamudi S (2023) Organic farming automation to revolutionize the agricultural industry than traditional farming practices using IOT and technological development. In: International conference on robotics, control, automation and artificial intelligence. Springer Nature Singapore, Singapore, pp 333–350

Gill R, Chawla P (2021) A review on various techniques to transform traditional farming to precision agriculture. Turk J Comput Math Educ 12(2):131–135

Guragain DP, Shrestha B, Bajracharya I (2024) A low-cost centralized IoT ecosystem for enhancing oyster mushroom cultivation. J Agric Food Res 15:100952

Hendinata LK, Fikri AIR (2023) Development of IoT-based temperature and relative humidity monitoring system for mushroom cultivation house. J Electron Control Telecommun Inf Power Eng 10(1):95–102

Higgins C, Margot H, Warnquist S, Obeysekare E, Mehta K (2017) Mushroom cultivation in the developing world: a comparison of cultivation technologies. In: 2017 IEEE global humanitarian technology conference (GHTC). IEEE, pp 1–7

Irwanto F, Hasan U, Lays ES, De La Croix NJ, Mukanyiligira D, Sibomana L, Ahmad T (2024) IoT and fuzzy logic integration for improved substrate environment management in mushroom cultivation. Smart Agric Technol 7:100427

Karunathilake EMBM, Le AT, Heo S, Chung YS, Mansoor S (2023) The path to smart farming: innovations and opportunities in precision agriculture. Agriculture 13(8):1593

Kassim MRM, Mat I, Yusoff IM (2019) Applications of internet of things in mushroom farm management. In: 2019 13th international conference on sensing technology (ICST). IEEE, pp 1–6

Kawuryan R, Pratiwi S, Kandang T, Anthama LP, Nurramadani A (2025) Optimizing mushroom cultivation for SMEs: development of a frugal innovation-based IoT-enabled smart Kumbung system. In: E3S web of conferences, vol 622. EDP Sciences, p 03011

Kumar CA, Shrinivasan L, Kumar RA, Harish S, Vaishnavi DAL (2023) Intelligent monitoring of Grey oyster mushroom cultivation with IoT. Int J Intell Syst Appl Eng 12(4s):235–239

Miller T, Mikiciuk G, Durlik I, Mikiciuk M, Łobodzińska A, Śnieg M (2025) The IoT and AI in agriculture: the time is now—a systematic review of smart sensing technologies. Sensors 25(12):3583

Moysiadis V, Kokkonis G, Bibi S, Moscholios I, Maropoulos N, Sarigiannidis P (2023) Monitoring mushroom growth with machine learning. Agriculture 13(1):223

Nguyen HH, Shin DY, Jung WS, Kim TY, Lee DH (2024) An integrated IoT sensor-camera system toward leveraging edge computing for smart greenhouse mushroom cultivation. Agriculture 14(3):489

Omar S, Nik WMFW, Ahmad MI, Chow TS, Bakar MNA, Man SF et al (2024) IoT enabled mushroom farm automation with machine learning. Adv Sustain Technol 3(1):29–37

Rahman H, Faruq MO, Hai TBA, Rahman W, Hossain MM, Hasan M et al (2022) IoT enabled mushroom farm automation with machine learning to classify toxic mushrooms in Bangladesh. J Agric Food Res 7:100267

Rukhiran M, Sutanthavibul C, Boonsong S, Netinant P (2023) IoT-based mushroom cultivation system with solar renewable energy integration: assessing the sustainable impact of the yield and quality. Sustainability 15(18):13968

Sakamoto Y (2018) Influences of environmental factors on fruiting body induction, development and maturation in mushroom-forming fungi. Fungal Biol Rev 32(4):236–248

Santos AF, Gaspar PD, de Souza HJ (2023) Eco-efficiency in mushroom production: a study on HVAC equipment to reduce energy consumption and CO2 emissions. Appl Sci 13(10):6129

Singh S, Simran SA, Sushma SJ (2020) Smart mushroom cultivation using IoT. Int J Eng Res Technol 8(13):65–69

Thong-un N, Wongsaroj W (2022) Productivity enhancement using low-cost smart wireless programmable logic controllers: a case study of an oyster mushroom farm. Comput Electron Agric 195:106798

# Development of Mushroom-Based Functional Food and Nutraceuticals

7

Abzar, R. Mohammad-Syaril, F. M. N. A. Aida, and Siva Raseetha


**Abstract**

In ancient China, mushrooms were regarded as promoting longevity and health, while in ancient Greece, they were believed to provide strength to warriors. The global production of edible mushrooms has risen significantly over the years due to consumer demand. The therapeutic potential of mushrooms is attributed to their bioactive compounds, such as lectins, glucans, phenolic compounds, and terpenoids. They contribute powerful action toward inflammation and oxidative reactions that could heal diseases inflicted on consumers. The availability of various mushroom species eases food industrial processes such as the production of enzymes, vitamins, antibiotics, pigments, amino acids, fatty acids, glycolipids, and polysaccharides. Further, mushrooms were added as food ingredients and pharmaceutical products. This chapter provides an eye-opener for researchers on the overall nutritional profile of various mushrooms and their health benefits. Additionally, the usage of mushroom extracts as functional food and in nutraceutical products has been described. This chapter aims to give insight into the food and pharmaceutical industries and food developers to explore more innovative products to meet the needs of consumers, depending on their utilization and application.



Abzar
Faculty of Applied Sciences, Universiti Teknologi MARA, Shah Alam, Selangor, Malaysia

Department of Crop Science, Faculty of Agriculture, University Putra Malaysia, Serdang, Selangor, Malaysia

R. Mohammad-Syaril
Johor Bahru Public Health Laboratory, Tampoi, Johor Bahru, Johor, Malaysia

F. M. N. A. Aida · S. Raseetha (✉)
Faculty of Applied Sciences, Universiti Teknologi MARA, Shah Alam, Selangor, Malaysia
e-mail: raseetha@uitm.edu.my







B. Giri et al. (eds.), *Advances in Mushroom Research*,
https://doi.org/10.1007/978-981-95-7692-0_7

**Keywords**

Food products · Nutraceuticals · Nutrition · Health benefits

## 7.1 Introduction

Mushrooms are macro-fungi of the Basidiomycota phylum and have been valued since ancient times for their nutritional and medicinal properties, with archaeological evidence dating back 18,700 years (Kotowski 2019). Historically revered as "Food of the Gods" in Roman culture and considered divine gifts in ancient Egypt (Rahi and Malik 2016), mushrooms continue to be appreciated for their rich nutritional profile, being high in protein, vitamins, minerals, and fiber while low in calories and cholesterol. The cultivation history spans centuries, beginning with Auricularia species, followed by *Flammulina velutipes* and *Lentinula edodes*, with France pioneering *Agaricus bisporus* cultivation in the 1600s. Today, approximately 35 species are commercially cultivated, with *A. bisporus* and *Pleurotus ostreatus* being predominant in Egypt, containing 26.05%–33.85% crude protein, 2.79%–2.41% crude fats, and 42.56%–57.05% carbohydrates (Muhammad and Suleiman 2015; Atila et al. 2017).

The therapeutic properties of mushrooms are attributed to bioactive compounds, including lectins with anticancer and antidiabetic properties (Chang 2008), beta-glucans demonstrating anti-inflammatory and immune-stimulatory effects, phenolic compounds like gallic and caffeic acids that have antioxidant and anticancer properties (Atila et al. 2017), and terpenoids, including gadoteric acids and ergosterol, which offer antimicrobial benefits (Power et al. 2015). These compounds collectively contribute to mushrooms' documented antimicrobial, anti-inflammatory, antioxidant, anticancer, and hepatoprotective effects (Chang and Miles 1992; Das et al. 2010), validating their traditional use in folk medicine across cultures.

## 7.2 An Overview of Functional Foods and Nutraceuticals

Food production has evolved significantly in recent decades due to rising consumer demand. Consumers recognize that food satisfies hunger and plays a crucial role in health (Mollet and Rowland 2002), preventing nutrition-related diseases and enhancing mental and physical well-being. Reports from the Food and Agriculture Organization and World Health Organization identify dietary patterns and lifestyle habits as key modifiable risk factors for chronic diseases such as cancer, heart disease, obesity, type 2 diabetes, osteoporosis, and periodontal disease (WHO 2003).

Functional foods are essential for addressing health issues, providing benefits beyond basic nutrition. They can be natural or processed foods and contain active compounds that help manage or prevent chronic diseases (Raut 2019). Healthy compounds include sulfur, glucosinolates, and polyphenols like anthocyanins, flavones, isoflavones, and terpenoids. Diets high in refined sugar, fat, salt, and

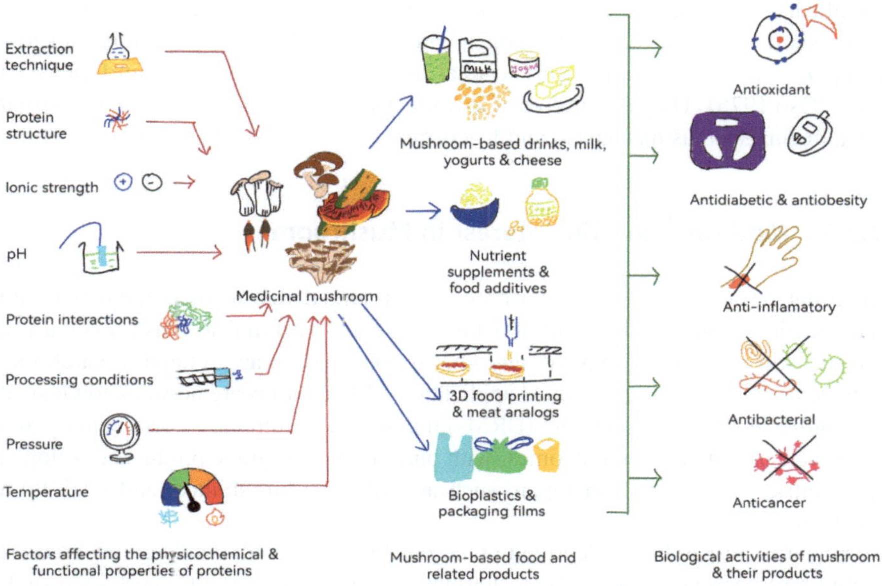


**Fig. 7.1** Mushroom products, factors affecting their functional properties, and their biological activities

cholesterol are linked to chronic diseases, especially in older adults. Side effects from medications have increased the need for safe, nutrient-rich foods (Guo et al. 2022). Eating foods with functional ingredients like minerals, vitamins, probiotics, antioxidants, and fiber can reduce the risk of chronic diseases and enhance physical and mental health.

Functional foods have gained attention from researchers and consumers, leading to the term "nutraceuticals," which combines nutrients and pharmaceuticals. Mushroom products, rich in bioactive peptides, vitamins, probiotics, and antioxidants, have captured interest due to their potential health benefits, as shown in Fig. 7.1. Nutraceuticals are food-derived products available in medicinal forms like capsules and powders, demonstrating physiological benefits and protection against chronic diseases. In Canada, these are known as "natural health products." Both functional foods and nutraceuticals help reduce healthcare burdens through prevention, driven by population trends and increasing consumer interest.

## 7.3 Mushrooms in Food and Medicine

### 7.3.1 Historical Use of Mushrooms in Traditional Medicine

Mushrooms have been used for food and medicine since ancient times, valued for their rich content of amino acids, protein, minerals, and other nutrients (Kalač 2009). In sub-Saharan Africa during the Paleolithic period, mushrooms held cultural significance, often linked to mysticism (Samorini 1992). The Yoruba tribe in Nigeria first recorded their use as hallucinogens and medicinally, with traditional doctors using *Termitomyces microcarpus* to treat diseases like gonorrhea (van Griensven 2009; Oso 1975). They also utilized *Calvatia cyathiformis* and *Daldinia concentrica* for ailments such as masomaso and leucorrhoea (El Enshasy et al. 2013).

### 7.3.2 Modern Scientific Interest in Mushrooms

In recent decades, interest in mushrooms has surged due to their medicinal and therapeutic potential. While traditional medicine has used mushrooms for centuries, advances in medicinal chemistry and pharmacology have accelerated research into their biological effects and clinical applications. Mushrooms are now acknowledged as biological response modifiers (BRMs) that enhance immune system activity and improve cancer treatment outcomes, with pharmaceutical-grade products developed in countries like China and Japan, such as lentinan from shiitake and PSK from turkey tail.

Scientific research has identified numerous bioactive compounds in mushrooms, including triterpenoids and polysaccharides, leading to over 50,000 published studies and over 15,000 significant patent activities. Organizations like the International Society for Medicinal Mushrooms are integrating traditional knowledge with modern research, particularly focusing on species like Cordyceps (Thakur 2020).

Beyond their therapeutic uses, mushrooms are nutritionally rich, with up to 35% protein, 19% crude fiber, and low fat, making them an attractive meat alternative and thereby increasing global demand (Cheung 2010; Thakur 2020).

## 7.4 Nutritional Profile of Mushrooms

Edible mushrooms are a good food source for human consumption in Eastern and Western countries. Researchers have explored mushrooms' chemical composition, providing medicinal properties and nutritional values.

### 7.4.1 Sterols and Polyphenols

Ergosterol is the most common sterol found in mushrooms, which can undergo photolysis to vitamin $D_2$ upon exposure to UV radiation. Zheng et al. (2013) reported that ergosterol demonstrates cytotoxicity toward acute promyelocytic leukemia and liver cancer cells and shows antimicrobial activity against selected fungi. Another sterol named H1-A, which shows similarity to testosterone and dehydroepiandrosterone in its structure, was extracted from Cordyceps. In vivo studies of this compound in mice have shown that it can effectively treat selected autoimmune diseases. Along with sterols, specific polyphenols, including phenolic acids represented by benzoic acid and cinnamic derivatives, have also been reported in mushrooms, which alternatively increase their antioxidant activities. These compounds inhibit free radicals, minimize peroxide decomposition, and block metals' action while catalyzing oxidation reactions (Sharpe et al. 2021). Another report by Peng and Shahidi (2022) showed that chaga ethanol extracts are detected with 111 different phenolic compounds, including coumarins, flavonoids, phenolic acids, styrylpyrones, and quinones. Flavonoids such as rutin, quercetin, myricetin, morin, hesperetin, and naringenin have also been reported in medicinal mushrooms.

### 7.4.2 Polysaccharides

Oyster mushrooms are a rich source of bioactive compounds, such as polysaccharides, mevinolin, and other statins, revealing hypocholesterolemic properties. In addition, proteoglycans derived from *Agaricus blazei* showed strong immunomodulatory effects and significantly played a therapeutic role in the management of cancer and immune-deficient conditions by promoting the maturation of dendritic cells (Gunde-Cimerman and Plemenitas 2001; Kim et al. 2005). The polysaccharides extracted from *Agaricus bisporus* have also shown strong inhibitory effects against breast cancer in humans (Jeong et al. 2012). Several β-glucans, like pleuran produced by *Pleurotus* species, lentinan from *Lentinula edodes*, schizophyllan from *Schizophyllum commune*, calocyban from *Calocybe indica*, and ganoderan and ganopoly from *Ganoderma lucidum*, have been discovered and isolated from several species of mushrooms (Villares et al. 2012). Additionally, polysaccharides from *Hericium erinaceus* have proved anti-fatigue properties, which support their potential application in sports nutrition (Liu et al. 2015). Overall, mushroom-derived polysaccharides are recognized for their diverse physiological activities, such as antitumor, antioxidant, antiviral, immunomodulatory, anti-inflammatory, and anticancer effects (Rathore et al. 2017).

### 7.4.3 Bioactive Protein Molecules

Essential nutrients for human health can be classified into seven groups based on their physicochemical composition and functions, such as carbohydrates, lipids, proteins, vitamins, minerals, dietary fiber, and water. It is essential to obtain these nutrients from the diet to support normal physiological functions (Wang and Zhao 2023). Furthermore, mushrooms provide a complete profile of amino acids, in which all eight essential amino acids, which are required by adults, and histidine, which is necessary for infants, are included. In many species of mushrooms, the amino acid composition closely aligns with the ideal amino acid pattern for human nutrition. Some species, such as *Agaricus bisporus*, *Flammulina velutipes*, *Tricholoma matsutake*, and *Pleurotus eryngii*, have an amino acid profile considered optimal (Gao et al. 2012; Zhang et al. 2017). The digestibility pattern of mushrooms typically ranges from 72% to 83%, which is comparable to or lower than the digestibility of other protein sources: 74% for soybeans, 82% for rice, 92%–94% for meat, 98% for eggs, and 97%–98% for dairy products (Wani et al. 2010). These properties related to digestibility mark mushrooms as an excellent alternative source of dietary protein. Furthermore, proteins obtained from cereals are often low in lysine, and legume proteins tend to lack sufficient sulfur-containing amino acids such as methionine and cysteine; that is why mushrooms offer a more balanced amino acid profile, making them a superior substitute in plant-based diets (Sá et al. 2020).

### 7.4.4 Terpenes and Terpenoids

Terpenes are a class of volatile, unsaturated hydrocarbons widely recognized for their anti-inflammatory properties and have been widely isolated from various mushroom species (Duru and Çayan 2015). These terpenoid compounds show various pharmacological activities, such as anticancer, anticholinesterase, antimalarial, anti-inflammatory, antiviral, and antibacterial effects (Duru and Çayan 2015). Monoterpenes, sesquiterpenoids, and triterpenoids are among the isolated terpenoids from various mushroom species. It has been found that sesquiterpenoids such as flammulinol and flammulinolides can be extracted from *Flammulina velutipes*. At the same time, triterpenoids like lanostane have been identified in *Ganoderma lucidum* and sesquiterpenoids from *Pleurotus cornucopiae* (Rathore et al. 2017). Due to their diverse bioactivities, these terpenes can be efficiently utilized to develop drugs for curing Alzheimer's and other degenerative diseases.

Furthermore, scientific reports revealed that terpenes and terpenoids are organic compounds commonly found in medicinal mushrooms and are acknowledged for their immunomodulatory effects (Zhao et al. 2020). These compounds can regulate immune function by enhancing or suppressing immune responses based on physiological needs. Terpenes are hydrocarbons that are widely distributed around the globe, synthesized via isopentenyl pyrophosphate, and produced by plants, some insects, and fungi such as mushrooms. Terpenoids, which are oxygenated derivatives of terpenes, further strengthen the biological activities of terpenes. Mushroom species like *Ganoderma* are reported to be rich in triterpenoids, which are potentially immunomodulatory and anti-infective by modulating gene transcription in the NF-κB and MAPK pathways (Gao et al. 2002; Ma et al. 2014; Chen et al. 2014).

Prastiyanto et al. (2020) confirmed the presence of terpenoids in *Pleurotus* species, particularly *P. ostreatus,* by thin-layer chromatography, which shows anticancer activity against Raji cells and antibacterial effects against MRSA. In addition to immune modulation, terpenes are also known for their broad pharmacological potential, such as antioxidant, antitumor, anti-inflammatory, antiviral, insecticidal, and nematicidal properties (Dudekula et al. 2020; Devi et al. 2024). Additionally, pleurospiroketals A–E, isolated from *Pleurotus cornucopiae*, were found to inhibit HeLa cell proliferation (Wang et al. 2013), while enokipodins B, D, and J from *Flammulina velutipes* displayed significant antioxidant capacity and moderate cytotoxicity against several human tumor cell lines (MCF-7, HepG2, A549, and SGC7901) (Wang et al. 2012).

### 7.4.5 Vitamins and Minerals

Mushrooms possess various bioactive properties, such as anticancer, antibiotic, antiviral, immunomodulatory, and hypolipidemic effects. Some species from the genus *Pleurotus* are specifically known for their medicinal properties. These therapeutic effects are due to the presence of certain biologically active compounds in mushrooms. Several studies have identified mushrooms as a rich source of proteins, essential amino acids, dietary fiber, vitamins, and certain minerals (Huang et al. 2015; Vishwakarma et al. 2017).

Several nutritional parameters were assessed by Gnanwa et al. (2021) for *Pleurotus ostreatus*, highlighting its value as a functional food with a rich nutritional profile. *Pleurotus ostreatus* showed high concentrations of essential minerals, such as copper (36.357 ± 0.327%), iron (52.653 ± 0.061%), and magnesium (47.840 ± 1.126%). These minerals are critical in enzymatic functions, oxygen transport, and neuromuscular activity. Additionally, the same study's results showed that *P. ostreatus* flour was a significant source of key vitamins besides its mineral content. The variable amount of several vitamins, including vitamin A (0.1922 ± 0.0206%), thiamine (0.2978 ± 0.0458%), folic acid (2.7924 ± 0.0076%), pyridoxine (5.0699 ± 0.0006%), and vitamin C (17.4833 ± 2.1605%), was assessed. Furthermore, trace levels of vitamin E, riboflavin, and niacin were also detected in the *Pleurotus ostreatus* flour. These nutrients contribute to antioxidant defense, metabolic regulation, and overall physiological health, further strengthening the nutritional and therapeutic potential of *P. ostreatus* in human diets.

### 7.4.6 Dietary Fiber

The carbohydrates not digested or absorbed by the human body are known as dietary fiber (DF). DF consists of β-glucans, cellulose, hemicellulose, pectin, lignin, and numerous oligosaccharides (Fuller et al. 2016). DF is central in maintaining human health and is recognized as the "seventh nutrient." An increasing body of research highlights its diverse health benefits, such as anticancer, anti-diabetes, anti-obesity, and anti-inflammatory properties (Nie and Luo 2021). DF is generally found in vegetables, fruits, grains, legumes, and edible fungi (He et al. 2022). Edible fungi are large edible fruiting bodies of macro-fungi, which are nutritionally rich, particularly in DF, with contents ranging from 14.4 g to 70.2 g per 100 g of mushrooms (Yu et al. 2020). On a solubility basis, DF is classified into soluble dietary fiber (SDF) and insoluble dietary fiber (IDF). The previous studies have primarily focused on SDF's physiological activities in mushrooms, while IDF has received comparatively less attention. Various applications have been shown by SDF; for example, polysaccharides from *Tremella fuciformis* showed neuroprotective effects against cellular damage, those from *Lentinula edodes* are used as adjuncts in cancer therapy, and *Cordyceps militaris* polysaccharides demonstrate antioxidant and antiaging properties (Zhang et al. 2021; Gu et al. 2022). Recently, research on IDF has been gaining attention through mechanisms such as adsorption and fermentation. IDF can contribute to hypoglycemic effects by absorbing lipids, bile acids, and cholesterol. At the same time, its fermentation by gut microbiota enhances the abundance of beneficial bacteria, further promoting the production of short-chain fatty acids. IDF's content is reported to be 3–10.5 times higher than SDF in mushrooms, indicating its greater economic value. However, the importance of SDF extraction commonly leads to a large amount of IDF waste, highlighting the necessity of improving IDF generated from mushrooms to facilitate the production of functional foods and enable practical resource usage (Tian et al. 2022).

The mushroom cell walls are a significant source of dietary fiber (DF), composed of nondigestible carbohydrates (NDCs) that resist digestion by human enzymes. The components of cell wall structure include chitin, a straight-chain (1 → 4)-β-linked polymer of N-acetyl-glucosamine, as well as polysaccharides such as (1 → 3)-β-D-glucans and mannans. Carbohydrates account for approximately 35%–70% of dry weight and are considered the major constituents of mushrooms; however, this percentage depends on the mushroom species. Mostly, these carbohydrates are NDCs, including oligosaccharides and key cell wall polysaccharides like chitin, β-glucans, and mannans. Among these, the content of chitin is generally lower than that of β-glucans in various mushroom species (Cheung 2010; Ghosh 2016).

## 7.5 Health Benefits of Mushrooms

### 7.5.1 Immune System Modulation of Mushrooms

Bioactive mushroom polysaccharides are known as "Biological Response Modifiers" (BRMs) due to their ability to stimulate the immune system against infections, tumors, and inflammation, while also enhancing hormone and immune cell synthesis (Ohno et al. 2001; Giavasis 2013). The immunomodulatory effects depend on the properties of their biopolymers, such as molecular weight, conformation, branching degree, and sugar composition, all of which influence their medicinal benefits (Bohn and BeMiller 1995; Giavasis 2014). Additionally, the methods used for extraction, like water or alkali extraction and alcohol precipitation, can also affect their size and functionality (Giavasis 2013).

These compounds enhance the immune response, rather than directly killing cancer cells. They aid in preventing metastasis, reduce chemotherapy and radiotherapy side effects, and promote bone marrow recovery (Ikekawa 2001). Mushroom glucans, such as lentinan and schizophyllan, stimulate immune activity through macrophage activation and tumor necrosis factor (TNF-α) secretion, often associated with their high molecular weight and triple helix structure (Chang and Wasser 2012). In contrast, scleroglucan, which has a random coil structure, may effectively stimulate human blood monocytes (Kulicke et al. 1997). Lentinan and schizophyllan are usually given via injection due to low oral absorption, while scleroglucan can be taken orally, making it advantageous for mushroom-based nutraceuticals (Chan et al. 2009).

Medicinal mushrooms, like *Grifola frondosa* and *Trametes versicolor,* have shown benefits in clinical trials, improving survival rates in cancer patients when used alongside chemotherapy or surgery. The *Ganoderma lucidum* is particularly notable; its polysaccharide ganoderan has been found to extend survival in tumor-bearing mice, enhance cytotoxic drug effects, and stimulate key immune factors such as TNF-α, IFN-γ, and natural killer (NK) cells. Additionally, Ganoderma polysaccharides can help restore immune function in immunosuppressed animals and may act prophylactically on macrophages (Vannucci et al. 2013; Giavasis 2014). The immunomodulatory effects of Ganoderma may also stem from other bioactive compounds like ganoderic acids and terpenoids (Lindequist et al. 2005).

Polysaccharides from *Agaricus blazei* mushrooms enhance IFN-γ production and suppress allergic reactions, acting as antiallergic immunomodulators. Similarly, β-glucans from Saccharomyces cerevisiae (a fungal yeast) exhibit strong mitogenic activity, stimulating cytokine production and enhancing monocyte and neutrophil phagocytosis (Giavasis 2013). Chemically modified mushroom glucans (e.g., glucan sulfate, carboxymethyl glucan) also show strong antitumor effects. Edible mushrooms such as *Pleurotus ostreatus* and *Pleurotus tuber-regium* are also reported to induce apoptosis in prostate cancer cells and inhibit breast and colon cancer cell proliferation (Šandula et al. 1999; Khan and Tania 2012).

The immune-enhancing and antitumor effects of mushroom-derived glucans result from various mechanisms, including mitogenic stimulation, activation of immune cells, and enhanced phagocytic activity, highlighting their potential as natural immunomodulators (Ohno et al. 2001; Giavasis 2014).

## 7.5.2 Anti-Inflammatory and Anticancer Properties

Inflammation is a natural immune response that helps defend against health issues like heart disease, chronic pain, age-related disorders, memory decline, and autoimmune diseases. It also contributes to tumor development, as many cancers arise from chronic infection and irritation. Therefore, targeting inflammation and apoptosis is crucial for effective cancer treatment (Iman et al. 2016). A proper diet can help reduce inflammation, with functional foods like mushrooms, berries, vegetables, legumes, and fish particularly beneficial (Joseph et al. 2014; Taofiq et al. 2016; Zhu et al. 2018). Various mushrooms with anti-inflammatory properties are shown in Fig. 7.2.

Several mushrooms exhibit chemopreventive and anti-inflammatory properties, as highlighted by in vitro studies. Ethyl acetate extracts have shown more potent inhibition of COX-1 and COX-2 enzymes compared to 70% ethanol extracts, especially from *Ganoderma applanatum, Naematoloma sublateritium, Pleurotus eryngii, and P. salmoneostramineus*, with higher COX-2 inhibition. Additionally, ergosterol and peroxide from edible and medicinal mushrooms significantly reduced LPS-induced inflammatory responses by inhibiting NF-κB and C/EBPβ activity and downregulating MAPK phosphorylation (Kobori et al. 2007).

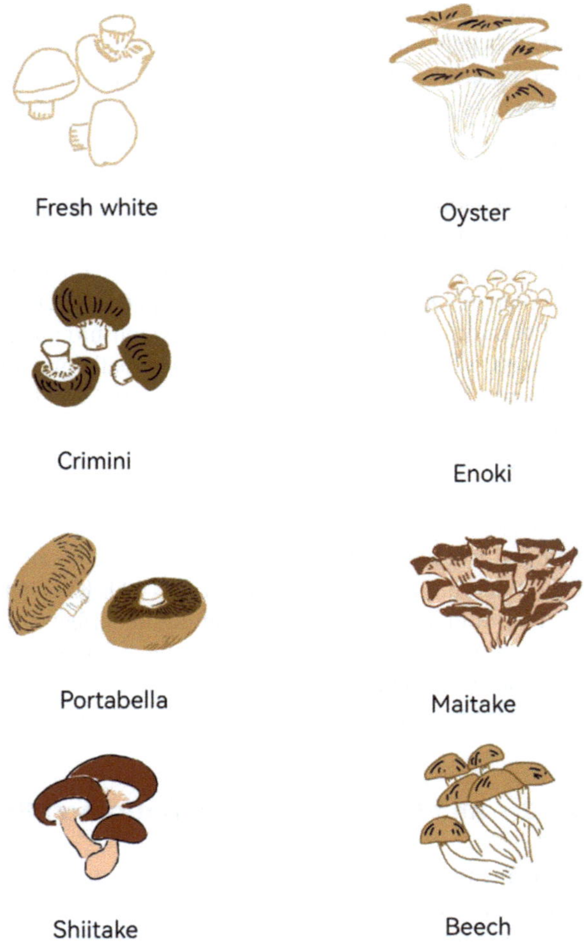


**Fig. 7.2** Mushroom varieties with anti-inflammatory activity. (Fresh white mushroom: *Agaricus bisporus*; oyster mushroom: *Pleurotus ostreatus*; crimini: *Portabella*; enoki: *Flammulina velutipes* (Fr.) Sing; maitake: *Grifola frondosa*; beech: *Hypsizygus tessellatus*)

Dudhgaonkar et al. (2009) reported that a triterpene extract from *Ganoderma lucidum* significantly reduced the secretion of pro-inflammatory cytokines TNF-α and IL-6, along with the inflammatory mediators nitric oxide (NO) and prostaglandin $E_2$ ($PGE_2$), in LPS-stimulated murine RAW264.7 macrophage cells. The extract also downregulated the expression of inducible nitric oxide synthase (iNOS) and COX-2. Jo et al. (2010) evaluated the anti-inflammatory potential of a hot water extract from *Cordyceps militaris* fruiting bodies. They showed that treatment of RAW264.7 cells with increasing concentrations of the extract significantly suppressed LPS-induced production of NO, TNF-α, and IL-6 in a dose-dependent manner.

Additionally, the anti-inflammatory activity of oyster mushroom concentrate was evaluated using RAW264.7 murine macrophage cells and murine splenocytes treated with the concentrate, which was prepared by resuspending dried mushrooms in distilled water and concentrating the solution. In both LPS-stimulated and unstimulated conditions, the oyster mushroom concentrate significantly inhibited the LPS-induced secretion of TNF-α, IL-6, IL-12p40, prostaglandin $E_2$ ($PGE_2$), and nitric oxide (NO) in RAW264.7 macrophages. Moreover, it repressed the LPS-induced DNA-binding activity of the transcription factors AP-1 and NF-κB in these cells (Jedinak et al. 2011).

Moro et al. (2012) investigated the anti-inflammatory effects of methanolic extracts from *Agaricus bisporus*, *Cantharellus cibarius*, and *Lactarius deliciosus* in activated macrophages. Their findings confirmed that these extracts exhibited vigorous anti-inflammatory activity by inhibiting NO production and downregulating the mRNA expression of iNOS, IL-1β, and IL-6 in response to LPS stimulation. In another study, Hur et al. (2012) assessed the anti-inflammatory potential of *Russula virescens* ethanol extracts in RAW264.7 macrophage cells. Treatment with concentrations ranging from 1 to 5 mg/mL significantly reduced LPS-induced NO production, with the highest concentration (5 mg/mL) notably decreasing TNF-α mRNA expression. Kamiyama et al. (2013) examined the anti-inflammatory properties of various extracts derived from *Trametes versicolor*. Among the tested fractions, the acetone extract demonstrated a dose-dependent inhibition of 15-lipoxygenase (15-LOX), indicating considerable anti-inflammatory potential.

### 7.5.3 Cardiovascular and Metabolic Health Benefits

Cardiometabolic health conditions encompass the overall state of cardiovascular and metabolic health, including factors like blood pressure (BP), blood lipids (cholesterol and triglycerides [TG]), and blood glucose, along with conditions such as metabolic syndrome (MetS), type 2 diabetes (T2D), cardiovascular disease (CVD), and stroke (de Waard et al. 2019). These conditions have traditionally been linked to unhealthy lifestyles, making lifestyle modification essential in their prevention and management (Sakakibara et al. 2019).

Among the healthy diets, mushrooms have been used for medicinal purposes for over two millennia, with growing evidence supporting their beneficial effects onvarious health parameters, including oxidative stress, inflammation, immune function, and lipid profiles linked to atherosclerosis (Hong et al. 2024). They are rich in vitamins (such as B vitamins and vitamin D) and essential minerals (including iron, magnesium, and zinc). Their bioactive compounds, like beta-glucan, ergosterol, ergothioneine, and numerous polysaccharides, may further enhance their cardiometabolic benefits (Kirbağ and Akyüz 2010; Muszyńska et al. 2018).

### 7.5.4 Hypolipidemic and Prebiotic Potential

Mushrooms contain nondigestible polysaccharides that serve as dietary fiber, slowing glucose absorption and reducing blood cholesterol levels. These biopolymers may enhance their beneficial effects (Giavasis 2013; Chen et al.

2013). The hypolipidemic action is linked to increased fecal excretion of bile acids, while the hypoglycemic effects arise from their binding to the intestinal lining. For instance, lentinan reduces LDL and HDL levels, making it an effective cholesterol-lowering agent. Glucan-rich extracts from *P. ostreatus* have also shown promise in lowering atherogenic plaques in animal studies. Additionally, other compounds like lovastatin (mevinolin) from *Pleurotus* species may aid in lowering cholesterol (Bobek and Galbavý 1999; Lakhanpal and Rana 2005; Giavasis 2013).

Glucans from *S. cerevisiae* have also been shown to lower total cholesterol and raise HDL levels in obese men by consumption of 15 g daily (Waszkiewicz-Robak 2013). Even if native pullulan is not inherently bioactive, its chemically modified forms may exhibit bile acid-binding and cholesterol-lowering properties. Pullulan has been proposed as a potential prebiotic due to its ability to promote the growth of *Bifidobacteria* in the gut (Izydorczyk 2006; Giavasis 2013). A glucuronoxylomannan from *Tremella mesenterica* has been found to regulate blood glucose levels and reduce serum fructosamine concentrations in diabetic and normal rats, supporting its potential use as an oral hypoglycemic agent or as a functional food ingredient for diabetic individuals (Lo et al. 2006; Giavasis 2014).

### 7.5.5 Antidiabetic Effects of Mushrooms

Diabetes mellitus (DM) is a significant global health issue, especially type II diabetes, which often requires progression from lifestyle changes to antihyperglycemic medications due to hyperglycemia (Zimmet et al. 2014). While numerous treatment options exist, diabetes management is achieved through the use of antihyperglycemic agents. These mushrooms exhibit antidiabetic effects through mechanisms such as antioxidant action, blood glucose reduction, and stimulation of insulin secretion (Elkhateeb et al. 2020; Wa and Daba 2021). Several medicinal fungi, including *Ganoderma lucidum*, *Auricularia auricula-judae*, *Inonotus obliquus*, *Morchella conica*, *Coprinus comatus*, *Tremella fuciformis*, *Laetiporus sulphureus*, *Lentinula edodes*, *Agaricus blazei*, *Lignosus rhinocerotis*, *Grifola frondosa*, and various *Phellinus* spp., have demonstrated hypoglycemic effects in studies (De Silva et al. 2012; Elkhateeb et al. 2018).

The genus *Agaricus* is one of the most cultivated edible mushrooms worldwide, and it contains bioactive compounds that might play a central role in DM management. *Agaricus bisporus* (white button mushroom) is mainly rich in dietary fibers and antioxidants like vitamins C, D, and $B_{12}$; folates; and polyphenols, which may play a significant role in the improvement of cardiovascular and diabetic health (Jeong et al. 2010; Niwa et al. 2011). Ekowati et al. (2018) reported that administration of *A. bisporus* extract remarkably reduced blood glucose and malondialdehyde levels while increasing superoxide dismutase activity in alloxan-induced diabetic rats. Additionally, *A. bisporus* extract has shown potential as a natural antidiabetic agent suitable for industrial application (Ekowati et al. 2018). Abou et al. (2017) showed that *A. bisporus* exhibits antidiabetic effects by protecting pancreatic islets from streptozotocin-induced damage, promoting β-cell repair and regeneration, and stimulating insulin synthesis and secretion, likely due to its antioxidant properties. *Coprinus comatus*, which is commonly known as shaggy mane or lawyer's wig, is an edible mushroom widely cultivated in Asia and valued for its numerous physiological benefits, including antidiabetic, antioxidant, anticancer, hepatoprotective, and anti-inflammatory activities. It possesses antitumor, antimutagenic, and liver-protective properties, making it a nutritionally beneficial and therapeutically promising species (Cao et al. 2020; Nowakowski et al. 2020).

### 7.5.6 Antioxidant Activity of Mushrooms

Mushrooms possess potent antioxidant properties due to bioactive compounds like phenolics, beta-glucans, and ergothioneine. These compounds help combat oxidative stress, which is crucial for managing inflammation, cancers, and cellular health. As functional foods, mushrooms support immune function and overall well-being, showcasing their therapeutic potential (Al Qutaibi and Kagne 2024). Among the Throoms from Agaricomycetes—*Pleurotus columbinus*, *Lentinus sajor-caju*, and *Agaricus bisporus*—were systematically evaluated for antioxidant properties. *P. columbinus* exhibited the highest antioxidant activity in the DPPH radical scavenging assay. All extracts demonstrated superior activity compared to Trolox in the ABTS assay, indicating their significant potential for disease management and prevention (Elhusseiny et al. 2021).

Chowdhury et al. (2015) evaluated the antioxidant activity of several edible mushrooms cultivated in Bangladesh. Findings from their study indicated that total phenolic compounds were the predominant bioactive constituents in the mushroom extracts, quantified as milligrams of gallic acid equivalents (GAEs) per gram of fruiting body, with values ranging from 3.20 to 10.66 mg/ml. The coflavonoid concentration was also notable, ranging between 2.50 mg/ml and 4.76 mg/ml. At the same time, the ascorbic acid content was present only in trace amounts, with concentrations ranging from 0.06 to 0.21 mg/ml across all isolates. Most isolates showed high levels of phenolic and flavonoid compounds, excluding *Pleurotus ostreatus*, which showed comparatively lower values. Similarly, Kosanić et al. (2013) assessed the in vitro antioxidant properties of acetone and methanol extracts from *Amanita rubescens*, *Cantharellus cibarius*, *Lactarius piperatus*, and *Russula cyanoxantha*. Antioxidant potential was determined using four assays: free radical scavenging activity, reducing power, total phenolic content, and total flavonoid content. Among the tested extracts, the acetone extract of *Russula cyanoxantha* demonstrated the most vigorous potentiating activity. Furthermore, all tested extracts exhibited significant reducing power, indicating their potential as natural antioxidant sources.

## 7.6 The Use of Mushroom Extracts and Compounds

### 7.6.1 Mushroom Extracts and Compounds as a Functional Food

International health bodies recommend reducing overall fat intake by replacing saturated fats with unsaturated fats and avoiding trans fats (Alkerwi 2014). Diets rich in fruits, vegetables, legumes, whole grains, and nuts, while limiting salt and free sugar intake, are advised to help maintain a healthy weight and reduce disease risk. A balanced diet promotes overall health, safety, and well-being (Alkerwi 2014). Functional foods have gained attention for their potential health benefits, as they contain higher concentrations of beneficial components that improve bodily functions and may reduce disease risk (El Sohaimy 2012).

Plants and a plant-based diet, including mushrooms, are often considered healthy options. Their popularity has surged recently due to growing awareness of their nutritional and medicinal benefits. Rich in bioactive compounds, mushrooms offer numerous health benefits, including immunomodulatory, antitumor, cholesterol-lowering, antibacterial, antifungal, anti-inflammatory, antiviral, antidiabetic, and cardiovascular effects (Wasser 2014; Valverde et al. 2015), and are highly recommended to be considered as a functional food. Several mushroom species, such as *Agaricus bisporus* and *Lentinula edodes* (shiitake), have shown effectiveness against a broad spectrum of pathogens, including Gram-positive and Gram-negative bacteria, as well as fungi like *Candida albicans*, *Aspergillus*, and *Penicillium* species. These natural antimicrobial effects highlight mushrooms' potential use in health-promoting diets and support their classification as functional foods (Reis et al. 2017).

Additionally, mushrooms can be considered functional foods due to their strong immunomodulatory properties. The presence of bioactive compounds, especially polysaccharides like β-glucans, lectins, and FIPs, enhances the immune responses of mushrooms by activating key immune cells and promoting cytokine production. Species like *Lentinula edodes* and *Ganoderma lucidum* have shown immune-boosting and antitumor effects. Compounds such as lentinan, PSK, and PSP are even used as adjuvants in cancer therapy, highlighting mushrooms' potential in supporting immune health and disease prevention (Reis et al. 2017).

### 7.6.2 Mushroom Extracts and Compounds in Nutraceuticals

A nutraceutical is a food or natural product used to enhance health or treat disease, available in extracts or tablets (Howlett 2008; El Sohaimy 2012). Mushrooms can be consumed as nutraceuticals, or "mushroom pharmaceuticals," which are extracts or dried biomass from their mycelium or fruiting body. These are often offered in capsules or tablets and may boost the immune system, increasing disease resistance and potentially aiding disease regression (Giavasis 2014; Wasser 2014). The bioactive compounds in mushrooms, regarded as immunoceuticals, possess antitumor and immunomodulatory effects that target disease development (Morris et al. 2016).

Bioactive nutraceuticals have been mainly studied and developed as dietary supplements that help inhibit certain life-threatening diseases and protect the body from harmful environmental influences (Ma et al. 2018). Several antitumor polysaccharides have been developed from the fruiting bodies, mycelia, and culture media of various medicinal mushrooms, such as *Agaricus bisporus*, *A. brasiliensis*, *Auricularia auricula*, *Flammulina velutipes*, *Ganoderma lucidum*, *Grifola frondosa*, *Inonotus obliquus*, *Lentinula edodes*, *Pleurotus ostreatus*, *Schizophyllum commune*, *Trametes versicolor*, and *Tricholoma matsutake*, which show anticancer, antibacterial, antiviral, antifungal, antidiabetic, anti-inflammatory, and cardiovascular protective activities. Several pharmaceutical companies worldwide, including Zhejiang Fangge Pharmaceutical (China), FineCo Ltd. (South Korea), ecoNugenics, Mushroom Wisdom, MycoFormulas (USA), and others, are developing and marketing mushroom-based nutraceuticals for anticancer uses (Reis et al. 2017). In China, commercial nutraceutical products include extracts from *Ganoderma lucidum*, *Lentinula edodes*, *Poria cocos*, and *Cordyceps sinensis*, marketed for anticarcinogenic, antiviral, hypolipidemic, antiaging, and cardiovascular benefits (Xu 2001). Forms of mushroom-based dietary supplements on the market include fruiting body powders and extracts, dried preparations from mycelium and substrates, biomass from submerged culture, and spores or spore extracts (Wasser 2014). Commercial products such as Lentinan, Concord Sunchih, Reishi Plus, Grifon, and Didanosine are examples of mushroom-derived nutraceuticals available on the market, with additional clinical studies evaluating compounds like *Agaricus blazei* extract, AHCC, Hispidin, and Hispolon (Reis et al. 2017). However, challenges such as standardization, safety, regulatory acceptance, and clinical validation have remained (Wasser 2014; Rathore et al. 2017), and while the Asian market has capitalized on mushroom nutraceuticals, the Western market still requires more investment and development (Giavasis 2014). In Europe, these products are sold as dietary supplements under food law, prohibiting medicinal claims (Chang and Wasser 2012).

## 7.7 Mushroom-Based Functional Food Products

With advancements in food science, people are becoming more aware of the chemical composition of foods to address nutritional needs and manage metabolic diseases like gout. To avoid high purine levels, patients typically steer clear of seafood, red meat, and high-fructose fruit like mango.

Many functional foods contain bioactive compounds with therapeutic properties, such as anti-gout and anti-inflammatory effects. For instance, individuals with vitamin A deficiency can consume carrot juice, which is rich in beta-carotene, to enhance night vision. Biotin supplements may also benefit those with brittle nails and hair thinning. Fish oil, high in omega-3 fatty acids, can help reduce cholesterol levels. Despite around 1600 mushroom species worldwide, only about 33 are cultivated, with just three widely grown for consumption (Kumar et al. 2021).

### 7.7.1 Processed Forms of Mushrooms (Powders, Extracts, Tinctures)

Mushrooms have been processed into powders, extracts, and tinctures. Mushroom powder is produced by cutting, drying, and grinding into fine particles and can be easily added to food such as soups, baked goods, and smoothies as a seasoning to give umami flavor. Traditionally, mushroom extracts are obtained by solvent-assisted extraction, soaking mushroom bodies into water or organic solvents at various concentrations to optimize the recovery of bioactive compounds in mushrooms, including polysaccharides (beta-glucans) and proteins. Other common conventional techniques are percolation, the Soxhlet method, and pressurized liquid extraction. This old method is not sustainable due to the required long extraction time, high volume of solvent, and high temperature, which may result in the breakdown of heat-sensitive compounds. Kumar et al. (2021) reported advanced extraction techniques to recover bioactive compounds in mushrooms, including enzyme extraction, ultra-sound assisted, microwave-assisted, pulsed electric field-assisted, and subcritical water. Tinctures are alcohol-based extracts where alcohol is the primary solvent. Some tinctures require dual extraction using water and alcohol, especially for tougher mushrooms such as chaga and reishi, which have both alcohol-soluble and water-soluble bioactive compounds. The combination of dual extraction using water and alcohol (ethanol) ensures both water-soluble (β-glucans, polysaccharides, proteoglycans, glycoproteins, and peptides) and alcohol-soluble (triterpenes, sterols, flavonoids, phenolic compounds, cordycepin, hericenones, and ergosterol peroxide) compounds are recovered. Tinctures are extensively used in herbal medicine due to the quick absorption of mushroom compounds. In China, the split gill mushroom was made into tea and folk medicine, or mushroom tincture, that can cure gynecological issues, enhance immunity, and increase vitality.

### 7.7.2 Mushroom-Based Snacks and Supplements

Oyster and white buttom mushrooms had been utilized in bakery products for enrichment or replacement such as noodles (as wheat flour substitution, increase protein and fiber content), enrichment of maize flour (increase protein, fiber, minerals (zinc and iron), foam capacity and stability, water retention and absorption capacity, reduction in fat content and energy, compact and bulk density), substitution of wheat flour (increase protein, fiber, ash and minerals including Zn, K, Na, Ca, Mg, Mn, Fe, and Cu), substitute wheat flour for biscuits production (increase in protein, volume and spread ratio), replacement of wheat flour for cake production (enhance texture, flavor and color of the cake). In addition, supplementation of mushroom extract in dairy products, including yogurt, showed fat reduction, antimicrobial properties against coliform bacteria, yeast, and molds, and an increase in CFU counts for probiotic strains of both *Streptococcus thermophilus* and *Lactobacillus bulgaricus*. Split gill mushrooms have been made into many dishes around the world. For example, in the northeastern Indian state of Manipur, it is known as kanglayen and is one of the preferred ingredients for paaknam, the Manipuri-style pancakes. Pasi is the native name for one of Mizoram's most revered edible mushrooms (pa means mushroom and si means little).

Cookies significantly increase the amount of iron, calcium, protein, calories, fiber, and some of the B vitamins in our diet. The staple baked good of the human diet, biscuits are frequently paired with beverages and served to young children as weaning foods (Ikuomola et al. 2017). One example of mushroom products is fortified cookies, where the mushroom is dried into a stable form, which is known as composite flour. A composite flour is a blend of starch-rich flours from tubers like cassava, yams, and potatoes, as well as protein-rich flour and cereals, with or without wheat flour, that was developed to meet specific functional requirements and nutrient content (Hasmadi et al. 2020).

Cosmetics are a group of personal care products used to clean, brighten up, or change how the skin or body looks, or to make it healthier. Cosmetics are made up of chemical compounds from natural or artificial sources. Mushrooms, also known as natural source ingredients, are used in the cosmeceutical industry. This idea was already well known in Asian countries like China, Korea, and Japan, where certain mushrooms supplemented it. Hyaluronic acid consists of a polysaccharide that is mainly found in moisturizers. Therefore, if the mushroom-based moisturizer contains a high volume of polysaccharides, it will positively impact the skin, which can retain moisture and prevent skin dehydration. Hence, the skin will be maintained moist for the entire day (Kanaporn et al. 2021).

### 7.7.3 Mushroom Proteins and Their Application in Plant-Based Foods

Oyster mushrooms (*Pleurotus ostreatus*) and white button mushrooms (*Agaricus bisporus L.*) have been used to improve the quality, flavor, and nutritional values of a few plant-based foods such as beef salami, frankfurter, and liver pate. The addition of mushrooms aids in inhibiting lipid oxidation, increasing moisture content, reducing fat content, increasing juiciness, increasing protein content, and increasing antimicrobial effects toward *Pseudomonas* sp. (Bulam et al. 2022).

The need for easier-to-use, healthier food products that customers can readily incorporate into their diets is rising. The food industry is reformulating and developing new products to make its products healthier and more sustainable. The development of meals for the flexitarian market is the main topic of this essay. It is done by substituting a portion of meat or fish products with mushrooms. A study about the effect of beef patties fortified with mushrooms on lipid profile and lipid enzymes in rats by El Zainy and Adrabou (2011) has shown that after 30 days, as opposed to 15 days, hypercholesterolemic rats showed significantly decreased levels of total cholesterol, triglycerides, total lipids, phospholipids, LDL-c and VLDL-c, atherogenic index, and the liver enzymes GOT and GPT.

Additionally, the reduction of hypercholesterolemic impact was enhanced by increasing the amount of dry mushroom fortification added to beef patties. This further proved that beef burgers made with mushrooms can become a high-protein snack that improves renal blood flow. It was also demonstrated that at 0, during, and after 6 months of frozen storage, fortification with dried mushrooms at levels of 4%, 8%, and 12% positively impacted all beef patties' chemical and physical attributes. Compared to beef patties fortified with 12% dried mushrooms, those treated with 4% or 8% dried mushrooms or none exhibited better organoleptic characteristics (El-Refai et al. 2014).

## 7.8 Conclusion

The availability of various mushroom species worldwide and their nutritional profiles have been presented as well. It has been shown that mushrooms possess many bioactive components such as sterols, polyphenols, polysaccharides, glucan, terpenes, terpenoids, vitamins, and dietary fiber. This goodness contributes toward immune system modulation and anti-inflammatory and anticancer properties, cardiovascular and metabolic health benefits, hypolipidemic and prebiotic potential, and antioxidant and antidiabetic potential for consumers. The notion of functional food drew consideration by boosting more evidence linking a balanced diet to overall health quality with developments in food science and technology. The bioactive components from mushrooms can be added to functional food and nutraceutical products targeting certain biological processes to alleviate the development or minimize existing disease. Many commercial products are available after clinical studies. However, challenges such as safety and regulatory acceptance are still ambiguous due to limitations in medicinal claims imposed by food industrialists.

**Acknowledgments** The authors would like to thank the lab staff at Universiti Teknologi MARA for the facilities and assistance provided during the completion of this research.

**Declaration** The authors do not have any competing interests.

## References

Abou ZOA, Sonbaty SE, Neama MA (2017) Anti-diabetic activity of *Agaricus bisporus*: a biochemical and pathological study. Int J Pharm Sci Res 7(2):1740–1745

Al Qutaibi M, Kagne SR (2024) Exploring the phytochemical compositions, antioxidant activity, and nutritional potentials of edible and medicinal mushrooms. Int J Microbiol 2024(1):6660423

Alkerwi AA (2014) Diet quality concept. Nutrition 30(6):613–618

Atila F, Owaid MN, Shariati MA (2017) The nutritional and medical benefits of *Agaricus bisporus*: a review. J Microbiol Biotechnol Food Sci 7:281–286

Bobek P, Galbavý Š (1999) Hypocholesterolemic and antiatherogenic effect of oyster mushroom (*Pleurotus ostreatus*) in rabbits. Food/Nahrung 43(5):339–342

Bohn JA, BeMiller JN (1995) (1→3)-β-D-glucans as biological response modifiers: a review of structure–functional activity relationships. Carbohydr Polym 28(1):3–14

Bulam S, Ustun NS, Peksen A (2022) Oyster mushroom (*Pleurotus ostreatus*) as a healthy ingredient for sustainable functional food production. J Fungus 13:131–143
Cao H, Qin D, Guo H, Cui X, Wang S, Wu Y, Han C (2020) The shaggy ink cap medicinal mushroom, *Coprinus comatus* (Agaricomycetes), a versatile functional species: a review. Int J Med Mushrooms 22(3):245–255
Chan GCF, Chan WK, Sze DMY (2009) The effects of β-glucan on human immune and cancer cells. J Hematol Oncol 2(1):25
Chang ST (2008) Overview of mushroom cultivation and utilization as functional foods. In: Mushrooms as functional foods. Wiley, pp 1–33
Chang ST, Miles PG (1992) Mushroom biology—a new discipline. Mycologist 6:64–65
Chang ST, Wasser SP (2012) The role of culinary medicinal mushrooms on human welfare with a pyramid model for human health. Int J Med Mushrooms 1:95–134
Chen PX, Wang S, Nie S, Marcone M (2013) Properties of cordyceps sinensis: a review. J Funct Foods 5(2):550–569
Chen HP, Dong WB, Feng T, Yin X, Li ZH, Dong ZJ, Liu JK (2014) Four new sesquiterpenoids from fruiting bodies of the fungus *Inonotus rickii*. J Asian Nat Prod Res 16(6):581–586
Cheung PC (2010) The nutritional and health benefits of mushrooms. Nutr Bull 35(4):292–299
Chowdhury MMH, Kubra K, Ahmed SR (2015) Screening of antimicrobial, antioxidant properties and bioactive compounds of some edible mushrooms cultivated in Bangladesh. Annals Clin Microbiol Antimicrob 14(1):8
Das K, Das NK, Rathore VS, Kundu S, Choudhury S, Gharami RC, Datta PK (2010) Keratoacanthoma centrifugum marginatum: unresponsive to oral retinoid and successfully treated with wide local excision. Dermatol Rep 2(1):e4
De Silva DD, Rapior S, Hyde KD, Bahkali AH (2012) Medicinal mushrooms in prevention and control of diabetes mellitus. Fungal Divers 56(1):1–29
de Waard AM, Hollander M, Korevaar JC, Nielen MM, Carlsson AC, Lionis C, Seifert B, Thilsing T, de Wit NJ, Schellevis FG (2019) Selective prevention of cardiometabolic diseases: activities and attitudes of general practitioners across Europe. Eur J Pub Health 29:88–93
Devi PV, Islam J, Narzary P, Sharma D, Sultana F (2024) Bioactive compounds, nutraceutical values and its application in food product development of oyster mushroom. J Future Foods 4(4):335–342
Dudekula UT, Doriya K, Devarai SK (2020) A critical review on submerged production of mushroom and their bioactive metabolites. 3 Biotech 10(8):337
Dudhgaonkar S, Thyagarajan A, Sliva D (2009) Suppression of the inflammatory response by triterpenes isolated from the mushroom *Ganoderma lucidum*. Int Immunopharmacol 9(11): 1272–1280
Duru ME, Çayan GT (2015) Biologically active terpenoids from mushroom origin: a review. Rec Nat Prod 9(4):456–474
Ekowati N, Yuniati NI, Hernayanti H, Ratnaningtyas NI (2018) Antidiabetic potentials of button mushroom (*Agaricus bisporus*) on alloxan-induced diabetic rats. Biosaintifika 10(3):655–662
El Enshasy H, Elsayed EA, Aziz R, Wadaan MA (2013) Mushrooms and truffles: historical biofactories for complementary medicine in Africa and in the Middle East. Evid Based Complement Alternat Med 2013:620451
El Sohaimy SA (2012) Functional foods and nutraceuticals-modern approach to food science. World Appl Sci J 20(5):691–708
El Zainy R, Adrabou E (2011) Effect of feeding on beef patties fortifited with mushroom on serum lipids profile and liver enzymes of hypercholesterolemic rats. J Food Dairy Sci Mansoura Univ 2(8):411–428
Elhusseiny SM, El-Mahdy TS, Awad MF, Elleboudy NS, Farag MM, Aboshanab KM, Yassien MA (2021) Antiviral, cytotoxic, and antioxidant activities of three edible agaricomycetes mushrooms: *Pleurotus columbinus*, *Pleurotus sajor-caju*, and *Agaricus bisporus*. J Fungi 7(8):645

Elkhateeb WA, El-Hagrassi AM, Fayad W, El-Manawaty MA, Zaghlol GM, Daba GM, Ahmed EF (2018) Cytotoxicity and hypoglycemic effect of the Japanese jelly mushroom *Auricularia auricula-judae*. Chem Res J 3:123–133

Elkhateeb W, ElDien AN, Fadl E, Elhagrasi A, Fayad W, Wen TC (2020) Therapeutic potentials of n-hexane extracts of three medicinal mushrooms regarding their anti-colon cancer, antioxidant, and hypocholesterolemic capabilities. Biodiversitas 21(6):2625–2634

El-Refai A, El-Zeiny AR, El Abd Rabo EA (2014) Quality attributes of mushroom-beef patties as a functional meat product. J Hyg Eng Des 6:49–62

Fuller S, Beck E, Salman H, Tapsell L (2016) New horizons for the study of dietary fiber and health: a review. Plant Foods Human Nutr 71(1):1–12

Gao Y, Zhou S, Chen G, Dai X, Ye J (2002) A phase I/II study of a Ganoderma lucidum (Curt.: Fr.) P. Karst. extract (Ganopoly) in patients with advanced cancer. Int J Med Mushr 4(3):207–214

Gao GS, Zhang T, Wu SR, Sang L (2012) Quality evaluation of protein in the edible fungus and utilizing the complementary principle of amino acid to improve the nutritional value of protein. Edible Fungi China 31:35–38

Ghosh K (2016) A review: edible mushrooms as source of dietary fiber and its health effects. J Drug Deliv Ther 6(6):94–99

Giavasis I (2013) Production of microbial polysaccharides for use in food. In: Juneja VK, Sofos JN (eds) Microbial production of food ingredients, enzymes and nutraceuticals. Woodhead Publishing, pp 413–468

Giavasis I (2014) Bioactive fungal polysaccharides as potential functional ingredients in food and nutraceuticals. Curr Opin Biotechnol 26:162–173

Gnanwa JM, Soro LC, Fagbohoun JB, Yorou NS, Kouame LP (2021) Assessment of minerals, vitamins, amino and fatty acids components of Pleurotus ostreatus mushrooms cultivated and sold in the village of M'Badon (Abidjan, Côte d'Ivoire). Int J Curr Microbiol Appl Sci 10(9): 276–283

Gu CM, Zhang DB, Zhai WJ, Zhang HP, Wang SD, Lv SY (2022) Research progress on cordyceps militaris polysaccharides. Food Biosci 45:101503

Gunde-Cimerman N, Plemenitas A (2001) Hypocholesterolemic activity of the genus Pleurotus (Jacq.: Fr.) P. Kumm. (Agaricales s. I., Basidiomycetes). Int J Med Mushrooms 3(4):395–399

Guo J, Huang X, Dou L, Yan M, Shen T, Tang W, Li J (2022) Aging and aging-related diseases: from molecular mechanisms to interventions and treatments. Signal Transduct Target Ther 7(1): 391

Hasmadi M, Noorfarahzilah M, Noraidah H, Jahurul MHA (2020) Functional properties of composite flour: a review. Food Res 4(6):1820–1831

He Y, Wang BX, Wen LK, Wang FZ, Yu HS, Chen DX (2022) Effects of dietary fiber on human health. Food Sci Human Wellness 11(1):1–10

Hong JY, Kim MK, Yang N (2024) Mushroom consumption and cardiometabolic health outcomes in the general population: a systematic review. Nutr Res Pract 8(2):165–179

Howlett J (2008) Functional foods: from science to health and claims. ILSI Europe, Bruxelles

Huang SJ, Lin CP, Tsai SY (2015) Vitamin $D_2$ content and antioxidant properties of fruit body and mycelia of edible mushrooms by UV-B irradiation. J Food Compos Anal 42:38–45

Hur SJ, Cho SY, Lim BO (2012) In vitro anti-inflammatory activity of Russula virescens in the macrophage-like cell line RAW 264.7 activated by lipopolysaccharide. J Nutr Food Sci 2(4):154

Ikekawa T (2001) Beneficial effects of edible and medicinal mushrooms on health care. Int J Med Mushrooms 3(4):291–298

Ikuomola DS, Otutu OL, Oluniran DD (2017) Quality assessment of cookies produced from wheat flour and malted barley (*Hordeum vulgare*) bran blends. Cogent Food Agric 3(1). https://doi.org/10.1080/23311932.2017.1293471

Iman V, Mohan S, Abdelwahab SI, Karimian H, Nordin N, Fadaeinasab M, Noor SM (2016) Anticancer and anti-inflammatory activities of girinimbine isolated from *Murraya koenigii*. Drug Des Dev Ther 10:103–121

Izydorczyk MS (2006) Functional food carbohydrates. CRC Press, Boca Raton

Jedinak A, Dudhgaonkar S, Wu QL, Simon J, Sliva D (2011) Anti-inflammatory activity of edible oyster mushroom is mediated through the inhibition of NF-κB and AP-1 signaling. Nutr J 10(1): 52

Jeong SC, Jeong YT, Yang BK, Islam R, Koyyalamudi SR, Pang G, Song CH (2010) White button mushroom (*Agaricus bisporus*) lowers blood glucose and cholesterol levels in diabetic and hypercholesterolemic rats. Nutr Res 30(1):49–56

Jeong SC, Koyyalamudi SR, Jeong YT, Song CH, Pang G (2012) Macrophage immunomodulating and antitumor activities of polysaccharides isolated from *Agaricus bisporus* white button mushrooms. J Med Food 15(1):58–65

Jo WS, Choi YJ, Mm HJ, Lee JY, Nam BH, Lee JD, Jeong MH (2010) The anti-inflammatory effects of water extract from *Cordyceps militaris* in murine macrophage. Mycobiology 38(1): 46–51

Joseph SV, Edirisinghe I, Burton-Freeman BM (2014) Berries: anti-inflammatory effects in humans. J Agric Food Chem 62(18):3886–3903

Kalač P (2009) Chemical composition and nutritional value of European species of wild growing mushrooms: a review. Food Chem 113(1):9–16

Kanaporn S, Nakarin S, Jaturong K, Thanasak L (2021) Mushrooms: splendid gifts for the cosmetic industry. Chiang Mai J Sci 48(3):699–725

Kamiyama T, Yokoo H, Furukawa JI, Kurogochi M, Togashi T, Miura N, Nakanishi K, Kamachi H, Kakisaka T, Tsuruga Y, Fujiyoshi M, Taketomi A, Nishimura S and Todo S (2013) Identification of novel serum biomarkers of hepatocellular carcinoma using glycomic analysis. Hepatology 57(6):2314–2325

Khan MA, Tania M (2012) Nutritional and medicinal importance of Pleurotus mushrooms: an overview. Food Rev Int 28(3):313–329

Kim YW, Kim KH, Choi HJ, Lee DS (2005) Anti-diabetic activity of β-glucans and their enzymatically hydrolyzed oligosaccharides from *Agaricus blazei*. Biotechnol Lett 27:483–487

Kirbağ S, Akyüz M (2010) Nutritive value of edible wild and cultured mushrooms. Turk J Biol 34(1):97–102

Kobori M, Yoshida M, Ohnishi-Kameyama M, Shinmoto H (2007) Ergosterol peroxide from an edible mushroom suppresses inflammatory responses in RAW264.7 macrophages and growth of HT29 colon adenocarcinoma cells. Br J Pharm 150(2):209–219

Kosanić M, Ranković B, Dasić M (2013) Antioxidant and antimicrobial properties of mushrooms. Bulg J Agric Sci 19(5):1040–1046

Kotowski MA (2019) History of mushroom consumption and its impact on traditional view on mycobiota—an example from Poland. Microb Biosyst 4(1):1–13

Kulicke WM, Lettau AI, Thielking H (1997) Correlation between immunological activity, molar mass, and molecular structure of different (1→3)-β-D-glucans. Carbohydr Res 297(2):135–143

Kumar K, Mehra R, Guiné RPF, Lima MJ, Kumar N, Kaushik R, Ahmed N, Yadav AN, Kumar H (2021) Edible mushrooms: a comprehensive review on bioactive compounds with health benefits and processing aspects. Foods 10(12):2996

Lakhanpal TN, Rana M (2005) Medicinal and nutraceutical genetic resources of mushrooms. Plant Genet Resour 3(2):288–303

Lindequist U, Niedermeyer TH, Jülich WD (2005) The pharmacological potential of mushrooms. Evid Based Complement Alternat Med 2(3):285–299

Liu J, Du C, Wang Y, Yu Z (2015) Anti-fatigue activities of polysaccharides extracted from *Hericium erinaceus*. Exp Ther Med 9(2):483–487

Lo HC, Tsai FA, Wasser SP, Yang JG, Huang BM (2006) Effects of ingested fruiting bodies, submerged culture biomass, and acidic polysaccharide glucuronoxylomannan of tremella mesenterica Retz.: Fr. on glycemic responses in normal and diabetic rats. Life Sci 78(17): 1957–1966

Ma K, Bao L, Han J, Jin T, Yang X, Zhao F, Liu H (2014) New benzoate derivatives and hirsutane type sesquiterpenoids with antimicrobial activity and cytotoxicity from the solid-state fermented rice by the medicinal mushroom *Stereum hirsutum*. Food Chem 143:239–245

Ma G, Yang W, Zhao L, Pei F, Fang D, Hu Q (2018) A critical review on the health promoting effects of mushroom nutraceuticals. Food Sci Human Wellness 7:125–133

Mollet B, Rowland I (2002) Functional foods: at the frontier between food and pharma. Curr Opin Biotechnol 13(5):483–485

Moro C, Palacios I, Lozano M, D'Arrigo M, Guillamón E, Villares A, García-Lafuente A (2012) Anti-inflammatory activity of methanolic extracts from edible mushrooms in LPS activated RAW 264.7 macrophages. Food Chem 130(2):350–355

Morris HJ, Llaurado G, Beltran Y, Lebeque Y, Bermudez RC, Garcia N, Gaime-Perraud I, Moukha S (2016) The use of mushrooms in the development of functional foods, drugs or nutraceuticals. In: Wild plants, mushrooms and nuts: functional food properties and applications. Wiley, pp 123–157

Muhammad BL, Suleiman B (2015) Global development of mushroom biotechnology. Int J Emerg Trends Sci Technol 2(6):2660–2669

Muszyńska B, Grzywacz-Kisielewska A, Kała K, Gdula-Argasińska J (2018) Anti-inflammatory properties of edible mushrooms: a review. Food Chem 243:373–381

Nie Y, Luo F (2021) Dietary fiber: An opportunity for a global control of hyperlipidemia. Oxid Med Cell Longev 2021:5542342

Niwa A, Tajiri T, Higashino H (2011) *Ipomoea batatas* and *Agaricus blazei* ameliorate diabetic disorders with therapeutic antioxidant potential in streptozotocin-induced diabetic rats. J Clin Biochem Nutr 48(3):194–202

Nowakowski P, Naliwajko SK, Markiewicz-Żukowska R, Borawska MH, Socha K (2020) The two faces of *Coprinus comatus*—functional properties and potential hazards. Phytother Res 34(11): 2932–2944

Ohno N, Miura T, Miura NN, Adachi Y, Yadomae T (2001) Structure and biological activities of hypochlorite oxidized zymosan. Carbohydr Polym 44(4):339–349

Oso BA (1975) Mushrooms and the Yoruba people of Nigeria. Mycol 67(2):311–319

Peng H, Shahidi F (2022) Qualitative analysis of secondary metabolites of chaga mushroom (*Inonotus obliquus*): phenolics, fatty acids, and terpenoids. J Food Bioact 17. https://doi.org/10.31665/JFB.2022.17304

Power RC, Salazar-García DC, Straus LG, Morales MRG, Henry AG (2015) Microremains from El Mirón cave human dental calculus suggest a mixed plant–animal subsistence economy during the Magdalenian in northern Iberia. J Archaeol Sci 60:39–46

Prastiyanto ME, Rukmana RM, Saraswati DK, Darmawati S, Maharani ETW, Tursinawati Y (2020) Anticancer potential of methanolic extracts from Pleurotus species on raji cells and antibacterial activity against methicillin-resistant *Staphylococcus aureus*. Biodiversitas 21(12). https://doi.org/10.13057/biodiv/d211221

Rahi DK, Malik D (2016) Diversity of mushrooms and their metabolites of nutraceutical and therapeutic significance. J Mycol 2016:1–18

Rathore H, Prasad S, Sharma S (2017) Mushroom nutraceuticals for improved nutrition and better human health: a review. PharmaNutrition 5(2):35–46

Raut JK (2019) Current status, challenges and prospects of mushroom industry in Nepal. Int J Agric Econ 4(4):154–160

Reis FS, Martins A, Vasconcelos MH, Morales P, Ferreira IC (2017) Functional foods based on extracts or compounds derived from mushrooms. Trends Food Sci Technol 66:48–62

Sá AGA, Moreno YMF, Carciofi BAM (2020) Plant proteins as high-quality nutritional source for human diet. Trends Food Sci Technol 97:170–184

Sakakibara BM, Obembe AO, Eng JJ (2019) The prevalence of cardiometabolic multimorbidity and its association with physical activity, diet, and stress in Canada: evidence from a population-based cross-sectional study. BMC Public Health 19(1):1–9

Samorini G (1992) The oldest representations of hallucinogenic mushrooms in the world (Sahara Desert, 9000–7000 BP). Integration 2(3):69–78

Šandula J, Kogan G, Kačuráková M, Machová E (1999) Microbial (1→3)-β-D-glucans, their preparation, physico-chemical characterization and immunomodulatory activity. Carbohydr Polym 38(3):247–253

Sharpe E, Farragher-Gnadt AP, Igbanugo M, Huber T, Michelotti JC, Milenkowic A, Bou-Abdallah F (2021) Comparison of antioxidant activity and extraction techniques for commercially and laboratory prepared extracts from six mushroom species. J Agric Food Res 4:100130

Taofiq O, Martins A, Barreiro MF, Ferreira IC (2016) Anti-inflammatory potential of mushroom extracts and isolated metabolites. Trends Food Sci Technol 50:193–210

Thakur MP (2020) Advances in mushroom production: key to food, nutritional and employment security: a review. Indian Phytopathol 73:377–395

Tian BM, Pan YZ, Wang J, Cai M, Ye BW, Yang K (2022) Insoluble dietary fibers from by-products of edible fungi industry: basic structure, physicochemical properties, and their effects on energy intake. Front Nutr 9:851228

Valverde ME, Hernández-Pérez T, Paredes-López O (2015) Edible mushrooms: improving human health and promoting quality life. Int J Microbiol 2015(1):376387

van Griensven LJLD (2009) Mushrooms, must action be taken? In: The 5th international medicinal mushroom conference, Nantong, China, pp 407–412

Vannucci L, Krizan J, Sima P, Stakheev D, Caja F, Rajsiglova L, Saieh M (2013) Immunostimulatory properties and antitumor activities of glucans. Int J Oncol 43(2):357–364

Villares A, Mateo-Vivaracho L, Guillamón E (2012) Structural features and healthy properties of polysaccharides occurring in mushrooms. Agri 2(4):452–471

Vishwakarma K, Upadhyay N, Kumar N, Yadav G, Singh J, Mishra RK, Sharma S (2017) Abscisic acid signaling and abiotic stress tolerance in plants: a review on current knowledge and future prospects. Front Plant Sci 8:161

Wa A, Daba GM (2021) Highlights on the golden mushroom *Cantharellus cibarius* and unique shaggy ink cap mushroom *Coprinus comatus* and smoky bracket mushroom *Bjerkandera adusta* ecology and biological activities. Open Access J Mycol Mycol Sci 4:1–8

Wang M, Zhao R (2023) A review on nutritional advantages of edible mushrooms and its industrialization development situation in protein meat analogues. J Future Foods 3(1):1–7

Wang Y, Bao L, Yang X, Li L, Li S, Gao H, Liu HW (2012) Bioactive sesquiterpenoids from the solid culture of the edible mushroom *Flammulina velutipes* growing on cooked rice. Food Chem 132(3):1346–1353

Wang SJ, Bao L, Han JJ, Wang QX, Yang XL, Wen HA, Liu HW (2013) Pleurospiroketals A–E, perhydrobenzannulated 5,5-spiroketal sesquiterpenes from the edible mushroom *Pleurotus cornucopiae*. J Nat Prod 76(1):45–50

Wani BA, Bodha RH, Wani AH (2010) Nutritional and medicinal importance of mushrooms. J Med Plants Res 4(24):2598–2604

Wasser SP (2014) Medicinal mushroom science: current perspectives, advances, evidences, and challenges. Biomed J 37:345–356

Waszkiewicz-Robak B (2013) Spent brewer's yeast and beta-glucans isolated from them as diet components modifying blood lipid metabolism disturbed by an atherogenic diet. In: Lipid metabolism. IntechOpen, pp 261–290

WHO J, Consultation FE (2003) Diet, nutrition and the prevention of chronic diseases. world Health Organ Tech Rep Ser 916(1–8):1–149

Yu QN, Guo MJ, Zhang B, Wu H, Zhang Y, Zhang LT (2020) Analysis of nutritional composition in 23 kinds of edible fungi. J Food Qual 2020:1–9

Zhang L, Gong ZQ, Wang WL, Cao H, Yu MM, Ge LL (2017) Analysis of flavor components and evaluation on umami of seven kinds of edible fungi. Food Sci Technol 42(3):274–278

Zhang YR, Wang DW, Chen YT, Liu TT, Zhang SS, Fan HX (2021) Healthy function and high valued utilization of edible fungi. Food Sci Human Wellness 10(4):408–420

Zheng J, Wang Y, Wang J, Liu P, Li J, Zhu W (2013) Antimicrobial ergosteroids and pyrrole derivatives from halotolerant *Aspergillus flocculosus* PT05-1 cultured in a hypersaline medium. Extremophiles 17:963–971

Zhu F, Du B, Xu B (2018) Anti-inflammatory effects of phytochemicals from fruits, vegetables, and food legumes: a review. Crit Rev Food Sci Nutr 58(8):1260–1270

Zimmet PZ, Magliano DJ, Herman WH, Shaw JE (2014) Diabetes: a 21st century challenge. Lancet Diabetes Endocrinol 2(1):56–64

# Functional Roles of Mushroom-Derived Polysaccharides in Human Health

8

Juliana Garcia, Irene Gouvinhas, Jani Silva, Eva Olo-Fontinha, and Maria José Alves

**Abstract**

Edible and medicinal mushrooms provide structurally distinctive polysaccharides, dominated by β-glucans with β(1 → 3) backbones and β(1 → 6) branches, whose molecular weight, branching, conformation, and solubility regulate bioactivity and vary with species and extraction, motivating rigorous chemotyping across studies. These polymers engage conserved pattern-recognition circuitry (notably dectin-1 → spleen tyrosine kinase (Syk) → CARD9–BCL10–MALT1(CBM) → nuclear factor kappa B (NF-κB),

Irene Gouvinhas, Jani Silva and Eva Olo-Fontinha contributed equally with all other contributors.

J. Garcia (✉)
AquaValor – Centro de Valorização e Transferência de Tecnologia da Água – Associação, Chaves, Portugal

LiveWell – Research Centre for Active Living & Wellbeing, Instituto Politécnico de Bragança, Bragança, Portugal

CITAB – Centre for the Research and Technology of Agro-Environment and Biological Sciences, Inov4Agro, University of Trás-os-Montes e Alto Douro, Vila Real, Portugal
e-mail: juliana.garcia@aquavalor.pt

I. Gouvinhas
Animal and Veterinary Research Center (CECAV), University of Trás-os-Montes and Alto Douro, Vila Real, Portugal, Vila Real, Portugal

Associate Laboratory of Animal and Veterinary Sciences (AL4AnimalS), Vila Real, Portugal, Vila Real, Portugal

J. Silva
AquaValor – Centro de Valorização e Transferência de Tecnologia da Água – Associação, Chaves, Portugal

LiveWell – Research Centre for Active Living & Wellbeing, Instituto Politécnico de Bragança, Bragança, Portugal

 
B. Giri et al. (eds.), *Advances in Mushroom Research*,
https://doi.org/10.1007/978-981-95-7692-0_8

with complement receptor 3 (CR3)/toll-like receptor 2 (TLR2) crosstalk) and, beyond acute signaling, can imprint trained immunity; ex vivo work with *Agaricus bisporus* β-glucans shows canonical chromatin and metabolic reprogramming of human monocytes. Across indications, the most mature oncology data concern *Trametes versicolor* products (polysaccharopeptides (PSP) and polysaccharide-K (PSK)) and lentinan from *Lentinula edodes*: meta-analyses and cohorts in resected or advanced gastric cancer report survival benefits when lentinan is added to chemotherapy, and historical randomized trials with schizophyllan (sizofiran) plus radiotherapy improved tumor control in Stage II cervical cancer. In infectious-disease contexts, parenteral lentinan is used as an immunostimulatory adjuvant in parts of Asia, while oral pleuran (*Pleurotus ostreatus*) is positioned for respiratory support; recent multicenter pediatric data indicate improved asthma control and fewer respiratory infections, whereas a pediatric gastroenteritis randomized controlled trial (RCT) was negative, emphasizing indication-specific effects and disease-formulation matching. At the gut interface, mechanistic colitis models delineate epithelial (dectin–1–tumor necrosis factor receptor 1 (TNFR1)) and myeloid checkpoints targeted by oral lentinan. Convergent evidence indicates that β-glucans act as prebiotic-like substrates, increasing short-chain fatty acid (SCFA) production, reinforcing tight junctions, and attenuating endotoxemia. Neurocognitive signals include randomized trials of *Hericium erinaceus* in mild cognitive impairment (MCI)/early Alzheimer's disease (AD) alongside cohort and biomarker data that implicate neurotrophic and microbiota–SCFAs pathways. However, key translational gaps persist in product heterogeneity, small underpowered trials, and inconsistent endpoints. This chapter therefore prioritizes (i) chemistry-anchored standardization, (ii) matrix-aware delivery (including colon-targeted strategies), and (iii) biomarker-linked clinical designs (trained immunity, barrier, and microbiome readouts) to enable dose–response modeling and responder stratification across oncology, infection, metabolism, gut, and brain health.

E. Olo-Fontinha
AquaValor – Centro de Valorização e Transferência de Tecnologia da Água – Associação, Chaves, Portugal

LiveWell – Research Centre for Active Living & Wellbeing, Instituto Politécnico de Bragança, Bragança, Portugal

Universidade Católica Portuguesa, CBQF – Centro de Biotecnologia e Química Fina, Laboratório Associado, Escola Superior de Biotecnologia, Porto, Portugal

M. J. Alves
AquaValor – Centro de Valorização e Transferência de Tecnologia da Água – Associação, Chaves, Portugal

LiveWell – Research Centre for Active Living & Wellbeing, Instituto Politécnico de Bragança, Bragança, Portugal

CIMO – Centro de Investigação de Montanha, Instituto Politécnico de Bragança, Bragança, Portugal

**Keywords**

Mushroom β-glucans · Immunomodulation · Oncology · Gut–brain axis · Microbiota

## 8.1 Introduction

Polysaccharides from edible and medicinal mushrooms, especially β-glucans with a β(1 → 3) backbone and β(1 → 6) branches, are increasingly recognized as biologically active compounds capable of modulating key physiological systems. These macromolecules exert multiple biological effects on mucosal and systemic immunity, host–microbe interactions, and metabolic homeostasis, positioning them as potential nutraceutical agents with relevance across immunological, metabolic, and inflammatory disorders. The integrative nature of these effects is summarized in Fig. 8.1. Mushroom β-glucans are structurally characterized by a β(1 → 3)-linked D-glucose backbone with β(1 → 6)-linked side chains. Their biological effects depend on factors such as molecular weight, degree of branching, solubility, and

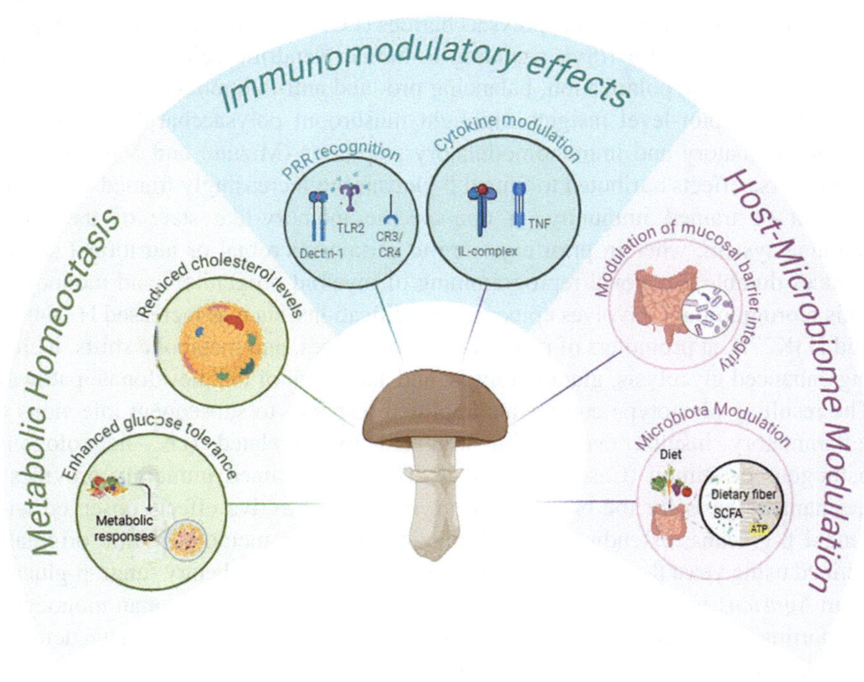


**Fig. 8.1** Biological activities of mushroom-derived β-glucans. Through their structural features, these polysaccharides contribute to metabolic homeostasis, exert immunomodulatory effects, and promote host–microbiome modulation

higher-order conformation (e.g., triple helix versus random coil), which also account for their functional differences from cereal β-glucans, typically composed of β(1 → 3,1 → 4)-linked glucans. These structure–function relationships, together with heterogeneity introduced by extraction/processing, motivate the field's current emphasis on rigorous chemotyping and standardization when comparing studies or developing products (van Steenwijk et al. 2021; Mizuno and Minato 2024; Netea et al. 2020; Case et al. 2024).

Following oral ingestion, particulate fungal β-glucans are translocated by M cells over Peyer's patches to the lamina propria, where they engage pattern-recognition receptors (PRRs) on dendritic cells, macrophages, and neutrophils. A central signaling cascade involves dectin-1 (CLEC7A) activation leading to Syk phosphorylation and downstream engagement of the CARD9–BCL10–MALT1 (CMB) complex, culminating in nuclear factor kappa B (NF-κB) activation. This pathway integrates with co-receptors such as complement receptor 3 (CR3; CD11b/CD18) and toll-like receptor 2 (TLR2) to manage phagocytosis, reactive oxygen species generation, antigen presentation, and the induction of a context-dependent cytokine milieu, including tumor necrosis factor-alpha (TNF-α), interleukin 6 (IL-6), and interleukin 10 (IL-10) (van Steenwijk et al. 2021). Comparative immunology across species underscores the importance of linkage patterns: lentinan from *Lentinula edodes* (β-1,3/1,6-glucan, ~300–800 kDa) robustly activates innate effectors yet can suppress colitic inflammation at the epithelial/myeloid interface, whereas a β-glucan from *Pleurotus citrinopileatus* polysaccharides (PCPs, ~450 kDa), enriched in β-1,6 linkages, drives dectin-1/Syk signaling in human dendritic cells and dynamically tunes macrophage polarization, balancing pro- and anti-inflammatory outputs.

These receptor-level insights highlight mushroom polysaccharides' conserved anti-inflammatory and immunomodulatory responses (Mizuno and Minato 2024). The diverse effects attributed to fungal β-glucans are increasingly framed within the concept of trained immunity—a non-specific memory-like state of the innate immune system, wherein prior exposure to certain microbial or nutritional stimuli induces durable functional reprogramming of myeloid progenitors and monocytes. This reprogramming involves epigenetic modifications, such as increased H3K4me3 and H3K27ac at promoters of pro-inflammatory genes, and metabolic shifts, including enhanced glycolysis, glutaminolysis, and activation of the mevalonate pathway. The resulting phenotype confers an amplified response to subsequent infections or inflammatory insults, even when triggered by unrelated (i.e., heterologous) pathogens or stimuli (Case et al. 2024). Therefore, trained immunity provides a mechanistic basis for the broad-spectrum immunoprotective effects observed with fungal β-glucans, extending beyond classical adaptive memory. While originally defined using yeast β-glucans, this paradigm now extends to dietary fungi: β-glucans from *Agaricus bisporus* induced trained immunity signatures in human monocytes, supporting the idea that common edible mushrooms can prime innate defenses beyond acute exposure (Case et al. 2024; Netea et al. 2020). Recent overviews of fungal β-glucans, likewise, position trained immunity programming as a mechanistic connection between chemistry and clinical response (Yin et al. 2024). Translationally, the two ends of a food–medicine continuum are instructive. On the pharmacological side, intravenous lentinan (from *L. edodes*) is an approved immunomodulatory adjuvant in oncology in parts of Asia, with evidence of systemic immunobiology, although not a demonstration of anti-infective efficacy in the general population. On the nutritional side, pleuran (from *Pleurotus ostreatus*) exemplifies oral formulations developed to support immune function; the broader regulatory literature emphasizes that such uses must ultimately be tied to standardized preparations and clinically meaningful endpoints. In parallel, an RCT with yeast β(1,3/1,6)-glucans (structurally analogous to fungal β-glucans) reported reduced early upper respiratory tract symptom burden and modulation of innate cytokine responses, including in older adults, albeit with heterogeneity across products, again highlighting the need for chemical specification and harmonized outcomes (Dharsono et al. 2019; Fuller et al. 2017). In addition to host defense, mushroom polysaccharides influence metabolic pathways and barrier-tissue biology, both of which are central to healthy aging. Reviews list antidiabetic, antihyperlipidemic, and anti-inflammatory activities across genera, with mechanistic links to viscosity-mediated carbohydrate handling, bile acid signaling, and prebiotic fermentation to short-chain fatty acids (Krittanawong et al. 2021). Preclinical studies with *P. citrinopileatus* extracts show reduced weight gain, lower triglycerides/ cholesterol/ Low-Density Lipoprotein (LDL), and improved glucose tolerance in diet-induced obesity, findings that situate mushroom matrices within cardiometabolic nutrition research (Hu et al. 2006).

At the gut interface, oral lentinan protects against colitis, with evidence for epithelial tumor necrosis factor receptor 1 (TNFR1) down-modulation via apical dectin-1 signaling and for ileal and colonic T-cell trafficking, delineating epithelial and myeloid checkpoints through which defined β-glucans can promote barrier integrity and immune resolution (Liu et al. 2019).

Finally, as the COVID-19 era refocused attention on innate immunity influence, consolidated evidence summarizes a broad preclinical antiviral spectrum for mushroom extracts and purified polysaccharides (e.g., activity reported against influenza A, including H1N1pdm; Respiratory syncytial virus (RSV); Herpes simplex virus (HSV)-1 and HSV-2; enterovirus; Hepatitis C virus (HCV); and coronavirus), while cautioning that clinical validation remains limited and product-specific. Taken together, current mechanistic and translational evidence, spanning β-glucan molecular characterization, conserved pattern-recognition receptor (PRR)-mediated signaling cascades, induction of trained immunity, and early-phase clinical findings, provides a coherent rationale for the structure of this chapter. Specifically, we will examine: (i) the structural and physicochemical diversity of mushroom-derived β-glucans and their implications for receptor binding and biological activity; (ii) the molecular mechanisms underlying their immunomodulatory effects, including PRR engagement and downstream signaling; (iii) core therapeutic applications supported by preclinical and clinical evidence, such as oncology, infectious diseases, and metabolic and neurocognitive disorders; and (iv) current challenges and future directions for clinical translation, emphasizing the need for rigorous standardization, biomarker-based stratification of responders, and optimized delivery strategies.

## 8.2 Chemical Diversity and Natural Sources

Polysaccharides derived from medicinal mushrooms, most notably β-glucans, particularly β-(1 → 3)-glucans with β-(1 → 6) branches, alongside a variety of heteropolysaccharides and proteoglycans, are generally located within the fungal cell wall, where they fulfill important structural and physiological functions such as maintenance of osmotic balance, defense mechanisms, and intercellular communication (Garcia et al. 2022; Maity et al. 2021). The taxonomic and ecological breadth of medicinal fungi further yields substantial biochemical variability: species including *Ganoderma lucidum*, *Lentinula edodes*, *Grifola frondosa*, *Trametes versicolor*, *Cordyceps militaris,* and *Hericium erinaceus* are among the most intensively studied, both for their traditional uses and for well-characterized polysaccharide profiles (Wasser 2011; Zhang et al. 2007) (Fig. 8.2). Structurally, fungal polysaccharides range from homopolymeric glucose chains (linear or branched β-glucans) to more complex heteroglycans composed of galactose, mannose, fucose, rhamnose, xylose, and other monosaccharides; these heteropolymers commonly display high degrees of branching, molecular weights often exceeding 500 kDa, and solubility differences

**Fig. 8.2** Representative edible and medicinal mushrooms are commonly studied as sources of bioactive polysaccharides, particularly β-glucans. Species illustrated include (**a**): *Pleurotus eryngii*; (**b**): *Lentinula edodes*; (**c**): *Grifola frondosa;* (**d**): *Pleurotus citrinopileatus*; (**e**): *Pleurotus ostreatus*; (**f**): *Hericium erinaceus*; and (**g**): *Ganoderma lucidum*, all of which are recognized for their nutritional value and immunomodulatory properties

that depend on linkage type and substitution patterns, while the presence of sulfate or carboxyl groups can further augment antioxidant and immunomodulatory activities (Gunasekaran et al. 2021; Guo et al. 2017; Zhang et al. 2010).

This structural heterogeneity is modulated by intrinsic and extrinsic factors: species and strain determine core biosynthetic capacities, whereas cultivation conditions (substrate composition, pH, temperature, oxygenation, and light), together with the choice of extraction methodology, strongly influence monosaccharide composition, glycosidic linkages, branching patterns, and molecular weight distribution (Wang et al. 2022; Wu et al. 2024). Consequently, each medicinal fungus tends to produce species-specific polysaccharide ensembles; for example, *Cordyceps militaris* yields structurally diverse polysaccharides associated with antitumor, antioxidant, and immunomodulatory effects, whereas *Ganoderma* spp. typically synthesizes β-glucans and a range of heteropolysaccharides with distinct immunological activities (Wu et al. 2024). Advances in chromatographic and mass spectrometric techniques have increasingly enabled detailed compositional and structural profiling, revealing complex mixtures of neutral sugars, uronic acids, and amino sugars in organisms such as *Polyporus umbellatus*, thereby improving our ability to link molecular features to bioactivity (Guo et al. 2019).

Among the most functionally and clinically relevant fungal polysaccharides are protein-conjugated forms. Polysaccharopeptides (PSPs) and polysaccharide-K (PSK) exemplify protein-carbohydrate complexes in which the sugar backbone and the attached protein moiety jointly determine biological effects. PSP, typically isolated from *Coriolus versicolor*, comprises a polysaccharide backbone with α-1,4 and β-1,3 glucosidic linkages and distinctive branching; purified fractions such as PSP-1b1 possess alternating α-galactose and α-mannose residues with side chains (e.g., α-1,6-Manp, β-1,6-Glcp, β-1,3,6-Glcp) attached principally at O-6, plus secondary branches that increase structural complexity (Habtemariam 2020; Ooi and Liu 2000). Biological activity correlates with molecular weight and branching; higher molecular weight and relatively lower branching often align with enhanced immunomodulatory effects (Fritz et al. 2015; Habtemariam 2020; Ooi and Liu 2000), and the protein component of PSP, enriched in glutamic and aspartic acids, includes a small peptide ($\approx$16–18 kDa) with reported antiproliferative and antitumor properties (Ooi and Liu 2000). The carbohydrate composition further differentiates these complexes: PSP frequently contains rhamnose and arabinose, whereas PSK contains fucose, reflecting distinct biosynthetic and structural signatures (Fritz et al. 2015; Habtemariam 2020; Ooi and Liu 2000). This evidence shows that a complex interplay of genetic and environmental factors determines the biosynthesis and structural diversity of mushroom-derived polysaccharides. Both intrinsic and extrinsic influences shape their monosaccharide composition, molecular weight, branching patterns, and overall bioactivity, leading to substantial variability even within the same species.

**Genetic Determinants and Intraspecific Variation**

The genetic background of each species, and, in many cases, of individual strains, dictates the enzymatic machinery responsible for polysaccharide biosynthesis.

Variations in gene expression, metabolic regulation, and biosynthetic pathway efficiency result in distinct structural configurations and functional properties. Intra-specific variability is further amplified by epigenetic regulation and metabolic plasticity, leading to differences in the yield, structure, and biological effects of polysaccharides obtained from the same species but under different conditions (Luan et al. 2022; Su et al. 2016; Wang et al. 2017).

### 8.2.1 Environmental and Cultivation Factors

Environmental conditions play a pivotal role in modulating fungal metabolism and polysaccharide production. Parameters such as the cultivation system (e.g., static vs. agitated cultures), nutrient composition of the growth medium, pH, oxygen availability, temperature, and light directly influence enzyme activity and gene regulation, ultimately shaping the branching degree, molecular weight, and solubility of the polysaccharides. Studies on white-rot fungi have demonstrated that variations in culture conditions can lead to markedly different biosynthetic profiles, altering both structural complexity and biological activity (Monteiro et al. 2025; Rämä and Quandt 2021).

#### 8.2.1.1 Stress Responses and Cocultivation Effects

Environmental stressors, including exposure to heavy metals, nutrient limitation, or oxidative stress, can trigger adaptive metabolic responses and stimulate the synthesis of specific polysaccharide fractions and associated secondary metabolites (Hasani Zadeh et al., 2023). Likewise, cocultivation with other microorganisms may induce cross-signaling pathways that diversify the polysaccharide repertoire and, in some cases, enhance bioactive potential (Rämä and Quandt 2021).

The production of mushroom polysaccharides is tightly regulated by an integrated network of genetic, metabolic, and environmental signals. This adaptability enables fungi to modulate polysaccharide composition in response to ecological pressures, which not only has evolutionary relevance but also practical implications for optimizing cultivation strategies aimed at maximizing yield and functionality.

### 8.2.2 Extraction Strategies for Mushroom Polysaccharides: Methods, Efficiency, and Impact on Bioactivity

The extraction method significantly influences both the yield and the structural integrity of polysaccharides, thereby impacting their biological activity. While hot water extraction (HWE) remains the conventional method, a variety of modern techniques have been explored in recent years to overcome its limitations, such as thermal degradation, long extraction times, and low selectivity (Fig. 8.3).

Conventional extraction protocols generally start with HWE, capitalizing on the water solubility of many polysaccharides. This is often followed by ethanol precipitation to concentrate high molecular weight fractions (Lu 2023). Deproteinization

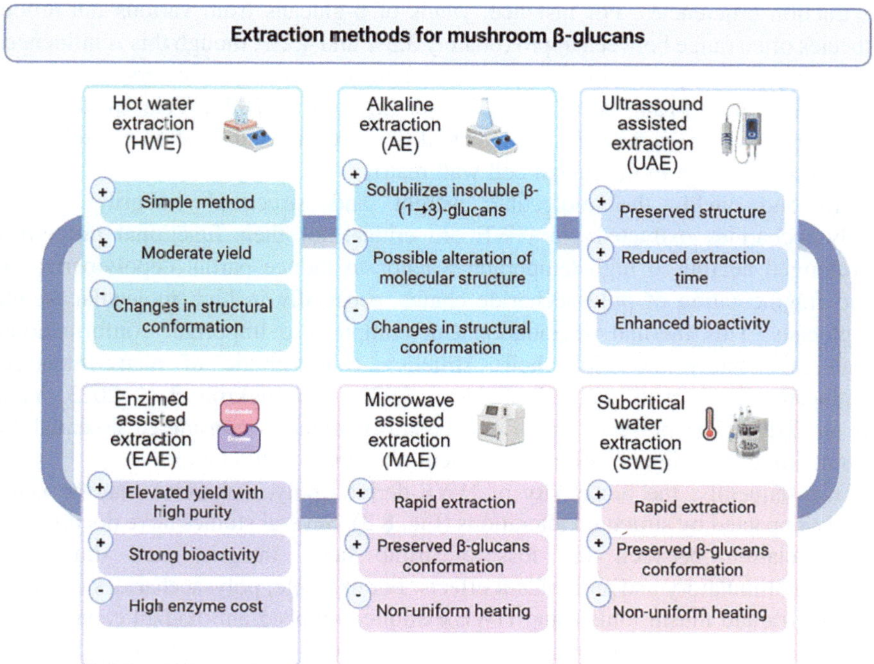


**Fig. 8.3** Comparative overview of extraction methods for mushroom β-glucans. Different techniques, including hot water, alkaline, ultrasound-assisted, enzymatic-assisted, microwave-assisted, and subcritical water extraction, present specific advantages and limitations

(e.g., Sevag method) and dialysis are frequently used to remove contaminants and small molecules. However, HWE can degrade heat-sensitive components and may not efficiently extract cell wall-bound β-glucans (Huang et al. 2021). Therefore, alkaline extraction (e.g., using NaOH) is employed to solubilize the insoluble fractions, particularly β-(1 → 3)-glucans associated with chitin (Maity et al. 2021).

Emerging extraction strategies such as ultrasound-assisted extraction (UAE), microwave-assisted extraction (MAE), enzyme-assisted extraction (EAE), and sub-critical water extraction (SWE) aim to improve yield, reduce extraction time, and preserve polysaccharide structure and bioactivity (Wang et al. 2022). These methods offer advantages in terms of scalability and sustainability and are increasingly used in industrial applications (Leong et al. 2021).

#### 8.2.2.1 Hot Water Extraction (HWE)

HWE remains one of the most widely employed techniques for isolating polysaccharides from mushrooms and plants due to its simplicity, cost-effectiveness, and suitability for food-grade applications. However, the conditions typically used in HWE, which involve prolonged exposure to elevated temperatures (commonly 80–100 °C), can have significant effects on both the quantity and quality of the extracted polysaccharides. Regarding yield, HWE generally provides moderate extraction efficiencies. For instance, yields of β-glucans from various mushroom species often range between approximately 3.5% and 4.2%, though this is influenced by the particular species, tissue type, and specific extraction parameters such as time and temperature (Tang et al. 2023; Tepsongkroh et al. 2023; Yuan et al. 2019). Variations in these parameters can lead to differences in solubilization and release of polysaccharides from the fungal cell wall matrix.

Beyond yield, the molecular weight and structural integrity of the polysaccharides extracted via HWE are critical for their functional properties. Prolonged heating at high temperatures tends to induce partial depolymerization and fragmentation of polysaccharide chains, especially in high molecular weight β-glucans. This thermal degradation can compromise important conformational features, notably the triple helix structure characteristic of many bioactive β-glucans (Chen et al. 2014; Tang et al. 2023; Tepsongkroh et al. 2023; Yuan et al. 2019). The preservation of such conformations is considered essential for maintaining the immunomodulatory potential of these compounds.

Consequently, the bioactivity of HWE-derived polysaccharides may be negatively impacted by structural alterations (Fig. 8.3). Several studies have documented a correlation between the loss of triple helix integrity and diminished immunostimulatory and antioxidant effects. For example, polysaccharides extracted from okra and mushrooms using HWE exhibited reduced antioxidant capacity and lower stimulation of immune responses compared to those obtained through enzyme-assisted extraction or pressurized water extraction techniques, which tend to preserve molecular size and conformation more effectively (Chen et al. 2014; Tang et al. 2023; Tepsongkroh et al. 2023; Yuan et al. 2019).

#### 8.2.2.2 Ultrasound-Assisted Extraction (UAE)

UAE has emerged as an effective technique for isolating polysaccharides from natural sources, outperforming conventional methods such as hot water or boiling extraction. The application of ultrasonic waves induces cavitation, which disrupts cell walls and enhances the release of polysaccharides, thereby improving extraction efficiency (Fig. 8.3).

Regarding extraction efficiency and yield, numerous studies report that UAE consistently achieves higher polysaccharide yields. For instance, *Cercis chinensis* leaves yielded 20.02 mg/g using UAE compared to 16.09 mg/g obtained via boiling extraction (Wei and Zhang 2023). Similar improvements have been documented for mushroom by-products, *Typha domingensis*, *Ginkgo biloba* leaves, and various other plant materials, often under UAE conditions optimized through response surface methodology (Aguiló-Aguayo et al. 2017; Sorourian et al. 2020).

In terms of structural integrity and bioactivity, UAE effectively preserves or enhances key structural features essential for biological activity. Research indicates that polysaccharides extracted via UAE maintain or improve β-glucan conformation, molecular weight distribution, and functional group integrity, attributes critical to antioxidant and immunomodulatory effects (Guo et al. 2017, 2024; Wang et al. 2022). Moreover, some studies suggest that UAE may induce beneficial structural modifications, such as decreased molecular weight and increased uronic acid content, potentially further enhancing bioactivity (Guo et al. 2024; Wang et al. 2022).

Concerning antioxidant and functional properties, polysaccharides obtained through UAE exhibit superior radical scavenging activities, including DPPH, hydroxyl, and ABTS assays, relative to those extracted by conventional methods (Wei and Zhang 2023; Yang et al. 2024). Enhanced prebiotic and immunomodulatory activities have also been observed, supporting the application of UAE-derived polysaccharides in functional foods and pharmaceutical formulations (Gharibzahedi et al. 2022; Sorourian et al. 2020).

#### 8.2.2.3 Microwave-Assisted Extraction (MAE)

MAE has gained prominence as an efficient technique for isolating bioactive polysaccharides from medicinal mushrooms (Fig. 8.3). It offers distinct advantages, including enhanced extraction yield, markedly reduced extraction times, and effective preservation of the structural and functional integrity of the polysaccharides. MAE significantly increases the yield of polysaccharides compared to conventional extraction methods. For instance, in *Fomitopsis ulmaria*, MAE under optimized conditions (400 W, 2.5 min) achieved an extraction yield of 8.36%, demonstrating rapid and effective polysaccharide recovery (Zhao et al. 2015). Similarly, in *Hericium erinaceum*, MAE at 140 °C for 5 min yielded comparable polysaccharide quantities to those obtained after 6 h of traditional heating at 100 °C, highlighting considerable time efficiency (Ookushi et al. 2006). Moreover, in *Cordyceps sinensis*, optimal MAE parameters (1300 W, 5 min, 1:60 solid-to-water ratio) resulted in the highest reported yield of 6.2% polysaccharides (Cheong et al. 2016). Polysaccharides extracted by MAE generally maintain key structural characteristics vital to their biological functions, such as β-glucan content and specific chain conformations (Cheong et al. 2016; Jia et al. 2022; Smiderle et al. 2017). In *Ganoderma lucidum*, MAE not only enhanced polysaccharide yield but also preserved the semi-rigid chain conformation, correlating with improved in vitro anti-inflammatory activity (Jia et al. 2022). Comparable structural preservation and potent bioactivities have been observed in polysaccharides derived from other mushroom species using MAE. These extracts exhibit significant antioxidant capabilities and α-glucosidase inhibitory effects, underscoring their potential applications in antidiabetic therapies and broader health promotion (Leong et al. 2021; Lin et al. 2019; Mirzadeh et al. 2020).

#### 8.2.2.4 Enzyme-Assisted Extraction (EAE)

EAE has emerged as an advanced technique for isolating polysaccharides, offering significant improvements in yield, purity, and biological activity over conventional extraction methods (Fig. 8.3). By employing hydrolytic enzymes such as cellulase, pectinase, and hemicellulase, EAE effectively disrupts cell wall matrices, facilitating enhanced release of intracellular polysaccharides while preserving their molecular integrity. The application of EAE markedly increases polysaccharide extraction yields. For example, in *Ulva lactuca*, EAE elevated polysaccharide yield from 6.43% using traditional methods to 26.68%, with further enhancement up to 30.14% when combined with ultrasound-assisted extraction (Wang et al. 2024). Similar yield improvements have been reported for *Potentilla anserina* (19.8%)

(Guo et al. 2023), *Schizochytrium limacinum* (11.86%) (Zhang et al. 2024), and *Cornus officinalis* (9.29%) (You et al. 2013). The synergistic integration of EAE with physical techniques such as ultrasound or microwave irradiation also reduces extraction time and energy consumption, improving overall process efficiency (Lin et al. 2022). Operating under mild extraction conditions, typically lower temperature and controlled pH, EAE minimizes polysaccharide degradation, thereby preserving molecular weight distribution and biofunctional properties. Numerous studies have demonstrated that polysaccharides extracted via EAE exhibit superior antioxidant activities, including enhanced DPPH, ABTS, and hydroxyl radical scavenging capacities, relative to those obtained by hot water extraction (Guo et al. 2023; Wang et al. 2024). Furthermore, enzyme-extracted polysaccharides from ginseng have shown improved immunostimulatory effects in both in vitro and in vivo models, highlighting their therapeutic potential (Song et al. 2018; Zhou et al. 2025). Polysaccharides isolated through EAE frequently display a more diverse monosaccharide composition and improved solubility, attributes that contribute to their enhanced functional properties in food, nutraceutical, and pharmaceutical applications (Nadar et al. 2018; Song et al. 2018). Beyond efficacy, EAE is recognized for its environmental sustainability and scalability, especially when coupled with enzyme immobilization technologies, which facilitate enzyme reuse and process optimization (Nadar et al. 2018).

#### 8.2.2.5 Subcritical Water Extraction (SWE)

SWE has gained recognition as a sustainable and efficient method for isolating polysaccharides from fungi, outperforming conventional extraction techniques in both yield and biofunctional properties (Fig. 8.3). This environmentally friendly process utilizes water at elevated temperatures (typically 100–374 °C) under high pressure to maintain its liquid state, thereby enhancing solvent properties and extraction efficiency without the need for organic solvents. SWE consistently achieves higher polysaccharide yields compared to traditional HWE. For instance, yields from *Morchella sextelata* increased from 4.95% with HWE to 18.09% under SWE conditions, while *Dendrobium nobile* extracts reached up to 21.88% polysaccharide content (Li et al. 2022; Liu et al. 2019). This significant improvement is attributed to the enhanced penetration and solubilization capacity of subcritical water, which facilitates the disruption of fungal cell walls and the efficient release of polysaccharides. Moreover, SWE induces notable structural modifications in extracted polysaccharides, including a reduction in molecular weight and a conformational shift from triple helix to random coil configurations. These changes typically result in increased solubility and may potentiate biological activity (Li et al. 2022). Importantly, SWE preserves, and in some cases enriches, the concentration of key bioactive constituents such as β-glucans and polyphenols, which are critical to the health-promoting effects of fungal extracts (Kodama et al. 2015; Ma et al. 2024). Polysaccharides obtained via SWE exhibit markedly enhanced antioxidant capacity, as evidenced by superior performance in 2,2-diphenyl-1-picryl-hydrazyl (DPPH), 2,2-azino-bis-3-ethylbenzothiazoline-6-sulphonic acid (ABTS), and ferric reducing antioxidant power (FRAP) radical scavenging assays (Ma et al. 2024). These extracts also demonstrate significant immunomodulatory activity, including stimulation of RAW264.7 macrophage proliferation and nitric oxide production, indicative of immune system activation (Ma et al. 2024).

## 8.3 Structure–Function Relationships and Mechanisms of Action

The bioactive properties of these compounds depend essentially on their structural characteristics, including molecular weight, degree of branching, solubility, and conformation in solution. These variables explain the correlation between structural variability and respective bioactivities (Sánchez 2017). Thus, depending on the structural conformation, there is variability in the composition of monosaccharides and, therefore, an ability to interact with various receptors present on the cell surface, both dectin-1 and Toll-like receptors (TLRs), which, when activated, trigger the inflammatory response through the activation of macrophages, natural killer cells, dendritic cells, and cytokines (Chakraborty et al. 2021).

Molecular weight is one of the commonly reported chemical characteristics that influences the biological properties of polysaccharides. Numerous bioactivities have been associated with high molecular weight polysaccharides (200–500 kDa), as these have been shown to be more biologically active than low molecular weight polysaccharides. This fact can be explained by the greater affinity that exists between them and immune cell receptors such as dectin-1, which is found on the surface of immune cells. On the other hand, the compounds with lower molecular weight have fewer direct effects, since they depend on the action of cytokines to modulate the immune response (Han et al. 2020).

In addition to molecular weight, the immunological activity of β-glucans is strongly influenced by their branching architecture. Structures composed of a linear β-(1 → 3)-glucan backbone with β-(1 → 6) side branches exhibit superior interaction with immune receptors, resulting in a stronger modulation of the innate immune systems (Chakraborty et al. 2021).

The biological activity is also influenced by side-chain features. Two aspects deserve attention: first, their length, and second, the frequency of branching. Branches that are too short, with only one glucose unit, have less capacity to act on the immune response than macromolecules with longer side chains. Furthermore, according to the literature, there is an ideal branching interval to ensure that polysaccharides have immunomodulatory properties, which should be between 0.2 (1:5 branching) and 0.33 (1:3 branching). Outside this interval, the activity of polysaccharides will tend to decrease (Han et al. 2020).

Mushroom polysaccharides have different degrees of branching; for example, PSK, Lentinan, and schizophyllan have 0.2 (1:5), 0.4 (2:5), and 0.33 (1:3), respectively. Despite these differences, all polysaccharides exhibit relevant bioactivities, demonstrating that the branching profile, molecular mass, and conformation in solution are variables that contribute to their recognition by receptors and influence the magnitude of the response (Chakraborty et al. 2021).

However, it is important to note that there are exceptions: polysaccharides with similar branching frequencies may have different affinities for receptors, which highlights the importance of analyzing other variables such as the type of bonds, their number, and their positions, which are also determining factors that affect the activity of polysaccharides derived from mushrooms. It is also important to recognize that excessively branched macromolecules may lose their immunomodulatory or antitumor capacity due to steric hindrance, which prevents affinity with specific receptors (Yang et al. 2022).

Fungal β-glucans can acquire three types of conformations, single helix, triple helix, or random coil, which result from the balance between molecular weight, the pattern of β-glycosidic bonds, and hydrogen bonds. In vivo, the triple helix is the most common and dominant organization in the cell wall, and its recognition by immune receptors is relevant for triggering subsequent responses (Han et al. 2020). Of the three conformations mentioned, the triple helix is the one that represents greater antitumor and immunomodulatory activity in relation to the single helix conformation (add references). Nowadays, this is not an absolute concept, as there are studies that show that the single helix or random coil also promotes the bioactivity of polysaccharides, with efficacy depending on the combination of various structural parameters (Zhang et al. 2022).

Polysaccharides are classified as bioactive macromolecules, owing to their structural characteristics that give them numerous bioactivities. However, they do not cross the plasma membrane but act through external recognition via PRRs located on the surface of immune system cells (Yang et al. 2016; Shen et al. 2024).

Among the main receptors involved are dectin-1 and the CR3, both of which are widely recognized as primary targets of fungal polysaccharides, namely β-glucans. TLRs, especially TLR2 and TLR4, also contribute to the recognition process. This initial interaction between the polysaccharide and the receptor triggers intracellular signaling cascades, most notably the nuclear factor kappa B (NF-kB) and mitogen-activated protein kinases (MAPKs) pathways (Konusova et al. 2021; Yin et al. 2024).

Dectin-1 is a C-type lectin receptor expressed on immune cells, such as macrophages and dendritic cells. Its activation is strongly influenced by the structural characteristics of the polysaccharide. This receptor has a higher affinity for polysaccharides formed by β-(1 → 3) glucan chains with β-(1 → 6) branches. Dectin-1 is particularly responsive to organized conformations, such as triple helix conformations, which are typically associated with higher molecular weight (Konusova et al. 2021).

The binding of β-glucans to their receptors promotes the activation of multiple signaling pathways involved in the innate immune response, leading to increased production of reactive oxygen species (ROS), enhanced phagocytic activity, and increased synthesis of inflammatory cytokines (Camilli et al. 2018).

Mushroom polysaccharides activate two main pathways, NF-kB and MAPKs, which regulate immune responses by converging on transcription factors, including NF-kB and AP-1, ultimately controlling the expression of genes related to inflammation and host defense (Kim et al. 2011; Yin et al. 2021).

Following recognition of β-glucans, dectin-1 undergoes dimerization, leading to phosphorylation of tyrosine residues within its hemITAM motif by kinases of the Src family. This creates a docking site for Syk kinase, which is subsequently recruited and activated. Syk then initiates the downstream CARD9–BCL10–MALT1 signaling axis by recruiting the adapter protein CARD9, which forms a complex with BCL10 and MALT1. This complex is essential for the activation of the IκB kinase (IKK) complex, resulting in phosphorylation and ubiquitination-mediated degradation of IκB. As a consequence, NF-κB is released and translocates into the nucleus, where it induces transcription of pro-inflammatory cytokine genes such as TNF-α, IL-6, and IL-12 (Kim et al. 2011; Wagener et al. 2018).

In addition to activating NF-κB, dectin-1 also promotes the phosphorylation of MAPKs, including ERK, JNK, and p38, which play crucial roles in regulating inflammation, cytokine production, and cell survival. Upon recognition of β-glucans, dectin-1 undergoes hemITAM phosphorylation, recruiting the Syk kinase, which in turn activates the CARD9–BCL10–MALT1 complex. This complex functions as a key signaling hub that facilitates the activation of MAPKs, although the precise intermediates involved in MAPK activation remain under investigation. Notably, MAPK activation leads to the phosphorylation and activation of transcription factors such as AP-1, which acts synergistically with NF-κB to regulate the expression of immune effector genes (Huang et al. 2015; Ruland and Hartjes 2019).

The ability of dectin-1 to trigger both NF-κB and MAPK pathways underscores its central role in orchestrating innate immune responses to fungal polysaccharides.

In addition to dectin-1, β-glucans are also recognized by TLRs, namely TLR2 and TLR4. These receptors are expressed in macrophages, dendritic cells, T cells, and B cells, acting as membrane receptors that detect pathogen-associated molecular patterns (PAMPs). However, it is important to note that TLRs are not always directly activated by polysaccharides alone. Their activation may depend on the presence of complex glycoproteins or, on the other hand, result from receptor cooperation, whereby activation of a receptor such as dectin-1 enhances TLR signaling (Brown 2006).

When specific ligands bind to TLRs, signaling pathways are activated, with signaling mediated by MyD88, which recruits signaling complexes capable of activating signaling pathways such as NF-kB and MAPKs (Kim et al. 2011).

TLRs initiate a MyD88-dependent signaling cascade, beginning with the recruitment of the adapter protein, MyD88, followed by activation of IL-1 receptor-associated kinases (IRAKs) and TNF receptor-associated factor 6 (TRAF6). This leads to the phosphorylation and activation of transforming growth factor–βactivated kinase 1 (TAK1), a central node that bifurcates the signal into two key pathways: the NF-κB pathway, via IκB kinase (IKK)-mediated degradation of IκB, and the MAPK pathway, which activates mitogen-activated protein kinases such as ERK, JNK, and p38 (Kim et al. 2011; Pereira and Gazzinelli 2023).

These MAPKs contribute to the phosphorylation and activation of activator protein-1 (AP-1), a transcription factor complex that, alongside NF-κB, regulates the expression of genes involved in inflammation, cytokine production, and host

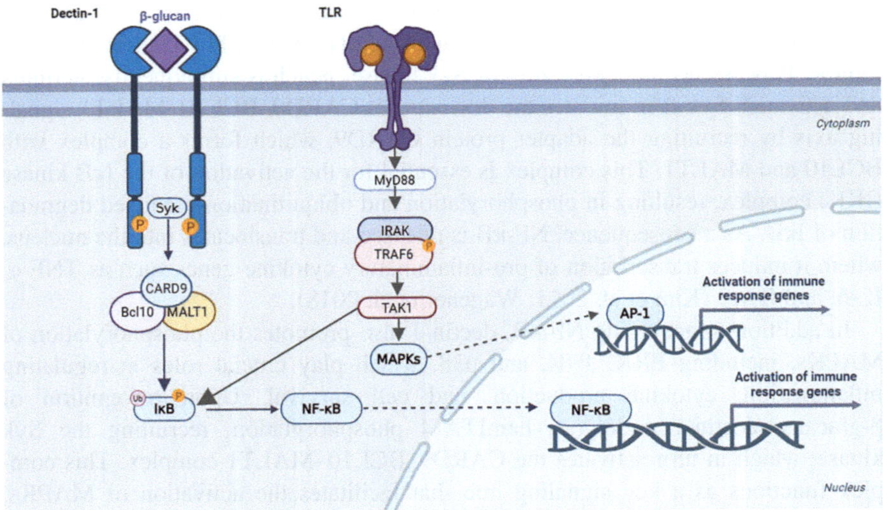


**Fig. 8.4** Signaling pathways activated by mushroom polysaccharides through dectin-1 and Toll-like receptors (TLRs). Recognition of polysaccharides by dectin-1 triggers signaling via Syk and the CARD9–BCL10–MALT1 complex, leading to NF-κB and MAPK activation. In parallel, TLRs signal through the MyD88–IRAK–TRAF6–TAK1 axis, also activating NF-κB and MAPKs. The convergence of these pathways results in the activation of transcription factors NF-κB and AP-1, which regulate the expression of immune response genes

defense. The convergence of both dectin-1 and TLR signaling on these transcriptional regulators illustrates a coordinated immune response and highlights the synergistic potential of these pathways in innate immunity (Yang et al. 2016). An overview of these signaling pathways is presented in Fig. 8.4.

A molecular weight above 100 kDa is particularly relevant, as it contributes to the promotion of organized conformations, particularly triple helices, which promote stability of the binding with dectin-1, contributing to the activation of the NF-kB pathway (Camilli et al. 2018). PSK, an extract from *Trametes versicolor*, exemplifies this mechanism. It directly engages TLR receptors, specifically TLR2. Its activation in vivo promotes an antitumor response mediated by dendritic cells and CD8+ T lymphocytes (Lu et al. 2011). Moreover, PSK also interacts with the TLR4 receptor, promoting the secretion of pro-inflammatory factors such as IL-6 and TNF-α (Price et al. 2010).

According to the literature, other polysaccharides, such as EPA-1 from *Pleurotus eryngii*, described as branched structures with a high molecular weight, have been shown to induce the release of pro-inflammatory factors, such as TNF-α and IL-6, with the increased expression of these genes resulting from the activation of the two pathways mentioned above (Yin et al. 2021).

Lentinan, a β-(1 → 3)-glucan from *Lentinula edodes*, is among the best-characterized examples of structure–function relationships in fungal polysaccharides. It consists of a β-(1 → 3)-glucan backbone with β-(1 → 6) branches, forming a triple helix conformation and a high molecular weight, key features for its biological activity. These structural characteristics are essential for the bioactivity of macromolecules. As a result of these characteristics, lentinan is favorably recognized by receptors such as dectin-1, located on the surface of macrophages and dendritic cells. This binding triggers the activation of intracellular pathways, particularly NF-kB, promoting the transcription of immunomodulatory genes, notably IL-10, which, with increased expression, plays a key role in balancing the inflammatory response (Balakrishnan et al. 2021).

## 8.4 Therapeutic Roles of Mushroom-Derived Polysaccharides in Human Health

The therapeutic roles of mushroom-derived polysaccharides, with β-glucans as the best-characterized class, are reviewed across major clinical domains, emphasizing evidence that is anchored in chemistry, delivery, and adjudicated outcomes. β-Glucans activate evolutionarily conserved innate immune pathways, most notably those associated with trained immunity, and concurrently modulate metabolic processes and epithelial barrier function. This multifunctional activity positions them within a translational continuum, ranging from parenterally administered immunoadjuvants (e.g., lentinan, approved in parts of Asia) to standardized oral formulations such as pleuran, used as functional food supplements (Mizuno and Minato 2024; Netea et al. 2020).

This section synthesizes translational signals in: (i) oncology, where compounds such as PSK/PSP and lentinan have been used as immunochemotherapy adjuncts; (ii) infectious-disease settings, where human data are modest but consistent when β-glucans are well-characterized and endpoints prespecified; (iii) cardiometabolic health, where randomized trials with Pleurotus and mixed mushroom interventions report improvements in fasting glucose and atherogenic lipids with matrix- and dose-dependence; (iv) neurocognitive health, where *Hericium erinaceus* trials in mild cognitive impairment MCI and early Alzheimer's disease (AD) show promising but heterogeneous results; and (v) gut inflammation, where defined β-glucans such as lentinan target epithelial and myeloid checkpoints in dextran sulfate sodium (DSS) colitis models. Throughout, attention is given to product heterogeneity and study design, reporting of linkage/branching, molecular weight distributions, solubility/conformation, and bridging in vitro pharmacology to clinical endpoints because these factors determine comparability and effect size in forthcoming trials.

### 8.4.1 Oncology: Adjuvant and Supportive Roles of Mushroom Polysaccharides

Over the past decades, both preclinical and clinical studies have progressively unveiled β-glucans' potential in oncology, positioning them as promising candidates for integrative and precision medicine approaches. PSK and PSP from *Trametes versicolor* exemplify the most advanced oncology evidence. At the cellular level, PSK is a selective TLR2 agonist that activates dendritic cells (DCs) and promotes $CD8^+$ T-cell and NK-cell-dependent antitumor activity. PSK also matures DCs in vivo and functions as a vaccine adjuvant in preclinical studies (Lu et al. 2011). Cooperative signaling between β-glucan receptors (notably dectin-1) and TLRs provides the mechanistic basis for these effects and helps explain the broad innate and adaptive immune engagement attributed to β-glucans. PSP likewise engages pattern-recognition pathways (TLR4/dectin-1 reported in several models), enhancing phagocytosis, antigen presentation, and a cytokine program that includes IL-1β, TNF-α, and IFN-γ (Kim et al. 2016). Collectively, these data position PSK/PSP as biological response modifiers that enhance tumor immune surveillance rather than directly targeting cytotoxins. Clinically, PSK has been used as an adjuvant immunochemotherapy after curative surgery and has repeatedly shown a survival benefit in gastrointestinal malignancies. Meta-analyses of centrally randomized trials in colorectal and gastric cancer demonstrate improvements in overall and disease-free survival when PSK is combined with fluoropyrimidine-based regimens versus chemotherapy alone (Sakamoto et al. 2006); subsequent pooled and cohort analyses in resected gastric cancer confirm the signal. Benefits are accompanied by immune biomarker improvements (e.g., NK activity, immunoglobulins) and acceptable tolerability. Evidence for lung cancer is less definitive but suggests potential gains in immune function, symptom control, and, in some series, survival, warranting modern randomized trials.

For PSP, human data are more limited but directionally consistent with immunorestorative activity. In a randomized double-blind study in advanced non-small-cell lung cancer, PSP slowed clinical deterioration and improved leukocyte/neutrophil counts and humoral immunity, supporting its role as a supportive care adjunct rather than a stand-alone antineoplastic; current reviews concur and emphasize the need for larger, well-controlled trials (Fritz et al. 2015; Habtemariam 2020; Ooi and Liu 2000). Evidence from drug-mushroom-derived polysaccharide interaction studies, such as those involving tamoxifen in vivo, highlights the critical need to assess pharmacokinetic parameters when mushroom polysaccharides are administered concomitantly with conventional therapeutics (Tsang et al. 2003).

Lentinan (derived from *Lentinula edodes),* when administered intravenously alongside chemotherapy in patients with advanced gastric cancer, has been shown to significantly prolong overall survival compared to chemotherapy alone (Oba et al., 2009). For example, a prospective study reported a median survival of 689 days in the lentinan-chemotherapy group versus 565 days in controls ($p = 0.0406$), along with improved granulocyte-to-lymphocyte ratios, suggestive of immunomodulatory benefits (Datta et al. 2019). A patient-level meta-analysis further corroborated these findings, demonstrating a survival advantage across multiple studies of unresectable or recurrent gastric cancer (Oba et al. 2009). Additionally, a more recent retrospective cohort involving metastatic gastric cancer patients treated with S-1-based chemotherapy plus lentinan reported tumor control in 83% of cases and a median overall survival of approximately 407 days, highlighting the potential of lentinan as a chemo-immunotherapeutic adjunct (Ina et al. 2016). In the context of cervical cancer, historical randomized trials of schizophyllan (sizofiran), an immunomodulatory β-glucan from *Schizophyllum commune*, administered in combination with radiotherapy, demonstrated significant increases in time to disease recurrence and survival in patients with Stage II disease, whereas outcomes in Stage III patients were not significantly different from controls (Okamura et al. 1986). A similar pattern was observed across multiple institutions in Japan, reinforcing the notion of stage-dependent efficacy with adjuvant β-glucan treatments. Although oncological standards have evolved considerably, these results support the concept that select, well-characterized β-glucans can enhance antitumor immunity and improve clinical outcomes when used in adjuvant regimens. Their effects appear mediated via immune potentiation rather than direct tumoricidal activity, justifying further mechanistically informed exploration in integrated oncology protocols.

### 8.4.2 Immune Support and Infectious Diseases: Clinical Signals and Prophylaxis

Across the food–medicine continuum, fungal β-glucans illustrate how chemistry and route of delivery shape clinical intent. Intravenous lentinan is used in parts of Asia as an immunostimulatory adjuvant (predominantly in oncology settings), whereas oral pleuran from *Pleurotus ostreatus* has been developed as a nutraceutical for recurrent respiratory tract infections (RTIs) in children and other stress-exposed groups; these applications highlight that parenteral dosing targets systemic immunotherapy while oral formulations are positioned for mucosal/respiratory support and convalescence (add reference please). Recent pediatric data now extend beyond prevention-only outcomes: in a multicenter, randomized, double-blind trial in perennial asthma ($n \approx 300$; 24-week treatment + follow-up), pleuran (with vitamin C) improved asthma control, lowered RTI counts overall, and reduced exacerbations in children $<12$ years, with good tolerability, an effect profile consistent with reduced respiratory morbidity rather than acute bronchodilation (Jesenak et al. 2025). By contrast, a pediatric RCT in acute infectious gastroenteritis (β-glucan pleuran vs. placebo; 2–10 years) found no benefit on diarrhea duration or severity, underscoring indication-specific performance and the importance of matching formulation to disease biology (intestinal fluid loss vs. upper airway infection) (Wzorek-Łyczko et al. 2025).

Beyond oral capsules and beverages, interest in mucosa-targeted delivery has grown. Intranasal exposure to β-glucan (preclinical) reprograms tissue-resident alveolar macrophages toward a memory-like state that heightens early control of respiratory pathogens, evidence that local dosing can imprint the lung myeloid compartment and may complement oral strategies focused on gut–lung crosstalk (Theobald et al. 2024). Evidence for trained immunity, defined as durable epigenetic and metabolic reprogramming of innate immune cells, now extends beyond yeast β-glucans to edible mushrooms. β-Glucans purified from *Agaricus bisporus* products induced canonical trained immunity signatures in human monocytes ex vivo (e.g., increased H3K4me3/H3K27ac, a glycolytic shift, and augmented secondary cytokine responses), indicating that common edible mushrooms can prime host defenses beyond the immediate exposure. Crucially, trained immunity is not confined to enhanced *resistance* (reducing pathogen burden). In influenza models, prior β-glucan exposure reprograms neutrophils toward a disease-tolerance phenotype—limiting tissue injury and inflammatory collateral damage even when viral loads are similar. This state is marked by restrained excessive degranulation/reactive oxygen species (ROS) production and NETosis, improved resolution cues, and preservation of epithelial barrier function, yielding better pulmonary function and survival without necessarily accelerating viral clearance. Because morbidity and mortality in severe viral infections are often driven by immune-mediated pathology, this tolerance arm is as therapeutically relevant as resistance (Khan et al. 2025).

Clinical signals in community-acquired infections remain modest but directionally consistent when β-glucans are well-characterized and endpoints are adjudicated. In adults and older cohorts, randomized trials with yeast β(1,3/1,6)-glucans, structural analogs to mushroom insoluble glucans, repeatedly show reductions in upper respiratory tract infection (URTI) symptom severity and medication use, even when incidence is unchanged; a recent pediatric RCT with a yeast-fermentate postbiotic similarly lowered overall cold/flu symptom scores and analgesic use over 84 days without safety concerns (Horneck Johnston et al. 2024).

Systematic and narrative reviews from 2023 to 2025 converge on this pattern, attenuated symptom burden and improved well-being rather than sterilizing protection, while calling for product chemotyping (linkage pattern, branching, molecular mass/solubility) and harmonized outcomes (e.g., laboratory-confirmed respiratory tract infection, predefined symptom scales, and healthcare utilization) (Zhong et al. 2023).

Across preclinical models, polysaccharide-enriched extracts from *Pleurotus, Lentinula, Grifola, Inonotus,* and *Ganoderma* reduce viral load or disease-severity endpoints in influenza A, RSV, HSV-1 and HSV-2, enterovirus, and coronavirus infections. In many of these studies, decreases in viral burden coincide with strong cytoprotective effects, attenuated oxidative stress, preserved epithelial integrity, and dampened immunopathology, which makes it difficult to disentangle direct antiviral actions from host-directed benefits. The most conservative interpretation is that these preparations frequently act as host-modulators that improve tissue tolerance and early control, with any genuine virucidal or entry-blocking effects being context- and preparation-specific (Jesenak et al. 2025).

In children with allergic airways disease, adjunctive β-glucans may also reduce wheeze frequency when delivered as a nasal solution combined with barrier-supportive agents, again pointing to route- and matrix-specific opportunities to reduce respiratory morbidity (Indolfi et al. 2024).

A major barrier to translation is the marked heterogeneity among mushroom β-glucan preparations. Comparative studies of *Lentinula edodes* lentinan (β-1,3/1,6; ~300–800 kDa) and *Pleurotus citrinopileatus* polysaccharides (β-1,6-enriched; ~450 kDa) show that linkage architecture and molecular size produce distinct pharmacodynamic "fingerprints" in human myeloid cells and in vivo, affecting the balance of dendritic cell programs, macrophage polarization, NK-cell engagement, and downstream cytokine profiles. These differences are expected to yield nonidentical clinical effects across indications and routes (e.g., IV vs. oral or intranasal). Accordingly, trials should report full chemotyping, species/strain and tissue, extraction/processing, quantitative β-glucan content, β(1 → 6):β(1 → 3) ratio and branching degree, molecular weight distribution, solubility, conformational state (e.g., triple helix), and residuals (protein, endotoxin, and phenolics). Wherever possible, in vitro pharmacology in primary human cells should be explicitly bridged to prespecified clinical endpoints (e.g., URTI symptom-days or asthma exacerbations for respiratory indications; DFS/OS for oncology) via intermediate biomarkers (e.g., dectin-1 engagement, cytokine signatures, and trained immunity marks). This alignment enables meaningful dose–response modeling, comparison across products, and identification of responder subgroups.

### 8.4.3 Cardiometabolic Health: Glycemic Control and Lipid Modulation

β-glucans exert metabolic effects through complementary mechanisms that converge on glycemic control and lipid homeostasis. Through complementary physicochemical and microbiota-dependent routes, mushroom β-glucans modulate glycemic and lipid homeostasis. When present at sufficient molecular weight and solution viscosity, β-glucans slow gastric emptying and reduce the effective diffusion/uptake of luminal glucose, attenuating postprandial excursions—an effect quantified in meta-analyses of β-glucan meals and supported by the European Food Safety Authority (EFSA)'s dose–response opinion (Zurbau et al. 2021; EFSA Panel on Dietetic Products and Allergies 2011). In the colon, fermentation to SCFAs signals via epithelial G protein-coupled receptors (GPCRs) and reshapes enterohepatic/bile acid pathways, changes linked to lower intestinal cholesterol reabsorption and reduced hepatic de novo lipogenesis. SCFAs also tighten epithelial junctions and dampen low-grade inflammation, supporting insulin sensitivity (Chambers et al. 2018; Koh et al. 2016). Translationally, effects are product and matrix-specific: a 3-month RCT using a *Pleurotus eryngii* snack (UV-B–enhanced vitamin $D_2$) improved fasting glucose and LDL cholesterol in metabolically unhealthy adults, while a parallel trial adding daily mushrooms to a Mediterranean-style diet improved fasting glucose vs. the same diet without mushrooms. Conversely, an encapsulated *Ganoderma lucidum* intervention in type 2 diabetes/metabolic syndrome was neutral for hemoglobin A1c and fasting glucose (Kleftaki et al. 2022; Uffelman et al. 2024). A parallel randomized study embedding one daily serving of *Agaricus bisporus* and *Pleurotus ostreatus* mushrooms within a Mediterranean-style diet reported a greater fall in fasting serum glucose versus the same diet without mushrooms, while the overall diet, independent of mushrooms, lowered total cholesterol; this design isolates an incremental glycemic benefit attributable to mushrooms within a best-practice dietary pattern (Uffelman et al. 2024). By contrast, a double-blind, placebo-controlled trial in type 2 diabetes with metabolic syndrome using encapsulated *Ganoderma lucidum* (3 g/day, with/without *Cordyceps sinensis*) for 16 weeks found no effect on HbA1c or fasting glucose, emphasizing that not all mushroom interventions, sources, or formats yield glycemic benefits (Klupp et al. 2016). Compelling evidence suggests that human trials still vary in their characterization of β-glucans and endpoints but support the plausibility of improved glycemic control via viscosity-mediated carbohydrate handling and microbiota-dependent fermentation (Moniruzzaman et al. 2025).

For insulin resistance, small clinical trials with *Agaricus* extracts have reported improvements in the homeostatic model assessment of insulin resistance (HOMA-IR) when used as adjuncts to standard oral antidiabetic therapy, though these studies predate current trial-quality standards and require confirmation in RCTs; the signal nonetheless aligns with the innate immune and barrier effects noted above (Hsu et al. 2007). Complementary observational analyses suggest that higher habitual mushroom intake is associated with lower occurrence of type 2 diabetes in cohort settings, but causality remains to be established (Kim et al. 2024).

Lipid outcomes are supported by both controlled feeding studies and randomized trials, with source and preparation materially affecting results. Systematic reviews focused on *Pleurotus ostreatus* identify decreases in total and LDL cholesterol and triglycerides across several clinical trials, albeit with risk-of-bias concerns; individual trials in specific populations (e.g., HIV patients with antiretroviral therapy (ART)-induced dyslipidemia) have been null, likely reflecting dose, matrix, and cultivation-dependent variability in the active polysaccharide fraction (Dicks and Ellinger 2020). More recently, the *Pleurotus eryngii* randomized trial cited above documented reductions in LDL and oxidized LDL alongside improvements in inflammatory markers, an integrated lipid/oxidative profile consistent with β-glucan actions on bile acid turnover and lipoprotein handling (Kleftaki et al. 2022). Earlier controlled human studies with *Pleurotus* species have reported drops in triglycerides and cholesterol, though effect sizes vary and some results are pilot scale (Schneider et al. 2011). Recent reviews across edible mushrooms converge on a prudent conclusion: consistent lipid-lowering is most reproducible when β-glucan content, molecular traits, and dosing are adequately standardized and when outcomes include low-density lipoprotein (LDL)-C, non-high-density lipoprotein (HDL)-C, triglycerides, and oxidized LDL measured under controlled diets (Uffelman et al. 2023). Integrating these strands, mushroom β-glucans appear to support glycemic control (fasting glucose, occasionally HOMA-IR) and atherogenic lipid profiles (LDL-C, triglycerides, and oxidized LDL) in at-risk adults, with effect size and consistency contingent on (i) species/tissue and extraction (which determine solubility/viscosity and fermentability), (ii) delivery matrix (whole food vs. isolated fraction; co-nutrients such as vitamin $D_2$), and (iii) trial design (dietary background, duration, and adjudicated endpoints). Mechanistic reviews emphasize that higher molecular weight, more viscous mushroom β-glucans tend to yield greater postprandial glucose attenuation and cholesterol lowering, plausibly via delayed absorption and SCFA-mediated bile acid and hepatic lipid signaling, while prebiotic remodeling and barrier restoration reduce metabolic endotoxemia and low-grade inflammation that drive insulin resistance (Moniruzzaman et al. 2025).

### 8.4.4 Neurological Health and Cognition: Neurotrophic Actions and the Microbiota–Brain Axis

Lion's mane (*Hericium erinaceus*) concentrates two families of neuroactive molecules, hericenones from the fruiting body and erinacines from the mycelium, that upregulate neurotrophic signaling. Recent mechanistic work shows *Hericium erinaceus* constituents stimulate nerve growth factor (NGF) and brain-derived neurotrophic factor (BDNF) pathways and may act "pan-neurotrophically," promoting neurite outgrowth and synaptic plasticity while dampening neuroinflammation and oxidative stress, properties consistent with disease-modifying neuroprotection rather than mere symptomatic effects (Szućko-Kociuba et al. 2023). A double-blind RCT in older adults with MCI found that 16 weeks of HE fruiting body improved cognition on the Hasegawa Dementia Scale-Revised, with scores regressing after wash-out, an effect that underscores both efficacy and the need for continued exposure (Mori et al. 2009). A second parallel-group RCT reported Mini Mental State Examination (MMSE) improvement after 12 weeks of *Hericium erinaceus* in community-dwelling adults ≥50 years, albeit without concordant gains on two other memory tests, highlighting outcome-measure sensitivity and the heterogeneity of preparations used across trials (Saitsu et al. 2019). Moving upstream to an earlier disease stage, a 49-week double-blind pilot in mild AD using erinacine-A-enriched *Hericium erinaceus* mycelia (EAHE) showed higher MMSE and Instrumental Activities of Daily Living (IADL) scores vs. placebo, attenuation of plasma neurodegeneration/inflammation signals, and diffusion-tensor imaging changes consistent with preserved white-matter integrity; tolerability was good, with few withdrawals for minor adverse events (Li et al. 2020). Not all results are positive: an acute, well-controlled study in healthy young adults detected no overall cognitive or mood advantage from a single *Hericium erinaceus* dose, suggesting that chronic administration and/or at-risk populations are more plausible targets for benefit (Surendran et al. 2025). Observational evidence converges with these trials. In the Singapore Diet and Healthy Aging cohort, higher habitual mushroom intake (≥ ~300 g/week, mixed species commonly including *Hericium erinaceus*'s culinary analogs) was associated with markedly lower odds of MCI, a relationship the authors hypothesized could be mediated by ergothioneine (ET), a mushroom-specific antioxidant transported into the brain (Feng et al. 2019). That mechanistic link is now supported by longitudinal clinic-based data showing low plasma ET predicts faster cognitive and functional decline and by a pilot RCT in older adults with MCI in which 1 year of ET supplementation improved verbal learning and stabilized plasma neurofilament light versus placebo, signals of memory benefit and slower neuroaxonal injury (Wu et al. 2022). Together, these findings strengthen a biologically coherent narrative in which mushroom-borne actives (erinacines/hericenones/ET) engage neurotrophic, antioxidant, and anti-inflammatory axes relevant to age-related cognitive trajectories.

HE also interfaces with the microbiota–gut–brain axis. A human pilot (7-day supplementation) documented increased alpha diversity and enrichment of SCFAs, producing taxa (*Faecalibacterium prausnitzii, Eubacterium rectale*, *Gemmiger* spp.), alongside reductions in putative pathobionts; shifts correlated with serum biochemistry, indicating rapid, host-relevant modulation (Xie et al. 2021). Preclinical studies using HE polysaccharides show complementary changes in microbial composition and SCFAs levels with signaling through GPR41/43, reductions in colitis and systemic inflammation, and improved barrier integrity, mechanisms mechanistically linked to microglial priming, vagal signaling, and hippocampal plasticity in gut–brain biology (Diling et al. 2017). Recent studies of the gut–brain axis further situate SCFAs and tryptophan-pathway metabolites as key mediators that may bridge *Hericium erinaceus*-driven microbial remodeling to central effects on neuroinflammation and neurotrophism (Scuto et al. 2024).

Overall, *Hericium erinaceus* offers a mechanistically plausible, increasingly human-supported strategy for cognitive health: NGF/BDNF-oriented neurotrophic signaling from hericenones/erinacines; antioxidant and anti-inflammatory effects (with *Hericium erinaceus* as a notable, mushroom-specific contributor), and microbiota-mediated SCFAs and barrier-immune pathways converging on the brain. The clinical literature remains modest and heterogeneous in products and endpoints, but evidence now spans RCTs in MCI and early AD, a null acute study that refines expectations toward chronic use, and convergent cohort and biomarker data. Future trials should employ chemotyped *Hericium erinaceus* preparations (fruiting body vs. mycelium; erinacine A content), adjudicated cognitive and functional endpoints, and integrated microbiome/biomarker readouts to resolve dose–response and target-population questions (Li et al. 2020; Saitsu et al. 2019; Surendran et al. 2025).

### 8.4.5 Gut Health and Inflammatory Disease: Barrier Integrity and Immune Resolution

Mushroom β-glucans act at the intestinal interface where epithelial, microbial, and immune cues converge. After oral intake, particulate fungal glucans are sampled by M cells in Peyer's patches and presented to macrophages and dendritic cells in the lamina propria, where they engage pattern-recognition receptors such as dectin-1 (predominantly for insoluble β-glucans) and, for some soluble fractions, CR3/TLR2, initiating NF-κB–linked programs that coordinate chemokinesis, cytokine release, and crosstalk with the epithelium. This mucosal route explains how fungal glucans can modulate barrier function and local inflammation differently from cereal β-glucans, whose effects are largely physicochemical; the structural distinction (β-(1 → 3)/(1 → 6) in fungi versus (1 → 3)/(1 → 4) in cereals) is a key determinant of bioactivity (van Steenwijk et al. 2021).

Within inflammatory bowel disease (IBD) models, *Lentinula edodes* β-glucan (lentinan) provides a mechanistic template. In dextran sulfate sodium (DSS) colitis, oral lentinan was more effective than rectal administration and operated via two complementary mechanisms: in the ileum, dendritic cell sensing in Peyer's patches increased IL-12 and promoted Th1 differentiation (upregulation of *Il12b* and *Tbx21*), with Th1 cells subsequently migrating to the inflamed colon; in parallel, once in the colon, lentinan was recognized through dectin-1 on intestinal epithelial cells, which suppressed basolateral TNFR1 surface delivery and thereby blunted TNF-α–driven NF-κB activation and IL-8 output, culminating in reduced colonic inflammation (Minato et al. 2023; Nishitani et al. 2013). Suppression of pMLKL-mediated necroptosis has been reported as an additional epithelial-protective axis, and combining lentinan with *Lactobacillus reuteri* plus *Bacillus coagulans* outperformed either agent alone, consistent with β-glucan–microbiota synergy (Mizuno and Minato 2024).

Anti-inflammatory actions are not limited to *Lentinula.* A β-1,6/1,3-glucan from *Pleurotus citrinopileatus* (PCPs; ~450 kDa) drives dectin-1–Syk–CARD9 signaling in human dendritic cells and can also signal via TLR2/4, producing a tailored cytokine milieu that includes both pro- and anti-inflammatory mediators; during monocyte-to-macrophage differentiation, PCPs reduced TNF-α and IL-6 secretion while leaving IL-10 unchanged and primed CCL2/CCL8 expression with reciprocal dampening of CCR2 transcripts, features that align with context-appropriate immune resolution in the gut (Mizuno and Minato 2024). Beyond *Pleurotus*, other mushroom extracts relevant to colitis include *Inonotus obliquus* (Chaga). Its water-soluble β-glucans have been shown to lower colonic TNF-α and STAT1 signaling in DSS colitis and exhibit complementary anti-inflammatory effects in LPS-challenged macrophages. Additionally, β-glucans from *Pleurotus pulmonarius* have been found to reduce clinical signs of colitis (diarrhea, bleeding) and improve fecal quality in chronic DSS models. Although human data remain limited, a small trial of a mixed mushroom extract ("AndoSan"; *Agaricus blazei*, *Hericium erinaceus*, and *Grifola frondosa*) in ulcerative colitis and Crohn's disease reported reductions in circulating pro-inflammatory cytokines and a fall in fecal calprotectin in UC, supporting translational plausibility (Moniruzzaman et al. 2025).

These immunological effects are intertwined with microbiota ecology and barrier integrity. Recent studies conclude that mushroom β-glucans act as prebiotic-like substrates that are fermented to SCFAs, strengthening tight-junction complexes (e.g., ZO-1), enhancing mucin production, and reducing endotoxemia; higher molecular weight and viscosity generally correlate with greater benefits (Araújo-Rodrigues et al. 2024; Zhao et al. 2023). SCFAs signaling through GPCRs (notably GPR41/43) links these microbial shifts to epithelial and metabolic responses that dampen inflammation and restore homeostasis (Moniruzzaman et al. 2025). In IBD pathobiology, characterized by barrier failure and overexpression of TNF-α, IL-1β, IL-6, and IL-8 (Dunleavy et al. 2023), interventions that both reinforce junctional integrity and attenuate PRR-driven cytokine cascades are protective; β-glucans meet this profile by simultaneously reconditioning innate signaling (e.g., dectin-1/TNFR1 axis) and reshaping the luminal milieu toward SCFA-rich, less pro-inflammatory communities (Mizuno and Minato 2024; Sakaguchi et al. 2018).

Taken together, the gut evidence-based positions mushroom β-glucans as credible adjuncts for intestinal health and inflammatory disease control: rigorous mechanistic data in DSS colitis define epithelial and immune checkpoints targeted by lentinan; convergent results with *Pleurotus* and other genera extend generalizability; and human observations, though modest, suggest anti-inflammatory activity compatible with current care. Translation now depends on chemotyping and standardization (linkage pattern, branching degree, molecular weight/solubility), matrix-aware dosing strategies that favor colonic delivery, and trials that integrate clinical activity indices with biomarker panels (e.g., cytokines, fecal calprotectin), epithelial integrity readouts, and microbiome/SCFAs profiling to resolve dose–response and responder phenotypes (Mizuno and Minato 2024).

## 8.5 Challenges to Clinical Translation and Paths Forward

Despite a coherent mechanistic framework linking mushroom-derived polysaccharides, especially β-glucans, to innate and mucosal immunomodulation, the present evidence base remains constrained by chemical heterogeneity, variable preparations, and uneven clinical phenotyping. Across species and even within a given species, molecular weight distributions, $\beta(1 \rightarrow 3)/\beta(1 \rightarrow 6)$ linkage patterns, branching degree, and higher-order conformation are strongly molded by cultivation conditions and extraction methods; these parameters, in turn, determine solubility, viscosity, fermentability, and receptor engagement. As this chapter notes, aligning basic chemistry with biological readouts will require chemotyping and standardization of β-glucan fractions (e.g., linkage analysis, nuclear magnetic resonance profiles, molecular weight characterization) across studies and products, together with matrix-aware dosing strategies that deliver actives to the intended intestinal niche. Without this, cross-trial comparisons and meta-analyses will continue to blur the true signal with manufacturing noise. A second limitation is the translation gap from preclinical to clinical settings. Murine colitis models (e.g., dextran sulfate sodium) have delineated epithelial and myeloid checkpoints for defined β-glucans such as lentinan, including epithelial dectin-1–TNFR1 crosstalk and Peyer's patch-driven Th1 trafficking, yet human data in inflammatory bowel disease remain modest, often limited to small pilot trials or mixed extracts; robust efficacy trials with adjudicated activity indices, barrier integrity measures, fecal calprotectin, and integrated microbiome/SCFAs profiling are still missing. Similarly, in infectious-disease contexts, the clinical literature specific to mushroom β-glucans is sparse; much of the interventional signal derives from yeast β-(1,3/1,6)-glucans used as analogs, and regulatory evaluations continue to emphasize the need for standardized preparations and prespecified endpoints before population-level "immune defense" claims can be justified.

The trained immunity paradigm provides a unifying mechanistic model, but it is early in nutritional translation. While ex vivo human data show that β-glucans from *Agaricus bisporus* products can hallmark monocyte programs consistent with trained immunity, prospective trials that link defined dietary β-glucans to durable epigenetic/metabolic signatures (e.g., H3K4me3/H3K27ac, glycolysis/mevalonate rewiring) and infection- or vaccine-relevant outcomes in vivo are still needed. Establishing a minimal biomarker set, combining chromatin marks, cytokine responses, and metabolite fingerprints, would enable dose–response modeling and responder stratification. For metabolic disorders, heterogeneity of species, delivery matrices (whole foods vs. encapsulated extracts), co-nutrients (e.g., vitamin $D_2$), dose, and trial design continues to yield mixed results. Signals for improved fasting glucose and atherogenic lipids are most consistent when β-glucan content and physicochemistry are specified; in contrast, some encapsulated preparations (e.g., *Ganoderma lucidum* in type 2 diabetes) have been neutral, underscoring that not all mushroom interventions have been successfully applied. Future studies should incorporate standardized β-glucans, controlled dietary backgrounds, and harmonized lipid endpoints (LDL-C, non-HDL-C, triglycerides, and oxidized LDL) to reduce variance and clarify effect sizes.

In neurocognitive health, promising randomized trials with *Hericium erinaceus* in MCI and early AD sit alongside an acute null study in healthy young adults and heterogeneous product profiles (fruiting body vs. mycelium; erinacine vs. hericenone content). The field requires chemotyped interventions with quantified erinacine/hericenone loads, longer durations, and adjudicated cognitive/functional endpoints, ideally coupled to mechanistic intermediates (e.g., NGF/BDNF signaling, plasma neurofilament light, and microbiome/SCFAs correlates) to bridge biology and outcomes.

Across domains, trial methodology represents a persistent gap. Many studies are small, underpowered, or rely on composite or surrogate endpoints without prespecified minimal clinically important differences. Head-to-head comparisons across species (e.g., *Pleurotus* vs. *Lentinula*) and across formats (whole food vs. purified fraction) are rare, making it difficult to construct equivalence matrices that guide substitution or dose conversions. Safety appears favorable overall, but systematic pharmacovigilance in populations with autoimmunity, immunosuppression, or polypharmacy is limited and should be embedded prospectively.

Future directions should therefore emphasize (i) reference standards for mushroom β-glucans, molecular weight distributions, and conformation metrics, reported alongside viscosity/solubility and fermentation indices; (ii) matrix-aware delivery and colon-targeted formulations to enhance local availability where barrier and immune effects are expressed; (iii) multi-omics endpoints (chromatin, metabolome, microbiome) anchored to clinical outcomes to resolve mechanisms and responders; (iv) vaccine-adjunct and human challenge models to test trained immunity hypotheses with clear, time-bounded readouts; and (v) precision-nutrition frameworks that stratify by baseline microbiome, metabolic status, age (including immunosenescence), and genetic polymorphisms in PRR pathways. These programs, if executed with regulatory-grade product characterization and harmonized endpoints, can convert a mechanistically rich but heterogeneous literature into actionable, indication-specific guidance for the clinical and public-health deployment of mushroom polysaccharides (Fig. 8.5).

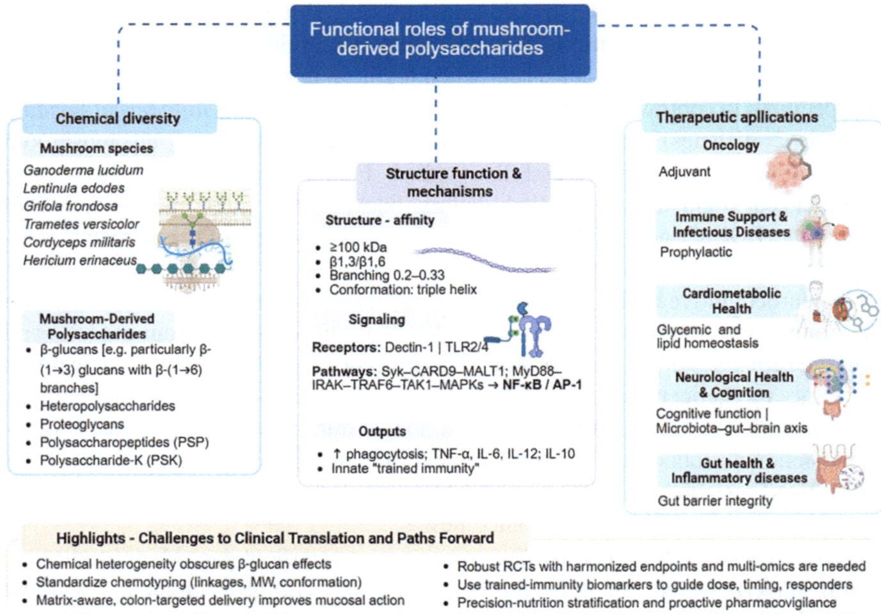


**Fig. 8.5** Integrated view of mushroom-derived polysaccharides from chemistry to clinic. *Left*: representative species and classes (β-1,3/1,6-glucans, heteropolysaccharides, proteoglycans, PSP/PSK). Center: Structural determinants (≥100 kDa, β-1,3 backbone with β-1,6 branches, branching 0.2–0.33, triple helix) govern receptor binding (dectin-1, TLR2/4) and signaling (Syk–CARD9–MALT1; MyD88–IRAK–TRAF6–TAK1–MAPKs → NF-κB/AP-1), producing phagocytosis/ROS, cytokines (TNF-α, IL-6, IL-12), IL-10, and trained immunity programs. *Right*: therapeutic domains—oncology adjuvant use, immune/infectious support, cardiometabolic regulation, neurocognitive health, and gut-barrier/inflammatory disease. *Bottom*: translation priorities—standardized chemotyping, matrix-aware mucosal delivery, robust RCTs with harmonized multi-omics endpoints, biomarker-guided dosing, and precision-nutrition stratification

## 8.6 Conclusion

Mushroom-derived polysaccharides, particularly β-(1 → 3,1 → 6)-glucans and related heteropolysaccharides, occupy a distinctive position at the interface between nutrition, immunology, and clinical medicine. Across species, strains, and extraction strategies, these macromolecules display a high degree of structural diversity in molecular weight, branching architecture, solubility, and higher-order conformation, and these parameters collectively dictate receptor engagement, downstream signaling, and ultimately biological effect. At the mechanistic level, fungal β-glucans are now firmly established as ligands for dectin-1, CR3, and selected Toll-like receptors, converging on the Syk–CARD9–MALT1 and MyD88–IRAK–TRAF6–TAK1 axes and activating NF-κB/MAPK-dependent transcriptional programs. In parallel, trained immunity paradigms and prebiotic-like actions at the gut interface provide unifying frameworks that link polysaccharide chemistry to innate immune memory, barrier integrity, and microbiota-dependent metabolism.

Translational evidence, although heterogeneous, supports a multi-domain role for mushroom polysaccharides in human health. In oncology, well-characterized preparations such as PSK, PSP, lentinan, and schizophyllan function as biological response modifiers that enhance antitumor immunity and, in specific settings, improve survival when used as adjuvants to chemotherapy or radiotherapy. In infectious and respiratory disease, standardized β-glucan formulations, most notably pleuran and yeast β-(1,3/1,6)-glucans as structural analogs, recurrently reduce upper respiratory tract symptom burden, improve asthma control in children, and support convalescence, consistent with host-directed rather than pathogen-specific benefits. Cardiometabolic trials using Pleurotus and other edible mushrooms indicate improvements in fasting glucose, insulin resistance proxies, and atherogenic lipid profiles when β-glucan content and physicochemical traits are adequately specified. In neurocognitive health, *Hericium erinaceus* exemplifies how mushroom matrices can interface with neurotrophic pathways and the microbiota–gut–brain axis, with early randomized controlled trials in mild cognitive impairment and Alzheimer's disease suggesting clinically relevant effects. Finally, in gut inflammation, mechanistic models with lentinan and other β-glucans delineate epithelial and myeloid checkpoints through which defined polysaccharides restore barrier integrity and promote immune resolution, with emerging human data in inflammatory bowel disease pointing in the same direction.

At the same time, this literature is constrained by substantial heterogeneity in products, dosing regimens, and clinical endpoints. Species and strain selection, cultivation conditions, and extraction technologies generate polysaccharide ensembles that differ in linkage patterns, branching degree, molecular weight distributions, and conformational states, with direct consequences for solubility, viscosity, fermentability, and receptor binding. Many clinical studies remain small, underpowered, or insufficiently chemotyped, limiting cross-trial comparability and formal synthesis. For trained immunity, gut–microbiota modulation, and cardiometabolic outcomes in particular, the gap between elegant preclinical work and robust, biomarker-anchored randomized trials remains wide.

Looking forward, the functional roles of mushroom-derived polysaccharides in human health will be defined less by single "panacean" claims and more by chemistry-anchored, indication-specific applications. Priority areas include (i) the establishment of reference standards for mushroom β-glucans and related polysaccharides, including linkage analysis, branching metrics, molecular weight profiles, and conformation; (ii) matrix-aware formulation and colon-targeted delivery strategies that align the site of action with desired outcomes in immunity, metabolism, and barrier biology; and (iii) clinical programs that integrate validated endpoints with multi-omics readouts, epigenetic and metabolic signatures of trained immunity, microbiome and SCFA profiles, and barrier/inflammatory biomarkers to enable dose–response modeling and responder stratification. Within this framework, valorization of mushroom by-products as sustainable sources of β-glucans and complementary polysaccharides can support circular bioeconomy goals while supplying rigorously characterized ingredients for next-generation nutraceuticals and adjuvant therapies.

In summary, mushroom-derived polysaccharides constitute a mechanistically rich and translationally promising class of bioactive compounds. By coupling structural standardization and advanced extraction with biomarker-driven clinical trials and precision-nutrition approaches, the field is well positioned to convert a diverse and historically descriptive evidence base into actionable strategies for promoting immune competence, metabolic health, gut and brain homeostasis, and integrative oncology care.

**Acknowledgments** We are thankful to the Fundação para a Ciência e a Tecnologia (Portugal) through project 2024.04884.RESTART—PrebioFungi - Exploring the Role of β-Glucans Derived from Mushroom By-Products in Gut Microbiota Modulation and Prebiotic Effects. Irene Gouvinhas thanks FCT (Fundação para a Ciência e Tecnologia) for funding through the Scientific Employment Stimulus—Individual Call (2022.00498.CEECIND), https://doi.org/10.54499/2022.00498.CEECIND/CP1749/CT0001 (accessed on 17 May 2024). Juliana Garcia and Jani Silva are grateful for funding support from Portuguese public funding through Investimento RE-C05-i02-Missão Interface N.° 01/C05-i02/2022 and from FCT.

We used AI to assist with drafting and restructuring sections of the manuscript. All content was verified, edited, and approved by the authors, who take full responsibility for the final version.

**Funding** This work is supported by national funds from FCT—Portuguese Foundation for Science and Technology—under the projects UID/04033: Centro de Investigação e de Tecnologias Agro-Ambientais e Biológicas and LA/P/0126/2020 (10.54499/LA/P/0126/2020) (accessed on August 25, 2025). This work is funded by national funds from FCT—Fundação para a Ciência e a Tecnologia, I.P., under the project/support UID/6157/2023. This work is also supported by national funds from FCT – Fundação para a Ciência e a Tecnologia, I.P., under the project/support UID/6157/2025 and the projects UID/00772/2025, DOI https://doi.org/10.54499/UID/00772/2025 and LA/P/0059/2020 of the Portuguese Science and Technology Foundation (FCT). Eva Olo-Fontinha would like to acknowledge FCT - Fundação para a Ciência e Tecnologia, I.P. for the support and individual PhD grant 2024.06281.BDANA.

## References

Aguiló-Aguayo I, Walton J, Viñas I, Tiwari BK (2017) Ultrasound assisted extraction of polysaccharides from mushroom by-products. LWT 77:92–99. https://doi.org/10.1016/j.lwt.2016.11.043

Araújo-Rodrigues H, Sousa AS, Relvas JB, Tavaria FK, Pintado M (2024) An overview on mushroom polysaccharides: health-promoting properties, prebiotic and gut microbiota modulation effects and structure-function correlation. Carbohydr Polym 333:121978. https://doi.org/10.1016/j.carbpol.2024.121978

Balakrishnan B, Liang Q, Fenix K, Tamang B, Hauben E, Ma L, Zhang W (2021) Combining the anticancer and immunomodulatory effects of Astragalus and shiitake as an integrated therapeutic approach. Nutrients 13(8). https://doi.org/10.3390/nu13082564

Brown GD (2006) Dectin-1: a signalling non-TLR pattern-recognition receptor. Nat Rev Immunol 6(1):33–43. https://doi.org/10.1038/nri1745

Camilli G, Tabouret G, Quintin J (2018) The complexity of fungal β-Glucan in health and disease: effects on the mononuclear phagocyte system. Front Immunol 9:673. https://doi.org/10.3389/fimmu.2018.00673

Case S, O'Brien T, Ledwith AE, Chen S, Horneck Johnston CJ, Hackett EE, O'Sullivan M, Charles-Messance H, Dempsey E, Yadav S (2024) β-Glucans from Agaricus bisporus mushroom products drive trained immunity. Front Nutr 11:1346706

Chakraborty N, Banerjee A, Sarkar A, Ghosh S, Acharya K (2021) Mushroom polysaccharides: a potent immune-modulator. Biointerface Res Appl Chem 11(2):8915–8930

Chambers ES, Preston T, Frost G, Morrison DJ (2018) Role of gut microbiota-generated short-chain fatty acids in metabolic and cardiovascular health. Curr Nutr Rep 7(4):198–206. https://doi.org/10.1007/s13668-018-0248-8

Chen H, Zhou X, Zhang J (2014) Optimization of enzyme assisted extraction of polysaccharides from Astragalus membranaceus. Carbohydr Polym 111:567–575. https://doi.org/10.1016/j.carbpol.2014.05.033

Cheong KL, Wang LY, Wu DT, Hu DJ, Zhao J, Li SP (2016) Microwave-assisted extraction, chemical structures, and chain conformation of polysaccharides from a novel Cordyceps Sinensis fungus UM01. J Food Sci 81(9):C2167–C2174. https://doi.org/10.1111/1750-3841.13407

Datta NR, Stutz E, Gomez S, Bodis S (2019) Efficacy and safety evaluation of the various therapeutic options in locally advanced cervix cancer: a systematic review and network meta-analysis of randomized clinical trials. Int J Radiat Oncol Biol Phys 103(2):411–437. https://doi.org/10.1016/j.ijrobp.2018.09.037

Dharsono T, Rudnicka K, Wilhelm M, Schoen C (2019) Effects of yeast (1, 3)-(1, 6)-beta-glucan on severity of upper respiratory tract infections: a double-blind, randomized, placebo-controlled study in healthy subjects. J Am Coll Nutr 38(1):40–50

Dicks L, Ellinger S (2020) Effect of the intake of oyster mushrooms (Pleurotus ostreatus) on cardiometabolic parameters-a systematic review of clinical trials. Nutrients 12(4). https://doi.org/10.3390/nu12041134

Diling C, Xin Y, Chaoqun Z, Jian Y, Xiaocui T, Jun C, Ou S, Yizhen X (2017) Extracts from Hericium erinaceus relieve inflammatory bowel disease by regulating immunity and gut microbiota. Oncotarget 8(49):85838–85857. https://doi.org/10.18632/oncotarget.20689

Dunleavy KA, Raffals LE, Camilleri M (2023) Intestinal barrier dysfunction in inflammatory bowel disease: underpinning pathogenesis and therapeutics. Dig Dis Sci 68(12):4306–4320. https://doi.org/10.1007/s10620-023-08122-w

EFSA Panel on Dietetic Products N, Allergies (2011) Scientific opinion on the substantiation of health claims related to beta-glucans from oats and barley and maintenance of normal blood LDL-cholesterol concentrations (ID 1236, 1299), increase in satiety leading to a reduction in energy intake (ID 851, 852), reduction of post-prandial glycaemic responses (ID 821, 824), and "digestive function"(ID 850) pursuant to Article 13 (1) of Regulation (EC) No 1924/2006. EFSA J 9(6):2207

Feng L, Cheah IK, Ng MM, Li J, Chan SM, Lim SL, Mahendran R, Kua EH, Halliwell B (2019) The association between mushroom consumption and mild cognitive impairment: a community-based cross-sectional study in Singapore. J Alzheimer's Dis 68(1):197–203. https://doi.org/10.3233/jad-180959

Fritz H, Kennedy DA, Ishii M, Fergusson D, Fernandes R, Cooley K, Seely D (2015) Polysaccharide K and Coriolus versicolor extracts for lung cancer: a systematic review. Integr Cancer Ther 14(3):201–211

Fuller R, Moore MV, Lewith G, Stuart BL, Ormiston RV, Fisk HL, Noakes PS, Calder PC (2017) Yeast-derived β-1,3/1,6 glucan, upper respiratory tract infection and innate immunity in older adults. Nutrition 39–40:30–35. https://doi.org/10.1016/j.nut.2017.03.003

Garcia J, Rodrigues F, Saavedra MJ, Nunes FM, Marques G (2022) Bioactive polysaccharides from medicinal mushrooms: a review on their isolation, structural characteristics and antitumor activity. Food Biosci 49:101955. https://doi.org/10.1016/j.fbio.2022.101955

Gharibzahedi SMT, Marti-Quijal FJ, Barba FJ, Altintas Z (2022) Current emerging trends in antitumor activities of polysaccharides extracted by microwave- and ultrasound-assisted methods. Int J Biol Macromol 202:494–507. https://doi.org/10.1016/j.ijbiomac.2022.01.088

Gunasekaran S, Govindan S, Ramani P (2021) Sulfated modification, characterization and bioactivities of an acidic polysaccharide fraction from an edible mushroom Pleurotus eous (Berk.) Sacc. Heliyon 7(1):e05964. https://doi.org/10.1016/j.heliyon.2021.e05964

Guo Q, Wang C, Hu X, Ai L (2017) Polysaccharides: structure and solubility. In: Xu Z (ed) Solubility of polysaccharides. IntechOpen, Rijeka. https://doi.org/10.5772/intechopen.71570

Guo N, Bai Z, Jia W, Sun J, Wang W, Chen S, Wang H (2019) Quantitative analysis of polysaccharide composition in Polyporus umbellatus by HPLC–ESI–TOF–MS. Molecules 24(14):2526

Guo P, Chen H, Ma J, Zhang Y, Chen H, Wei T, Gao D, Li J (2023) Enzyme-assisted extraction, characterization, and in vitro antioxidant activity of polysaccharides from Potentilla anserina L. Front Nutr 10:1216572. https://doi.org/10.3389/fnut.2023.1216572

Guo Y, Nan S, Qiu C, Song C, Wu B, Tang Y, Cheng L, Ma H (2024) Ultrasound-assisted enzymatic extraction of jujube (Ziziphus jujuba Mill.) polysaccharides: extraction efficiency, antioxidant activity, and structure features. Ultrason Sonochem 111:107088. https://doi.org/10.1016/j.ultsonch.2024.107088

Habtemariam S (2020) Trametes versicolor (Synn. Coriolus versicolor) polysaccharides in cancer therapy: targets and efficacy. Biomedicine 8(5). https://doi.org/10.3390/biomedicines8050135

Han B, Baruah K, Cox E, Vanrompay D, Bossier P (2020) Structure-functional activity relationship of β-glucans from the perspective of immunomodulation: a mini-review. Front Immunol 11:658

Horneck Johnston CJH, Ledwith AE, Lundahl MLE, Charles-Messance H, Hackett EE, O'Shaughnessy SD, Clegg J, Prendeville H, McGrath JP, Walsh AM, Case S, Austen Byrne H, Gautam P, Dempsey E, Corr SC, Sheedy FJ (2024) Recognition of yeast β-glucan particles triggers immunometabolic signaling required for trained immunity. iScience 27(3):109030. https://doi.org/10.1016/j.isci.2024.109030

Hsu CH, Liao YL, Lin SC, Hwang KC, Chou P (2007) The mushroom Agaricus Blazei Murill in combination with metformin and gliclazide improves insulin resistance in type 2 diabetes: a randomized, double-blinded, and placebo-controlled clinical trial. J Altern Complement Med 13(1):97–102. https://doi.org/10.1089/acm.2006.6054

Hu S, Liang Z-C, Chiu C-C, Lien J, Chen K, Lee M, Wang J (2006) Antihyperlipidemic and antioxidant effects of extracts from Pleurotus citrinopileatus. J Agric Food Chem 54:2103–2110. https://doi.org/10.1021/jf052890d

Huang J-H, Lin C-Y, Wu S-Y, Chen W-Y, Chu C-L, Brown GD, Chuu C-P, Wu-Hsieh BA (2015) CR3 and Dectin-1 collaborate in macrophage cytokine response through association on lipid rafts and activation of Syk-JNK-AP-1 pathway. PLoS Pathog 11(7):e1004985. https://doi.org/10.1371/journal.ppat.1004985

Huang G, Chen F, Yang W, Huang H (2021) Preparation, deproteinization and comparison of bioactive polysaccharides. Trends Food Sci Technol 109:564–568. https://doi.org/10.1016/j.tifs.2021.01.038

Ina K, Furuta R, Kataoka T, Kayukawa S, Ina H, Yoneda M (2016) Chemo-immunotherapy using Lentinan for the treatment of gastric cancer with liver metastases. Med Sci 4(2):8

Indolfi C, Mignini C, Valitutti F, Bizzarri I, Dinardo G, Klain A, Miraglia Del Giudice M, Di Cara G (2024) Effects of nasal solution incorporating resveratrol and Carboxymethyl-B-Glucan in preschool non-atopic children with wheezing. Nutrients 16(14). https://doi.org/10.3390/nu16142197

Jesenak M, Hrubisko M, Chudek J, Bystron J, Rennerova Z, Diamant Z, Majtan J, Emeryk A (2025) Beneficial effects of pleuran on asthma control and respiratory tract-infection frequency in children with perennial asthma. Sci Rep 15(1):7146. https://doi.org/10.1038/s41598-025-91830-2

Jia X, Ma B, Xue F, Xing Y, Wu P, Li T, Shi F, Xu C (2022) Structure characterization and anti-inflammatory activity of polysaccharides from Lingzhi or Reishi medicinal mushroom Ganoderma lucidum (Agaricomycetes) by microwave-assisted freeze-thaw extraction. Int J Med Mushrooms 24(11):49–61. https://doi.org/10.1615/IntJMedMushrooms.2022045268

Khan N, Tran KA, Chevre R, Locher V, Richter M, Sun S, Sadeghi M, Pernet E, Herrero-Cervera A, Grant A, Saif A, Downey J, Kaufmann E, Khader SA, Joubert P, Barreiro LB, Yipp BG, Soehnlein O, Divangahi M (2025) β-Glucan reprograms neutrophils to promote disease tolerance against influenza a virus. Nat Immunol 26(2):174–187. https://doi.org/10.1038/s41590-024-02041-2

Kim HS, Hong JT, Kim Y, Han S-B (2011) Stimulatory effect of β-glucans on immune cells. Immune Netw 11(4):191–195

Kim HS, Park KH, Lee HK, Kim JS, Kim YG, Lee JH, Kim KH, Yun J, Hwang BY, Hong JT, Kim Y, Han S-B (2016) Curdlan activates dendritic cells through dectin-1 and toll-like receptor 4 signaling. Int Immunopharmacol 39:71–78. https://doi.org/10.1016/j.intimp.2016.07.013

Kim Y-M, Woo HW, Shin M-H, Koh SB, Kim HC, Kim MK (2024) A prospective association between dietary mushroom intake and the risk of type 2 diabetes: the Korean Genome and Epidemiology Study–Cardiovascular Disease Association Study. Epidemiol Health 46: e2024017

Kleftaki S-A, Amerikanou C, Gioxari A, Lantzouraki DZ, Sotiroudis G, Tsiantas K, Tsiaka T, Tagkouli D, Tzavara C, Lachouvaris L, Zervakis GI, Kalogeropoulos N, Zoumpoulakis P, Kaliora AC (2022) A randomized controlled trial on Pleurotus eryngii mushrooms with antioxidant compounds and vitamin D2 in managing metabolic disorders. Antioxidants 11(11):2113

Klupp NL, Kiat H, Bensoussan A, Steiner GZ, Chang DH (2016) A double-blind, randomised, placebo-controlled trial of Ganoderma lucidum for the treatment of cardiovascular risk factors of metabolic syndrome. Sci Rep 6(1):29540. https://doi.org/10.1038/srep29540

Kodama S, Shoda T, Wahyudiono W, Machmudah S, Kanda H, Goto M (2015) Subcritical water extraction of polysaccharides using a semi-batch extractor. Mod Appl Sci 9:220–224

Koh A, De Vadder F, Kovatcheva-Datchary P, Bäckhed F (2016) From dietary fiber to host physiology: short-chain fatty acids as key bacterial metabolites. Cell 165(6):1332–1345. https://doi.org/10.1016/j.cell.2016.05.041

Konusova V, Vorbeychikov E, Shamtsyan M (2021) Potential role of mushroom beta-glucans in immunity and inflammation in viral infections and COVID-19. J Food Bioact 16:8–18

Krittanawong C, Isath A, Hahn J, Wang Z, Fogg SE, Bandyopadhyay D, Jneid H, Virani SS, Tang WW (2021) Mushroom consumption and cardiovascular health: a systematic review. Am J Med 134(5):637–642. e632

Leong YK, Yang F-C, Chang J-S (2021) Extraction of polysaccharides from edible mushrooms: emerging technologies and recent advances. Carbohydr Polym 251:117006. https://doi.org/10.1016/j.carbpol.2020.117006

Li IC, Chang HH, Lin CH, Chen WP, Lu TH, Lee LY, Chen YW, Chen YP, Chen CC, Lin DP (2020) Prevention of early Alzheimer's disease by Erinacine A-enriched Hericium erinaceus mycelia pilot double-blind placebo-controlled study. Front Aging Neurosci 12:155. https://doi.org/10.3389/fnagi.2020.00155

Li F, Lei H, Xu H (2022) Influences of subcritical water extraction on the characterization and biological properties of polysaccharides from Morchella sextelata. J Food Process Preserv 46(9):e16024. https://doi.org/10.1111/jfpp.16024

Lin Y, Zeng H, Wang K, Lin H, Li P, Huang Y, Zhou S, Zhang W, Chen C, Fan H (2019) Microwave-assisted aqueous two-phase extraction of diverse polysaccharides from Lentinus edodes: process optimization, structure characterization and antioxidant activity. Int J Biol Macromol 136:305–315. https://doi.org/10.1016/j.ijbiomac.2019.06.064

Lin Y, Pi J, Jin P, Liu Y, Mai X, Li P, Fan H (2022) Enzyme and microwave co-assisted extraction, structural characterization and antioxidant activity of polysaccharides from purple-heart radish. Food Chem 372:131274. https://doi.org/10.1016/j.foodchem.2021.131274

Liu Y, Zhao J, Zhao Y, Zong S, Tian Y, Chen S, Li M, Liu H, Zhang Q, Jing X, Sun B, Wang H, Sun T, Yang C (2019) Therapeutic effects of lentinan on inflammatory bowel disease and colitis-associated cancer. J Cell Mol Med 23(2):750–760. https://doi.org/10.1111/jcmm.13897

Lu X (2023) Changes in the structure of polysaccharides under different extraction methods. eFood 4(2):e82. https://doi.org/10.1002/efd2.82

Lu H, Yang Y, Gad E, Wenner CA, Chang A, Larson ER, Dang Y, Martzen M, Standish LJ, Disis ML (2011) Polysaccharide krestin is a novel TLR2 agonist that mediates inhibition of tumor growth via stimulation of CD8 T cells and NK cells. Clin Cancer Res 17(1):67–76. https://doi.org/10.1158/1078-0432.Ccr-10-1763

Luan F, Peng X, Zhao G, Zeng J, Zou J, Rao Z, Liu Y, Zhang X, Ma H, Zeng N (2022) Structural diversity and bioactivity of polysaccharides from medicinal mushroom Phellinus spp.: a review. Food Chem 397:133731. https://doi.org/10.1016/j.foodchem.2022.133731

Ma Y, Zheng X, Chu Z, Nan W, Zhao Y, Bai Y, Ma R, Jia Y, Lü X, Wang X (2024) Polysaccharides from Inonotus obliquus employing subcritical water extraction: extraction optimization, physiochemical properties and bioactivities analysis. Ind Crop Prod 222: 119638. https://doi.org/10.1016/j.indcrop.2024.119638

Maity P, Sen IK, Chakraborty I, Mondal S, Bar H, Bhanja SK, Mandal S, Maity GN (2021) Biologically active polysaccharide from edible mushrooms: a review. Int J Biol Macromol 172: 408–417. https://doi.org/10.1016/j.ijbiomac.2021.01.081

Minato KI, Oura K, Mizuno M (2023) The inhibitory effect of oral administration of lentinan on DSS-induced inflammation is exerted by the migration of T cells activated in the ileum to the colon. Eur J Pharmacol 946:175631. https://doi.org/10.1016/j.ejphar.2023.175631

Mirzadeh M, Arianejad MR, Khedmat L (2020) Antioxidant, antiradical, and antimicrobial activities of polysaccharides obtained by microwave-assisted extraction method: a review. Carbohydr Polym 229:115421. https://doi.org/10.1016/j.carbpol.2019.115421

Mizuno M, Minato K-i (2024) Anti-inflammatory and immunomodulatory properties of polysaccharides in mushrooms. Curr Opin Biotechnol 86:103076. https://doi.org/10.1016/j.copbio.2024.103076

Moniruzzaman M, Tuong DTC, Chin S, Anjana S, Karthikeyan A, Min T, Kim I-J (2025) A review on pharmacological insights of edible and medicinal mushroom based β-glucans. Appl Biol Chem 68(1):41. https://doi.org/10.1186/s13765-025-01006-9

Monteiro LMO, del Cerro C, Kijpornyongpan T, Yaguchi A, Bennett A, Donohoe BS, Ramirez KJ, Benson AF, Mitchell HD, Purvine SO, Markillie LM, Burnet MC, Bloodsworth KJ, Bowen BP, Harwood TV, Louie K, Northen T, Salvachúa D (2025) Metabolic profiling of two white-rot fungi during 4-hydroxybenzoate conversion reveals biotechnologically relevant biosynthetic pathways. Commun Biol 8(1):224. https://doi.org/10.1038/s42003-025-07640-9

Mori K, Inatomi S, Ouchi K, Azumi Y, Tuchida T (2009) Improving effects of the mushroom Yamabushitake (Hericium erinaceus) on mild cognitive impairment: a double-blind placebo-controlled clinical trial. Phytother Res 23(3):367–372. https://doi.org/10.1002/ptr.2634

Nadar SS, Rao P, Rathod VK (2018) Enzyme assisted extraction of biomolecules as an approach to novel extraction technology: a review. Food Res Int 108:309–330. https://doi.org/10.1016/j.foodres.2018.03.006

Netea MG, Domínguez-Andrés J, Barreiro LB, Chavakis T, Divangahi M, Fuchs E, Joosten LAB, van der Meer JWM, Mhlanga MM, Mulder WJM, Riksen NP, Schlitzer A, Schultze JL, Stabell Benn C, Sun JC, Xavier RJ, Latz E (2020) Defining trained immunity and its role in health and disease. Nat Rev Immunol 20(6):375–388. https://doi.org/10.1038/s41577-020-0285-6

Nishitani Y, Zhang L, Yoshida M, Azuma T, Kanazawa K, Hashimoto T, Mizuno M (2013) Intestinal anti-inflammatory activity of Lentinan: influence on IL-8 and TNFR1 expression in intestinal epithelial cells. PLoS One 8(4):e62441. https://doi.org/10.1371/journal.pone.0062441

Oba K, Kobayashi M, Matsui T, Kodera Y, Sakamoto J (2009) Individual patient based meta-analysis of Lentinan for unresectable/recurrent gastric cancer. Anticancer Res 29(7):2739–2745

Okamura K, Suzuki M, Chihara T, Fujiwara A, Fukuda T, Goto S, Ichinohe K, Jimi S, Kasamatsu T, Kawai N et al (1986) Clinical evaluation of schizophyllan combined with irradiation in patients with cervical cancer. A randomized controlled study. Cancer 58(4): 865–872. https://doi.org/10.1002/1097-0142(19860815)58:4<865::aid-cncr2820580411>3.0.co;2-s

Ooi VE, Liu F (2000) Immunomodulation and anti-cancer activity of polysaccharide-protein complexes. Curr Med Chem 7(7):715–729. https://doi.org/10.2174/0929867003374705

Ookushi Y, Sakamoto M, Azuma J-i (2006) Optimization of microwave-assisted extraction of polysaccharides from the fruiting body of mushrooms. J Appl Glycosci 53(4):267–272. https://doi.org/10.5458/jag.53.267

Pereira M, Gazzinelli RT (2023) Regulation of innate immune signaling by Irak proteins. Front Immunol 14:1133354. https://doi.org/10.3389/fimmu.2023.1133354

Price LA, Wenner CA, Sloper DT, Slaton JW, Novack JP (2010) Role for toll-like receptor 4 in TNF-alpha secretion by murine macrophages in response to polysaccharide Krestin, a Trametes versicolor mushroom extract. Fitoterapia 81(7):914–919. https://doi.org/10.1016/j.fitote.2010.06.002

Rämä T, Quandt CA (2021) Improving fungal cultivability for natural products discovery. Front Microbiol 12:706044. https://doi.org/10.3389/fmicb.2021.706044

Ruland J, Hartjes L (2019) CARD–BCL-10–MALT1 signalling in protective and pathological immunity. Nat Rev Immunol 19(2):118–134

Saitsu Y, Nishide A, Kikushima K, Shimizu K, Ohnuki K (2019) Improvement of cognitive functions by oral intake of Hericium erinaceus. Biomed Res 40(4):125–131. https://doi.org/10.2220/biomedres.40.125

Sakaguchi K, Shirai Y, Itoh T, Mizuno M (2018) Lentinan exerts its anti-inflammatory activity by suppressing TNFR1 transfer to the surface of intestinal epithelial cells through Dectin-1 in an in vitro and mice model. Immunome Res 14(3):1–11

Sakamoto J, Morita S, Oba K, Matsui T, Kobayashi M, Nakazato H, Ohashi Y (2006) Efficacy of adjuvant immunochemotherapy with polysaccharide K for patients with curatively resected colorectal cancer: a meta-analysis of centrally randomized controlled clinical trials. Cancer Immunol Immunother 55(4):404–411. https://doi.org/10.1007/s00262-005-0054-1

Sánchez C (2017) Bioactives from mushroom and their application. In: Food bioactives: extraction and biotechnology applications. Springer, pp 23–57

Schneider I, Kressel G, Meyer A, Krings U, Berger RG, Hahn A (2011) Lipid lowering effects of oyster mushroom (Pleurotus ostreatus) in humans. J Funct Foods 3(1):17–24. https://doi.org/10.1016/j.jff.2010.11.004

Scuto M, Rampulla F, Reali GM, Spanò SM, Trovato Salinaro A, Calabrese V (2024) Hormetic nutrition and redox regulation in gut-brain axis disorders. Antioxidants 13(4). https://doi.org/10.3390/antiox13040484

Shen Y, Zhao H, Wang X, Wu S, Wang Y, Wang C, Zhang Y, Zhao H (2024) Unraveling the web of defense: the crucial role of polysaccharides in immunity. Front Immunol 15:1406213

Smiderle FR, Morales D, Gil-Ramírez A, de Jesus LI, Gilbert-López B, Iacomini M, Soler-Rivas C (2017) Evaluation of microwave-assisted and pressurized liquid extractions to obtain β-d-glucans from mushrooms. Carbohydr Polym 156:165–174. https://doi.org/10.1016/j.carbpol.2016.09.029

Song Y-R, Sung S-K, Jang M, Lim T-G, Cho C-W, Han C-J, Hong H-D (2018) Enzyme-assisted extraction, chemical characteristics, and immunostimulatory activity of polysaccharides from Korean ginseng (Panax ginseng Meyer). Int J Biol Macromol 116:1089–1097. https://doi.org/10.1016/j.ijbiomac.2018.05.132

Sorourian R, Khajehrahimi AE, Tadayoni M, Azizi MH, Hojjati M (2020) Ultrasound-assisted extraction of polysaccharides from Typha domingensis: structural characterization and functional properties. Int J Biol Macromol 160:758–768. https://doi.org/10.1016/j.ijbiomac.2020.05.226

Su CH, Lai MN, Lin CC, Ng LT (2016) Comparative characterization of physicochemical properties and bioactivities of polysaccharides from selected medicinal mushrooms. Appl Microbiol Biotechnol 100(10):4385–4393. https://doi.org/10.1007/s00253-015-7260-3

Surendran G, Saye J, Binti Mohd Jalil S, Spreadborough J, Duong K, Shatwan IM, Lilley D, Heinrich M, Dodd GF, Surendran S (2025) Acute effects of a standardised extract of Hericium erinaceus (Lion's Mane mushroom) on cognition and mood in healthy younger adults: a double-blind randomised placebo-controlled study. Front Nutr 12. https://doi.org/10.3389/fnut.2025.1405796

Szućko-Kociuba I, Trzeciak-Ryczek A, Kupnicka P, Chlubek D (2023) Neurotrophic and neuroprotective effects of Hericium erinaceus. Int J Mol Sci 24(21):15960

Tang Y, He X, Liu G, Wei Z, Sheng J, Sun J, Li C, Xin M, Li L, Yi P (2023) Effects of different extraction methods on the structural, antioxidant and hypoglycemic properties of red pitaya stem polysaccharide. Food Chem 405:134804. https://doi.org/10.1016/j.foodchem.2022.134804

Tepsongkroh B, Thaihuttakij C, Supawong S, Jangchud K (2023) Impact of high pressure pre-treatment and hot water extraction on chemical properties of crude polysaccharide extract obtained from mushroom (Volvariella volvacea). Food Chem X 19:100864. https://doi.org/10.1016/j.fochx.2023.100864

Theobald H, Bejarano DA, Katzmarski N, Haub J, Schulte-Schrepping J, Yu J, Bassler K, Ament AL, Osei-Sarpong C, Piattini F, Vornholz L, T'Jonck W, Györfi AH, Hayer H, Yu X, Sheoran S, Al Jawazneh A, Chakarov S, Haendler K, Brown GD, Williams DL, Bosurgi L, Distler JHW, Ginhoux F, Ruland J, Beyer MD, Greter M, Bain CC, Vazquez-Armendariz AI, Kopf M, Schultze JL, Schlitzer A (2024) Apolipoprotein E controls Dectin-1-dependent development of monocyte-derived alveolar macrophages upon pulmonary β-glucan-induced inflammatory adaptation. Nat Immunol 25(6):994–1006. https://doi.org/10.1038/s41590-024-01830-z

Tsang KW, Lam CL, Yan C, Mak JC, Ooi GC, Ho JC, Lam B, Man R, Sham JS, Lam WK (2003) Coriolus versicolor polysaccharide peptide slows progression of advanced non-small cell lung cancer. Respir Med 97(6):618–624. https://doi.org/10.1053/rmed.2003.1490

Uffelman CN, Chan NI, Davis EM, Wang Y, McGowan BS, Campbell WW (2023) An assessment of mushroom consumption on cardiometabolic disease risk factors and morbidities in humans: a systematic review. Nutrients 15(5):1079

Uffelman CN, Schmok JN, Campbell RE, Hartman AS, Olson MR, Anderson NL, Reisdorph NA, Tang M, Krebs NF, Campbell WW (2024) Consuming mushrooms when adopting a healthy Mediterranean-style dietary pattern does not influence short-term changes of most cardiometabolic disease risk factors in healthy middle-aged and older adults. J Nutr 154(2): 574–582. https://doi.org/10.1016/j.tjnut.2023.12.026

van Steenwijk HP, Bast A, de Boer A (2021) Immunomodulating effects of fungal beta-glucans: from traditional use to medicine. Nutrients 13(4). https://doi.org/10.3390/nu13041333

Wagener M, Hoving JC, Ndlovu H, Marakalala MJ (2018) Dectin-1-Syk-CARD9 signaling pathway in TB immunity. Front Immunol 9:225. https://doi.org/10.3389/fimmu.2018.00225

Wang Q, Wang F, Xu Z, Ding Z (2017) Bioactive mushroom polysaccharides: a review on monosaccharide composition, biosynthesis and regulation. Molecules 22(6). https://doi.org/10.3390/molecules22060955

Wang W, Tan J, Nima L, Sang Y, Cai X, Xue H (2022) Polysaccharides from fungi: a review on their extraction, purification, structural features, and biological activities. Food Chem X 15: 100414. https://doi.org/10.1016/j.fochx.2022.100414

Wang W, Li J, Lu F, Liu F (2024) Ultrasound-assisted multi-enzyme extraction for highly efficient extraction of polysaccharides from Ulva lactuca. Foods 13(6). https://doi.org/10.3390/foods13060891

Wasser SP (2011) Current findings, future trends, and unsolved problems in studies of medicinal mushrooms. Appl Microbiol Biotechnol 89(5):1323–1332. https://doi.org/10.1007/s00253-010-3067-4

Wei Q, Zhang Y-h (2023) Ultrasound-assisted polysaccharide extraction from Cercis chinensis and properites, antioxidant activity of polysaccharide. Ultrason Sonochem 96:106422. https://doi.org/10.1016/j.ultsonch.2023.106422

Wu LY, Kan CN, Cheah IK, Chong JR, Xu X, Vrooman H, Hilal S, Venketasubramanian N, Chen CP, Halliwell B, Lai MKP (2022) Low plasma ergothioneine predicts cognitive and functional decline in an elderly cohort attending memory clinics. Antioxidants (Basel) 11(9). https://doi.org/10.3390/antiox11091717

Wu N, Ge X, Yin X, Yang L, Chen L, Shao R, Xu W (2024) A review on polysaccharide biosynthesis in Cordyceps militaris. Int J Biol Macromol 260(Pt 1):129336. https://doi.org/10.1016/j.ijbiomac.2024.129336

Wzorek-Łyczko K, Piwowarczyk A, Woźniak W, Kuchar E, Szymański H (2025) A randomised trial of pleuran in paediatric acute gastroenteritis. Sci Rep 15(1):16912. https://doi.org/10.1038/s41598-025-94893-3

Xie XQ, Geng Y, Guan Q, Ren Y, Guo L, Lv Q, Lu ZM, Shi JS, Xu ZH (2021) Influence of short-term consumption of Hericium erinaceus on serum biochemical markers and the changes of the gut microbiota: a pilot study. Nutrients 13(3). https://doi.org/10.3390/nu13031008

Yang Y, Lv J, Jiang S, Ma Z, Wang D, Hu W, Deng C, Fan C, Di S, Sun Y, Yi W (2016) The emerging role of toll-like receptor 4 in myocardial inflammation. Cell Death Dis 7(5):e2234–e2234. https://doi.org/10.1038/cddis.2016.140

Yang M, Ren W, Li G, Yang P, Chen R, He H (2022) The effect of structure and preparation method on the bioactivity of polysaccharides from plants and fungi. Food Funct 13(24): 12541–12560

Yang S, Li X, Zhang H (2024) Ultrasound-assisted extraction and antioxidant activity of polysaccharides from Tenebrio molitor. Sci Rep 14(1):28526. https://doi.org/10.1038/s41598-024-79482-0

Yin Z, Liang Z, Li C, Wang J, Ma C, Kang W (2021) Immunomodulatory effects of polysaccharides from edible fungus: a review. Food Sci Human Wellness 10(4):393–400. https://doi.org/10.1016/j.fshw.2021.04.001

Yin Z, Zhang J, Qin J, Guo L, Guo Q, Kang W, Ma C, Chen L (2024) Anti-inflammatory properties of polysaccharides from edible fungi on health-promotion: a review. Front Pharmacol 15: 1447677

You Q, Yin X, Zhao Y (2013) Enzyme assisted extraction of polysaccharides from the fruit of Cornus officinalis. Carbohydr Polym 98(1):607–610. https://doi.org/10.1016/j.carbpol.2013.06.036

Yuan Q, Lin S, Fu Y, Nie X-R, Liu W, Su Y, Han Q-H, Zhao L, Zhang Q, Lin D-R, Qin W, Wu D-T (2019) Effects of extraction methods on the physicochemical characteristics and biological activities of polysaccharides from okra (Abelmoschus esculentus). Int J Biol Macromol 127: 178–186. https://doi.org/10.1016/j.ijbiomac.2019.01.042

Zadeh PH, Fermoso FG, Collins G, Serrano A, Mills S, Abram F (2023) Impacts of metal stress on extracellular microbial products, and potential for selective metal Recovery. Ecotoxicol Environ Saf 252:114604. https://doi.org/10.1016/j.ecoenv.2023.114604

Zhang M, Cui SW, Cheung PCK, Wang Q (2007) Antitumor polysaccharides from mushrooms: a review on their isolation process, structural characteristics and antitumor activity. Trends Food Sci Technol 18(1):4–19. https://doi.org/10.1016/j.tifs.2006.07.013

Zhang K, Peschel D, Klinger T, Gebauer K, Groth T, Fischer S (2010) Synthesis of carboxyl cellulose sulfate with various contents of regioselectively introduced sulfate and carboxyl groups. Carbohydr Polym 82(1):92–99. https://doi.org/10.1016/j.carbpol.2010.04.027

Zhang S, Lei L, Zhou Y, Ye F-y, Zhao G-h (2022) Roles of mushroom polysaccharides in chronic disease management. J Integr Agric 21(7):1839–1866. https://doi.org/10.1016/S2095-3119(21)63871-6

Zhang N, Chen W, Li X, Chen X, Wang Y, Huang G, Wang J, Jia Z (2024) Enzyme-assisted ultrasonic extraction and antioxidant activities of polysaccharides from Schizochytrium limacinum meal. Foods 13(6). https://doi.org/10.3390/foods13060880

Zhao YT, Tang YQ, Liu KZ, Zhang Y (2015) A study about microwave-assisted extraction of polysaccharides form medicinal mushrooms Fomitopsis ulmaria (sor.: for.) bond. Et sing. Adv Mater Res 1073:1837–1840

Zhao J, Hu Y, Qian C, Hussain M, Liu S, Zhang A, He R, Sun P (2023) The interaction between mushroom polysaccharides and gut microbiota and their effect on human health: a review. Biology 12(1):122

Zhong X, Wang G, Li F, Fang S, Zhou S, Ishiwata A, Tonevitsky AG, Shkurnikov M, Cai H, Ding F (2023) Immunomodulatory effect and biological significance of β-Glucans. Pharmaceutics 15(6). https://doi.org/10.3390/pharmaceutics15061615

Zhou H, Dong Y, Wu Z, Peng X, Yan M, Chen S, Lv G (2025) Ultrasound-assisted enzyme extraction of dendrobium Officinale polysaccharides: extraction process, characterization, immunomodulatory effects. Ultrason Sonochem 114:107248. https://doi.org/10.1016/j.ultsonch.2025.107248

Zurbau A, Noronha JC, Khan TA, Sievenpiper JL, Wolever TMS (2021) The effect of oat β-glucan on postprandial blood glucose and insulin responses: a systematic review and meta-analysis. Eur J Clin Nutr 75(11):1540–1554. https://doi.org/10.1038/s41430-021-00875-9

# Therapeutic Potential of Medicinal Mushrooms

9

Damanpreet Kaur, Bhawana, and Kohinoor Kaur

**Abstract**

Mushrooms are recognized as functional foods that not only supply essential nutrients but also deliver a wide range of additional health benefits. These benefits are largely attributed to the presence of various bioactive compounds, including polysaccharides, phenolic compounds, triterpenoids, and lectins found in diverse mushroom species. These bioactive compounds, offering numerous health benefits, make medicinal mushrooms promising candidates for developing dietary supplements as well as for the treatment and control of chronic conditions such as diabetes, cardiovascular ailments, and neurodegenerative diseases. They can also play supportive roles in existing treatments, such as those for cancer. Their immunoregulatory properties can also be exploited to combat microbial infections and autoimmune diseases. This chapter summarizes the range of bioactive properties associated with medicinal mushrooms and their clinical significance, underscoring their role in contemporary therapeutic approaches while also discussing the current challenges and future research prospects.

**Keywords**

Medicinal mushrooms · Bioactive compounds · Neuroprotective · Antitumor · Immunomodulatory · Antioxidant

D. Kaur · K. Kaur (✉)
Department of Microbiology, Institute of Home Economics, University of Delhi, New Delhi, India
e-mail: kohinoor.kaur@ihe.du.ac.in

Bhawana
Department of Botany, Institute of Home Economics, University of Delhi, New Delhi, India





B. Giri et al. (eds.), *Advances in Mushroom Research*,
https://doi.org/10.1007/978-981-95-7692-0_9

## 9.1 Introduction

The nutritional and medicinal value of mushrooms has been recognized for centuries, and their cultivation has been well documented. Mushrooms have been consumed as part of human culture and diets since ancient times. Mushrooms have been known to provide strength and vigor to the Greeks and Romans, warranting their use on battlefields. In traditional Chinese culture, mushrooms have long been considered a health-promoting food, often referred to as an "elixir of life" (Miles and Chang 2004).

Mushrooms are highly nutritious, loaded with proteins (including all essential amino acids), dietary fiber, and various vitamins such as those in groups B, C, D, and E. They also contain vital minerals and trace elements. Their low carbohydrate and fat content, combined with beneficial fatty acids, make mushrooms a healthy, low-calorie, and cholesterol-free food ideal for those aiming for weight management. In addition to their nutritional benefits, mushrooms are appreciated for their culinary versatility and low levels of calories, fat, carbohydrates, and sodium.

Numerous mushroom species are known to produce enzymes and bioactive compounds with wide-ranging therapeutic effects, making them promising candidates for pharmaceutical and biotechnological applications. These bioactive compounds display tremendous biochemical diversity, including polysaccharides, lectins, quinones, phenolic compounds, glycoproteins, ergosterols, sterols, unsaturated fatty acids, nucleotides, and tocopherols. These pharmacologically active compounds have great potential to be developed as food additives, nutraceuticals, and novel drugs by the food and pharmaceutical industries (Bhambri et al. 2022; Fitsum et al. 2025).

Moreover, mushrooms have potent antioxidant properties due to the presence of key nutrients such as ergothioneine, glutathione, vitamin D, and selenium, which help alleviate oxidative stress (Bhambri et al. 2022). Mushrooms have also been shown to stimulate metabolism, fight obesity, and display anti-aging potential, largely due to their strong antioxidant properties. A wide range of therapeutic effects have been attributed to medicinal mushrooms, including antidiabetic, anticholesterolemic, neuroprotective, antimicrobial, anticancer, immunomodulatory, hepatoprotective, cytotoxic, antioxidant, anti-inflammatory, and anti-aging benefits. Mushrooms are a valuable reservoir of a number of active ingredients, several of which are yet to be discovered, that could offer further health benefits. This chapter highlights the diverse health-promoting properties of mushrooms, emphasizing their biologically active constituents that exhibit a broad array of pharmacological effects including neuroprotective, cardioprotective, antitumor, antioxidant, antidiabetic, anti-inflammatory, and immunomodulatory activities and explores the therapeutic capacity of selected medicinal mushrooms in human health and discusses current challenges and future directions for incorporating mushrooms into therapeutic and nutritional applications.

## 9.2 Diverse Health Benefits Conferred by Mushrooms

### 9.2.1 Neuroprotective Effects of Mushrooms

Mushrooms are emerging as valuable natural reservoirs of bioactive metabolites, many of which exhibit notable neuroprotective activities, mainly in terms of antioxidant properties, suppressive effects on neuroinflammation, and the ability to prevent neuronal apoptosis, all crucial mechanisms that underpin neurodegeneration in diseases like Alzheimer's disease, Parkinson's disease, and amyotrophic lateral sclerosis. Numerous species have been shown to produce structurally diverse bioactive compounds such as diterpenoids, triterpenoids, polysaccharides, flavonoids, and sterols that target multiple molecular descriptors mediating neuronal survival and cognitive function. The bioactive compounds with neuroprotective benefits are listed in Table 9.1.

One such highly studied mushroom showing neuroprotective effects is *Hericium erinaceus* (lion's mane mushroom), which produces erinacines (A-I) and hericenones (C-H). These cyathin diterpenoids, as well as benzyl alcohol derivatives, can traverse the blood-brain barrier and are able to induce nerve growth factor (NGF) synthesis in vitro and in vivo with corresponding neuronal differentiation and promotion of nerve cell survival (Spelman et al. 2017; Chong et al. 2019; Venturella et al. 2021).

Similarly, lanostane-type triterpenoids, such as lanosterol, inotodiol, and inonotsutriol A, were reported to be isolated from *Inonotus obliquus* (Chaga) together with new inonotusols H–N. These compounds have shown great potential as neuroprotectors, since they decreased the nitric oxide (NO) release from activated microglia. Inonotusols I and L bind inducible nitric oxide synthase (iNOS) to attenuate neuroinflammation as well as provide neuroprotection against NO-induced neuronal death (Kou et al. 2021; Tee et al. 2024). Additionally, phenolic compounds isolated from *I. obliquus*, including inoscavin A, have been shown to enhance cell viability in $H_2O_2$-injured human neuroblastoma cells, largely through their antioxidant capacity (Wei et al. 2022; Tee et al. 2024).

Diterpenes from the mushroom *Taiwanofungus camphoratus*, protein hydrolysates from *Pleurotus geesteranus*, and *Phallus indusiatus* produce dictyoquinazols A, B, and C, which have all been reported to exert neuroprotective effects, particularly through their activity in neuronal cell culture models (Fitsum et al. 2025). Ergosterol, an isolated ergosteroid from *Auricularia polytricha*, reduces neuroinflammation induced by overactivated microglia through the downregulation of NF-κB and upregulation of Nrf2/SOD-1 antioxidant pathways. The molecule also prevents hippocampal cells from incurring TNF-α-induced damage through the modulation of Akt/GSK-3 and EGR-1 pathways (Sillapachaiyaporn et al. 2022a, b; Rangsinth et al. 2023). Additionally, some mushrooms produce other polyphenols and phenolic compounds with similar effects.

Polysaccharides are a prevalent type of bioactive metabolite, widely studied for their potential neuroprotective properties. The β-glucan-rich extracts from *Coprinus comatus* and *Coprinellus truncorum* inhibit acetylcholinesterase (AChE), thereby

**Table 9.1** Bioactive compounds with neuroprotective activity

| Compound/Class | Source species | Pharmacological activity | Mechanism of action | References |
|---|---|---|---|---|
| Erinacines (A-I) | *Hericium erinaceus* (lion's mane) | Neuroprotective, neurotrophic | Cross the blood-brain barrier, induce nerve growth factor (NGF) synthesis, neuronal differentiation promotion, nerve cell survival enhancement | Spelman et al. (2017), Chong et al. (2019), and Venturella et al. (2021) |
| Hericenones (C-H) | *Hericium erinaceus* (lion's mane) | Neuroprotective, neurotrophic | Cross the blood-brain barrier, induce nerve growth factor (NGF) synthesis, neuronal differentiation promotion, nerve cell survival enhancement | Spelman et al. (2017), Chong et al. (2019), and Venturella et al. (2021) |
| Lanostane-type triterpenoids (lanosterol, inotodiol, inonotsutriol A) | *Inonotus obliquus* (Chaga) | Neuroprotective | Decreased nitric oxide (NO) release from activated microglia, neuroprotection against NO-induced neuronal death | Kou et al. (2021) and Tee et al. (2024) |
| Inonotusols (H–N, particularly I and L) | *Inonotus obliquus* (Chaga) | Neuroprotective, anti-neuroinflammatory | Bind inducible nitric oxide synthase (iNOS) functionally, attenuate neuroinflammation, provide neuroprotection against NO-induced neuronal death | Kou et al. (2021) and Tee et al. (2024) |
| Inoscavin A | *Inonotus obliquus* | Neuroprotective, antioxidant | Enhance cell viability in $H_2O_2$-injured human neuroblastoma cells through antioxidant capacity | Wei et al. (2022) and Tee et al. (2024) |
| Diterpenes | *Taiwanofungus camphoratus* | Neuroprotective | Neuroprotective properties in in vitro neuronal cell lines | Fitsum et al. (2025) |
| Protein hydrolysates | *Pleurotus geesteranus* | Neuroprotective | Neuroprotective properties in in vitro neuronal cell lines | Fitsum et al. (2025) |
| Dictyoquinazols | *Phallus indusiatus* | Neuroprotective | Neuroprotective properties in in vitro neuronal cell lines | Fitsum et al. (2025) |

| | | | | |
|---|---|---|---|---|
| Ergosterol | *Auricularia polytricha* | Anti-neuroinflammatory, neuroprotective | Reduces neuroinflammation induced by over-activated microglia through NF-κB downregulation and Nrf2/SOD-1 antioxidant pathway upregulation; prevents hippocampal TNF-α-induced damage via Akt/GSK-3 and EGR-1 pathway modulation | Sillapachaiyaporn et al. (2022a, b) and Rangsinth et al. (2023) |
| β-Glucan-rich extracts | *Coprinus comatus*, *Coprinellus truncorum* | Neuroprotective, cholinergic enhancement | Inhibit acetylcholinesterase (AChE) and enhance cholinergic neurotransmission (critical therapeutic strategy in Alzheimer's disease) | Tel-Çayan et al. (2020) |
| Flavonoids (kaempferol, epicatechin, formononetin) | *Flammulina velutipes* | Neuroprotective, antioxidant | Confer neuroprotection against $H_2O_2$-induced oxidative damage and regulate apoptosis through intrinsic mitochondrial pathways | Hu et al. (2017) |
| Ergothioneine | Various mushrooms | Neuroprotective, cognitive enhancement | Reduction in amyloid deposition, improved glucose metabolism, delayed cognitive impairment in transgenic Alzheimer's disease models | Ahn et al. (2017) and Cheah and Halliwell (2021) |

enhancing cholinergic neurotransmission, a critical therapeutic strategy in Alzheimer's disease (Tel-Çayan et al. 2020).

Flavonoids such as kaempferol, epicatechin, and formononetin, found in *Flammulina velutipes* mushrooms, were shown to confer neuroprotection to neuronal cells against $H_2O_2$-induced oxidative damage. The mechanisms involved in these protective properties were mainly related to the regulation of apoptosis through the intrinsic mitochondrial pathways (Hu et al. 2017). In another study, ergothioneine, a sulfur-containing amino acid with a high concentration in mushrooms, was identified as another potent neuroprotective molecule. Oral administration of dietary ergothioneine was linked with a reduction in amyloid deposition, improved glucose metabolism, and delayed development of cognitive dysfunction in a transgenic model of Alzheimer's disease (Cheah and Halliwell 2021; Whitmore et al. 2022).

Studies involving preclinical and clinical trials on Alzheimer's disease models consistently displayed improvements in cognitive functioning and behavioral impairments upon consumption of *H. erinaceus* extracts. The results of the clinical studies were similar to the preclinical results, which supported the effectiveness of *H. erinaceus* extracts in improving cognitive performance and counteracting the behavioral impairments in Alzheimer's disease models (Feng et al. 2019). In addition, epidemiological evidence indicated that the intake of two mushroom servings per week could reduce the risk of mild cognitive impairment by half in the studied population, further highlighting the potential translational application of medicinal mushrooms in neuroprotection (Yanshree et al. 2022).

### 9.2.2 Antitumor Effects of Medicinal Mushrooms

Mushroom extracts have been identified as a useful source of antitumor compounds, and an increasing body of evidence is accumulating in favor of their ability to inhibit the proliferation of cancer cells and tumor progression (Table 9.2). Extracts from *Albatrellus confluens* and *Auricularia polytricha*, rich in phenols, have been shown to possess good antiproliferative activity against gastric and ovarian cancer cell lines as well as in cancer models of the colon (COLO-205), breast (MCF-7), and kidney (ACHN). Similarly, extracts from *Inonotus obliquus*, *Coprinus atramentarius*, and *Ganoderma lucidum* displayed antileukemic, anti-lung, and anti-colon cancer properties, respectively, further confirming their antitumor potential (Zhou et al. 2024; Camilleri et al. 2024; Fitsum et al. 2025).

Another mushroom that has proven to have high cancer cell suppression ability is *Pleurotus ostreatus*, commonly known as the oyster mushroom. Purified water-soluble proteoglycan fractions isolated from *P. ostreatus* were able to significantly reduce tumor cell population and arrest cell division in the G0/G1 phase of the cell cycle (Sarangi et al. 2006). The process involves the increased lytic activity of natural killer cells as well as the priming of macrophages to make nitric oxide, thereby making the environment unfavorable to the tumor. *Pleurotus ostreatus* has been shown, in comparative studies, to exhibit the strongest inhibitory effects on breast and colon cancer cell growth among medicinal mushrooms, acting via both

**Table 9.2** Bioactive compounds with antitumor activity

| Compound/Class | Source species | Pharmacological activity | Mechanism of action | References |
|---|---|---|---|---|
| Phenolic-enriched fractions | *Albatrellus confluens*, *Auricularia polytricha* | Antitumor | Significant proliferation inhibitory action against gastric, ovarian, colon, breast, and kidney cancer cell lines | Zhou et al. (2024) and Fitsum et al. (2025) |
| Extracts (general) | *Inonotus obliquus* | Antileukemic | Antileukemic effects | Zhou et al. (2024) and Fitsum et al. (2025) |
| Extracts (general) | *Ganoderma lucidum* | Anti-lung cancer | Anti-lung cancer effects | Zhou et al. (2024) and Fitsum et al. (2025) |
| Extracts (general) | *Coprinus atramentarius* | Anti-colon cancer | Anti-colon cancer effects | Zhou et al. (2024) and Fitsum et al. (2025) |
| Polysaccharides (ethyl acetate fraction) | *Pleurotus ramosus* | Antitumor | Inhibition against Dalton's lymphoma ascites and esophageal adenocarcinoma cell lines | Sasikumar et al. (2023) |
| Polysaccharides | *Ganoderma lucidum* | Anticancer | Anticancerous effect against EAC-produced solid cancer in vitro | Sasikumar et al. (2023) |
| Polysaccharides (mycelium/ fruiting body) | *Pleurotus tuber-regium* | Chemopreventive | Significantly decreased tumor load in rats, chemopreventive agents of natural origin | Sivanesan et al. (2022) |
| Flavonoids | Various mushrooms | Antitumor | Inhibit cell division, induce cell death, prevent tumor growth, strengthen antioxidant defenses, and immunity | Camilleri et al. (2024) |
| Extracts (angiogenesis regulation) | *Agrocybe aegerita* | Anti-angiogenic, antitumor | Downregulation of vascular endothelial growth factor (VEGF) and bFGF, | Fitsum et al. (2025) |

(continued)

**Table 9.2** (continued)

| Compound/Class | Source species | Pharmacological activity | Mechanism of action | References |
|---|---|---|---|---|
| | | | limited capacity to vascularize tumors | |
| Phenolic acids (ferulic, chlorogenic, protocatechuic, gallic, erucic) | *Agrocybe aegerita* | Antitumor | Water extract with prominent antitumor activity attributed to phenolic acid content | Fitsum et al. (2025) |
| Extracts (antiproliferative) | *Hypsizygus tessellatus*, *Flammulina velutipes* | Antiproliferative | Marked antiproliferative effects on breast cancer cell lines (MCF-7 and MDA-MB-231), bioactivity linked to hydroxylated compounds, and mitochondrial cytochrome P450 enzyme interactions | Fitsum et al. (2025) |

p53-dependent and independent mechanisms (Jedinak and Sliva 2008; Venturella et al. 2021).

The most significant anticancer agents found in mushrooms are polysaccharides. Extracts of *Pleurotus ramosus* prepared using ethyl acetate, methanol, and aqueous solvents were shown to inhibit esophageal adenocarcinoma (EAC) and Dalton lymphoma ascites (DLA) cell lines in vivo, with the ethyl acetate fraction demonstrating the strongest effect (Sivanesan et al. 2022; Sasikumar et al. 2023). Angiogenesis plays a crucial role in cancer prevention. Angiogenesis was shown to be inhibited by *Agrocybe aegerita* by reducing the production of vascular endothelial growth factor (VEGF) and basic fibroblast growth factor (bFGF), suggesting its potential to enhance the efficacy of conventional cancer therapies (Venturella et al. 2021). Similar anti-angiogenic activity is demonstrated by ergosterol, which acts by inhibiting the activity of VEGFC and STAT3, which mediate angiogenesis (Tan et al. 2017).

*Inonotus obliquus*, traditionally used in Eastern European folk medicine, contains potent antitumor compounds, including triterpenes and ergosterol peroxide (Lindequist et al. 2005; Olusola Ogidi et al. 2020). Aqueous extracts from this species demonstrate dose-dependent growth inhibition in various cancer cell lines, including HCT-15, B16-F10 melanoma, and HT-29 colon cancer cells. The mechanism involves the G0/G1 phase arrest followed by caspase-3-mediated apoptosis. The in vivo studies have confirmed these effects, with consistent extract administration reducing metastatic tumor nodules by 25% and tumor growth by 60.3% in carcinoma-implanted mouse models (Youn et al. 2009; Tee et al. 2024).

### 9.2.3 Immunomodulatory Effects of Medicinal Mushrooms

One of the major pharmacological effects of the bioactive components of mushrooms is immunomodulation, which has potential in developing therapies for a variety of immune disorders through a series of complicated cellular signaling processes and molecular mechanisms (Table 9.3). The turkey tail mushroom, *Coriolus versicolor*, contains two structurally different protein-bound polysaccharide extracts, polysaccharide peptide (PSP) and polysaccharide-Krestin (PSK), which are complex β-glucans. These are known to display potent immunomodulatory and antitumor effects (Venturella et al. 2021). These compounds are shown to stimulate cytokine production and enhance the activity of natural killer cells as well as dendritic and T cell infiltration of the tumor microenvironment through their pattern recognition receptors like dectin-1 and toll-like receptors (TLRs), including TLR-2, TLR-4, and TLR-6. As such, they trigger immune activation downstream that boosts immune surveillance and clinical efficacy in the management of cancer (Chang et al. 2017; Saleh et al. 2017).

*Lentinula edodes* contains lentinan, a β-1, 3-D-glucan that has wide use in Japanese cancer therapy due to its potential ability to encourage Th1 responses to immunity and restore Th1/Th2 equilibrium. Lentinan amplifies the dendritic cell response and induces T and NK cell proliferation, which enhances dendritic cell-mediated NK cell cytotoxicity; increases Fc receptor expression to enhance the NK cell antitumor effect; and increases complement-dependent tumor lysis by complement receptor-3 involvement (Ina et al. 2013; Ahn et al. 2017; Zhang et al. 2018). This immunomodulatory activity is supported by clinical data that also show that lentinan enhances tumor-infiltrating CD86+ cells and restores the killer-to-survival cell ratios, reinforcing its potential as an immunomodulatory bioregulator, a physiologically active substance with therapeutic effects on gastric and other cancers (Venturella et al. 2021).

Cordycepin, the nucleoside analog from *Cordyceps militaris* and *C. sinensis*, is a bidirectional compound in the sense that it is both an immune stimulant and suppressant, dependent on the physiological context. It activates the release of cytokines by peripheral blood mononuclear cells and also regulates transcription factors in human immune cells in vitro; it has immunosuppressive effects, verified in heterotopic heart transplants and skin grafts, where it increased the graft survival duration. Such results make cordycepin a prospective drug in the development of autoimmune disorders, in improving recipient survival in transplants, and in augmenting anticancer therapies by regulating the immune system in a context-specific manner (Ashraf et al. 2020).

β-Glucans isolated from several mushrooms bind to a number of immune receptors, including dectin-1 found in monocytes, macrophages, neutrophils, and dendritic cells. The interaction of these polysaccharides triggers downstream events of cellular signaling, which leads ultimately to the activation of NF-κB, expression of immune-regulatory genes, and cytokines, nitric oxide, and reactive oxygen species. The mechanism behind the ability of β-glucans in mushrooms to increase innate and adaptive immunity is this receptor-mediated signaling pathway (Cerletti et al. 2021).

**Table 9.3** Bioactive compounds with immunomodulatory activity

| Compound/Class | Source species | Pharmacological activity | Mechanism of action | References |
|---|---|---|---|---|
| Polysaccharide peptide (PSP) | *Coriolus versicolor* | Immunomodulatory, anticancer | Stimulate cytokine production, enhance natural killer cell activity, and dendritic and T cell infiltration of the tumor microenvironment through pattern recognition receptors (dectin-1, TLR-2, TLR-4, and TLR-6); trigger immune activation downstream; boost immune surveillance | Venturella et al. (2021) |
| Polysaccharide-K (PSK) | *Coriolus versicolor* | Immunomodulatory, anticancer | 12% β-glucan component stimulates cytokine production, enhances natural killer cell activity, dendritic and T cell infiltration through pattern recognition receptors, and triggers immune activation for enhanced immune surveillance | Venturella et al. (2021) |
| Lentinan (β-1, 3 D-glucan) | *Lentinula edodes* | Immunomodulatory, anticancer | Encourage Th1 responses, restore Th1/Th2 equilibrium, amplify dendritic cell response, induce T and NK cell proliferation, enhance dendritic cell-related NK cell cytotoxicity, increase Fc receptor expression, enhance NK cell antitumor effect, and complement-dependent tumor lysis via complement receptor-3 | Ina et al. (2013), Ahn et al. (2017), and Zhang et al. (2018) |

| | | | | |
|---|---|---|---|---|
| Cordycepin | *Cordyceps militaris*, *C. sinensis* | Bidirectional immunomodulatory | Both immune stimulant and suppressant, depending on physiological context, activates cytokine release by peripheral blood mononuclear cells, regulates transcription factors in human immune cells, has immunosuppressive effects in transplant models, and increases graft survival | Ashraf et al. (2020) |
| β-Glucans | Various mushrooms | Immunomodulatory | Bind to immune receptors (primarily dectin-1 on monocytes, macrophages, neutrophils, and dendritic cells); trigger downstream cellular signaling; activate NF-κB; express immune-regulatory genes; produce cytokines, nitric oxide, and reactive oxygen species; and increase adaptive and innate immunity | Cerletti et al. (2021) |
| Bioactive proteins (lectins, fungal immunomodulatory proteins, ribosome-inactivating proteins) | Various mushrooms | Immunomodulatory, antiviral, antitumor, antifungal, and antibacterial | Act as biological response modifiers, activate host defense mechanisms, regulate cellular responses to infection and malignancy | Xu et al. (2011) and Valverde et al. (2015) |

In addition to polysaccharides, mushrooms also possess an array of other bioactive proteins that include ribosome-inactivating proteins, immunomodulatory compounds, and lectins that display a range of effects, including antibacterial, antiviral, antifungal, and antitumor activity. Such proteins improve immunity and shield against the invasion of pathogens, as well as accelerate enzyme functions and nutrient absorption (Xu et al. 2011; Valverde et al. 2015). Lectins attach themselves to cell surface carbohydrates, where they exhibit immunomodulatory as well as antimicrobial effects (Singh et al. 2014). Along with antifungal and antiviral (inhibition of HIV-1 reverse transcriptase) abilities, ribosome-inactivating proteins show immunomodulatory effects against a range of cancer cell lines, such as MCF-7, HepG2, and leukemic T cells. These proteins help regulate immunomodulatory activities by facilitating an active host immune system that can generate rigorous cellular responses to infection and malignancy (Puri et al. 2012; Bhambri et al. 2022; Dahiya et al. 2022). The recent body of literature highlights a list of several other pharmacologically significant bioactives, including polysaccharides, triterpenoids, proteins, and nucleoside analogs with promising immunoregulatory activity. These compounds represent a strong molecular basis of therapeutic intervention in cancer, transplantation medicine, autoimmune disease, and infectious disorders, which further supports the role in modern immunotherapy by controlling cytokine production, boosting cellular immune responses, and inducing situation-specific immune suppression when required (Ashraf et al. 2020; Venturella et al. 2021; Cerletti et al. 2021; Bhambri et al. 2022).

### 9.2.4 Cardiovascular Benefits of Medicinal Mushroom

Components of mushrooms have emerged as potential tools to regulate cardiovascular conditions, especially in regulating the cholesterol content and the prevention of vascular dysfunctions. Extracts from *Pleurotus ostreatus* possess polysaccharides that have powerful hypocholesterolemic properties and binding capabilities to non-absorbable bile acids, which reduce cholesterol absorption and lower the levels of low-density lipoprotein (LDL) in experimental models (Abotaleb et al. 2020; Venturella et al. 2021). An interesting study showed a 17.1% reduction in plasma glucose levels due to the hypoglycemic activity of exopolysaccharides derived from *Pleurotus geesteranus*, mediated by a reduction in streptozotocin-induced pancreatic β cell damage and insulin mimicry (Song et al. 2018). Along with the impact on lipid metabolism, mushroom-derived β-glucans are shown to assist in lowering cholesterol through preventing its production, increasing the formation of bile acids, decreasing the uptake of fats in the intestines, and implementing a balance of microbiota in the gut (Cerletti et al. 2021). In addition, vasoprotective effects through inhibition of HMG-CoA reductase and cholesterol synthesis have been reported for triterpenoids from *Ganoderma lucidum* (Liang et al. 2019). Cordycepin from *Cordyceps* improves cardiovascular outcomes via activation of AMPK to increase lipid metabolism and reduces total cholesterol, low-density lipoprotein (LDL)-C, and triglycerides, thereby decreasing the risks of hyperlipidemia and heart disease (Ashraf et al. 2020). Together, these bioactive compounds highlight mushrooms as promising nutraceuticals for cardiometabolic regulation.

Comprehensively, these facts highlight the contribution of mushroom metabolites to cardiovascular health. Mushrooms are a naturally occurring and biologically dense source of cardioprotective agents, offering the potential to supplement existing prevention efforts in the field of cardiovascular disease, including lowering cholesterol and preventing oxidative damage, and having the capacity to increase the resilience of the vascular system. The bioactive compounds with cardiovascular benefits are listed in Table 9.4.

### 9.2.5 Antioxidant Properties of Medicinal Mushrooms

Medicinal mushrooms are highly effective natural sources of antioxidants which provide significant protection against oxidative damage to the cells induced by reactive oxygen species. This protection is mediated through a wide range of bioactive components that sequentially respond to oxidative stress and prevent age-related pathologies (Table 9.5). In this respect, *Ganoderma lucidum*, *Cordyceps militaris*, *Phellinus linteus*, and *Inonotus obliquus* have been repeatedly studied in terms of their potential to reduce diverse harmful molecules that impair cellular integrity.

The antioxidant effects of mushrooms is due to bioactive compounds such as phenolic compounds, complex polysaccharides, and ergosterol derivatives. Metabolites shown to be notable are hispidin and hispolon in *T. linteus*, as well as cordycepin, inotilone, caffeic acid, and 4-(3,4-dihydroxyphenyl)-3-buten-2-one expressed in various species. These are natural defenders because they interfere with and eliminate the free radicals before they can cause the destruction of tissues. The most important factors affecting variability in efficacy are the mushroom species itself, the extraction procedure, and the concentration of active compounds available (Ashraf et al. 2020; Tee et al. 2024).

Extensive laboratory studies have demonstrated that numerous preparations possess significant antioxidant properties. In a DPPH assay, a 4 mg/mL alcohol extract of *Taiwanofungus camphoratus* achieved 83% DPPH radical scavenging activity, comparable to that of vitamin C (0.5 mg/mL). Notably, the same extract outperformed vitamin C in certain evaluations, particularly in the superoxide dismutase (SOD) assay, where it exhibited 71–87% efficacy compared to the 67% observed for vitamin C (0.5 mg/mL) (Tee et al. 2024).

*Inonotus obliquus* polysaccharides were found to be capable of scavenging DPPH radicals as well as neutralizing hydroxyl radicals (Tee et al. 2024). *Ganoderma* species polysaccharides provide protective effects against oxidative stress through multiple mechanisms. *Ganoderma atrum* polysaccharide (PSG-1) demonstrates protective effects against D-galactose-induced oxidative stress in mouse brain tissue,

**Table 9.4** Cholesterol-lowering and cardioprotective compounds from medicinal mushrooms

| Compound/Class | Source species | Pharmacological activity | Mechanism of action | References |
|---|---|---|---|---|
| Polysaccharides | *Pleurotus ostreatus* | Hypocholesterolemic | Powerful hypocholesterolemic properties, binding capability across nonabsorbable bile acids, reduce cholesterol absorption, and lower LDL levels | Venturella et al. (2021) |
| β-Glucans | *Pleurotus ostreatus*, *Grifola frondosa* | Cholesterol-lowering, antioxidant | Lessen oxidative stress and lipid peroxidation (central roles in atherogenesis) | Cerletti et al. (2021) |
| Triterpenoids | *Ganoderma lucidum* | Vasoprotective | Inhibition of nitric oxide pathways, resulting dilation and increased endothelial integrity, downregulates pro-inflammatory cytokines, alleviates vascular inflammation, reduces chance of plaque development | Liang et al. (2019) |
| Phenolic extracts | *Auricularia polytricha*, *Agrocybe aegerita* | Antioxidant, cardioprotective | High antioxidant activity, protection to LDL particles against oxidative modification (crucial event in cardiovascular disease development) | Abotaleb et al. (2020) |
| Ferulic acid derivatives | Various mushrooms | Antioxidant, cardioprotective | Inhibit initiators of lipid peroxidation, endothelial protection | Abotaleb et al. (2020) |
| Chlorogenic acid derivatives | Various mushrooms | Antioxidant, cardioprotective | Inhibit initiators of lipid peroxidation, endothelial protection | Abotaleb et al. (2020) |

**Table 9.5** Bioactive agents with antioxidant properties

| Compound class | Bioactive compound | Source species | Pharmacological activity | Mechanism of action | References |
|---|---|---|---|---|---|
| Phenolic compounds | Hispidin | *Phellinus linteus* | Antioxidant, dose-dependent protection | Free radical scavenging, up to 65% effectiveness at high dose | Ashraf et al. (2020) and Tee et al. (2024) |
| Phenolic compounds | Hispolon | *Phellinus linteus* | Antioxidant | Free radical elimination | Ashraf et al. (2020) and Tee et al. (2024) |
| Phenolic compounds | Caffeic acid | Various mushroom species | Antioxidant | Superior antioxidant activity compared to butylated hydroxyanisole | Ashraf et al. (2020) and Tee et al. (2024) |
| Phenolic compounds | Inotilone | Various mushroom species | Antioxidant | Superior antioxidant activity compared to standard reagents | Ashraf et al. (2020) and Tee et al. (2024) |
| Phenolic compounds | 4-(3,4-dihydroxyphenyl)-3-buten-2-one | Various mushroom species | Antioxidant | Superior antioxidant activity compared to standard reagents | Ashraf et al. (2020) and Tee et al. (2024) |
| Terpenoids and steroids | Ergosterol and oxide derivatives | *Pleurotus ostreatus*, various species | Antioxidant | Electron donation, metal chelation, enzyme inhibition | Lee et al. (2014) and Tee et al. (2024) |
| Terpenoids and steroids | Antcin A | *Taiwanofungus camphoratus* | Antioxidant | Electron donation, metal chelation, enzyme inhibition | Lee et al. (2014) and Tee et al. (2024) |
| Terpenoids and steroids | Antcin B | *Taiwanofungus camphoratus* | Antioxidant | Electron donation, metal chelation, enzyme inhibition | Lee et al. (2014) and Tee et al. (2024) |
| Polysaccharides and β-Glucans | *Inonotus obliquus* polysaccharides | *Inonotus obliquus* | Antioxidant | DPPH radical scavenging (82.3%), hydroxyl radical neutralization (81.3%) | Tee et al. (2024) |
| Polysaccharides and β-Glucans | Selenium-enriched polysaccharides | Various species | Antioxidant | Twofold activity compared to unenriched polysaccharides | Tee et al. (2024) |

(continued)

**Table 9.5** (continued)

| Compound class<br>Bioactive compound | | Source species | Pharmacological activity | Mechanism of action | References |
|---|---|---|---|---|---|
| Polysaccharides and β-Glucans | Mannoglucan polysaccharides | *Taiwanofungus camphoratus* | Antioxidant | Multiple antioxidant pathways | Tee et al. (2024) |
| Nucleosides | Cordycepin | *Cordyceps militaris* | Antioxidant | Free radical scavenging | Ashraf et al. (2020) |
| Amino acids and specialized compounds | Ergothioneine | Various mushroom species | Antioxidant | Mitochondrial protection, ROS scavenging | Olusola Ogidi et al. (2020) |

significantly reducing apoptosis while increasing endogenous antioxidant activity and attenuating intracellular calcium accumulation (Li et al. 2023).

The protective bioactive compounds include ergosterol and its oxide derivatives, selenium-enriched polysaccharides, and specific compounds such as ergothioneine and tocopherols. The presence of mannoglucan polysaccharide, antcin A, and antcin B, with an assortment of triterpenoids in *Taiwanofungus camphoratus*, coupled with a myriad of other compounds, further supports the overall antioxidant activity. Cellular protection against oxidative injury is exerted by a variety of mechanisms, such as electron donation, metal chelation, and enzyme inhibition (He et al. 2019; Bhambri et al. 2022; Rangsinth et al. 2023; Tee et al. 2024).

### 9.2.6 Antidiabetic Properties of Medicinal Mushrooms

Clinical research suggests that medicinal mushrooms hold potential for diabetes treatment, given the compositional diversity of bioactive compounds found in them, which display glucose- and lipid-lowering effects on the human body (Table 9.6). There are several fungal taxa with high effectiveness in plasma glucose level stability and reduction of diabetic complications. *Agaricus blazei*, *Pleurotus ostreatus*, *Ganoderma lucidum*, and *Inonotus obliquus* are among the most widely studied taxa that have demonstrated plenty of potential as antidiabetic agents.

A clinical study with *Pleurotus ostreatus* demonstrated that 89 diabetic subjects who consumed 50 g of cooked mushroom preparations three times daily over 24 days experienced a significant decline in blood pressure, plasma glucose, total cholesterol, and triglycerides, and after discontinuing the supplementation, the effects of these components vanished (Khatun et al. 2007).

Similarly, beta-glucan polysaccharides found in several mushroom species have the ability to restore the integrity of pancreatic tissues and stimulate the release of more insulin hormone by the beta cells (Olusola Ogidi et al. 2020). A collection of bioactive components, including a complex of polysaccharides, ergosterol derivatives, and structurally and functionally diverse compounds, such as cordycepin, davallialactone, hypholomine B, and ellagic acid, creates comprehensive glucoregulatory systems (Ashraf et al. 2020; Tee et al. 2024). The peripheral tissue insulin responsiveness to beta-glucans is raised due to PI3K/Akt-mediated signal transduction and GLUT4 translocation in insulin-sensitive tissue. As a result, the renormalization of molecular pathways occurs to resolve deficiencies in insulin production and peripheral insulin resistance (Lee et al. 2014; Wang et al. 2017; Rangsinth et al. 2023). Another randomized, double-blinded study with 72 Type 2 diabetic individuals indicated that supplementation of *Agaricus blazei* extract at the level of 1500 mg/day during the period of at least 12 weeks significantly increased insulin sensitivity through the rise of plasma adiponectin levels, and that the placebo-treated group reported a decrease instead (Hsu et al. 2007).

Molecular mechanisms have been unraveled in recent advanced preclinical studies. *Inonotus obliquus* polysaccharides demonstrated statistical significance in improving glucose tolerance, reducing fasting blood glucose levels, increasing

**Table 9.6** Bioactive compounds with antidiabetic activity

| Compound class | Bioactive compound | Source species | Pharmacological activity | Mechanism of action | References |
|---|---|---|---|---|---|
| Polysaccharides and β-Glucans | β-1,3/1,6-glucans | *Ganoderma lucidum*, *Pleurotus* spp. | Antidiabetic, immunomodulatory | Restore pancreatic tissue integrity, stimulate insulin release by beta cells, enhance peripheral tissue insulin responsiveness via PI3K/Akt-mediated signal transduction and GLUT4 translocation | Olusola Ogidi et al. (2020) and Rangsinth et al. (2023) |
| Polysaccharides and β-Glucans | HIOP1-S | *Inonotus obliquus* | Hypoglycemic | Inhibits α-glucosidase, enhances glucose consumption | Tee et al. (2024) |
| Polysaccharides and β-Glucans | HIOP2-S | *Inonotus obliquus* | Hypoglycemic | Enhances extracellular glucose consumption | Tee et al. (2024) |
| Polysaccharides and β-Glucans | IOEP1/IOEP2 | *Inonotus obliquus* | Hypoglycemic | Increases glucose consumption in insulin-resistant cells | Tee et al. (2024) |
| Polysaccharides and β-Glucans | *Inonotus obliquus* polysaccharides | *Inonotus obliquus* | Hypoglycemic, insulin sensitizing | Augment PI3K-p85 and phosphorylated Akt proteins, increase GLUT4 expression, and improve glucose tolerance | Wang et al. (2017) |
| Polysaccharides and β-Glucans | Water-soluble melanin complexes | *Inonotus obliquus* | Insulin sensitizing | Dose-dependent Akt phosphorylation and GLUT4 translocation | Lee et al. (2014), Wang et al. (2017), and Rangsinth et al. (2023) |
| Polysaccharides and β-Glucans | Branched glycan | *Tropicoporus linteus* | Antidiabetic | Reduces blood glucose levels | Tee et al. (2024) |
| Terpenoids and steroids | Ergosterol | *Pleurotus ostreatus* | Antidiabetic, nephroprotective | GLUT4 translocation through PI3K/Akt and PKC signaling, normalizes biochemical markers, decreases inflammatory cytokines (IL-6, TNF-α, and MCP-1) | Ashraf et al. (2020) and Tee et al. (2024) |
| Terpenoids and steroids | Davallialactone | *Tropicoporus linteus* | Antidiabetic, antioxidant | Inhibits aldose reductase enzyme | Lee et al. (2014) and Tee et al. (2024) |

| | | | | | |
|---|---|---|---|---|---|
| Proteins and peptides | Cordymin | *Cordyceps sinensis* | Antihyperglycemic | Glucose metabolism modulation | Wong et al. (2011) and Ashraf et al. (2020) |
| Nucleosides | Cordycepin | *Cordyceps militaris* | Anti-inflammatory, antidiabetic | Inhibits TNF-α, IL-1β, IL-6; modulates diabetes-regulatory 11β-HSD1 and PPAR-γ via NF-κB pathway | Shin et al. (2009) and Ji et al. (2009) |
| Phenolic compounds | Hypholomine B | *Tropicoporus linteus* | Antioxidant, hypolipidemic, hepatoprotective | Reduces diabetic complications | Tee et al. (2024) |

hepatic glycogen content, and ameliorating insulin resistance in diabetic mice via augmenting the PI3K-p85 and phosphorylated Akt proteins and, in parallel, increasing the GLUT4 expression (Rangsinth et al. 2023). Insulin-mediated glucose uptake was also stimulated by water-soluble melanin complexes originating in the same taxon through dose-dependent Akt phosphorylation and GLUT4 translocation.

Ergosterol, isolated from *Pleurotus ostreatus*, has antidiabetic properties related to the GLUT4 translocation through the signaling mediated by, e.g., PI3K/Akt and PKC, and anticomplementary effects in diabetic nephropathy shown by its ability to normalize biochemical markers and decrease inflammatory cytokines, including IL-6, TNF-?, and MCP-1 (Liu et al. 2020; Rangsinth et al. 2023). Cordycepin, the product of *Cordyceps militaris*, has multi-mechanism functioning since it inhibits the pro-inflammatory cytokines such as TNF-α, IL-1β, and IL-6 and modulates diabetes-regulatory 11β-HSD1 and PPAR-γ through an NF-κB-mediated mechanism (Youn et al. 2009; Tee et al. 2024). A further advantage comes in the form of additional specialized compounds like davallialactone and ellagic acid, which also inhibit aldose reductase enzyme action, making it less possible to develop any diabetic complications (Lee et al. 2008; Tee et al. 2024). Clinical evidence shows that pharmacological use is not limited to glycemic control but also covers the management of diabetic complications as a whole. SREBP-1, acetyl-CoA carboxylase expression, and fatty acid synthase mRNA are inhibited by polysaccharide treatment, which promotes improved lipid metabolism (Tee et al. 2024). The shared impact on glucose metabolism, insulin sensitivity, and prevention of complications makes mushroom-derived compounds a promising adjunct treatment where conventional treatments lack efficiency.

### 9.2.7 Anti-Inflammatory Properties of Medicinal Mushrooms

Recent studies have shown that multiple medicinal mushrooms have significant anti-inflammatory potential by regulating an array of bioactive factors and signaling processes (Table 9.7). *Ganoderma lucidum*, *Cordyceps militaris*, *Inonotus obliquus*, and *Phellinus linteus* have been shown to reduce inflammatory diseases mediated by TNF-α, IL-6, nitric oxide, and other inflammatory mediators. The key mechanisms can be broadly attributed to ergosterol-derived substances (Bhambri et al. 2022), β-glucans, as well as phenolic compounds, which affect NF-kB signal transduction and decrease the generation of pro-inflammatory cytokines. These results are, however, varying with the mushroom species, extraction methods, and models of inflammation and therefore require stringent controls of methods (Bawadekji et al. 2016; Cerletti et al. 2021; Rangsinth et al. 2023).

Clinical studies provide additional evidence that extracts with high concentrations of ergosterol can significantly inhibit inflammatory mechanisms through JAK-STAT and MAPK signal blockade. Cordycepin isolated from *Cordyceps militaris* exhibits potent effects on LPS-induced inflammation by preventing the degradation of I kB-κ and inhibition of NF-κB activation and reducing phosphorylation of ERK, JNK, and p38 kinases (Ashraf et al. 2020; Rangsinth et al. 2023). *Phellinus linteus* extract

**Table 9.7** Bioactive compounds with anti-inflammatory activity

| Compound/ Class | Source species | Pharmacological activity | Mechanism of action | References |
|---|---|---|---|---|
| Cordycepin | *Cordyceps militaris* | Anti-inflammatory, antitumor | Prevents IκB-α degradation, inhibits NF-κB activation, reduces ERK/JNK/p38 phosphorylation | Ashraf et al. (2020) and Rangsinth et al. (2023) |
| β-D-glucan | *Taiwanofungus camphoratus*, various species | Anti-inflammatory, immunomodulatory | Reduces LPS-induced inflammation, modulates immune response | Bawadekji et al. (2016), Cerletti et al. (2021), and Rangsinth et al. (2023) |
| Ergosterol derivatives | *Ganoderma lucidum*, *Inonotus obliquus* | Anti-inflammatory | JAK-STAT and MAPK pathway inhibition, NF-κB suppression | Ashraf et al. (2020) and Rangsinth et al. (2023) |
| Hispidin | *Phellinus linteus* | Anti-inflammatory, antioxidant | Regulates redox-sensitive pathways in macrophages, NF-κB inhibition | Galappaththi et al. (2022) and Tee et al. (2024) |
| Hispolon | *Phellinus linteus* | Anti-inflammatory | Anti-inflammatory signaling modulation | Galappaththi et al. (2022) and Tee et al. (2024) |
| Ganoderic acids (A–X) | *Ganoderma lucidum* | Anti-inflammatory, immunomodulatory | HMG-CoA reductase inhibition, cytokine regulation | Galappaththi et al. (2022) |
| Ganoluciduone B | *Ganoderma lucidum* | Anti-inflammatory | Nitric oxide suppression | Yang et al. (2022) and Galappaththi et al. (2022) |
| Butyl lucidenate analogs | *Ganoderma lucidum* | Anti-inflammatory | Multiple inflammatory pathway modulation | Galappaththi et al. (2022) |
| Sulfated polysaccharides | *Taiwanofungus camphoratus*, various species | Anti-inflammatory, anticancer | Enhanced activity vs. non-sulfated forms, cytokine modulation | Galappaththi et al. (2022) |
| Phenolic compounds | *Inonotus obliquus*, various species | Anti-inflammatory, antioxidant | NF-κB signal transduction inhibition | Rangsinth et al. (2023) |
| Polysaccharide-protein complexes | *Ganoderma lucidum*, various species | Immunomodulatory, anti-inflammatory | Immune cell signaling modulation, TNF-α/IL-6 reduction | Cerletti et al. (2021) |
| Triterpenoids (general) | *Ganoderma lucidum* | Anti-inflammatory, hepatoprotective | Multiple molecular targets, pro-inflammatory cytokine reduction | Galappaththi et al. (2022) |

containing hispidin can exert intense anti-inflammatory effects by regulating redox-sensitive pathways in macrophages (Bhambri et al. 2022).

The similar results were obtained using standardized extracts of *G. lucidum and I. obliquus* may be due to the strong nitric oxide-suppressive capability of the terpenoid ganoluciduone B (Yang et al. 2022; Galappaththi et al. 2022). These findings underscore the broad range of bioactive ingredients in medicinal mushrooms, including ergosterol derivatives, triterpenoids, and sulfated polysaccharides, which have proven useful across a vast array of inflammation-related disorders, including colitis and neuroinflammation (Ashraf et al. 2020; Galappaththi et al. 2022; Rangsinth et al. 2023). Hispolon and butyl lucidenate analogs have also been shown to be useful anti-inflammatory bioactives. Precise molecular signaling mechanisms through which these compounds inhibit chronic inflammatory diseases need to be better understood to develop these as effective supplements to conventional anti-inflammatory therapeutic approaches (Galappaththi et al. 2022; Tee et al. 2024).

## 9.3 Current Challenges and Future Prospects in Medicinal Mushroom Research

### 9.3.1 Clinical Research and Methodological Limitations

The medicinal mushroom application arena has the drawback of the nature of clinical research design and deployment. Many experiments lack proper definition, like the inadequate sample sizes that have yielded false negatives, a lack of proper randomization procedures, a placebo, and the use of double-blind protocols (Venturella et al. 2021). The absence of follow-up studies in pilot trials keeps their results tentative and far from clinical application (Venturella et al. 2021), as well as the challenge of production, since, in many species, there is a conflict between the accumulation of bioactive compounds, which may be stimulated by elicitors, and cellular growth inhibition by the same compounds (Chen 2020). These theoretical gaps are compounded by the fact that to date, the field has identified only 700 of the 2000 safe mushroom species with pharmacological potential, noting that there is immense research potential lying in wait (Galappaththi et al. 2022).

### 9.3.2 Standardization and Consistency Goals and Objectives

A significant drawback in mushroom research is the lack of standardized preparations, as the same study may produce contradictory or even antagonistic results with different preparations, as was the case with *Agaricus blazei Murrill*, where proven preparations failed or even become toxic (Yoshimura et al. 2010; Ohno et al. 2011). The inconsistency is brought about by various aspects that include the extraction method, strain variations, geographic origin, cultivation conditions, the composition of the vegetative medium, the part of the mushroom used, and the harvest during the phase, and all these have fundamental effects on the bioactive potential of the mushroom (Venturella et al. 2021). The implication that the precise concentrations of single metabolites in extracts cannot reasonably be ascertained easily also contributes to the difficulty of efforts toward standardization, and gaps in research infrastructure also persist despite having roughly identified 126 medicinal functions of mushrooms. Mushrooms' response mechanisms to various elicitor molecules remain unclear and cannot be further used to optimize the production of bioactive compounds (Wasser 2011; Venturella et al. 2021).

## 9.4 Future Prospects

### 9.4.1 Technological Innovation and Therapeutic Development

The discovery of new bioactive molecules could lead to the development of novel therapeutics for the treatment of neurodegenerative disorders, cardiovascular diseases, and cancers. Furthermore, nanobiotechnology-based targeted drug delivery systems for bioactive compounds such as cordycepin could be designed to improve their bioavailability (Ashraf et al. 2020). Research on metabolomics could also shed light on ways to alter metabolic pathways in order to achieve increased synthesis of these compounds in mushrooms. The pharmaceutical industry can focus efforts to develop strategies for improved cultivation and better extraction mechanisms for these active compounds from mushrooms to enhance the production of such health-promoting metabolites (Chen 2020; Fitsum et al. 2025).

### 9.4.2 Market Development and Biotechnological Innovations

The medicinal mushroom supplement industry has a promising future, currently generating 18 billion US dollars in annual revenue (Galappaththi et al. 2022). Mushrooms have been found to have excellent functional food qualities, such as rich nutrients and bioactive substances, and are capable of replacing meat as an ingredient in diverse food preparations. Mushrooms hold great potential for the future, especially in delivering personalized nutrition to patients suffering from common chronic ailments such as diabetes, high blood pressure, and raised cholesterol levels. The active component ergosterol from mushrooms can be employed to develop new antioxidant and anti-inflammatory formulations that may also be integrated with chemotherapeutic drugs in order to enhance their functionality and reduce their side effects (Rangsinth et al. 2023). Additionally, some of these mushrooms can substantially be used in the bioremediation of heavy metals and the transformation of waste into biofuels (Fitsum et al. 2025).

### 9.4.3 Research Improvements and Clinical Trials

Future advancements in this field would benefit from more experimental and clinical studies to understand the specific mechanisms of action of therapeutic effects. Safety and efficacy profiles of nanotherapy and combination therapies should be properly established (Ashraf et al. 2020). It is essential to standardize clinical trial methodologies, sample sizes, and randomization and blinding to achieve conclusive and significant results in the field. Safety and adverse effect profiles should also be studied before advancing to clinical applications (Venturella et al. 2021). Continual pharmacological studies of bioactive compounds like ergosterol would be necessary for furthering their development into potential drugs and nutraceuticals. There should also be formulation studies to increase the bioavailability of these compounds (Rangsinth et al. 2023). It is important that integrating new biotechnologies with these traditional applications of medicinal mushrooms could lead to novel advancements in the fields of health, pharmaceuticals, and environmental sustainability (Wasser 2011).

## 9.5 Conclusion

One of the most striking aspects of medicinal mushrooms is their ability to target multiple health concerns simultaneously. For instance, the neuroprotective compound erinacines in lion's mane can cross the blood-brain barrier and promote the synthesis of nerve growth factor. The turkey tail mushroom, on the other hand, is a potent immunomodulator that can enhance the activity of the natural killer cells due to the presence of β-glucans. Not only do they help to boost the immune system, but they also display anti-inflammatory and antioxidant potential. These mushrooms contain several bioactive compounds, including polysaccharides, triterpenoids, and phenolic compounds, among other metabolites, thereby explaining the broad range of health benefits offered by them. This integrated array of benefits such as this is akin to the practice of ancient medicine and offers an interesting opportunity for modern science.

However, to utilize this potential of medicinal mushrooms, certain important challenges need to be addressed. It is essential that the medicinal mushroom products be standardized since the concentration of active compounds in the product could vary considerably, based on the conditions in which the mushrooms are cultivated, the mode through which they are extracted, and the part of the mushroom used to prepare the extract. The inadequacy of the clinical trial design and the absence of studies that extend well into the future are also issues that ought to be considered. A combination of biotechnological innovations with the traditional mushroom cultivation protocols in the future can lead to the development of improved mushroom strains with increased production of functional compounds. Furthermore, it is evident that the research on the medicinal prospects of yet to be explored mushrooms has enormous possibilities for the discovery of new nutraceuticals in the future.

**AI Disclosure** This manuscript was prepared with the assistance of Claude Opus 4.1 and ChatGPT version 5, which were used solely for language and grammar corrections. All AI-generated suggestions were carefully reviewed and manually revised by the authors to ensure accuracy and clarity. The authors take full responsibility for the content of the manuscript.

## References


Abotaleb M, Liskova A, Kubatka P, Büsselberg D (2020) Therapeutic potential of plant phenolic acids in the treatment of cancer. Biomolecules 10(2):221. https://doi.org/10.3390/biom10020221

Ahn H, Jeon E, Kim J-C, Kang SG, Yoon S, Ko H-J, Kim P-H, Lee G-S (2017) Lentinan from shiitake selectively attenuates AIM2 and non-canonical inflammasome activation while inducing pro-inflammatory cytokine production. Sci Rep 7(1):1314. https://doi.org/10.1038/s41598-017-01462-4

Ashraf SA, Elkhalifa AEO, Siddiqui AJ, Patel M, Awadelkareem AM, Snoussi M, Ashraf MS, Adnan M, Hadi S (2020) Cordycepin for health and wellbeing: a potent bioactive metabolite of an entomopathogenic medicinal fungus Cordyceps with its nutraceutical and therapeutic potential. Molecules 25(12):2735. https://doi.org/10.3390/molecules25122735

Bawadekji A, Al Ali K, Al Ali M (2016) A review of the bioactive compound and medicinal value of Cordyceps Militaris. 76–69:(1)1 [illegible] [illegible] [illegible] الشمال. https://doi.org/10.12816/0021378

Bhambri A, Srivastava M, Mahale VG, Mahale S, Karn SK (2022) Mushrooms as potential sources of active metabolites and medicines. Front Microbiol 13:837266. https://doi.org/10.3389/fmicb.2022.837266

Camilleri E, Blundell R, Baral B, Karpinski TM, Aruci E, Atrooz OM (2024) A brief overview of the medicinal and nutraceutical importance of Inonotus obliquus (chaga) mushrooms. Heliyon 10(15):e35638. https://doi.org/10.1016/j.heliyon.2024.e35638

Cerletti C, Esposito S, Iacoviello L (2021) Edible mushrooms and Beta-Glucans: impact on human health. Nutrients 13(7):2195. https://doi.org/10.3390/nu13072195

Chang Y, Zhang M, Jiang Y, Liu Y, Luo H, Hao C, Zeng P, Zhang L (2017) Preclinical and clinical studies of Coriolus versicolor polysaccharopeptide as an immunotherapeutic in China. Discov Med 23(127):207–219

Cheah IK, Halliwell B (2021) Ergothioneine, recent developments. Redox Biol 42:101868. https://doi.org/10.1016/j.redox.2021.101868

Chen S (2020) The pharmacological effects of triterpenoids from Ganoderma lucidum and the regulation of its biosynthesis. Adv Biol Chem 10(02):55–65. https://doi.org/10.4236/abc.2020.102005

Chong PS, Fung M-L, Wong KH, Lim LW (2019) Therapeutic potential of Hericium erinaceus for depressive disorder. Int J Mol Sci 21(1):163. https://doi.org/10.3390/ijms21010163

Dahiya S, Dahiya R, Fuloria NK, Mourya R, Dahiya S, Fuloria S, Kumar S, Shrivastava J, Saharan R, Chennupati SV, Patel JK (2022) Natural bridged bicyclic peptide macrobiomolecules from Celosia argentea and amanita phalloides. Mini-Rev Med Chem 22(13):1772–1788. https://doi.org/10.2174/1389557522666220113122117

Feng L, Cheah IK-M, Ng MM-X, Li J, Chan SM, Lim SL, Mahendran R, Kua E-H, Halliwell B (2019) The association between mushroom consumption and mild cognitive impairment: a community-based cross-sectional study in Singapore. J Alzheimer's Dis 68(1):197–203. https://doi.org/10.3233/JAD-180959

Fitsum S, Sbhatu DB, Gebreyohannes G (2025) Harnessing the nutritional value, therapeutic applications, and environmental impact of mushrooms. Food Sci Nutr 13(7):e70611. https://doi.org/10.1002/fsn3.70611

Galappaththi MCA, Patabendige NM, Premarathne BM, Hapuarachchi KK, Tibpromma S, Dai D-Q, Suwannarach N, Rapior S, Karunarathna SC (2022) A review of Ganoderma triterpenoids and their bioactivities. Biomolecules 13(1):24. https://doi.org/10.3390/biom13010024

He R, Wu K, Zhang A, Xie Z, Sun P (2019) Mechanochemical-assisted extraction and pharmacological study of triterpenoids from Antrodia Camphorata. Appl Sci 9(20):4281. https://doi.org/10.3390/app9204281

Hsu C-H, Liao Y-L, Lin S-C, Hwang K-C, Chou P (2007) The mushroom *Agaricus Blazei* Murill in combination with metformin and Gliclazide improves insulin resistance in type 2 diabetes: a randomized, double-blinded, and placebo-controlled clinical trial. J Altern Complement Med 13(1):97–102. https://doi.org/10.1089/acm.2006.6054

Hu Q, Wang D, Yu J, Ma G, Pei F, Yang W (2017) Neuroprotective effects of six components from Flammulina velutipes on H2O2-induced oxidative damage in PC12 cells. J Funct Foods 37: 586–593. https://doi.org/10.1016/j.jff.2017.07.043

Ina K, Kataoka T, Ando T (2013) The use of Lentinan for treating gastric cancer. Anti Cancer Agents Med Chem 13(5):681–688. https://doi.org/10.2174/1871520611313050002

Jedinak A, Sliva D (2008) Pleurotus ostreatus inhibits proliferation of human breast and colon cancer cells through p53-dependent as well as p53-independent pathway. Int J Oncol 33(6): 1307–1313

Ji D, Ye J, Li C, Wang Y, Zhao J, Cai S (2009) Antiaging effect of *Cordyceps sinensis* extract. Phytother Res 23(1):116–122. https://doi.org/10.1002/ptr.2576

Khatun K, Mahtab H, Khanam P, Sayeed M, Khan K (2007) Oyster mushroom reduced blood glucose and cholesterol in diabetic subjects. Mymensingh Med J 16(1):94–99. https://doi.org/10.3329/mmj.v16i1.261

Kou R-W, Han R, Gao Y-Q, Li D, Yin X, Gao J-M (2021) Anti-neuroinflammatory polyoxygenated lanostanoids from Chaga mushroom Inonotus obliquus. Phytochemistry 184: 112647. https://doi.org/10.1016/j.phytochem.2020.112647

Lee YS, Kang Y-H, Jung J-Y, Kang I-J, Han S-N, Chung J-S, Shin H-K, Lim SS (2008) Inhibitory constituents of aldose reductase in the fruiting body of Phellinus linteus. Biol Pharm Bull 31(4): 765–768. https://doi.org/10.1248/bpb.31.765

Lee T-H, Chen C-C, Chen J-J, Liao H-F, Chang H-S, Sung P-J, Tseng M-H, Wang S-Y, Ko H-H, Kuo Y-H (2014) New and cytotoxic components from Antrodia camphorata. Molecules 19(12): 21378–21385. https://doi.org/10.3390/molecules191221378

Li N, Li H, Liu Z, Feng G, Shi C, Wu Y (2023) Unveiling the therapeutic potentials of mushroom bioactive compounds in Alzheimer's disease. Foods 12(15):2972. https://doi.org/10.3390/foods12152972

Liang C, Tian D, Liu Y, Li H, Zhu J, Li M, Xin M, Xia J (2019) Review of the molecular mechanisms of Ganoderma lucidum triterpenoids: ganoderic acids A, C2, D, F, DM, X and Y. Eur J Med Chem 174:130–141. https://doi.org/10.1016/j.ejmech.2019.04.039

Lindequist U, Niedermeyer THJ, Jülich W-D (2005) The pharmacological potential of mushrooms. Evid Based Complement Alternat Med 2(3):285–299. https://doi.org/10.1093/ecam/neh107

Liu C, Zhao S, Zhu C, Gao Q, Bai J, Si J, Chen Y (2020) Ergosterol ameliorates renal inflammatory responses in mice model of diabetic nephropathy. Biomed Pharmacother 128:110252. https://doi.org/10.1016/j.biopha.2020.110252

Miles PG, Chang S-T (2004) Mushrooms, 0th edn. CRC Press. https://doi.org/10.1201/9780203492086

Ohno S, Sumiyoshi Y, Hashine K, Shirato A, Kyo S, Inoue M (2011) Phase I clinical study of the dietary supplement, *Agaricus blazei* Murill, in cancer patients in remission. Evid Based Complement Alternat Med 2011(1):192381. https://doi.org/10.1155/2011/192381

Olusola Ogidi C, Olusegun Oyetayo V, Juliet Akinyele B (2020) Wild medicinal mushrooms: potential applications in Phytomedicine and functional foods. In: Kumar Passari A, Sánchez S (eds) An introduction to mushroom. IntechOpen. https://doi.org/10.5772/intechopen.90291

Puri M, Kaur I, Perugini MA, Gupta RC (2012) Ribosome-inactivating proteins: current status and biomedical applications. Drug Discov Today 17(13–14):774–783. https://doi.org/10.1016/j.drudis.2012.03.007

Rangsinth P, Sharika R, Pattarachotanant N, Duangjan C, Wongwan C, Sillapachaiyaporn C, Nilkhet S, Wongsirojkul N, Prasansuklab A, Tencomnao T, Leung GP-H, Chuchawankul S (2023) Potential beneficial effects and pharmacological properties of ergosterol, a common bioactive compound in edible mushrooms. Foods 12(13):2529. https://doi.org/10.3390/foods12132529

Saleh MH, Rashedi I, Keating A (2017) Immunomodulatory properties of Coriolus versicolor: the role of polysaccharopeptide. Front Immunol 8:1087. https://doi.org/10.3389/fimmu.2017.01087

Sarangi I, Ghosh D, Bhutia SK, Mallick SK, Maiti TK (2006) Anti-tumor and immunomodulating effects of Pleurotus ostreatus mycelia-derived proteoglycans. Int Immunopharmacol 6(8): 1287–1297. https://doi.org/10.1016/j.intimp.2006.04.002

Sasikumar V, Govindan S, Rajendran G, Rajendran A, Ramani P, Pateiro M, Lorenzo JM (2023) Protective effect of aqueous extract of Pleurotus eous against Dalton's lymphoma ascites tumor in mice. J Agric Food Res 13:100638. https://doi.org/10.1016/j.jafr.2023.100638

Shin S, Lee S, Kwon J, Moon S, Lee S, Lee C-K, Cho K, Ha N-J, Kim K (2009) Cordycepin suppresses expression of diabetes regulating genes by inhibition of lipopolysaccharide-induced inflammation in macrophages. Immune Netw 9(3):98. https://doi.org/10.4110/in.2009.9.3.98

Sillapachaiyaporn C, Chuchawankul S, Nilkhet S, Moungkote N, Sarachana T, Ung AT, Baek SJ, Tencomnao T (2022a) Ergosterol isolated from cloud ear mushroom (Auricularia polytricha) attenuates bisphenol A-induced BV2 microglial cell inflammation. Food Res Int 157:111433. https://doi.org/10.1016/j.foodres.2022.111433

Sillapachaiyaporn C, Mongkolpobsin K, Chuchawankul S, Tencomnao T, Baek SJ (2022b) Neuroprotective effects of ergosterol against TNF-α-induced HT-22 hippocampal cell injury. Biomed Pharmacother 154:113596. https://doi.org/10.1016/j.biopha.2022.113596

Singh S, Wang H, Chan Y, Pan W, Dan X, Yin C, Akkouh O, Ng T (2014) Lectins from edible mushrooms. Molecules 20(1):446–469. https://doi.org/10.3390/molecules20010446

Sivanesan I, Muthu M, Gopal J, Oh J-W (2022) Mushroom polysaccharide-assisted anticarcinogenic mycotherapy: reviewing its clinical trials. Molecules 27(13):4090. https://doi.org/10.3390/molecules27134090

Song X, Liu Z, Zhang J, Yang Q, Ren Z, Zhang C, Liu M, Gao Z, Zhao H, Jia L (2018) Anti-inflammatory and hepatoprotective effects of exopolysaccharides isolated from Pleurotus geesteranus on alcohol-induced liver injury. Sci Rep 8(1):10493. https://doi.org/10.1038/s41598-018-28785-0

Spelman K, Sutherland E, Bagade A (2017) Neurological activity of lion's mane ( *Hericium erinaceus* ). J Restor Med 6(1):19–26. https://doi.org/10.14200/jrm.2017.6.0108

Tan W, Pan M, Liu H, Tian H, Ye Q, Liu H (2017) Ergosterol peroxide inhibits ovarian cancer cell growth through multiple pathways. Onco Targets Ther 10:3467–3474. https://doi.org/10.2147/OTT.S139009

Tee PYE, Krishnan T, Cheong XT, Maniam SAP, Looi CY, Ooi YY, Chua CLL, Fung S-Y, Chia AYY (2024) A review on the cultivation, bioactive compounds, health-promoting factors and clinical trials of medicinal mushrooms Taiwanofungus camphoratus, Inonotus obliquus and Tropicoporus linteus. Fungal Biol Biotechnol 11(1):7. https://doi.org/10.1186/s40694-024-00176-3

Tel-Çayan G, Muhammad A, Deveci E, Duru ME, Öztürk M (2020) Isolation, structural characterization, and biological activities of galactomannans from Rhizopogon luteolus and Ganoderma adspersum mushrooms. Int J Biol Macromol 165:2395–2403. https://doi.org/10.1016/j.ijbiomac.2020.10.040

Valverde ME, Hernández-Pérez T, Paredes-López O (2015) Edible mushrooms: improving human health and promoting quality life. Int J Microbiol 2015:1–14. https://doi.org/10.1155/2015/376387

Venturella G, Ferraro V, Cirlincione F, Gargano ML (2021) Medicinal mushrooms: bioactive compounds, use, and clinical trials. Int J Mol Sci 22(2):634. https://doi.org/10.3390/ijms22020634

Wang J, Wang C, Li S, Li W, Yuan G, Pan Y, Chen H (2017) Anti-diabetic effects of Inonotus obliquus polysaccharides in streptozotocin-induced type 2 diabetic mice and potential mechanism via PI3K-Akt signal pathway. Biomed Pharmacother 95:1669–1677. https://doi.org/10.1016/j.biopha.2017.09.104

Wasser SP (2011) Current findings, future trends, and unsolved problems in studies of medicinal mushrooms. Appl Microbiol Biotechnol 89(5):1323–1332. https://doi.org/10.1007/s00253-010-3067-4

Wei Y-M, Yang L, Mei W-L, Chen H-Q, Cai C-H, Li W, Dong W-H, Chen Z-B, Dai H-F (2022) Phenolic compounds from the sclerotia of *Inonotus obliquus*. Nat Prod Res 36(9):2413–2417. https://doi.org/10.1080/14786419.2020.1833202

Whitmore CA, Haynes JR, Behof WJ, Rosenberg AJ, Tantawy MN, Hachey BC, Wadzinski BE, Spiller BW, Peterson TE, Paffenroth KC, Harrison FE, Beelman RB, Wijesinghe P, Matsubara JA, Pham W (2022) Longitudinal consumption of Ergothioneine reduces oxidative stress and amyloid plaques and restores glucose metabolism in the 5XFAD mouse model of Alzheimer's disease. Pharmaceuticals 15(6):742. https://doi.org/10.3390/ph15060742

Wong JH, Ng TB, Wang H, Sze SCW, Zhang KY, Li Q, Lu X (2011) Cordymin, an antifungal peptide from the medicinal fungus Cordyceps militaris. Phytomedicine 18(5):387–392. https://doi.org/10.1016/j.phymed.2010.07.010

Xu X, Yan H, Chen J, Zhang X (2011) Bioactive proteins from mushrooms. Biotechnol Adv 29(6):667–674. https://doi.org/10.1016/j.biotechadv.2011.05.003

Yang L, Kong D-X, Xiao N, Ma Q-Y, Xie Q-Y, Guo J-C, Ying Deng C, Ma H-X, Hua Y, Dai H-F, Zhao Y-X (2022) Antidiabetic lanostane triterpenoids from the fruiting bodies of Ganoderma weberianum. Bioorg Chem 127:106025. https://doi.org/10.1016/j.bioorg.2022.106025

Yanshree YWS, Fung ML, Lee CW, Lim LW, Wong KH (2022) The monkey head mushroom and memory enhancement in Alzheimer's disease. Cells 11(15):2284. https://doi.org/10.3390/cells11152284

Yoshimura K, Kamoto T, Ogawa O, Matsui S, Tsuchiya N, Tada H, Murata K, Yoshimura K, Habuchi T, Fukushima M (2010) Medical mushrooms used for biochemical failure after radical treatment for prostate cancer: an open-label study. Int J Urol 17(6):548–554. https://doi.org/10.1111/j.1442-2042.2010.02528.x

Youn M-J, Kim J-K, Park S, Kim Y, Park C, Kim ES, Park K-I, So HS, Park R (2009) Potential anticancer properties of the water extract of Inontus obliquus by induction of apoptosis in melanoma B16-F10 cells. J Ethnopharmacol 121(2):221–228. https://doi.org/10.1016/j.jep.2008.10.016

Zhang Y, Zhang M, Jiang Y, Li X, He Y, Zeng P, Guo Z, Chang Y, Luo H, Liu Y, Hao C, Wang H, Zhang G, Zhang L (2018) Lentinan as an immunotherapeutic for treating lung cancer: a review of 12 years clinical studies in China. J Cancer Res Clin Oncol 144(11):2177–2186. https://doi.org/10.1007/s00432-018-2718-1

Zhou Y, Chu M, Ahmadi F, Agar OT, Barrow CJ, Dunshea FR, Suleria HAR (2024) A comprehensive review on phytochemical profiling in mushrooms: occurrence, biological activities, applications and future prospective. Food Rev Int 40(3):924–951. https://doi.org/10.1080/87559129.2023.2202738

# The Role of Mushrooms in Cancer and Neurodegenerative Disease Treatment 10

Avrajit Manna, Merina Yasmin, Indranil Chakraborty, Shibdas Maity, and Prasenjit Maity

**Abstract**

Mushrooms have long been recognized for their medicinal properties, particularly in the treatment of cancer and neurodegenerative diseases. Bioactive compounds derived from various mushroom species exhibit potent anticancer, anti-inflammatory, antioxidant, hepatoprotective, immunomodulatory, antibacterial, and neuroprotective effects. In cancer therapy, polysaccharides such as β-D-glucans, Lentinan, and PSK have demonstrated the ability to increase the immune system's cell-mediated response, helping to inhibit tumor growth and improve patient outcomes. These compounds function as biological response modifiers, stimulating immune cells like macrophages, natural killer cells, and T lymphocytes. Regarding neurodegenerative diseases, mushrooms produce neuroprotective compounds, including antioxidants, phenolics, and ergothioneine, that help mitigate oxidative stress and inflammation, both of

A. Manna
Department of Chemistry, Raja Narendra Lal Khan Women's College (Autonomous), Gope Palace, Paschim Medinipur, West Bengal, India

M. Yasmin
Department of Human Physiology, Vidyasagar University, Midnapore, Paschim Medinipur, West Bengal, India

I. Chakraborty
Department of Chemistry, Kharagpur College, Kharagpur, Paschim Medinipur, West Bengal, India

S. Maity
Department of Botany, Sabang Sajanikanta Mahavidyalaya, Lutunia, Midnapore, Paschim Medinipur, West Bengal, India

P. Maity (✉)
Department of Chemistry, Sabang Sajanikanta Mahavidyalaya, Lutunia, Paschim Medinipur, West Bengal, India

 
B. Giri et al. (eds.), *Advances in Mushroom Research*,
https://doi.org/10.1007/978-981-95-7692-0_10

which are critical factors in disorders such as Alzheimer's and Parkinson's disease. Collectively, these findings underscore the promising potential of mushrooms as natural agents in adjunct therapies for cancer and neurodegenerative disorders, offering a foundation for the development of innovative functional foods and pharmaceuticals.

**Keywords**

Medicinal mushrooms · Bioactive compounds · Anticancer properties · Neuroprotective effects

## 10.1 Introduction

The mushroom is an important natural food and a potent source of medicinal compounds and nutraceuticals (Wasser 2002). High in protein and low in calories, mushrooms are also a good source of vitamins and minerals. They play a potential role in preventing diabetes, asthma, allergies, stress, cancer, and insomnia (Shi et al. 2013). Mushrooms are rich in polysaccharides, dietary fibers, oligosaccharides, triterpenoids, peptides, proteins, alcohols, phenols, and minerals like iron, zinc, copper, iodine, selenium, and vitamins (Cör et al. 2018). They reduce the negative effects of radiation and chemotherapy, boost the immune system, and have anticancer properties (Yao et al. 2024). Polysaccharides and polysaccharide-protein complexes with anticancer characteristics have been found in a variety of mushrooms. β-D-glucans derived from medicinal mushrooms, including Lentinan from *Lentinus edodes*, Ganoderan from *Ganoderma lucidum*, Schizophyllan from *Schizophyllum commune*, and Grifolan from *Grifola frondosa*, as well as protein-bound polysaccharides such as PSK (Krestin) and polysaccharide-protein complexes from *Trametes versicolor*, have been extensively studied for their strong immunomodulatory and anticancer activities (Cui and Chisti 2003; Moradali et al. 2007; Taguchi et al. 1983). Polysaccharides and polysaccharide-protein complexes are believed to enhance cell-mediated immune responses in both in vitro and in vivo settings by serving as biological response modifiers. They can cause the expression of various cytokine receptor genes and immunomodulatory cytokines, which is why they are considered multi-cytokine inducers. The ability of bioactive polysaccharides and polysaccharide-bound proteins to influence so many important immune cells may be due to their structural diversity and unpredictable nature.

Mushrooms such as shiitake (*Lentinula edodes*) and button mushrooms (*Agaricus bisporus*) are biologically relevant due to their nutritional value and potential health benefits (Wasser and Weis 1999). New-generation foods that are becoming more and more popular include medicinal mushrooms like *Lentinula edodes*, *Ganoderma lucidum*, *Inonotus obliquus*, *Ophiocordyceps sinensis*, and *Trametes versicolor* (Sohretoglu and Huang 2018). They are distinguished due to their high concentration of physiologically active ingredients, including phenolic compounds, triterpenes, sterols, and (1,3), (1,6)-β-D-glucans (Arunachalam et al. 2022). These physiologically active compounds exhibit anticancer, antidiabetic, antioxidant, and immunomodulatory effects (Maity et al. 2014; Nandi et al. 2013). Medicinal mushrooms show great promise in the development of new functional foods (Bohn and BeMiller 1995); however, the vast array of therapeutic advantages of mushrooms is still unknown.

Cancer is becoming the leading cause of mortality worldwide. In 2018, it accounted for about 9.6 million deaths. Approximately 70% of cancer-related deaths occur in low- and middle-income countries, highlighting a significant global health disparity (Ye et al. 2011). Among the cancer chemotherapy drugs commonly used to treat particular cancer cases are Taxol, cisplatin, cyclophosphamide, mitomycin, and 5-fluorouracil derivatives. Hepatotoxicity, renal toxicity, immunological suppression, diarrhea, and gastrointestinal toxicity are just a few of the adverse side effects that these drugs might cause. These chemicals have been shown to be specific agents for killing cancer cells, and they also poison normal cells. Adjuvant chemotherapy or radiation therapy may be curative after a malignant tumor has been surgically removed. However, several cancer types, such as osteogenic sarcoma, breast cancer, and colon cancer, have the potential to grow quickly and spread to other vital organs after surgical excision of malignant tumors. Therefore, one of the main goals of biomedical science research is to uncover and identify new, safe anticancer drugs that have antitumor and antimetastasis capabilities (Dan et al. 2023). Increasing the energy of the host defense mechanism is one possible way to halt tumor growth without putting the host in danger. From this angle, the polysaccharides extracted from therapeutic mushrooms have been the subject of extensive research. This chapter focuses on the anticancer properties of polysaccharides and polysaccharide-protein complexes that have been isolated from medicinal mushrooms (Krishnakumar et al. 2023).

## 10.2 Biologically Significant Mushroom Species

Several mushroom species are recognized for their significant biological and therapeutic properties, particularly in cancer and neurodegenerative disease treatment (Gokce et al. 2015; Li et al. 2023; Zhang et al. 2019) (Fig. 10.1). Mushrooms are a suitable ingredient to produce functional foods with enhanced health benefits because of their nutritional richness and bioactive compounds (Kaur et al. 2022). Numerous studies suggest that mushrooms may be useful in the treatment and prevention of a variety of conditions, including cancer, diabetes, and heart disease (Ambhore et al. 2024). Over the past three decades, scientific and medical research conducted in China, Korea, Japan, and, more recently, the USA has confirmed the unique properties and compounds extracted from mushrooms for the prevention and treatment of cancer and other chronic diseases (Figueiredo and Régis 2017). Among these species are Xerocomus, Albatrellus, Calvatia, Clitocybe, Cordyceps, Flammulina, Fomes, Funalia, Ganoderma, Inocybe, Inonotus, Lactarius, Phellinus, Pleurotus, Russula, Schizophyllum, Suillus, Trametes, and Agaricus (Table 10.1). They might have potent anticancer compounds and show promising anticancer

**Fig. 10.1** Different mushrooms in anticancer and antineurodegenerative potentials

properties. Wasser (2017) stated that one can even prevent cancer by eating more mushrooms. In addition to their anticancer effects, mushrooms include a variety of non-neurodegenerative compounds that are mostly known for their antioxidant capabilities and potential benefits for overall health. Neurodegenerative diseases, including Parkinson's, Alzheimer's, and Huntington's disease, may be slowed down in progression, and neurological health may be enhanced by mushrooms, particularly the medicinal and edible varieties (Cha et al. 2024).

## 10.3 Anticancer and Non-Neurodegenerative Compounds

Mushrooms contain a diverse array of bioactive compounds, including alkaloids, phenolics, steroids, terpenoids, and a variety of polysaccharides that offer numerous health benefits (Ahn et al. 2004). Among these, the most studied are alkaloids, sterols, β-glucans, polysaccharide–peptides, polysaccharide-protein complexes, lectins, and phenolic compounds (Table 10.1). Polysaccharides and polysaccharide-protein complexes are potent anticancer molecules isolated from various mushroom species. Prominent β-D-glucans known for their strong immunomodulatory and anticancer properties include Lentinan from *Lentinus edodes*,

**Table 10.1** Biologically significant mushrooms and their bioactive compounds

| Mushroom species | Bioactive compounds | Bioactivity of mushroom | References |
|---|---|---|---|
| *Grifola frondosa* | Grifolan, Maitake D fraction | Anticancer activity | Shi et al. (2007) |
| *Coprinus comatus* | Cinnamic acid, P-coumaric acid, and P-hydroxybenzoic acid | Anticancer activity | Zaidman et al. (2008) |
| *Armillaria mellea* | Armillarikin | Anticancer activity | Standish et al. (2008) |
| *Piptoporous betulinus* | Betulin, betulinic acid, polyporenic acid A, C, ergosterol peroxides | Anticancer activity | Lemieszek et al. (2009) |
| *Agaricus blazei* | Agaricus polysaccharides | Anticancer activity | Endo et al. (2010) |
| *Ganoderma lucidum* | Polysaccharide-peptide, Ganoderma polysaccharides, β-glucan | Anticancer activity and neurodegenerative diseases | Valdez-Morales et al. (2010) |
| *Fomes fomentarius* | Exopolysaccharide | Anticancer activity | Chen et al. (2011) |
| *Flammulina velutipes* | Flammulin | Anticancer activity | Song et al. (2013) |
| *Albatrellus confluens* | Grifolin | Anticancer activity | Luo et al. (2015) |
| *Agaricus bisporus* | Polysaccharide, lectin | Anticancer activity | Gąsecka et al. (2018) |
| *Antrodia camphorata* | Antroquinonol | Anticancer activity | Zhu et al. (2018) |
| *Cantharellus cibarius* | Protocatechuic acid, p-hydroxybenzoic acid, ferullic acid, gallic acid, and homogentisic acid Myrecetin | Anticancer activity | Lemieszek et al. (2019) |
| *Hericium erinaceus* | Hericenones, erinacines, ergosterol peroxide, cerevisterol | Anticancer activity and neurodegenerative disease treatment | Chong et al. (2020a) |
| *Pleurotus citrinopileatus* | Polysaccharides | Anticancer activity | Wang et al. (2021) |
| *Lignosus rhinocerotis* | Polyphenols, terpenoids, alkaloid, sesquiterpenes, polysaccharides | Neurodegenerative disease treatment | Rai et al. (2021) |
| *Lentinula edodes* | Lentinan, polysaccharides, glycoproteins, and phenolic compounds (p-hydroxybenzoic, p-coumaric, and cinnamic acids) | Anticancer activity | Spim et al. (2021) |
| *Boletus edulis* | Phenol, flavonoids, and vitamin C | Anticancer activity | Zhang et al. (2021) |
| *Amanita caesarea* | Catechin, ferulic acid, p-coumaric acid, and cinnamic acid | Neurodegenerative disease treatment (Alzheimer's disease) | Li et al. (2023) |

Ganoderan from *Ganoderma lucidum*, Schizophyllan from *Schizophyllum commune*, and Grifolan from *Grifola frondosa* (Bhanja et al. 2012; Maity et al. 2014). Furthermore, protein-bound polysaccharides, including PSK (Krestin) and other polysaccharide-protein complexes derived from *Trametes versicolor*, have exhibited notable therapeutic efficacy (Krishnakumar et al. 2023). Medicinal mushrooms such as shiitake (*Lentinula edodes*), reishi (*Ganoderma lucidum*), and turkey tail (*Trametes versicolor*) are emerging as key ingredients in next-generation functional foods. These fungi are distinguished by their high concentrations of bioactive substances, including phenolics, triterpenes, sterols, and (1 → 3), (1 → 6)-β-D-glucans, which are classified as dietary fibers and contribute to their health-promoting properties (Ye et al. 2011). Terpenoids exhibit potent anti-inflammatory and anticancer activities by inducing apoptosis in cancer cells and inhibiting metastasis. Phenolic compounds and antioxidants such as ergothioneine protect neural cells by mitigating oxidative stress and inflammation, critical factors in neurodegenerative diseases like Alzheimer's and Parkinson's.

## 10.4 Uses of Mushrooms in Cancer Therapy

According to the WHO, cancer accounted for around 10 million deaths in 2020, thus becoming one of the leading causes of mortality globally (WHO 2022). Since each kind of cancer has a different treatment plan, a precise diagnosis is essential to successful treatment. Surgery, anticancer medication treatment, radiotherapy, and systemic therapy (chemotherapy and targeted biological therapies) are some of the possible cancer treatments. The current cancer treatments, however, are not target-specific, which can result in drug resistance and several adverse effects during clinical chemotherapy (Joseph et al. 2018). As a result, developing new, potent anticancer drugs with minimal side effects is crucial. In this regard, natural substances have been emphasized as anticancer medications. Because of their anti-inflammatory, anticancer, and immune-modulatory properties, mushrooms have long been utilized in East Asian traditional medicine (Joseph et al. 2018). Approximately 700 of the 14,000 species of mushrooms have been identified as possessing therapeutic potential (Han et al. 1999).

### 10.4.1 Liver Cancer

Liver cancer represents a significant global health challenge (Friedman 2015). This high mortality-to-incidence ratio underscores its aggressive nature and the urgent need for more effective treatments. Polysaccharides significantly enhance p27Kip expression in HepG2 and Bel-7404 cells, suppressing cyclin D1/CDK4 and/or cyclin E/CDK activity. They also suppress AKT activity by inhibiting AKT phosphorylation at Thr308 and/or Ser473 and up-regulate a subunit of PI3K and phospho-PTEN, which modulate AKT activity. *Ganoderma lucidum* polysaccharides activate the mitochondria-mediated apoptosis pathway, causing cell cycle arrest and apoptosis

(OuYang et al. 2013). Sulfated polysaccharide from *Grifola frondosa* inhibits HepG2 cell proliferation in a dose-dependent manner (Wang et al. 2013). *Pleurotus pulmonarius* contains a polysaccharide-protein complex (PP) that has shown potential in liver cancer treatment (Joseph et al. 2018). Studies indicate that PP not only suppresses the PI3K/AKT signaling pathway but also enhances the effectiveness of cisplatin, a common chemotherapy drug.

### 10.4.2 Colorectal Cancer

Mushrooms are not a cure for colorectal cancer, but they may offer benefits when used with standard medical treatments. Chronic gut inflammation is linked to increased intestinal permeability, leading to microbial dysbiosis and increased antigen flow. This leads to inflammatory bowel diseases and colorectal cancer (Bugajewski et al. 2025). MSCE and lignosulfonic acid improve the intestinal barrier in human colorectal adenocarcinoma cell lines. These compounds suppress claudin-2 expression and attenuate TNF-α/IFN-γ-induced NF-κB activation (Matsuhisa et al. 2018). *Lentinula edodes* active extracts stimulate NK cell activity, improve survival and quality of life, and may enhance chemo/radiotherapy efficacy (Dan et al. 2023). *Inonotus obliquus* compounds, rich in beta-glucans, betulinic acid, and antioxidants, have immune-modulating, anti-inflammatory, and antitumor properties across various cancer models (Kim et al. 2020). *Inonotus obliquus* demonstrates anticancer properties against colorectal cancer through its ability to downregulate the β-catenin signaling pathway (Kang et al. 2015). A bioactive compound called 4-Acetylantroquinonol B, isolated from the mycelia of the Taiwanese mushroom *Antrodia camphorata*, inhibits colorectal cancer tumorigenesis and suppresses the cancer stem-like phenotype (Chang et al. 2015).

### 10.4.3 Ovarian Cancer

Though ovarian cancer is less common than other major cancers in women, such as breast or lung cancer, it is the deadliest of the gynecologic cancers and has a disproportionately high mortality rate (Jemal et al. 2011). *Ganoderma lucidum* is a promising therapeutic agent for ovarian cancer due to its multifaceted antitumor effects (Dai et al. 2014). It inhibits the growth of ovarian cancer cells by modulating several key molecular pathways. Cordycepin, a bioactive substance from *Ophiocordyceps sinensis*, has been found to have both pro-apoptotic and antiapoptotic ovarian cancer properties. It has been shown to increase the susceptibility of cells to TNF-α-induced apoptosis, suppressing NF-κB, a key pro-survival component in TNF-α signaling. Cordycepin induces the phosphorylation of eukaryotic translation initiation factor 2α, and this activation subsequently contributes to the suppression of the NF-κB pathway. This activation sensitizes cells to TNF-α-induced apoptosis. Cordycepin enhances the sensitivity of cells to apoptosis induced by TNF-α, a powerful inflammatory signaling molecule, and achieves it by activating the eIF2α pathway, which in turn influences the mTORC1 pathway and ultimately suppresses the pro-survival NF-κB pathway (Kadomatsu et al. 2012).

### 10.4.4 Breast Cancer

Improved mammography screenings and new treatments have greatly increased breast cancer survival rates over the last few decades. While it remains the most commonly diagnosed invasive cancer and a leading cause of death for women worldwide, the outlook is much better than in the past (Arora et al. 2015). Ganoderma lucidum exhibits potent anticancer effects as it specifically targets and inhibits various functions of breast cancer cells, including their ability to grow, adhere to other cells, and spread throughout the body (Jin et al. 2016). This inhibitory effect has been observed in different types of breast cancer, including both hormone receptor-positive and triple-negative cells (Bulam et al. 2019). *Pleurotus ostreatus* also exhibited anticancer properties, particularly against breast cancer (Andrej and Daniel 2008). Its extract inhibits the proliferation of human breast cancer cell lines, such as MCF-7 and MDA-MB-231. Polysaccharide-K (PSK) derived from *Trametes versicolor* has been used for decades as an approved adjunctive treatment for various cancers (Standish et al. 2008).

### 10.4.5 Lung Cancer

Cancer imposes a massive economic burden globally, and lung cancer is the single largest contributor to this cost (Minna et al. 2002). This burden includes not only the high expenses of medical care but also significant indirect costs from lost productivity and premature deaths. Active compounds present in the extract of *Agaricus lanipes* possess significant anticancer properties, as it was effective for inhibiting the rapid proliferation of A549 lung cancer cells and also induced apoptosis, the process of programmed cell death, in these same cells (Kaygusuz et al. 2017). The extracts obtained from *Calvatia gigantea* have also been shown to stop the proliferation of A549 human lung cancer cells (Eroğlu et al. 2016). Moreover, the cytotoxic constituents from *Inonotus obliquus* mushrooms induce apoptosis in human lung adenocarcinoma cells (Baek et al. 2018).

## 10.5 Uses of Mushrooms in Neurodegenerative Disease Treatment

Neurodegenerative syndromes are noncommunicable disorders that are most frequently observed in aged people. Alzheimer's disease (AD), Parkinson's disease (PD), other dementias, motor neuron diseases, Huntington's disease, prion disease, spinocerebellar ataxia, and spinal muscular atrophy belong to the category of neurodegenerative illnesses (Skovronsky et al. 2006). For the last few decades, scientists have been using mushrooms and mushroom-derived substances for the treatment and management of innumerable peripheral nervous system (PNS) and central nervous system (CNS) syndromes (Barnham et al. 2004). As per the evidence received earlier, the use of culinary medicinal mushrooms has been used as a preventive measure for age-related neurodegenerative disorders like Parkinson's and Alzheimer's diseases. The mushrooms, including *Ganoderma lucidum, Hericium erinaceus, Pleurotus giganteus, Lignosus rhinocerotis, Antrodia camphorata, Sarcodon scabrosus, Mycoleptodonoides aitchisonii, Paxillus panuoides*, and several other species, prevent the development of neurodegeneration and dementia (Lin and Beal 2006).

### 10.5.1 Alzheimer's Disease (AD)

Alzheimer's disease (AD), a leading cause of dementia, is the most common age-related neurodegenerative illness. Accumulated hyperphosphorylated tau protein is also accountable for the progression of AD (Weller and Budson 2018). Mushrooms are a good source of anti-inflammatory and antioxidative substances. These are classified into three groups: nonedible, edible, and medicinal mushrooms (Chaturvedi et al. 2018). The investigators evaluate the effect of erinacine A-augmented *Hericium erinaceus* (EAHE) mycelia of mushrooms in male and female rats, which shows absolutely no toxic effect for extended use. An edible medicinal mushroom, *Hericium erinaceus* (HE), displays an immunomodulatory effect by regulating the microbiota of the gut. The risk of progression of AD may be decreased by introducing oyster mushrooms into the Mediterranean diet (Sabaratnam et al. 2013; Diling et al. 2017). The polysaccharide extract of edible mushrooms *Coprinellus truncorum* and *Coprinus comatus* inhibits the AChE in the AD disorder (Šimić et al. 2016; Ganeshpurkar et al. 2010; Cör et al. 2018; Agrawal and Dhansekaran 2019).

### 10.5.2 Parkinson's Disease

Parkinson's disease (PD), a persistent neurodegenerative disease, is a disorder related to difficulty in movement categorized by motor and nonmotor structures. Apart from the major motor indicators of the disorder, such as postural unsteadiness, resting tremor, and sluggishness of movements, other nonmotor symptoms like cognitive impairment, mood disorder, anxiety, and depression appear much before the motor indications (Valverde et al. 2015; DeMaagd and Philip 2015; Mazzoni et al. 2012). Medicinal and edible mushrooms have been applied in recent years for the treatment and well-being of persons suffering from PD. Neuroinflammation and oxidative stress are the primary agents accountable for the initiation and progression of PD. Different bioactive components of edible mushrooms help to alleviate the advanced degeneration of dopaminergic neurons by exerting anti-inflammatory and antioxidant effects. Extracts of *Ganoderma lucidum* (GLE) have been reported to attenuate microglial activation, the key contributor to the progression of neuroinflammation in Parkinson's disease. GLE can also be utilized as an anti-inflammatory medication for the treatment of PD (Ciulla et al. 2019). Niacin-rich mushrooms also display pretty good beneficial efficacy in the treatment of PD. *Agaricus blazei* extract (ABE) displays neuroprotective activity in the rotenone-induced PD. Rotenone persuaded motor and ABE suggestively enhanced nonmotor irregularities. A nourishing enhancement of *Agaricus blazei* might be employed for PD treatment (Alexander 2004; Goldman and Postuma 2014).

### 10.5.3 Motor Neuron Diseases and Multiple Sclerosis

Reports of several investigations reveal that many mushroom species positively influence different motor neuron diseases and multiple sclerosis (MS). This disease is a central nervous system-based syndrome, where the myelin sheath-covered neurons are confronted by the immune system. It is an advanced disorder that commences because of the annihilation of axons' fatty coatings and sporadic electric whims. The disease shows symptoms like fatigue, paralysis, and muscle spasms. Medicinal mushrooms exhibited wonderful results in numerous other neurodegenerative diseases and were investigated in the MS context (Warner 2009). *H. erinaceus* is the leading class that regulates the development of this pathology's effects by preserving the nerve growth factor. In Japan, the *H. erinaceus* is considered a "nerve of steel," associated with properties that arouse renaissance and nerve cell growth. Some studies on other animal models have shown that *H. erinaceus* can arouse myelin sheath production and encourage growth of nerves with the increased stride. It has been confirmed that it can accomplish and recuperate the MS degeneration stage (Li et al. 2014). Piwep is the mushroom excerpt from *Phellinus igniarius*, which upgrades the disease conditions in the animal model of multiple sclerosis (MS). MS is also a unique example of NDs, which are enhanced by the dietary mushrooms and their biologically potent components (Vasile et al. 2017).

### 10.5.4 Huntington's Disease (HD)

Huntington's disease (HD) is a complex, neurodegenerative disorder originating from substantial neuron bereavement in the brain. It is a genetic disorder resultant of the over-reiteration of the CAG categorization in the gene that ciphers for the huntingtin protein (Phan et al. 2015). To date, there are no clinically effective therapies for slowing down the progression of this disorder, making it a thrust area for therapeutic research. Several studies propose the usage of psilocybin in the treatment of HD due to its neurotrophic, neuroprotective, and neuroplastic effects, ensuing in a diminished rate of neuron loss and amplified synaptic density. HD is associated with emotional disturbances, motor dysfunction, memory loss, etc. A number of medicinal mushrooms are utilized to resist or diminish long-term functional dysfunctions correlated to a diminution in the impression of inflammatory evolution and oxidative tension that conveys the neuroprotective effect. According to several studies, the species *Hericium erinaceus* (HE) and *Ganoderma lucidum* (GL) consist of bioactive components that act to normalize the effects of HD. The most common cause of neurodegenerative dysfunction is oxidative stress, measured by harmonizing methods that embrace medicinal plants and mushrooms (Choi 2018). Hericenones in the maturing body and erinacines extracted from the mushroom's mycelium have neuroprotective effects accountable for preserving neuronal veracity. Other bioactive components, such as cordysinin extracted from species of the genus Cordyceps, are multi-targeted components but are restrained from having the same advantageous effect. Classes of GL and HE are most recurrently quantified to comprise bioactive components with a role in normalizing HD effects. Typically, for GL, neuroprotective activity (oxidative stress) is meticulously connected to the immune response's inflection, mitochondrial protection, which diminishes with age, and antioxidant protection, enhanced by dysfunctions of enzymes with antioxidant effects (Bell et al. 2019; Wang et al. 2017).

### 10.5.5 Spinocerebellar Ataxia

Cerebellar ataxia is an advanced motor syndrome categorized by dysmetria, dyssynergia, gait instability, and balance impairment. The etiology of this disorder can be infrequent or genetic, with the common cases showing neurodegeneration in the cerebellum, leading to ataxia. The ataxic motor dysfunctions are accompanied by adjustments in cerebellar circuit connectivity and Purkinje cell functions (Klockgether and Paulson 2011; Marsden and Harris 2011). Due to its composite pathophysiology, presently, there are no operative actions for cerebellar ataxia. Existing physical therapy or training offers only limited therapeutic effects against motor and balance deterioration. Therefore, effective initiatives are required to improve or alleviate the ataxic symptoms. *Hericium erinaceus* (HE), famously known as the lion's mane mushroom, is a medicinal mushroom that has been utilized in Chinese folk medicine since ancient times. It is famous for its diverse advantageous effects on the brain, including neurotrophic, antiaging, and antioxidative activities. Furthermore, HE has also been recommended to have neuroprotective activity against cerebellar ataxia (Jaques et al. 2021; Chong et al. 2020a, b). A recent study displayed that HE had neuroprotective effects in a rat model of neurotoxin 3-acetylpyridine (3-AP)-encouraged cerebellar ataxia (Ormerod et al. 1994; Sarva and Shanker 2014).

## 10.6 Different Mechanistic Pathways to Prevent Disease

Owing to the versatile usefulness of the active ingredients, mushrooms are widely appreciated for their generous role in the treatment of cancer and neurodegenerative diseases.

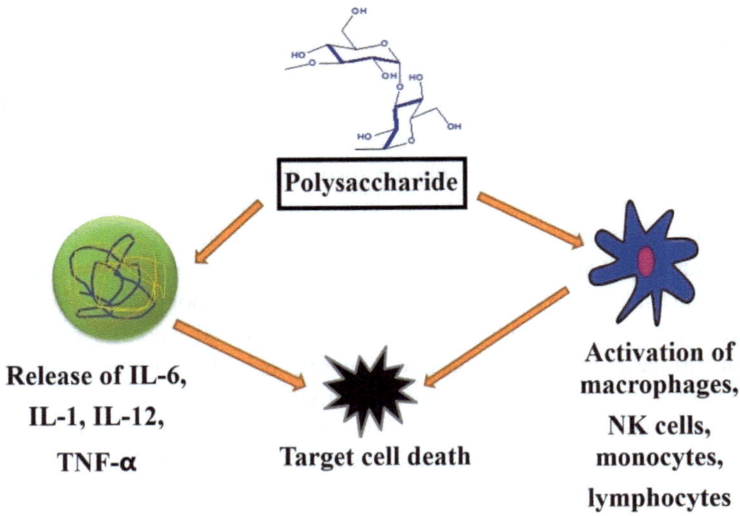


**Fig. 10.2** Schematic diagram showing the antitumor activity of mushroom polysaccharides

Mushroom-derived terpenoids and different types of polysaccharides, including potentially active glucans and other heteroglycans, proteins, polysaccharide-protein complexes, lectins, and phenolic compounds, are the active ingredients responsible for showing considerable antitumor and anticancer activities in numerous ways. Since the last few decades or so, medicinal mushrooms have been used for treating cancer in China and Japan either alone or combined with radiation or chemotherapy (Ayeka 2018). The active ingredients in mushrooms stimulate the host's immune system, such as monocytes, NK cells, T cells, B cells, and cytotoxicity, to make them able to easily recognize and fight cancer cells. Effectively modify the cell signal pathways, resulting in potentially inhibiting the cancer cell proliferations through cell cycle arrest & controlled growth of cancer cells, improved apoptotic pathway, etc. (Fig. 10.2). The active ingredients in mushrooms, either alone or used in combination with conventional chemotherapeutic drugs, have also been shown to improve the effectiveness of the drugs in cancer therapy.

Mushrooms are now being explored as potential therapeutic agents for the treatment of neurodegenerative diseases. Several active biomolecules like erinacines, hericenones, and polysaccharides, particularly β-glucans, along with several biomolecules like polyphenols, terpenoids, and alkaloids, isolated from different mushrooms, have been found to show promising health benefits in this regard (Demain and Vaishnav 2011). Neurodegenerative diseases are the progressive degeneration or loss of neurons, leading to a waning in the functions of the entire nervous system. These diseases are generally incurable, and the treatment is simply focused on managing the symptoms for the improvement of quality of life. Different therapies, supportive care, and modification of lifestyle are the general approaches to managing the disease. In certain cases, surgery can lead to relief from movement-related symptoms. Extensive research is going on to find any remedy that can stop the progression of the disease. In the recent past, mushrooms have been emphasized as a very rich and largely unexplored source of biomolecules for the treatment of neurodegenerative disease, both in vitro and in vivo models (Abitbol et al. 2022).

Active ingredients present in the mushrooms are found to show neuroprotective effects that reduce neuroinflammation and improve cognitive function, neurogenesis, and neuroprotection through antioxidant and anti-inflammatory effects, modulation of neuroinflammation, anticholinesterase activity, and regulation of neurotrophin synthesis (Bhambri et al. 2022; Roszczenko et al. 2024). Mushrooms exhibit neuroprotective effects through several mechanisms that include but are not limited to stimulating neurogenesis, reducing oxidative damage, modulating inflammatory pathways, and promoting nerve growth factor synthesis, and regulating inflammatory processes. *Hericium erinaceus*, a medicinal mushroom, has been reported to stimulate nerve growth factor synthesis, stimulating the development and endurance of neurons. These also improve cognitive function and potentially enhance memory and concentration (Bhambri et al. 2022; Roszczenko et al. 2024). Owing to its neuroprotective benefits, *Hericium erinaceus*, by stimulating nerve growth factor, can also be used for treating neurodegenerative diseases like Alzheimer's and Parkinson's (Skovronsky et al. 2006). Studies show that natural products from mushrooms like *H. erinaceus*, *C. africanus*, *C. militaris*, *I. obliquus*, and many *Ganoderma* species have the potential to alleviate neuroinflammation-mediated cell death, increase abnormal protein clearance, and promote anti-neuroinflammatory effects in both animal and cell models without displaying cytotoxicity (Bai et al. 2019; Yang et al. 2011, 2015). Although mushrooms are considered a promising source for treating neurodegenerative diseases, more investigation is required to completely understand their mode of action so that they can be potentially and effectively used for the treatment of neurodegenerative diseases. Clinical trials using mushroom extracts or specific biomolecules alone or in conjunction with other therapies are therefore needed.

## 10.7 Conclusion and Future Scope

Mushrooms have emerged as a valuable source of bioactive compounds with significant potential in the treatment of cancer and neurodegenerative diseases. Their unique constituents, including β-glucans, terpenoids, polysaccharide-protein complexes, and antioxidants, contribute to their immunomodulatory, anticancer, and neuroprotective effects. By enhancing immune responses, inhibiting tumor progression, and protecting neural cells from oxidative stress and inflammation, mushrooms offer promising complementary approaches to conventional therapies. However, to fully realize their therapeutic potential, more extensive clinical trials and mechanistic research are needed to confirm efficacy, safety, and optimal dosing. Innovations in biotechnology and extraction methods hold promise for producing more potent and standardized mushroom-based pharmaceuticals and functional foods. Incorporating medicinal mushrooms into mainstream medical practice could offer safer, natural, and cost-effective alternatives, ultimately improving outcomes for patients with cancer and neurodegenerative disorders.

**Acknowledgment** The authors are thankful to Prof. Syed Sirajul Islam, former professor, Department of Chemistry and Chemical Technology, Vidyasagar University, West Bengal, India.

**Declaration of AI Usage** AI was not used for the research described in the article.

## References

Abitbol A, Mallard B, Tiralongo E, Tiralongo J (2022) Mushroom natural products in neurodegenerative disease drug discovery. Cells 11(23):3938

Agrawal DC, Dhansekaran M (2019) Medicinal mushrooms. Recent progress in research and development. Springer Nature Singapore Pte Ltd., Springer, Singapore

Ahn WS, Kim DJ, Chae GT, Lee JM, Bae SM, Sin JI, Kim YW, Namkoong SE, Lee IP (2004) Natural killer cell activity and quality of life were improved by consumption of a mushroom extract, Agaricus blazei Murill Kyowa, in gynecological cancer patients undergoing chemotherapy. Int J Gynecol Cancer 14:589–594

Alexander GE (2004) Biology of Parkinson's disease: pathogenesis and pathophysiology of a multisystem neurodegenerative disorder. Dialogues Clin Neurosci 6(3):259–280

Ambhore JP, Adhao VS, Rafique SS, Telgote AA, Dhoran RS, Shende BA (2024) A concise review: edible mushroom and their medicinal significance. Explor Foods Foodomics 2(3): 183–194

Andrej J, Daniel S (2008) Pleurotus ostreatus inhibits proliferation of human breast and colon cancer cells through p53-dependent as well as p53-independent pathway. Int J Oncol 33:1307–1313

Arora R, Schmitt D, Karanam B, Tan M, Yates C, Dean-Colomb W (2015) Inhibition of the Warburg effect with a natural compound reveals a novel measurement for determining the metastatic potential of breast cancers. Oncotarget 6:662–678

Arunachalam K, Sasidharan SP, Yang X (2022) A concise review of mushrooms antiviral and immunomodulatory properties that may combat against COVID-19. Food Chem Adv 1:100023

Ayeka PA (2018) Potential of mushroom compounds as immunomodulators in cancer immunotherapy: a review. eCAM 2018(1):7271509

Baek J, Roh HS, Baek KH, Lee S, Lee S, Song SS, Kim KH (2018) Bioactivity-based analysis and chemical characterization of cytotoxic constituents from Chaga mushroom (Inonotus obliquus) that induce apoptosis in human lung adenocarcinoma cells. J Ethnopharmacol 224:63–75

Bai Y, Chen L, Chen Y, Chen X, Dong Y et al (2019) A Maitake (Grifola frondosa) polysaccharide ameliorates Alzheimer's disease-like pathology and cognitive impairments by enhancing microglial amyloid-β clearance. RSC Adv 9:37127–37135

Barnham KJ, Masters CL, Bush AI (2004) Neurodegenerative diseases and oxidative stress. Nat Rev Drug Discov 3:205–214

Bell V, Ferr˜ao J, Chaquisse E, Manuel B, Fernandes T (2019) Role of mushrooms in autism. Austin J Nutr Food Sci 7:1128

Bhambri A, Srivastava M, Mahale VG, Mahale S, Karn SK (2022) Mushrooms as potential sources of active metabolites and medicines. Front Microbiol 13:837266

Bhanja SK, Nandan CK, Mandal S, Bhunia B, Maiti TK, Mondal S, Islam SS (2012) Isolation and characterization of the immunostimulating β-glucans of an edible mushroom *Termitomyces robustus* var. Carbohydr Res 357:83–89

Bohn JA, BeMiller JN (1995) (1→ 3)-β-D-Glucans as biological response modifiers: a review of structure-functional activity relationships. Carbohydr Polym 28:3–14

Bugajewski M, Angerhoefer N, Pączek L, Kaleta B (2025) Lentinula edodes as a source of bioactive compounds with therapeutical potential in intestinal inflammation and colorectal cancer. Int J Mol Sci 26(7):3320

Bulam S, Üstün NŞ, Pekşen A (2019) Health benefits of Ganoderma lucidum as a medicinal mushroom. Turk J Agric Food Sci Technol 7:84–93

Cha S, Bell L, Shukitt-Hale B, Williams CM (2024) A review of the effects of mushrooms on mood and neurocognitive health across the lifespan. Neurosci Biobehav Rev 158:105548

Chang TC, Yeh CT, Adebayo BO, Lin YC, Deng L et al (2015) 4-Acetylantroquinonol B inhibits colorectal cancer tumorigenesis and suppresses cancer stem-like phenotype. Toxicol Appl Pharmacol 288:258–268

Chaturvedi VK, Agarwal S, Gupta KK, Ramteke PW, Singh MP (2018) Medicinal mushroom: boon for therapeutic applications. Biotech 8(8):334

Chen W, Zhao Z, Li Y (2011) Simultaneous increase of mycelial biomass and intracellular polysaccharide from Fomes fomentarius and its biological function of gastric cancer intervention. Carbohydr Polym 85(2):369–375

Choi JH (2018) Biologically functional molecules from mushroom-forming fungi. Biosci Biotechnol Biochem 82:372–382

Chong PS, Khairuddin S, Tse ACK, Hiew LF, Lau CL, Tipoe GL, Fung M-L, Wong KH, Lim LW (2020a) Hericium erinaceus potentially rescues behavioural motor deficits through ERK-CREB-PSD95 neuroprotective mechanisms in rat model of 3- acetylpyridine-induced cerebellar ataxia. Sci Rep 10:14945

Chong PS, Fung ML, Wong KH, Lim LW (2020b) Therapeutic potential of Hericium erinaceus for depressive disorder. Int J Mol Sci 21(1):163

Ciulla M, Marinelli L, Cacciatore I, Stefano AD (2019) Role of dietary supplements in the management of Parkinson's disease. Biomolecules 9(7):271

Cör D, Knez Ž, Knez Hrnčič M (2018) Antitumour, antimicrobial, antioxidant and antiacetylcholinesterase effect of Ganoderma lucidum terpenoids and polysaccharides: a review. Molecules 23(3):649/1–649/21

Cui J, Chisti Y (2003) Polysaccharopeptides of Coriolus versicolor: physiological activity, uses, and production. Biotechnol Adv 21:109–122

Dai S et al (2014) Ganoderma lucidum inhibits proliferation of human ovarian cancer cells by suppressing VEGF expression and up-regulating the expression of connexin 43. BMC Complement Altern Med 14:434

Dan A, Swain R, Jacobs RJ, Belonce SE (2023) Therapeutic effects of medicinal mushrooms on gastric, breast, and colorectal cancer: a scoping review. Cureus 15:e37574

DeMaagd G, Philip A (2015) Parkinson's disease and its management: part 1: disease entity, risk factors, pathophysiology, clinical presentation, and diagnosis. P T 40(8):504–532

Demain AL, Vaishnav P (2011) Natural products for cancer chemotherapy. Microb Biotechnol 4(6):687–699

Diling C, Chaoqun Z, Jian Y, Jian L, Jiyan S, Yizhen X, Guoxiao L (2017) Immunomodulatory activities of a fungal protein extracted from *Hericium erinaceus* through regulating the gut microbiota. Front Immunol 8:666

Endo M, Beppu H, Akiyama H, Wakamatsu K, Ito S, Kawamoto et al (2010) Agaritine purified from Agaricus blazei Murrill exerts anti-tumor activity against leukemic cells. BBA-Gen Subj 1800(7):669–673

Eroğlu C, Seçme M, Atmaca P, Kaygusuz O, Gezer K, Bağcı G, Dodurga Y (2016) Extract of Calvatia gigantea inhibits proliferation of A549 human lung cancer cells. Cytotechnology 68: 2075–2081

Figueiredo L, Régis WCB (2017) Medicinal mushrooms in adjuvant cancer therapies: an approach to anticancer effects and presumed mechanisms of action. Nutrire 42(1):28

Friedman M (2015) Chemistry, nutrition, and health-promoting properties of Hericium erinaceus (Lion's Mane) mushroom fruiting bodies and mycelia and their bioactive compounds. J Agric Food Chem 64:7108–7123

Ganeshpurkar A, Rai G, Jain AP (2010) Medicinal mushrooms: towards a new horizon. Pharmacogn Rev 4(8):127–135

Gąsecka M, Magdziak Z, Siwulski M, Mleczek M (2018) Profile of phenolic and organic acids, antioxidant properties and ergosterol content in cultivated and wild growing species of Agaricus. Eur Food Res Technol 244:259–268

Gokce EC, Kahveci R, Atanu OM, Gurer B et al (2015) Neuroprotective effects of Ganoderma lucidum polysaccharides against traumatic spinal cord injury in rats. Int J Care Inj 46:2146–2155

Goldman JG, Postuma R (2014) Premotor and nonmotor features of Parkinson's disease. Curr Opin Neurol 27(4):434–441

Han SB, Lee CW, Jeon YJ, Hong ND, Yoo ID, Yang KH, Kim HM (1999) The inhibitory effect of polysaccharides isolated from Phellinus linteus on tumor growth and metastasis. Immunopharmacology 41:157–164

Jaques CS, Escorcio-Bezerra ML, Pedroso JL, Barsottini OGP (2021) The intersection between cerebellar ataxia and neuropathy: a proposed classification and a diagnostic approach. Cerebellum 21:497–513

Jemal A, Bray F, Center MM, Ferlay J, Ward E, (2011) Forman D: Global cancer statistics. CA Cancer J Clin 61(2): 69–90

Jin X, Ruiz BJ, Sze DM, Chan GC (2016) Ganoderma lucidum (Reishi mushroom) for cancer treatment. Cochrane Database Syst Rev 4(4):007731

Joseph TP, Chanda W, Padhiar AA, Batool S, LiQun S, Zhong M, Huang MA (2018) Preclinical evaluation of the antitumor activities of edible and medicinal mushrooms: a molecular insight. Int Cancer Ther 17:200–209

Kadomatsu M, Nakajima S, Kato H, Gu L, Chi Y, Yao J, Kitamura M (2012) Cordycepin as a sensitizer to tumour necrosis factor (TNF)-α-induced apoptosis through eukaryotic translation initiation factor 2α (eIF2α)- and mammalian target of rapamycin complex 1 (mTORC1)-mediated inhibition of nuclear factor (NF)-κB. Clin Exp Immunol 168:325–332

Kang JH, Jang JE, Mishra SK, Lee HJ, Nho CW et al (2015) Ergosterol peroxide from Chaga mushroom (Inonotus obliquus) exhibits anti-cancer activity by down-regulation of the β-catenin pathway in colorectal cancer. J Ethnopharmacol 173:303–312

Kaur K, Sharma R, Srivastava I, Kaur S, Mehrotra R (2022) Edible mushrooms: nature's superfood for health and wellbeing. Int J Innov Multidiscip Res 1(1):40–53

Kaygusuz O, Kaygusuz M, Dodurga Y, Seçme M, Herken EN, Gezer K (2017) Assessment of the antimicrobial, antioxidant and cytotoxic activities of the wild edible mushroom Agaricus lanipes (F.H. Møller & Jul. Schäff.) Hlaváček. Cytotechnology 69:135–144

Kim TI, Choi JG, Kim JH, Li W, Chung HS (2020) Blocking effect of Chaga mushroom (Inonotus oliquus) extract for immune checkpoint CTLA-4/CD80 interaction. Appl Sci 10:5774

Klockgether T, Paulson H (2011) Milestones in ataxia. Mov Disord 26:1134–1141

Krishnakumar NM, Ramesh BT, Ceasar SA (2023) Medicinal mushrooms as potential sources of anticancer polysaccharides and polysaccharide–protein complexes. Stud Nat Prod Chem 76: 113–148

Lemieszek MK, Langner E, Kaczor J, Kandefer-Szerszen M, Sanecka B, Mazurkiewicz W, Rzeski W (2009) Anticancer effect of fraction isolated from medicinal Birch polypore mushroom, Piptoporus betulinus (Bull.: Fr.) P. Karst. (Aphyllophoromycetideae): in vitro studies. Int J Med Mushrooms 11(4):351–364

Lemieszek MK, Nunes FM, Rzeski W (2019) Branched mannans from the mushroom Cantharellus cibarius enhance the anticancer activity of natural killer cells against human cancers of lung and colon. Food Funct 10(9):5816–5826

Li L, Wu G, Choi BY, Jang BG, Kim JH, Sung GH, Cho JY, Suh SW, Park HJ (2014) Cardiac arrest and cardiopulmonary resuscitation: starting from basic science and bioengineering research to improve resuscitation outcome. BioMed Res Int:1–2

Li N, Li H, Liu Z, Feng G, Shi C, Wu Y (2023) Unveiling the therapeutic potentials of mushroom bioactive compounds in Alzheimer's disease. Foods 12(15):2972

Lin MT, Beal MF (2006) Mitochondrial dysfunction and oxidative stress in neurodegenerative diseases. Nature 443:787–795

Luo X, Yang L, Xiao L, Xia X, Dong X, Zhong J et al (2015) Grifolin directly targets ERK1/2 to epigenetically suppress cancer cell metastasis. Oncotarget 6(40):42704

Maity P, Nandi AK, Sen IK, Pattanayak M, Chattopadhyay S, Dash SK et al (2014) Heteroglycan of an edible mushroom Entoloma lividoalbum: structural characterization and study of its protective role for human lymphocytes. Carbohydr Polym 114:157–165

Marsden J, Harris C (2011) Cerebellar ataxia: pathophysiology and rehabilitation. Clin Rehabil 25: 195–216

Matsuhisa K, Watari A, Iwamoto K, Kondoh M, Yagi K (2018) Lignosulfonic acid attenuates NF-κB activation and intestinal epithelial barrier dysfunction induced by TNF-α/IFN-γ in Caco-2 cells. J Nat Med 72:448–455

Mazzoni P, Shabbott B, Cortés JC (2012) Motor control abnormalities in Parkinson's disease. Cold Spring Harb Perspect Med 2(6):a009282

Minna JD, Roth AJ, Gazdar AF (2002) Focus on lung cancer. Cancer Cell 1:49–52

Moradali MF, Mostafavi H, Ghods S, Hedjaroude GA (2007) Immunomodulating and anticancer agents in the realm of macromycetes fungi (macrofungi). Int Immunopharmacol 7:701–724

Nandi AK, Samanta S, Sen IK, Devi KSP, Maiti TK, Acharya K, Islam SS (2013) Structural elucidation of an immunoenhancing heteroglycan isolated from Russula albonigra (Krombh.) Fr. Carbohydr Polym 94:918–926

Ormerod IE, Harding AE, Miller DH, Johnson G, MacManus D, Du Boulay EP, Kendall BE, Moseley IF, McDonald WI (1994) Magnetic resonance imaging in degenerative ataxic disorders. J Neurol Neurosurg Psychiatry 57:51–57

OuYang F, Wang G, Guo W, Zhang Y, Xiang W, Zhao M (2013) AKT signalling and mitochondrial pathways are involved in mushroom polysaccharide-induced apoptosis and G1 or S phase arrest in human hepatoma cells. Food Chem 138:2130–2139

Phan CW, David P, Naidu M, Wong KH, Sabaratnam V (2015) Therapeutic potential of culinary-medicinal mushrooms for the management of neurodegenerative diseases: diversity, metabolite, and mechanism. Crit Rev Biotechnol 35(3):355–368

Rai SN, Mishra D, Singh P, Vamanu E, Singh MP (2021) Therapeutic applications of mushrooms and their biomolecules along with a glimpse of in silico approach in neurodegenerative diseases. Biomed Pharmacother 137:111377

Roszczenko P, Szewczyk-Roszczenko OK, Gornowicz A, Iwańska IA, Bielawski K, Wujec M, Bielawska A (2024) The anticancer potential of edible mushrooms: a review of selected species from Roztocze, Poland. Nutrients 16(17):2849

Sabaratnam V, Kah-Hui W, Naidu M, Rosie David P (2013) Neuronal health – can culinary and medicinal mushrooms help? J Tradit Complement Med 3(1):62–68

Sarva H, Shanker VL (2014) Treatment options in degenerative cerebellar ataxia: a systematic review. Mov Disord Clin Pract 1:291–298

Shi BJ, Nie XH, Chen LZ, Liu YL, Tao WY (2007) Anticancer activities of a chemically sulfated polysaccharide obtained from Grifola frondosa and its combination with 5-fluorouracil against human gastric carcinoma cells. Carbohydr Polym 68(4):687–669

Shi M, Zhang Z, Yang Y (2013) Antioxidant and immunoregulatory activity of Ganoderma lucidum polysaccharide. Carbohydr Polym 95:200–206

Šimić G, Babić Leko M, Wray S, Harrington C, Delalle I, Jovanov-Milošević N, Bažadona D, Buée L, de Silva R, Di Giovanni G, Wischik C, Hof PR (2016) Tau protein hyperphosphorylation and aggregation in Alzheimer's disease and other tauopathies, and possible neuroprotective strategies. Biomolecules 6(1):6

Skovronsky DM, Lee VMY, Trojanowski JQ (2006) Neurodegenerative diseases: new concepts of pathogenesis and their therapeutic implications. Annu Rev Pathol Mech Dis 1:151–170

Sohretoglu D, Huang SL (2018) Ganoderma lucidum polysaccharides as an anti-cancer agent. Anti-Cancer Agents Med Chem 18(5):667–674

Song FQ, Liu Y, Kong XS, Chang W, Song G (2013) Progress on understanding the anticancer mechanisms of medicinal mushroom: Inonotus obliquus. Asian Pac J Cancer Prev 14(3): 1571–1578

Spim SRV, Castanho NRCM, Pistila AMH, Jozala AF, Oliveira Júnior JM, Grotto D (2021) Lentinula edodes mushroom as an ingredient to enhance the nutritional and functional properties of cereal bars. J Food Sci Technol 58:1349–1357

Standish LJ, Wenner CA, Sweet ES, Bridge C, Nelson A, Martzen M, Novack J, Torkelson C (2008) Trametes versicolor mushroom immune therapy in breast cancer. J Soc Integr Oncol 6(3):122–128

Taguchi T, Furue H, Kimura T, Kondo T, Hattori T, Ogawa N (1983) Clinical efficacy of lentinan on neoplastic diseases. Adv Exp Med Biol 166:181–187

Valdez-Morales M, Barry K, Fahey GC Jr, Domínguez J, de Mejia EG, Valverde ME, Paredes-López O (2010) Effect of maize genotype, developmental stage, and cooking process on the nutraceutical potential of huitlacoche (Ustilago maydis). Food Chem 119(2):689–697

Valverde ME, Hernández-Pérez T, Paredes-López O (2015) Edible mushrooms: improving human health and promoting quality life. Int J Microbiol:1–14

Vasile D, Dincă L, Enescu CM (2017) Impact of collecting mushrooms from the spontaneous flora on forest ecosystems in Romania. Agro Life Sci J 6:268–275

Wang CL, Meng M, Liu SB, Wang LR, Hou LH, Cao XH (2013) A chemically sulfated polysaccharide from Grifola frondos induces HepG2 cell apoptosis by notch1–NF-κB pathway. Carbohydr Polym 95:282–287

Wang J, Cao B, Zhao H, Feng J (2017) Emerging roles of Ganoderma lucidum in anti-aging. Aging Dis 8:691–707

Wang Q, Niu LL, Liu HP, Wu YR, Li MY, Jia Q (2021) Structural characterization of a novel polysaccharide from Pleurotus citrinopileatus and its antitumor activity on H22 tumor-bearing mice. Int J Biol Macromol 168:251–260

Warner TT (2009) Chapter 10 – Motor neuron diseases, in: T.T. Warner, Simon R. Hammans (Eds.), Practical guide to neurogenetics WB Saunders, pp. 150–174

Wasser SP (2002) Medicinal mushrooms as a source of antitumor and immunomodulating polysaccharides. Appl Microbiol Biotechnol 60:258–274

Wasser SP (2017) Medicinal mushrooms in human clinical studies. Part I. Anticancer, oncoimmunological, and immunomodulatory activities: a review. Int J Med Mushrooms 19(4):279–317

Wasser SP, Weis AL (1999) Medicinal properties of substances occurring in higher Basidiomycetes mushrooms: current perspectives (Review)*. Int J Med Mushrooms 1:31–62

Weller J, Budson A (2018) Current understanding of Alzheimer's disease diagnosis and treatment. F1000Res 7:1161

World Health Organization Cancer (2022) Available online: https://www.who.int/news-room/fact-sheets/detail/cancer

Yang W, Fang Y, Liang J, Hu Q (2011) Optimization of ultrasonic extraction of Flammulina velutipes polysaccharides and evaluation of its acetylcholinesterase inhibitory activity. Food Res Int 44:1269–1275

Yang W, Yu J, Zhao L, Ma N, Fang Y, Pei F, Mariga AM, Hu Q (2015) Polysaccharides from Flammulina velutipes improve scopolamine-induced impairment of learning and memory of rats. J Funct Foods 18:411–422

Yao P, Liang S, Liu Z, Xu C (2024) A review of natural products targeting tumor immune microenvironments for the treatment of lung cancer. Front Immunol 15:1343316

Ye L, Zheng X, Zhang J, Tang Q, Yang Y, Wang et al (2011) Biochemical characterization of a proteoglycan complex from an edible mushroom Ganoderma lucidum fruiting bodies and its immunoregulatory activity. Food Res Int 44(1):367–372

Zaidman BZ, Wasser SP, Nevo E, Mahajna J (2008) Coprinus comatus and Ganoderma lucidum interfere with androgen receptor function in LNCaP prostate cancer cells. Mol Biol Rep 35:107–117

Zhang M, Zhang Y, Zhang L, Tian Q (2019) Mushroom polysaccharide lentinan for treating different types of cancers: a review of 12 years clinical studies in China. Prog Mol Biol Transl Sci 163:297–328

Zhang Y, Zhou R, Liu F, Ng TB (2021) Purification and characterization of a novel protein with activity against non-small-cell lung cancer in vitro and in vivo from the edible mushroom Boletus edulis. Int J Biol Macromol 174:77–88

Zhu PL, Fu XQ, Li JK, Tse AKW, Guo H, Yin CL et al (2018) Antrodia camphorata mycelia exert anti-liver cancer effects and inhibit STAT3 signaling in vitro and in vivo. Front Pharmacol 9:1449

# Comprehensive Characterization of *Hericium erinaceus*: Linking Sensory Evaluation, Nutritional Value, and Bioactive Potential

11

Mafalda Silva, Manuela Lageiro, Luisa Cristina Roseiro, Ana Cristina Ramos, Fernando C. Lidon, Fernando H. Reboredo, and Elsa M. Gonçalves

**Abstract**

Mushrooms are increasingly recognized not only as nutritious foods but also as sources of bioactive compounds with protective effects against diabetes, cardiovascular diseases, cancer, and neurodegenerative disorders. Their high fiber content, valuable micronutrients, and diverse bioactive molecules, combined with their culinary versatility, have fueled a global rise in consumption. This trend is further reinforced by the shift toward plant-based diets, where mushrooms contribute high-quality protein, umami flavor, and a fibrous texture, making them appealing alternatives to meat. Among edible mushrooms, *Hericium erinaceus* (Lion's Mane) is distinguished by its unique morphology, delicate flavor, and remarkable medicinal potential. Rich in erinacines, hericenones, polysaccharides, and ergothioneine, it represents a promising candidate for both gastronomy and nutraceutical applications. This paper integrates a

M. Silva
Faculdade de Ciências e Tecnologia, Universidade Nova de Lisboa, Campus da Caparica, Caparica, Portugal

M. Lageiro · L. C. Roseiro · A. C. Ramos · E. M. Gonçalves (✉)
UEIS-TSA - Unidade Estratégica de Investigação e Serviços de Tecnologia e Segurança Alimentar, Instituto Nacional de Investigação Agrária e Veterinária, Oeiras, Portugal

GeoBiotec—GeoBioTec Research Institute, Faculdade de Ciências e Tecnologia, Universidade Nova de Lisboa, Campus da Caparica, Caparica, Portugal
e-mail: elsa.goncalves@iniav.pt

F. C. Lidon · F. H. Reboredo
Faculdade de Ciências e Tecnologia, Universidade Nova de Lisboa, Campus da Caparica, Caparica, Portugal

GeoBiotec—GeoBioTec Research Institute, Faculdade de Ciências e Tecnologia, Universidade Nova de Lisboa, Campus da Caparica, Caparica, Portugal



B. Giri et al. (eds.), *Advances in Mushroom Research*,
https://doi.org/10.1007/978-981-95-7692-0_11

critical review of current knowledge with new experimental findings. Original data are presented on proximate composition, phenolic compounds, organic acids, amino acid profiles, and polysaccharides. In addition, a descriptive sensory lexicon was developed to address the lack of standardized evaluation for this species. By combining literature analysis with novel laboratory insights, this work underscores the potential of *H. erinaceus* as a functional food ingredient, particularly in the context of sustainable plant-based diets and health-promoting strategies. Future perspectives are outlined regarding clinical validation, technological innovation, and the broader role of mushrooms in advancing nutrition and food system sustainability.

**Keywords**

Edible mushrooms · Plant-based diets · *Hericium erinaceus* · Sensory analysis · Bioactive compounds

## 11.1 Introduction

Mushrooms have long been valued worldwide not only as food but also as a natural medicine. They are nutrient-dense, providing fiber, high-quality proteins, essential vitamins, and bioavailable minerals, while being naturally low in fat and calories. Beyond their nutritional value, mushrooms are important sources of bioactive compounds with reported antioxidant, anti-inflammatory, immunomodulatory, and anticancer activities. Numerous studies have highlighted their beneficial effects against chronic conditions such as diabetes, cardiovascular diseases, cancer, and neurodegenerative disorders (Ba et al. 2021; Szućko-Kociuba et al. 2023).

Global mushroom consumption has risen nearly fivefold in recent decades and continues to expand. This increase is driven by consumer interest in health-promoting foods, the search for sustainable protein alternatives, and the integration of mushrooms into a wide variety of food products (Grimm and Wösten 2018). Their role is particularly relevant in the context of plant-based diets, where they provide essential amino acids, umami flavor, and a fibrous texture that can replace or complement meat (Bach et al. 2017; Singh et al. 2023). Additionally, mushrooms have been incorporated into bakery products, such as bread, to improve nutritional value and functional properties (Lu et al. 2018). By contributing to nutritional quality while supporting environmentally responsible food production, mushrooms play a central role in advancing sustainable diets (Myrdal Miller et al. 2014; Banerjee et al. 2020).

Within this diverse group, *Hericium erinaceus* (commonly known as Lion's Mane) has attracted special attention. In China, it is called "Monkey Head Mushroom" (Houtougu), while in Japan it is known as "Yamabushitake." Traditionally used in East Asian medicine, it was prescribed for digestive health and as a tonic for overall vitality. Today, scientific interest has focused on its neuroprotective potential, particularly due to erinacines and hericenones—compounds that stimulate nerve

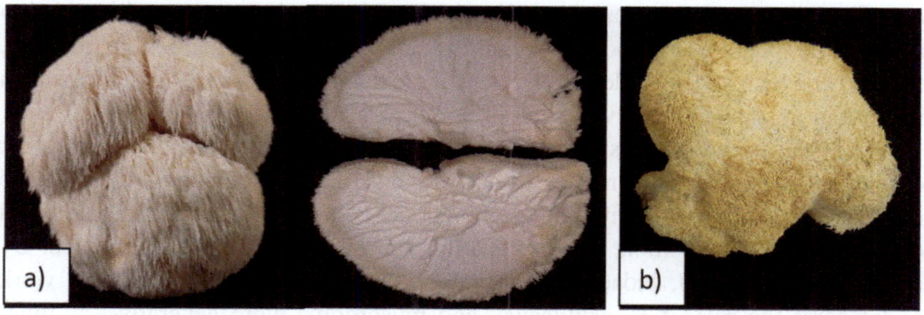


**Fig. 11.1** (**a**) Fresh *Hericium erinaceus* morphology; (**b**) *H. erinaceus* maturation

growth factor (NGF) and support cognitive function (Ratto et al. 2019). Culinary interest is equally strong: its seafood-like flavor and fibrous texture make it a sought-after ingredient in gastronomy and a promising component of plant-based meat or seafood analogs

From a biological perspective, *H. erinaceus* belongs to the phylum Basidiomycota, class Agaricomycetes, order Russulales, and family *Hericiaceae*. Its fruiting body is typically globular to semi-spherical and initially white to cream, gradually turning yellowish to brownish with age (Fig. 11.1). The most distinctive characteristic, and the origin of its common name, is its cascading "mane" of long, pendulous spines, 1–5 cm in length, which extend uniformly across the surface of the basidiocarp. This striking visual appearance not only makes it easy to identify in nature but also has contributed to its symbolic and gastronomic appeal.

Ecologically, *H. erinaceus* is a saprotrophic fungus, thriving on decaying hardwoods such as beech, oak, and maple. By decomposing lignocellulosic material, it plays a key role in nutrient recycling and forest regeneration within temperate ecosystems of Asia, Europe, and North America (Contato and Conte-Junior 2025). Naturally rare in the wild, it has been the focus of commercial cultivation since the 1980s, particularly in China, where controlled indoor systems using hardwood sawdust substrates allow consistent, high-yield production. This technological advance not only ensures a stable supply for the food and nutraceutical industries but also reduces harvesting pressure on wild populations.

Economically, *H. erinaceus* has become one of the most commercially valuable functional mushrooms. The global market for Lion's Mane extracts and related nutritional products was valued at several hundred million euros in 2023 and is projected to surpass €420 million by 2025, with a compound annual growth rate (CAGR) estimated between 15% and 23%. North America and the Asia-Pacific dominate this expansion, the former driven by the nutraceutical industry and the latter by traditional consumption and rising health awareness. This rapid growth highlights the mushroom's dual importance: as a sustainable agricultural product and as a high-value ingredient in the global bioeconomy.

The present work seeks to provide a comprehensive understanding of *Hericium erinaceus* by combining a critical review of the scientific literature with new experimental data. Original analyses are presented on proximate composition, amino acid profile, phenolic compounds, organic acids, and polysaccharides. In addition, the study develops the first descriptive sensory lexicon for *H. erinaceus*, addressing a major gap in standardized evaluation of its sensory attributes.

While the primary focus is on *H. erinaceus*, comparative insights from other edible and medicinal mushrooms are also included. These comparisons place the findings into a broader nutritional and functional context, allowing similarities and distinctions in composition, sensory quality, and bioactivity to be highlighted. By integrating nutritional, bioactive, sensory, and economic perspectives, this work underscores the role of *H. erinaceus*—and mushrooms more generally—as functional food ingredients and model species for sustainable, health-promoting diets within a rapidly expanding global market.

## 11.2 Materials and Methods

### 11.2.1 Materials

Fresh fruiting bodies of *Hericium erinaceus* were obtained from the Portuguese mushroom producer Serra da Lua (Portugal). Samples were harvested at commercial maturity, carefully packed, and transported under controlled conditions of temperature and humidity to ensure freshness and prevent quality loss. Upon arrival at the laboratory, mushrooms were immediately processed or stored at 4 °C until analysis.

### 11.2.2 Bibliometric Analysis

To map the global research landscape on *H. erinaceus*, a bibliometric analysis was performed using VOSviewer software (v.1.6.19). This software makes it possible to create maps of authors or journals based on co-citation data or to construct maps of keywords based on co-occurrence data (van Eck and Waltman 2010). The search was conducted in the Web of Science Core Collection database using the query: "*Hericium erinaceus*" AND ("nutritional" OR "bioactive properties"). The search covered publications between 2015 and 2025, yielding 73 records, including research articles, reviews, and conference proceedings. Data were analyzed to assess publication trends, research hotspots, and collaborative networks. Keyword co-occurrence maps were generated to identify thematic clusters, while co-citation analysis highlighted the most influential authors and journals. This approach provided a comprehensive overview of how scientific interest in *H. erinaceus* has evolved over the past decade.

A term map was constructed to visualize the co-occurrence of key terms, as shown in Fig. 11.2. The network analysis identified the five most frequent author keywords (occurring more than five times), with the most prominent being Hericium erinaceus, mushrooms, antioxidant activity, and bioactive compounds. Author keywords represent the terms that researchers consider most relevant to their work.

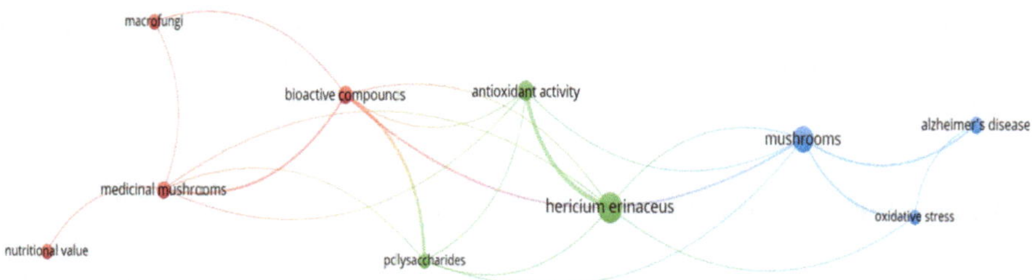


**Fig. 11.2** Bibliometric study: keyword co-occurrence analysis

In the map, node size reflects the frequency of each keyword, while the thickness of the connecting lines represents the strength of co-occurrence between keyword pairs.

The keyword co-occurrence analysis identified three major clusters. The first cluster is related to general studies on medicinal and edible mushrooms, with emphasis on nutritional value, macrofungi, and bioactive compounds. The second cluster centers on *Hericium erinaceus*, strongly linked to polysaccharides and antioxidant activity, reflecting the biochemical and functional characterization of this species. The third cluster connects mushrooms with biomedical applications, particularly through oxidative stress and Alzheimer's disease, suggesting increasing attention to the neuroprotective potential of *H. erinaceus*.

Overall, the structure of the network demonstrates a continuum of research, starting from general studies on medicinal and nutritional aspects of mushrooms, progressing toward the characterization of specific bioactive compounds, and culminating in targeted biomedical applications, particularly in the context of neurodegenerative diseases.

### 11.2.3 Proximate Composition

The proximate composition of mushroom fruiting bodies was determined following AOAC standard methods. Moisture content, ash, crude protein (Kjeldahl method), crude fat (Soxhlet extraction), and total carbohydrates (by difference) were quantified based on Nicolai et al.'s (2025) methodology. Energy values were estimated based on standard conversion factors.

### 11.2.4 Amino Acid Profile

Amino acids were determined after acid and alkaline hydrolysis of dried mushroom samples. The hydrolysates were derivatized and analyzed by high-performance liquid chromatography (HPLC) with fluorescence detection (Quaresma et al. 2023). Essential and nonessential amino acids were quantified using calibration curves prepared with certified standards. In addition, the Essential Amino Acid Index (EAAI) was calculated according to Oser's method [13], which compares the proportion of essential amino acids in the sample to those in a reference protein pattern. This index provides a comprehensive measure of protein quality and allows the nutritional value of Hericium erinaceus proteins to be evaluated relative to human dietary requirements.

### 11.2.5 Phenolic Compounds

Phenolic compounds were extracted using aqueous methanol and analyzed by HPLC-DAD, following the Nelson et al. (Pereira et al. 2023) method. Identification and quantification were based on retention time and UV–Vis spectra compared to authentic standards, including gallic acid, caffeic acid, ferulic acid, rutin, and quercetin.

### 11.2.6 Organic Acids

Organic acids were determined by HPLC-UV. Mushroom extracts were prepared in metaphosphoric acid, filtered, and injected for separation and quantification, based on Panda et al.'s (2022) methodology. Calibration curves were prepared with citric, malic, succinic, oxalic, lactic, and other standard acids.

### 11.2.7 Polysaccharides

Total polysaccharides were obtained by hot water extraction, ethanol precipitation, and freeze-drying, according to Panda et al.'s (2022) methodology. The monosaccharide composition was determined after hydrolysis by HPLC with refractive index detection, allowing quantification of glucose, mannose, galactose, arabinose, and other sugars.

### 11.2.8 Sensory Lexicon Development

A descriptive sensory lexicon for *Hericium erinaceus* was developed using a trained panel of 25 members (67% women and 33% men) ranging in age from 23 to 64 years. All sessions were conducted in a dedicated sensory laboratory under controlled temperature, humidity, and lighting to ensure reproducibility and minimize external influences on perception. Panelists were selected based on their availability, motivation, and sensory acuity and were familiarized with the objectives of the study prior to formal evaluations.

The evaluation sessions were designed to characterize fresh mushrooms at harvest and again after 3 days of storage at 5 °C in order to capture potential changes in sensory quality during early postharvest life. Training consisted of multiple sessions in which panelists generated descriptors through open discussion, followed by refinement and consensus to produce a standardized vocabulary. This vocabulary included terms related to aroma, flavor, and mouthfeel, each defined clearly and anchored with reference standards wherever possible.

The methodology followed the principles of descriptive sensory analysis as outlined by Suwonsichon (2019). which is widely regarded as one of the most robust approaches for identifying and quantifying sensory attributes in foods. Lexicon development is a central step in this methodology, as it provides the linguistic and conceptual framework for describing complex sensory perceptions in an objective and reproducible way.

Using this vocabulary, panelists conducted descriptive tests to evaluate attributes including new leather, old leather, wood, grains, mold, earthy, paper, potato, umami, bitter, green, salty, sour, humidity, and astringent. Attribute intensity was rated on a 0–5 scale, where 0 indicated "extremely weak" and 5 "extremely strong."

Once the lexicon was established, it was applied systematically by the trained panel to evaluate the intensity of each descriptor across samples. This approach not only allowed the characterization of the sensory profile of *H. erinaceus* but also enabled the monitoring of how these attributes evolved during storage. Such evaluations are particularly relevant for highly perishable commodities like mushrooms, where shelf-life is often constrained by rapid changes in texture, flavor, and aroma.

The development of this lexicon has broader implications beyond quality control. By linking sensory attributes to chemical composition and nutritional properties, it offers a practical tool for product innovation, especially in the context of plant-based food development. For *H. erinaceus*, which is increasingly incorporated into functional foods and meat analogs, a structured sensory vocabulary provides the basis for correlating consumer perception with compositional data, thus facilitating the design of products that are both nutritionally valuable and sensorially appealing.

## 11.3 Results and Discussion

This section integrates the findings of the present study with existing literature, providing a comprehensive assessment of the nutritional, bioactive, and sensory characteristics of *Hericium erinaceus*. Unlike previous works limited to literature reviews, the present study combines bibliometric analysis with new experimental data on proximate composition, amino acid profile, phenolic compounds, organic acids, and polysaccharides. Furthermore, it introduces the first descriptive sensory lexicon for *H. erinaceus*, addressing a major gap in standardized evaluation of this species.

The results are presented in a structured sequence, beginning with sensory attributes, which represent the most immediate determinant of consumer acceptance. This is followed by the analysis of proximate composition and amino acid profile, providing insights into nutritional quality. Finally, the characterization of phenolic compounds, organic acids, and polysaccharides highlights the bioactive potential of the species. Together, these findings position *H. erinaceus* not only as a nutritionally valuable mushroom but also as a promising functional ingredient for sustainable diets and innovative food applications.

### 11.3.1 Development of a Descriptive Sensory Lexicon

The first stage of this study focused on the development of a descriptive sensory lexicon, since sensory attributes are the primary factors influencing consumer acceptance of mushrooms. While nutritional and bioactive properties are essential to their scientific and functional relevance, appearance, aroma, flavor, and texture largely determine market success. For *Hericium erinaceus*, which is increasingly incorporated into gastronomy and plant-based foods, the absence of a standardized sensory framework has limited both scientific research and product innovation.

Although descriptive lexicons have been established for other foods and commodities—including fruits, nuts, and teas—they remain underexplored, particularly Lion's Mane. As summarized in Table 11.1, lexicons in other food systems have proven useful for monitoring sensory changes during storage, shelf-life testing, and product standardization (Lee and Chambers 2010; Riveros et al. 2013; Rosales and Suwonsichon 2015). However, no equivalent approach has been reported for *H. erinaceus*, despite its growing commercial and nutritional relevance.

To address this gap, a trained panel of 25 members (67% women and 33% men, aged 23–64 years) developed a standardized sensory lexicon specific to *H. erinaceus*. Evaluations were conducted in a sensory laboratory under controlled environmental conditions, following the descriptive sensory analysis methodology described by Suwonsichon (2019). Fresh mushrooms were assessed at harvest and again after storage at 5 °C, allowing the lexicon to capture both initial quality and early postharvest changes.

Through consensus, the panel generated a total of 17 descriptors, comprising 15 aroma/flavor attributes and two mouthfeel attributes. These included both favorable descriptors, such as umami, earthy, woody, grains, and salty, as well as less desirable ones, such as mold, paper, bitter, and astringent. The complete lexicon, with definitions and reference anchors, is presented in Table 11.2.

At harvest, the descriptors with the highest intensity were earthy, moldy, and bitter, and the mouthfeel attribute was humidity, indicating that *H. erinaceus* presents a complex and distinctive flavor profile even at peak freshness. This

**Table 11.1** Lexicon development for products under storage condition

| Food product | Application of lexicon | Key findings | References |
|---|---|---|---|
| Fresh-cut pomelo cultivars | Monitoring sensory changes during refrigerated storage | Some cultivars remained stable; others exhibited changes in texture and flavor after several days | Rosales and Suwonsichon (2015) |
| Peanut-based products | Evaluation of storage stability | High-oleic peanuts and coated roasted peanuts developed undesirable flavors (oxidized, cardboard) more slowly than controls | Riveros et al. (2013) |
| Korean green teas | Long-term storage assessment (2 years) | Minimal sensory changes observed; reference standards minimized panelist drift | Lee and Chambers (2010) |

**Table 11.2** Lexicon for flavor and mouthfeel description for *Hericinum erinaceus* evaluation during storage

| Attribute | | Definition |
|---|---|---|
| Aroma/flavor | New leather | Musty, new leather (new shoes or purses) |
| | Old leather | Musty, old leather (old books bindings) |
| | Wood | Aromatic associated with dry cut wood |
| | Grains | Aromatics associated with grains such as rice, oats, and wheat. May be characterized as sweet, brown, and/or musty |
| | Mold | Aromatic characterist of mold or wet damp soil |
| | Earthy | Sweet, aromatics associated with decaying vegetation and black soil |
| | Paper | The flat aroma that may be associated with cardboard |
| | Potato | The starchy, slightly metallic, and cooked vegetable aroma |
| | Umami | Typical taste of monosodium glutamate |
| | Bitter | Taste associated typically with caffeine or quinine |
| | Green | Aroma characterized by cut grass and unripe or green fruit |
| | Salty | Taste associated with sodium chloride |
| | Sour | Taste associated with citric acid |
| Mouthfeel | Humidity | Degree of wet or dry |
| | Astringent | Feeling in the mouth characterized by drying or a tingling sensation on the tongue |

contrasts with milder mushrooms such as *Agaricus bisporus*, whose flavor is dominated by umami and mild earthy notes (Grosshauser and Schieberle 2013).

During storage at 5 °C, most descriptors decreased in intensity, reflecting the progressive loss of freshness and quality previously described in mushrooms (Silva et al. 2025). These changes in attribute intensity are particularly relevant from a sensory quality perspective, as some of them are associated with negative consumer perceptions. Several of these characteristics have also been widely reported in previous studies. Certain attributes, particularly *astringent* and *bitter ones,* should be carefully monitored, as they are generally perceived as undesirable by consumers in some food types (Suwonsichon 2019).

These sensory shifts are visually represented in Fig. 11.3, which compares the profiles of fresh and stored mushrooms. Importantly, the umami attribute remained stable, reinforcing its role as a defining sensory characteristic of *H. erinaceus.*

The persistence of umami can be linked to the biochemical composition of mushrooms, particularly the presence of glutamic acid, aspartic acid, and ribonucleotides (Rotola-Pukkila et al. 2019; Sun et al. 2020). This stability makes *H. erinaceus* especially relevant for plant-based food applications, where umami contributes to savory taste and reduces the need for added sodium. As noted by Guinard et al. (2016), mushrooms can serve as a viable meat alternative and may help reduce sodium usage without diminishing sensory appeal for consumers. This effect is likely related to the presence of those umami compounds, which play a crucial role in sodium reduction by enhancing the perception of saltiness (Harada-

| Attributes | t=0 | t=3 |
|---|---|---|
| Paper | 2,5 | 2,6 |
| Earthy | 3,5 | 2,8 |
| Mold | 3,3 | 4,0 |
| Humidity | 4,0 | 3,3 |
| Potato | 2,3 | 1,5 |
| Old leather | 1,3 | 1,3 |
| Moldy | 3,2 | 2,7 |
| Blue cheese | 2,0 | 2,3 |
| Mushroom | 4,8 | 4,2 |
| Wood | 3,0 | 2,2 |
| Green | 2,4 | 2,3 |
| Paper | 2,0 | 2,5 |
| Umami | 2,3 | 2,3 |
| Bitter | 3,2 | 3,0 |
| Salty | 2,5 | 1,5 |
| Sour | 2,0 | 1,8 |
| Astringent | 3,5 | 3,6 |

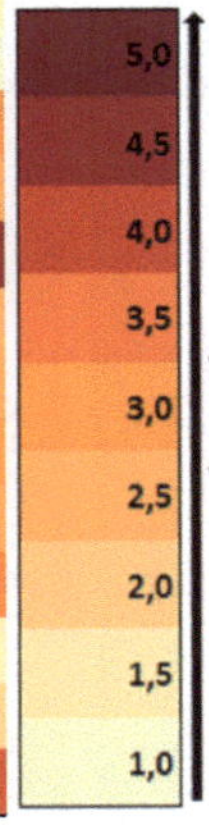


**Fig. 11.3** Sensory profile of fresh *Hericium erinaceus* mushrooms at the beginning and after 3 days of storage

Padermo et al. 2020). Other descriptors, such as earthy and moldy, can be associated with volatiles like 1-octen-3-ol, while green notes are linked to hexanal (Aisala et al. 2019; Sun et al. 2020). Bitter and astringent perceptions may be attributed to free amino acids such as histidine, arginine, and leucine (Chen et al. 2021).

Overall, this study presents the first standardized descriptive sensory lexicon for H. erinaceus. Comparable lexicons exist for *Lentinula edodes* (Politowicz et al. 2018a), *Agaricus bisporus* (Grosshauser and Schieberle 2013), and *Pleurotus ostreatus* (Politowicz et al. 2018b), but until now, Lion's Mane had not been characterized in this way. By filling this gap, the lexicon provides a critical tool for shelf-life studies, quality control, and the development of innovative food products. Moreover, by linking sensory descriptors with nutritional and chemical composition, it establishes a foundation for integrating *H. erinaceus* into sustainable plant-based diets while ensuring consumer acceptance.

It is also important to note that mushroom sensory attributes are inherently variable. As Chun et al. (2020) emphasize, a wide range of flavors—including woody, nutty, brown, bitter, sour, and astringent—are present across different mushroom species, though at varying intensity levels. This variability indicates that a sensory lexicon should not be regarded as fixed or final, but rather as a flexible tool that can be refined and adapted to reflect the unique characteristics of each mushroom type. The lexicon developed here for *H. erinaceus* should therefore be viewed as a starting framework that can be expanded and adjusted as new insights emerge from future studies and product applications.

## 11.3.2 Proximate Composition

Mushrooms are widely regarded as nutrient-dense foods, providing protein, dietary fiber, vitamins, and essential minerals while remaining low in fat and calories. This nutritional profile has positioned them as important contributors to both balanced diets and functional foods (González et al. 2020). The proximate composition of *Hericium erinaceus* determined in the present study is shown in Fig. 11.4.

The results confirm that *H. erinaceus* is composed primarily of carbohydrates, followed by protein and minor amounts of fat, which is consistent with previously reported data (Rodrigues et al. 2015). On a dry weight basis, carbohydrates account for about half of the mushroom's composition, largely in the form of polysaccharides and glycoproteins. Since many of these compounds are indigestible dietary fibers, *H. erinaceus* can be considered a low-calorie food, suitable for incorporation into diets designed to prevent obesity, cardiovascular disease, and diabetes (Dimopoulou et al. 2022).

The protein content observed in *H. erinaceus* falls within the range typically reported for edible mushrooms (19–35% dry weight) (Singh et al. 2025). While its protein levels are comparable to those of *Lentinula edodes* (20–23%) (Sheng et al. 2021), they are slightly lower than those of *Agaricus bisporus* (≈29%) (Atila et al. 2017). Nevertheless, the protein fraction of *H. erinaceus* is of particular nutritional interest because it contains a well-balanced amino acid profile (see Sect. 11.3.4).

Lipid levels in *H. erinaceus* are low, consistent with the general pattern of edible mushrooms, which typically contain only 2–6% fat on a dry weight basis (Singh et al. 2025). Although quantitatively minor, mushroom lipids are noteworthy because they are dominated by unsaturated fatty acids such as linoleic and oleic acids, which have been linked to beneficial effects on blood lipid regulation and cardiovascular health (den Hartigh 2019).

When compared to other mushroom species, *H. erinaceus* shows a proximate composition that is broadly similar but not identical. For example, *A. bisporus* generally contains higher protein and carbohydrate levels, while *L. edodes* is richer in carbohydrates and slightly fatter (Atila et al. 2017; Sheng et al. 2021). These

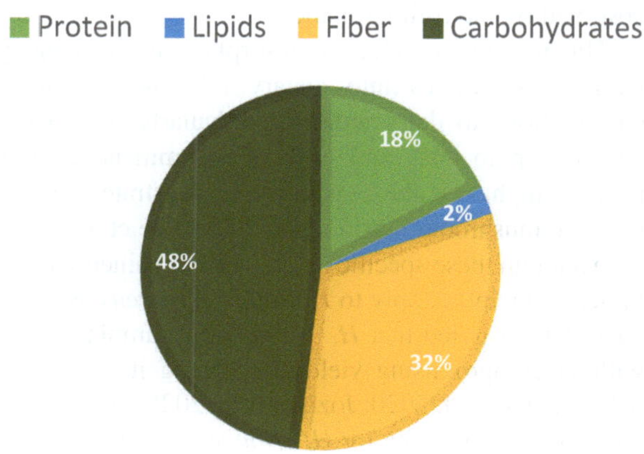


**Fig. 11.4** Proximal analysis of mushrooms *Hericium erinaceus*, per 100 g of dried weight *Original data

differences highlight the species-specific nature of mushroom nutritional profiles and underscore the importance of characterizing each mushroom type individually.

Taken together, the proximate composition results demonstrate that *H. erinaceus* offers a nutritional profile in line with other commonly consumed mushrooms while contributing unique bioactive components. Its low lipid content, moderate protein levels, and high proportion of polysaccharides make it not only a health-promoting food but also an excellent candidate for incorporation into functional food products and plant-based formulations.

### 11.3.3 Vitamins and Mineral Composition

Mushrooms are widely recognized as valuable sources of vitamins and essential minerals, making them unique contributors to human nutrition. They are especially rich in B-complex vitamins and are the only nonanimal food known to contain vitamin B12 in nutritionally relevant amounts. Among 16 mushroom species evaluated, *Hericium erinaceus* exhibited the highest vitamin B12 concentration, up to 1.04 mg/100 g dry weight (Dimopoulou et al. 2022). Mushrooms are also notable sources of ergosterol, which can be converted to vitamin $D_2$ upon exposure to ultraviolet (UV) radiation. This makes them a sustainable plant-based source of vitamin $D_2$, with direct benefits for bone health and calcium metabolism (Sławińska et al. 2016). Additionally, mushrooms contain vitamin C (ascorbic acid) and a wide range of minerals, including Fe, Mg, Zn, Se, and Cu, which contribute significantly to dietary requirements (Dimopoulou et al. 2022).

The mineral composition of mushrooms is strongly influenced by substrate and environmental conditions (Silva et al. 2024). In *Pleurotus ostreatus*, for example, mineral accumulation was significantly altered by substrate supplementation. As shown in Table 11.3, fruiting bodies cultivated on wheat straw supplemented with spent coffee grounds accumulated higher levels of potassium (K), iron (Fe), and arsenic (As) compared to those grown on wheat straw with bran. Interestingly, not all minerals present in high concentrations in the substrate were accumulated in the mushroom—for instance, aluminum (Al) was elevated in the substrate but not absorbed by the fruiting bodies.

This reflects the selective absorption mechanisms typical of mushrooms. Translating these values into dietary relevance (Table 11.4) highlights substantial contributions to daily intakes, particularly for potassium and manganese, which covered up to 44% and 68% of recommended allowances, respectively. These results emphasize the feasibility of substrate-driven biofortification as a tool to enhance mushroom nutritional quality (Lee et al. 2009; Koutrotsios et al. 2020).

Although these specific results were obtained with *P. ostreatus*, the same physiological principles apply to *Hericium erinaceus*. Selenium (Se) biofortification trials have demonstrated that *H. erinaceus* accumulates Se in a dose-dependent manner without compromising yield, confirming its suitability for targeted micronutrient delivery (Hu et al. 2020; Jozífek et al. 2025). To evaluate its dietary contribution, we compiled mineral data for *H. erinaceus* from literature sources and structured them

**Table 11.3** Effect of substrate composition on mineral content of *P. ostreatus* (Po) mushroom[c]

| Sample | Mineral content (% ppm) | | | | | | | | | | |
|---|---|---|---|---|---|---|---|---|---|---|---|
| | Al | Si | P | S | K | Ca | Mn | Fe | Cu | Zn | As |
| Sub_A | $0^{a}$ | $15941^{a}$ | $1507^{a}$ | $2037^{a}$ | $24917^{a}$ | $7584^{a}$ | $116^{a}$ | $578^{a}$ | $13^{a}$ | $44^{a}$ | $74^{a}$ |
| Sub_B | $2350^{b}$ | $19757^{b}$ | $428^{b}$ | $2251^{a}$ | $26038^{a}$ | $7404^{a}$ | $122^{a}$ | $992^{b}$ | $15^{a}$ | $30^{b}$ | $89^{b}$ |
| Po_A | $0^{a}$ | $2188^{a}$ | $9972^{a}$ | $3177^{a}$ | $55019^{a}$ | $439^{a}$ | $90^{a}$ | $171^{a}$ | $36^{a}$ | $136^{a}$ | $29^{a}$ |
| Po_B | $0^{a}$ | $2007^{a}$ | $8606^{b}$ | $3471^{a}$ | $58546^{b}$ | $341^{a}$ | $78^{a}$ | $205^{b}$ | $39^{a}$ | $130^{a}$ | $36^{b}$ |

Means within a column that do not share similar letters are significantly different ($p < 0.05$). Substrate composition: Sub_A- 80% wheat straw and 20% wheat bran; Sub_B- 80% wheat straw and 20% spent ground coffee (Authors' unpublished results). [c]Original data

**Table 11.4** Recommended Dietary Allowance (RDA) for mineral content for fresh *P. ostreatus* produced in each substrate

| Mineral | RDA (mg) | mg/100 g fw Po_SubA | %RDA (%) | mg/100 g fw Po_SubB | %RDA (%) |
|---|---|---|---|---|---|
| Ca | 800 | 6.6 | 1 | 5.1 | 1 |
| Fe | 14 | 2.6 | 18 | 3.1 | 22 |
| Mn | 2 | 1.4 | 68 | 1.2 | 59 |
| P | 700 | 149.6 | 21 | 129.1 | 18 |
| K | 2000 | 825.3 | 41 | 878.2 | 44 |
| Zn | 10 | 2.0 | 20 | 2.0 | 20 |

analogously to *P. ostreatus*. As shown in (Table 11.5), *H. erinaceus* is rich in potassium (2912 mg/100 g dw), phosphorus (771 mg/100 g dw), and magnesium (76 mg/100 g dw), while also providing significant amounts of iron (11.2 mg/100 g dw) and zinc (3.4 mg/100 g dw).

The estimated dietary contributions (Table 11.6) indicate that 100 g of dried *H. erinaceus* can provide over 80% of potassium's RDA, 62% of iron, and about 31% of zinc. Magnesium contributes a more modest 19% of the RDA but remains relevant for metabolic and neuromuscular function. These findings confirm the mushroom's value as a dense source of essential minerals.

From a health perspective, these minerals provide well-established benefits. Phosphorus (P) is vital for bone mineralization and energy metabolism (Serna and Bergwitz 2020). Iron (Fe), zinc (Zn), and manganese (Mn) play critical roles in neurocognitive function, with diets richer in these elements linked to improved brain performance (Zhao et al. 2023). Conversely, iron deficiency remains a widespread issue as a major cause of anemia and is a recognized risk factor for bone loss (Yang et al. 2023). The relatively high Fe content in *H. erinaceus*, therefore, adds to its nutritional importance.

Species-specific differences are also notable. According to Meyer et al. (2023), sodium (Na) levels in *H. erinaceus* (≈188 mg/100 g dw) are significantly higher than in *Pleurotus* spp. (≈11–24 mg/100 g dw). Although this sodium content remains low compared to processed foods, it is nutritionally meaningful and provides a plausible basis for the moderately salty taste detected by the sensory panel (Sect. 11.3.1). This alignment between mineral content and sensory perception illustrates the integration of compositional and sensorial quality in mushroom evaluation.

Overall, the combined evidence from *P. ostreatus* (Tables 11.3 and 11.4) and *H. erinaceus* (Tables 11.5 and 11.6) demonstrates two key points: (i) substrate engineering can effectively enhance the mineral profile of cultivated mushrooms, and (ii) *H. erinaceus* is inherently rich in several minerals that support bone health, neuroprotection, immune function, and sensory appeal. These features reinforce its potential as both a functional food and a model species for nutritional biofortification strategies.

**Table 11.5** Mineral content of *Hericium erinaceus* (mg per 100 g dw)[a]

| Mineral | Value (mg/100 g dw) |
|---|---|
| Iron (Fe) | ≈ 11.2 |
| Zinc (Zn) | ≈ 3.4 |
| Phosphorus (P) | ≈ 770.8 |
| Potassium (K) | ≈ 2912 |
| Magnesium (Mg) | ≈ 75.8 |

[a]Values derived from comparative mineral profiling across diferente published studies

**Table 11.6** Estimated RDA contribution of *H. erinaceus* (per 100 g dw)[a]

| Mineral | Value (mg) | RDA (%) |
|---|---|---|
| Iron (Fe) | ≈ 11.2 | ≈ 62% (18 mg) |
| Zinc (Zn) | ≈ 3.4 | ≈ 31% (11 mg) |
| Phosphorus (P) | ≈ 109 | ≈ 15.6% (700 mg) |
| Potassium (K) | ≈ 2.9 | ≈ 83% (3500 mg) |
| Magnesium (Mg) | ≈ 75.8 | ≈ 19% (400 mg) |

[a]Based on general adult RDAs

### 11.3.4 Amino Acid Profile

Proteins are an essential component of edible mushrooms, not only for their nutritional significance but also for their contribution to flavor development. The amino acid profile of *Hericium erinaceus*, presented in Table 11.7, confirms that this mushroom provides a broad spectrum of both essential and nonessential amino acids. The most abundant were alanine, aspartic acid, and glutamic acid, followed by leucine, glycine, and serine. These results are consistent with previous studies, which also reported high levels of glutamate- and aspartate-type amino acids, key contributors to the umami taste.

When compared to other edible mushrooms, the amino acid distribution of *H. erinaceus* exhibits some distinctive features. In *Agaricus bisporus*, aspartic acid, histidine, glutamic acid, lysine, and serine dominate the profile (Ramos et al. 2019), while *Lentinula edodes* shows particularly high levels of glutamic acid (Sheng et al. 2021). Although protein content in *H. erinaceus* is moderate relative to these species (see Sect. 11.3.2), its amino acid balance makes it a high-quality protein source, particularly in plant-based diets where essential amino acids are often limiting (Singh et al. 2025). In this regard, the Essential Amino Acid Index (EAAI) obtained for this study was 21.3, confirming that *H. erinaceus* contributes positively to protein quality, especially when consumed in combination with other plant foods.

Importantly, the amino acid profile contributes directly to sensory perception. Glutamic and aspartic acids are responsible for the umami attribute, while other amino acids modulate sweetness or bitterness. As summarized in Table 11.8, alanine, serine, glycine, and threonine contribute to sweetness, whereas histidine, arginine, valine, methionine, isoleucine, leucine, and phenylalanine contribute to bitterness (Chen et al. 2021). In *H. erinaceus*, the presence of high alanine levels

**Table 11.7** Amino acid profile of *Hericium erinaceus* (mg/100 g dw)[a]

| Amino acid | mg/100 g |
|---|---|
| Aspartic acid | 482.1 ± 52.3 |
| Glutamic acid | 350.8 ± 38.8 |
| Serine | 184.3 ± 6.5 |
| Histidine | 272.8 ± 21.2 |
| Tyrosine | 92.2 ± 7.5 |
| Glycine | 19.6 ± 2.3 |
| Arginine | 130.7 ± 15.5 |
| Alanine | 538.4 ± 94.5 |
| Threonine | 55 ± 8.3 |
| Methionine | 25.8 ± 5.2 |
| Valine | 243 ± 36.8 |
| Phenylalanine | 31.8 ± 6.3 |
| Isoleucine | 93.4 ± 12.1 |
| Leucine | 157.7 ± 18.0 |
| Lysine | 261.1 ± 92.2 |
| Tryptophan | 15.4 ± 0.3 |

[a]Original data

**Table 11.8** Amino acid contribution to profile sensory attributes (Chen et al. 2021)

| Sensory attributes | Amino acids |
|---|---|
| Umami | Aspartic and glutamic acid |
| Sweetness | Serine, glycine, threonine, alanine, proline |
| Bitterness | Histidine, arginine, valine, methionine, isoleucine, leucine, phenylalanine |
| Tasteless | Cysteine, lysine, and tyrosine |

explains its subtle sweet undertone, while glutamic and aspartic acids underpin the stable umami flavor observed in the sensory lexicon (Sect. 11.3.1). At the same time, the occurrence of histidine and arginine may explain the bitter and astringent notes that intensified slightly during storage.

Beyond sensory quality, amino acids in *H. erinaceus* also confer functional and health-related benefits. One of the most notable is ergothioneine (EGT), a histidine-derived amino acid with potent antioxidant properties. Mushrooms are the richest dietary source of EGT, with *H. erinaceus* showing particularly high concentrations compared to cereals, vegetables, and meat (Roda et al. 2022; Apparoo et al. 2022). Ergothioneine is actively transported into cells via the OCTN1 transporter, and evidence suggests it may cross the blood–brain barrier, making it highly relevant for neuroprotection and healthy aging (Roda et al. 2022). Moreover, EGT demonstrates stronger free radical scavenging activity than many conventional antioxidants (Apparoo et al. 2022), further supporting its role as a promising candidate in neuroprotection and longevity research. Beyond neuroprotection, higher EGT intake has been associated with broader health benefits, including reduced incidence of cardiovascular disease (Smith et al. 2020).

Interestingly, recent studies have also explored the use of *H. erinaceus* as an ingredient in seafood analogs. Meyer et al. (2023) compared the amino acid profiles of Lion's Mane and shrimp, noting similar levels of alanine, proline, and glycine. This compositional similarity supports the use of *H. erinaceus* in shrimp alternatives and broadens its application in plant-based food innovation.

Overall, the amino acid profile of *H. erinaceus* underscores its dual value: as a nutritionally balanced protein source and as a key driver of flavor. Its richness in glutamic acid and alanine explains the persistence of umami and subtle sweet notes in the sensory lexicon, while the presence of ergothioneine positions it as a unique mushroom with strong neuroprotective potential. Together, these findings reinforce the multifunctional role of *H. erinaceus* in sustainable, health-promoting diets.

### 11.3.5 Phenolic Compounds

Phenolic compounds are considered secondary metabolites of edible mushrooms. They play an important role in the development and defense of fungal structures. Analysis of these compounds provides insight into their antioxidant potential and contribution to the bioactive properties of the organism.

According to Singh et al. (2025), phenolic acids are mainly represented by two groups: hydroxybenzoic acids (HBAs) and hydroxycinnamic acids (HCAs). HBAs are often incorporated into complex structures such as tannins, lignin, and various organic acids, whereas HCAs are typically bound to cell wall constituents, including lignin, cellulose, and proteins. Among HCAs, ferulic, sinapic, caffeic, and coumaric acids are the most abundant, contributing significantly to lignin formation, defense mechanisms, and the regulation of growth. As reported by Ahmed et al. (2023), mushrooms contain various phenolic acids, including p-hydroxybenzoic, cinnamic, gallic, salicylic, p-coumaric, syringic, caffeic, ferulic, and chlorogenic acids, as well as a range of flavonoids. The latter group comprises compounds such as myricetin, rutin, naringenin, quercetin, and hesperetin, which are part of the polyphenol fraction and are known for their antiproliferative effects.

In the case of *H. erinaceus*, our analysis revealed a comparable qualitative composition, identifying gallic acid, protocatechuic acid, catechin, chlorogenic acid, p-hydroxybenzoic acid, caffeic acid, vanillic acid, p-coumaric acid, rutin, ellagic acid, ferulic acid, naringin, apigenin-7-glucoside, myricetin, resveratrol, and quercetin.

### 11.3.6 Organic Acids

Organic acids are key components of mushroom metabolism, influencing not only growth and development but also sensory attributes such as sourness, freshness, and umami enhancement. They also contribute to the overall bioactivity of mushrooms, with reported antioxidant, antimicrobial, and anti-inflammatory properties (Barros et al. 2013; Gąsecka et al. 2020).

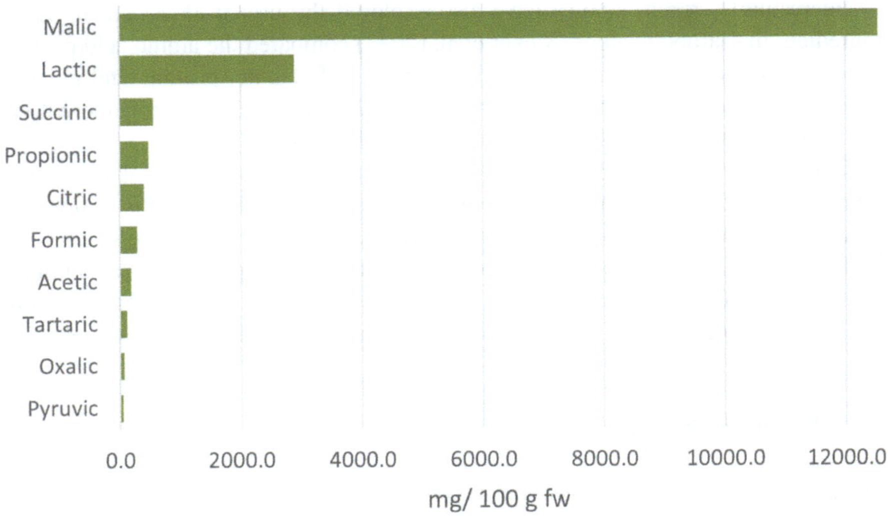


**Fig. 11.5** *Hericium erinaceus* organic acid composition (original data)

In the present study, the organic acid composition of *Hericium erinaceus* was determined by HPLC, revealing a total content of 18.7 mg/100 g fw (Fig. 11.5). The profile was dominated by malic acid (13,953.6 mg/100 g fw), followed by lactic acid (2848.5 mg/100 g fw). Intermediate concentrations were observed for succinic (616.6 mg/100 g fw), propionic (371.5 mg/100 g fw), and citric acid (297.5 mg/100 g fw). Minor contributors included formic, acetic, tartaric, oxalic, and pyruvic acids.

This composition is consistent with earlier reports on Portuguese edible mushrooms, where malic acid was also identified as the predominant organic acid (Barros et al. 2013). The high malic acid content in *H. erinaceus* is particularly relevant, as it not only contributes to a mild sourness but also plays roles in energy metabolism as part of the tricarboxylic acid cycle. Lactic and succinic acids may further enhance savory notes by interacting with amino acids and nucleotides, reinforcing the umami perception observed in the sensory lexicon (Sect. 11.3.1).

Beyond their sensory contribution, organic acids in *H. erinaceus* also exhibit functional properties. Succinic and citric acids act as natural antioxidants, protecting cells from oxidative damage, while oxalic and formic acids show antimicrobial and anti-inflammatory activities (Gąsecka et al. 2020). The detection of multiple acids in small but diverse quantities highlights the multifunctional role of organic acids in both nutrition and health.

Comparative studies with other mushrooms indicate species-specific differences in acid profiles. For instance, *Agaricus bisporus* is often richer in succinic acid (Gąsecka et al. 2018), while *Lentinula edodes* accumulates higher malic and acetic acid levels (Gao et al. 2020). The predominance of malic acid in *H. erinaceus* thus contributes to the flavor.

Overall, the organic acid composition of *H. erinaceus* underscores its dual relevance: as a determinant of sensory quality (notably sour and umami-enhancing notes) and as a source of bioactive compounds with antioxidant and antimicrobial effects. This positions organic acids alongside amino acids and polysaccharides as central contributors to both the flavor and functional potential of Lion's Mane mushrooms.

### 11.3.7 Polysaccharides

Polysaccharides are one of the most abundant and biologically active components of mushrooms, often regarded as key contributors to their functional food properties. In general, fungal polysaccharides encompass a wide variety of molecules, including chitin, β-glucans, heteropolysaccharides, and glycoproteins (Łysakowska et al. 2023). Among these, β-glucans are the most extensively studied, owing to their immunomodulatory, hypocholesterolemic, and antitumor effects (Leong et al. 2021; Boulifa et al. 2024).

In *Hericium erinaceus*, polysaccharides are particularly diverse. They include heteropolysaccharides composed of glucose, mannose, galactose, and arabinose (Contato and Conte-Junior 2025), as well as β-glucans with branching patterns that influence their biological activity. These compounds have been linked to the alleviation of oxidative stress, modulation of blood glucose levels, and improvements in gut health through prebiotic effects.

Our analysis confirmed this compositional diversity. As shown in (Fig. 11.6), mannitol was the most abundant sugar derivative, exceeding 1000 mg/100 g fw, followed by ribose (469 mg/100 g fw). Glucose and arabinose were present at intermediate levels (128 mg/100 g fw each), while sucrose was the least abundant (92 mg/100 g fw). The predominance of mannitol is consistent with its reported role as an osmotic regulator during fruiting body formation and maturation, with concentrations increasing progressively from mycelium to mature basidiocarps (Vetter 2023).

The functional relevance of these polysaccharides is supported by numerous studies. β-glucans with β-(1 → 3)/(1 → 6) linkages have demonstrated potent immunomodulatory activity, interacting with receptors such as dectin-1 and toll-like receptors to stimulate macrophages, natural killer cells, and cytokine release (Leong et al. 2021; Boulifa et al. 2024). Other structural variants, such as α-(1 → 3)-linked glucans, have been associated with selective cytotoxic effects against tumor cells (Boulifa et al. 2024). In addition, heteropolysaccharides of *H. erinaceus* have shown strong antioxidant capacity, contributing to cellular defense mechanisms.

Polysaccharides also influence texture and mouthfeel, indirectly linking them to sensory perception. The fibrous structure of *H. erinaceus*, noted in the sensory lexicon (Sect. 11.3.1), is partly attributable to its chitin and glucan content. This structural complexity explains its popularity in culinary applications, particularly in plant-based meat and seafood analogs, where a fibrous bite is desirable.

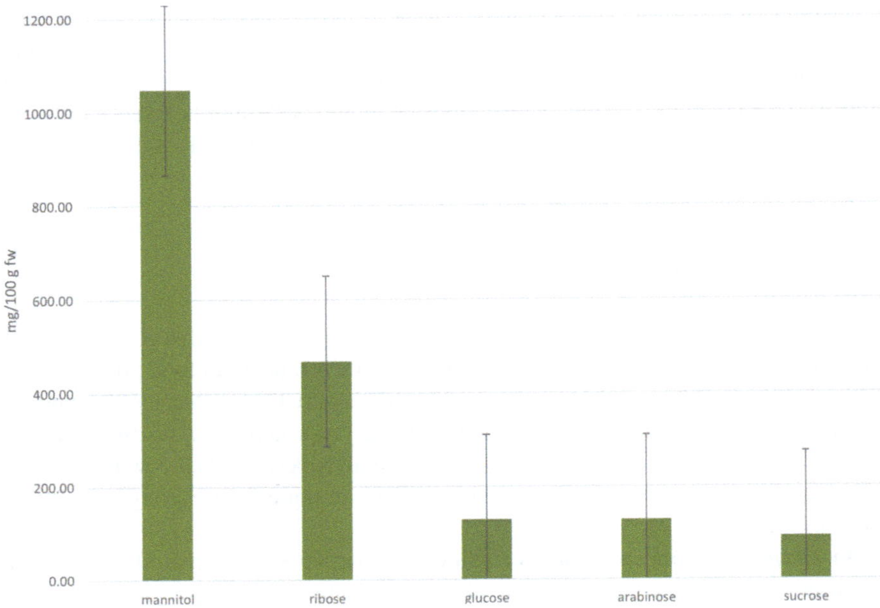


**Fig. 11.6** Polysaccharide composition of *Hericium erinaceus* (original data)

Overall, polysaccharides in *H. erinaceus* are multifunctional, combining nutritional, structural, and therapeutic roles. They reinforce the mushroom's status as a functional food by providing prebiotic fibers for gut health, supporting immune modulation, and offering potential antidiabetic and anticancer activities. When considered alongside its amino acids, organic acids, and phenolics, the polysaccharide profile completes the picture of *H. erinaceus* as a mushroom of both culinary value and bioactive significance.

### 11.3.8 Bioactive Compounds and Health Effects

The bioactive potential of *Hericium erinaceus* arises from a wide spectrum of compounds, ranging from common primary metabolites to highly specific secondary metabolites unique to this species. The results presented in previous sections already highlight this multifunctionality: minerals such as Fe, Zn, and Mn contribute to neurocognitive function (Zhao et al. 2023); organic acids such as citric, succinic, and malic exert antioxidant and antimicrobial effects (Gąsecka et al. 2020); phenolics and flavonoids provide radical scavenging and anti-inflammatory activities (Ahmed et al. 2023; Singh et al. 2025); and polysaccharides, particularly β-glucans and heteropolysaccharides, show strong immunomodulatory and antitumor properties (Leong et al. 2021; Boulifa et al. 2024). Taken together, these compounds illustrate how *H. erinaceus* integrates nutritional and bioactive roles, positioning it as both a food and a functional ingredient.

Beyond these shared mushroom metabolites, *H. erinaceus* is distinguished by unique diterpenoid derivatives, namely erinacines (from mycelium) and hericenones (from fruiting bodies). Both groups stimulate the synthesis of nerve growth factor (NGF), a neurotrophin critical for neuronal survival, differentiation, and regeneration (Szućko-Kociuba et al. 2023). Their capacity to cross the blood–brain barrier and promote NGF activity has been extensively studied, with promising results in models of Alzheimer's and Parkinson's diseases, as well as in mild cognitive impairment (Ratto et al. 2019; Szućko-Kociuba et al. 2023). These findings support the traditional use of Lion's Mane as a tonic for mental health and establish its unique status among medicinal mushrooms.

As previously reported, another important bioactive identified in *H. erinaceus* is ergothioneine (EGT), a histidine-derived amino acid with strong antioxidant properties, thus adding a further neuroprotective and cardioprotective dimension to its bioactive profile. Additionally, DLPE, a phospholipid isolated from the fruiting bodies of *H. erinaceus*, contains two unsaturated fatty acids (linoleic acids) that have been shown to protect neurons (Szućko-Kociuba et al. 2023).

The synergistic interplay between these compounds reinforces the health-promoting effects of *H. erinaceus*. Polysaccharides act primarily through immune modulation, activating macrophages and natural killer cells; phenolic acids and flavonoids enhance antioxidant defense; organic acids and amino acids contribute to metabolic regulation and taste-active properties, while erinacines, hericenones, and ergothioneine provide highly specific neuroprotective benefits. This multilevel activity explains the mushroom's wide range of reported benefits, from antioxidant and anti-inflammatory effects to immune modulation, metabolic regulation, and neuroprotection.

Collectively, the bioactive composition of *H. erinaceus* confirms its dual relevance: as a functional food that contributes essential nutrients and protective metabolites to the diet and as a nutraceutical candidate with applications in neurological and age-related disorders. By linking compositional data with health outcomes, this section underscores the value of Lion's Mane mushroom as both a culinary ingredient and a bioactive reservoir with significant potential for human health.

## 11.4 Applications in Plant-Based Diets

The demand for sustainable, health-conscious food options has elevated mushrooms as key contributors to plant-based diets. Efforts to reduce or replace meat, driven by environmental, ethical, and health considerations, have further strengthened their role as an essential dietary component. In vegetarian diets, mushrooms play a crucial role in enhancing amino acid intake, providing a valuable source of protein that complements plant-based nutrition (Bach et al. 2017). Their fibrous texture and naturally rich umami flavor make them an attractive alternative to animal products, positioning them as key ingredients in the development of meat substitutes (Singh et al. 2023). The food industry has capitalized on these attributes by incorporating mushrooms into a wide range of plant-based products, offering consumers sustainable and health-oriented alternatives to conventional meat (Myrdal Miller et al. 2014; Banerjee et al. 2020). Incorporation of mushrooms into meat analogs can modulate their sensory attributes, potentially enhancing flavor profiles (Fetriyuna et al. 2025). The high content of free amino acids in mushrooms contributes to several taste notes; thus, the majority of the free amino acids confer sweetness and bitterness, whereas glutamic acid and aspartic acid are responsible for umami flavor and replicate the flavor-enhancing effects of monosodium glutamate.

Additionally, mushrooms have been incorporated into bakery products, such as bread, to improve nutritional value and functional properties (Lu et al. 2018). However, they can also be an obstacle in some others by introducing a strong profile of mushroom aroma into food.

Among mushrooms, *Hericium erinaceus* stands out due to its unique combination of sensory qualities, nutritional richness, and bioactive potential. Its natural umami and moderate salty taste enhance flavor complexity, reducing the need for added sodium without diminishing sensory appeal for consumers of plant-based foods (Guinard et al. 2016; Harada-Padermo et al. 2020). Its fibrous structure mimics the texture of shredded meat or seafood, making it particularly suitable for meat analogs such as burgers, sausages, and vegan crab cakes.

Nutritionally, *H. erinaceus* provides high-quality proteins, essential amino acids, vitamins (including B12 and $D_2$), and key minerals such as potassium, phosphorus, zinc, and selenium (Sławińska et al. 2016; Hu et al. 2020; Dimopoulou et al. 2022; Łysakowska et al. 2023). These nutrients are often limited in strict plant-based diets, and their presence in *H. erinaceus* contributes to dietary adequacy. At the same time, polysaccharides such as β-glucans add prebiotic and immunomodulatory functions, while phenolics, organic acids, and ergothioneine contribute to antioxidant and anti-inflammatory protection (Gąsecka et al. 2020; Roda et al. 2022; Apparoo et al. 2022; Ahmed et al. 2023; Singh et al. 2025).

The integration of these bioactive compounds into plant-based foods opens opportunities not only for nutritional enrichment but also for the development of products with added health claims, especially in cognitive support and healthy aging.

To illustrate the multifunctional contributions of *H. erinaceus* to plant-based diets, Fig. 11.7 presents a conceptual diagram summarizing its main applications. The mushroom contributes to flavor enhancement (through umami and salty notes), texture improvement (via its fibrous structure), nutritional enrichment (proteins, vitamins, and minerals), bioactive benefits (neuroprotection and antioxidant activity), and direct incorporation into meat and seafood analogs (burgers, sausages, and shrimp alternatives).

In summary, *H. erinaceus* offers a rare convergence of sensory appeal, nutritional balance, and functional bioactivity. Its integration into plant-based diets not only improves product quality and consumer acceptance but also aligns with broader goals of sustainable food systems and health promotion.

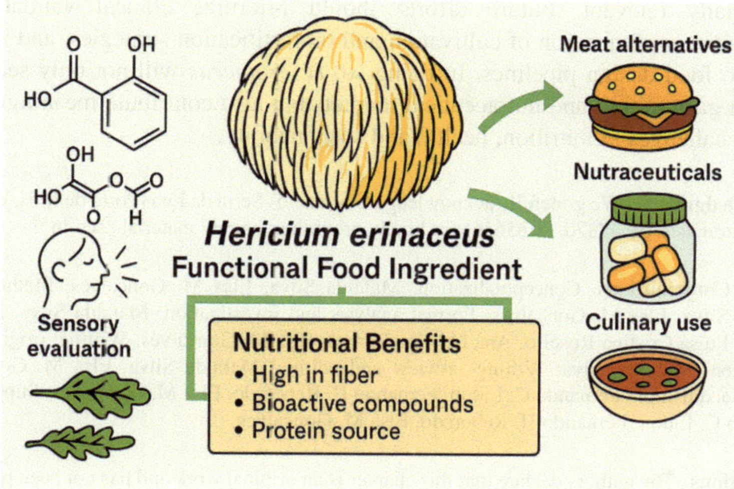


**Fig. 11.7** Conceptual diagram of the main applications of *Hericium erinaceus* in plant-based diets, highlighting its contributions to sensory quality, nutrition, bioactivity, and food innovation (image developed by artificial intelligence (AI))

## 11.5 Conclusions

The present work demonstrates that *Hericium erinaceus* is far more than a niche edible mushroom; it is an exemplary model for understanding how food biodiversity can contribute simultaneously to nutrition, sensory quality, and human health. By uniting original experimental data with a critical review of the literature, this study advances the knowledge of its nutritional composition, organic and phenolic acids, polysaccharides, amino acid profile, and bioactive metabolites such as erinacines, hericenones, and ergothioneine. The introduction of the first descriptive sensory lexicon for this species represents a significant step forward, providing a scientific language to describe its organoleptic identity and linking these attributes directly to its chemical composition.

The findings position *H. erinaceus* at the intersection of gastronomy, biotechnology, and preventive medicine. Its capacity to enrich plant-based diets with umami taste, fibrous texture, and essential nutrients addresses major sensory and nutritional limitations of current meat and seafood analogs. At the same time, its neuroprotective diterpenoids and antioxidant amino acid derivatives open avenues for nutraceutical development targeted at cognitive decline and age-related disorders. Few food species unite these diverse qualities in a single organism.

The broader implication of this work is that mushrooms like *H. erinaceus* can play a decisive role in the transition toward sustainable and health-promoting food systems. By valorizing such species, the food sector can innovate beyond imitation and offer products that are sensorially appealing, nutritionally balanced, and functionally relevant. Future efforts should prioritize clinical validation of bioactivities, optimization of cultivation and biofortification strategies, and integration into food design pipelines. In doing so, *H. erinaceus* will not only secure its place in gastronomy and nutraceutical markets but also contribute meaningfully to global challenges in nutrition, health, and sustainability.

**Acknowledgments** We gratefully acknowledge Cogumelos Serra da Lua (Porto de Mós, Portugal; GPS coordinates: 39.47870, -8.85636) for kindly providing the raw material used in this work.

**Author Contributions** Conceptualization: Mafalda Silva, Elsa M. Gonçalves; Methodology: Mafalda Silva, Elsa M. Gonçalves; Formal analysis and investigation: Mafalda Silva, Manuela Lageiro, Luisa Cristina Roseiro, Ana Cristina Ramos, Elsa M. Gonçalves; Writing: original draft preparation: Mafalda Silva; Writing—review and editing: Mafalda Silva, Elsa M. Gonçalves; Funding acquisition: Fernando C. Lidon, Fernando H. Reboredo, Elsa M. Gonçalves; Supervision: Fernando C. Lidon, Fernando H. Reboredo, Elsa M. Gonçalves.

**Declarations** The authors declare that this chapter is an original work and has not been published or submitted elsewhere, either in whole or in part. All sources of information and ideas from other works have been properly cited. The authors confirm that this chapter complies with Springer Nature's policies on originality, plagiarism, and ethical publication practices.

The authors declare that a generative artificial intelligence (AI) tool was used solely for the creation of the conceptual diagram (Fig. 11.7) associated with this chapter. The AI tool assisted in producing a visual representation of the authors' original concepts and data. All design inputs, verification, and final approval were performed by the authors. No AI tools were used to generate or modify the textual, analytical, or interpretative content of this chapter. The authors take full responsibility for the accuracy and integrity of all content presented.

The authors acknowledge financial support from FCT, through the strategic project UIDB/04035/2020 (https://doi.org/10.54499/UIDB/04035/2020) granted to the GeoBioTec Research Institute.

## References

Ahmed AF, Mahmoud GA-E, Hefzy M, Liu Z, Ma C (2023) Overview on the edible mushrooms in Egypt. J Future Foods 3(1):8–15

Aisala H, Sola J, Hopia A, Linderborg KM, Sandell M (2019) Odor-contributing volatile compounds of wild edible Nordic mushrooms analyzed with HS–SPME–GC–MS and HS–SPME–GC–O/FID. Food Chem 283:566–578

Apparoo Y, Phan CW, Kuppusamy UR, Sabaratnam V (2022) Ergothioneine and its prospects as an anti-ageing compound. Exp Gerontol 170:111982

Atila F, Nadhim Owaid M, Ali Shariati M (2017) The nutritional and medical benefits of *Agaricus bisporus*: a review. J Microbiol Biotechnol Food Sci 7(3):281–286

Ba DM, Ssentongo P, Beelman RB, Muscat J, Gao X, Richie JP (2021) Higher mushroom consumption is associated with lower risk of cancer: a systematic review and meta-analysis of observational studies. Adv Nutr 12(5):1691–1704

Bach F, Helm CV, Bellettini MB, Maciel GM, Haminiuk CWI (2017) Edible mushrooms: a potential source of essential amino acids, glucans and minerals. Int J Food Sci Technol 52(11):2382–2392

Banerjee DK, Das AK, Banerjee R, Pateiro M, Nanda PK, Gadekar YP, Biswas S, McClements DJ, Lorenzo JM (2020) Application of Enoki mushroom (Flammulina Velutipes) stem wastes as functional ingredients in goat meat nuggets. Foods 9(4):432

Barros L, Pereira C, Ferreira ICFR (2013) Optimized analysis of organic acids in edible mushrooms from Portugal by ultra fast liquid chromatography and photodiode Array detection. Food Anal Methods 6(1):309–316

Boulifa A, Raftery MJ, Franzén AS, Radecke C, Stintzing S, Blohmer J-U, Pecher G (2024) Role of beta-(1→3)(1→6)-D-glucan derived from yeast on natural killer (NK) cells and breast cancer cell lines in 2D and 3D cultures. BMC Cancer 24(1):339

Chen Z, Fang X, Wu W, Chen H, Han Y, Yang H, Gao H (2021) Effects of fermentation with Lactiplantibacillus plantarum GDM1.191 on the umami compounds in shiitake mushrooms (Lentinus edodes). Food Chem 364:130398

Chun SS, Chambers E, Han I (2020) Development of a sensory flavor lexicon for mushrooms and subsequent characterization of fresh and dried mushrooms. Foods 9(8):980

Contato AG, Conte-Junior CA (2025) Lion's mane mushroom (Hericium erinaceus): a neuroprotective fungus with antioxidant, anti-inflammatory, and antimicrobial potential—a narrative review. Nutrients 17:1307

den Hartigh L (2019) Conjugated linoleic acid effects on cancer, obesity, and atherosclerosis: a review of pre-clinical and human trials with current perspectives. Nutrients 11(2):370

Dimopoulou M, Kolonas A, Mourtakos S, Androutsos O, Gortzi O (2022) Nutritional composition and biological properties of sixteen edible mushroom species. Appl Sci 12(16):8074

Fetriyuna F, Rafi AZ, Zaida Z, Purwestri RC, Sikin AM (2025) From forest to fork: a systematic review of mushroom-based meat analogues. Int J Food Sci Technol 60:vvaf030

Gao S, Huang Z, Feng X, Bian Y, Huang W, Liu Y (2020) Bioconversion of rice straw agro-residues by Lentinula edodes and evaluation of non-volatile taste compounds in mushrooms. Sci Rep 10(1):1814

Gąsecka M, Magdziak Z, Siwulski M, Mleczek M (2018) Profile of phenolic and organic acids, antioxidant properties and ergosterol content in cultivated and wild growing species of Agaricus. Eur Food Res Technol 244(2):259–268

Gąsecka M, Siwulski M, Magdziak Z, Budzyńska S, Stuper-Szablewska K, Niedzielski P, Mleczek M (2020) The effect of drying temperature on bioactive compounds and antioxidant activity of Leccinum scabrum (Bull.) Gray and Hericium erinaceus (Bull.) Pers. J Food Sci Technol 57(2): 513–525

González A, Cruz M, Losoya C, Nobre C, Loredo A, Rodríguez R, Contreras J, Belmares R (2020) Edible mushrooms as a novel protein source for functional foods. Food Funct 11(9):7400–7414

Grimm D, Wösten HAB (2018) Mushroom cultivation in the circular economy. Appl Microbiol Biotechnol 102(18):7795–7803. https://doi.org/10.1007/s00253-018-9226-8

Grosshauser S, Schieberle P (2013) Characterization of the key odorants in Pan-fried white mushrooms (Agaricus bisporus L.) by means of molecular sensory science: comparison with the raw mushroom tissue. J Agric Food Chem 61(16):3804–3813

Guinard J-X, Myrdal Miller A, Mills K, Wong T, Lee SM, Sirimuangmoon C, Schaefer SE, Drescher G (2016) Consumer acceptance of dishes in which beef has been partially substituted with mushrooms and sodium has been reduced. Appetite 105:449–459

Harada-Padermo S d S, Dias-Faceto LS, Selani MM, Alvim ID, Floh EIS, Macedo AF, Bogusz S, Dias CT d S, Conti-Silva AC, Vieira TMF d S (2020) Umami ingredient: flavor enhancer from shiitake (Lentinula edodes) byproducts. Food Res Int 137:109540

Hu T, Hui G, Li H, Guo Y (2020) Selenium biofortification in Hericium erinaceus (lion's mane mushroom) and its in vitro bioaccessibility. Food Chem 331:127287

Jozífek M, Praus L, Matějka J, Jablonský I, Koudela M (2025) Selenium uptake by Hericium erinaceus Basidiocarps on various substrates and their effect on growth and yield. Agriculture 15(5):460

Koutrotsios G, Danezis G, Georgiou C, Zervakis G (2020) Elemental content in Pleurotus ostreatus and Cyclocybe cylindracea mushrooms: correlations with concentrations in cultivation substrates and effects on the production process. Molecules 25(9):2179

Lee J, Chambers DH (2010) Flavors of green tea change little during storage. J Sens Stud 25(4): 512–520

Lee C-Y, Park J-E, Kim B-B, Kim S-M, Ro H-S (2009) Determination of mineral components in the cultivation substrates of edible mushrooms and their uptake into fruiting bodies. Mycobiology 37(2):109

Leong YK, Yang F-C, Chang J-S (2021) Extraction of polysaccharides from edible mushrooms: emerging technologies and recent advances. Carbohydr Polym 251:117006

Lu X, Brennan MA, Serventi L, Brennan CS (2018) Incorporation of mushroom powder into bread dough—effects on dough rheology and bread properties. Cereal Chem 95(3):418–427

Łysakowska P, Sobota A, Wirkijowska A (2023) Medicinal mushrooms: their bioactive components, nutritional value and application in functional food production—a review. Molecules 28(14):5393

Meyer F, Hutmacher A, Lu B, Steiger N, Nyström L, Narciso JO (2023) Vegan shrimp alternative made with pink oyster and lion's mane mushrooms: nutritional profiles, presence of conjugated phenolic acids, and prototyping. Curr Res Food Sci 7:100572

Myrdal Miller A, Mills K, Wong T, Drescher G, Lee SM, Sirimuangmoon C, Schaefer S, Langstaff S, Minor B, Guinard J-X (2014) Flavor-enhancing properties of mushrooms in meat-based dishes in which sodium has been reduced and meat has been partially substituted with mushrooms. J Food Sci 79(9):S1795

Nicolai M, Palma ML, Reis R, Amaro R, Fernandes J, Gonçalves EM, Silva M, Lageiro M, Charmier A, Maurício E, Branco P, Palma C, Silva J, Nunes MC, Fernandes PCB, Pereira P (2025) Assessing the potential of brewer's spent grain to enhance cookie physicochemical and nutritional profiles. Foods 14(1):95

Panda A, Alvarenga N, Lita da Silva J, Partidário A, Lageiro M, Roseiro C, Dias J (2022) Influence of cocoa origin on the nutritional characterization of chocolate. Eur Food Res Technol 248(10): 2569–2577

Pereira N, Farrokhi M, Vida M, Lageiro M, Ramos AC, Vieira MC, Alegria C, Gonçalves EM, Abreu M (2023) Valorisation of wasted immature tomato to innovative fermented functional foods. Foods 12(7):1532

Politowicz J, Lech K, Lipan L, Figiel A, Carbonell-Barrachina ÁA (2018a) Volatile composition and sensory profile of shiitake mushrooms as affected by drying method. J Sci Food Agric 98(4):1511–1521

Politowicz J, Lech K, Sánchez-Rodríguez L, Figiel A, Szumny A, Grubor M, Carbonell-Barrachina ÁA (2018b) Volatile composition and sensory profile of oyster mushroom as affected by drying method. Dry Technol 36(6):685–696

Quaresma MAG, Pereira G, Nunes ML, Sponda C, Jardim A, Gonçalves H, Santos C, Roseiro LC (2023) Evaluating dried salted cod amino acid signature for nutritional quality assessment and discriminant analysis. Front Nutr 10:1144713

Ramos M, Burgos N, Barnard A, Evans G, Preece J, Graz M, Ruthes AC, Jiménez-Quero A, Martínez-Abad A, Vilaplana F, Ngoc LP, Brouwer A, van der Burg B, del Carmen Garrigós M, Jiménez A (2019) Agaricus bisporus and its by-products as a source of valuable extracts and bioactive compounds. Food Chem 292:176–187

Ratto D, Corana F, Mannucci B, Priori EC, Cobelli F, Roda E, Ferrari B, Occhinegro A, Di Iorio C, De Luca F, Cesaroni V, Girometta C, Bottone MG, Savino E, Kawagishi H, Rossi P (2019) Hericium erinaceus improves recognition memory and induces hippocampal and cerebellar neurogenesis in frail mice during aging. Nutrients 11(4):715

Riveros CG, Mestrallet MG, Quiroga PR, Nepote V, Grosso NR (2013) Preserving sensory attributes of roasted peanuts using edible coatings. Int J Food Sci Technol 48(4):850–859

Roda E, Ratto D, De Luca FD, Desiderio A, Ramieri M, Goppa L, Savino E, Bottone MG, Locatelli CA, Rossi P (2022) Searching for a longevity food, we bump into Hericium erinaceus primordium rich in Ergothioneine: the "longevity vitamin" improves locomotor performances during aging. Nutrients 14(6):1177

Rodrigues DMF, Freitas AC, Rocha-Santos TAP, Vasconcelos MW, Roriz M, Rodríguez-Alcalá LM, Gomes AMP, Duarte AC (2015) Chemical composition and nutritive value of Pleurotus citrinopileatus var cornucopiae, P. eryngii, P. salmoneo stramineus, Pholiota nameko and Hericium erinaceus. J Food Sci Technol 52(11):6927–6939

Rosales CK, Suwonsichon S (2015) Sensory lexicon of pomelo fruit over various cultivars and fresh-cut storage. J Sens Stud 30(1):21–32
Rotola-Pukkila M, Yang B, Hopia A (2019) The effect of cooking on umami compounds in wild and cultivated mushrooms. Food Chem 278:56–66
Serna J, Bergwitz C (2020) Importance of dietary phosphorus for bone metabolism and healthy aging. Nutrients 12(10):3001
Sheng K, Wang C, Chen B, Kang M, Wang M, Liu K, Wang M (2021) Recent advances in polysaccharides from Lentinus edodes (Berk.): isolation, structures and bioactivities. Food Chem 358:129833
Silva M, Ramos AC, Lidon FJ, Reboredo FH, Gonçalves EM (2024) Pre- and postharvest strategies for Pleurotus ostreatus mushroom in a circular economy approach. Foods 13(10):1464
Silva M, Vida M, Ramos AC, Lidon FJ, Reboredo FH, Gonçalves EM (2025) Storage temperature effect on quality and shelf-life of Hericium erinaceus mushroom. Horticulturae 11(2):158
Singh U, Tiwari P, Kelkar S, Kaul D, Tiwari A, Kapri M, Sharma S (2023) Edible mushrooms: a sustainable novel ingredient for meat analogs. eFood 4(6):e122
Singh A, Saini RK, Kumar A, Chawla P, Kaushik R (2025) Mushrooms as nutritional powerhouses: a review of their bioactive compounds, health benefits, and value-added products. Foods 14(5): 741
Sławińska A, Fornal E, Radzki W, Skrzypczak K, Zalewska-Korona M, Michalak-Majewska M, Parfieniuk E, Stachniuk A (2016) Study on vitamin D2 stability in dried mushrooms during drying and storage. Food Chem 199:203–209
Smith E, Ottosson F, Hellstrand S, Ericson U, Orho-Melander M, Fernandez C, Melander O (2020) Ergothioneine is associated with reduced mortality and decreased risk of cardiovascular disease. Heart 106(9):691–697
Sun L, Zhang Z, Xin G, Sun B, Bao X, Wei Y, Zhao X, Xu H (2020) Advances in umami taste and aroma of edible mushrooms. Trends Food Sci Technol 96:176–187
Suwonsichon S (2019) The importance of sensory lexicons for research and development of food products. Foods 8(1):27
Szućko-Kociuba I, Trzeciak-Ryczek A, Kupnicka P, Chlubek D (2023) Neurotrophic and neuroprotective effects of Hericium erinaceus. Int J Mol Sci 24(21):15960
van Eck NJ, Waltman L (2010) Software survey: VOSviewer, a computer program for bibliometric mapping. Scientometrics 84(2):523–538
Vetter J (2023) The mushroom Glucans: molecules of high biological and medicinal importance. Foods 12(5):1009
Yang J, Li Q, Feng Y, Zeng Y (2023) Iron deficiency and iron deficiency anemia: potential risk factors in bone loss. Int J Mol Sci 24(8):6891
Zhao D, Huang Y, Wang B, Chen H, Pan W, Yang M, Xia Z, Zhang R, Yuan C (2023) Dietary intake levels of iron, copper, zinc, and manganese in relation to cognitive function: a cross-sectional study. Nutrients 15(3):704

# Effect of Mushroom Nutraceuticals on Human Health

# 12

R. Rasyiqah-Nabihah, R. Mohammad-Syaril, F. M. N. A. Aida, Abzar, and Siva Raseetha

**Abstract**

Chemical constituents of mushrooms provide a wide range of therapeutic effects, enabling them to act as immunomodulatory, anticarcinogenic, antiviral, antioxidant, and anti-inflammatory agents. The accumulated secondary metabolites in medicinal mushrooms are widely accepted as sources of safe and effective nutraceuticals, cosmeceuticals, and pharmaceuticals. These compounds, working individually or synergistically, underscore mushrooms' vast potential as powerhouses of health-benefiting molecules. The cardioprotective effects of mushrooms are attributed to their rich composition of polysaccharides, sterols, antioxidants, and anti-inflammatory compounds alongside triterpenes, unique amino acids like ergothioneine, and fibrinolytic enzymes. This diverse array of bioactive works synergistically to improve lipid profiles, regulate blood pressure, and maintain vascular integrity. As such, mushrooms represent a powerful functional food with significant potential for inclusion in preventive dietary strategies aimed at reducing the global burden of cardiovascular disease. While the preclinical data are compelling, the path forward requires a continued commitment to rigorous scientific validation. This chapter delves into the goodness of mushroom nutraceuticals on human health from a holistic perspective.

R. Rasyiqah-Nabihah · F. M. N. A. Aida · S. Raseetha (✉)
Faculty of Applied Sciences, Universiti Teknologi MARA, Shah Alam, Selangor, Malaysia
e-mail: raseetha@uitm.edu.my

R. Mohammad-Syaril
Johor Bahru Public Health Laboratory, Tampoi, Johor Bahru, Johor, Malaysia

Abzar
Faculty of Applied Sciences, Universiti Teknologi MARA, Shah Alam, Selangor, Malaysia

Department of Crop Science, Faculty of Agriculture, University Putra Malaysia, Serdang, Selangor, Malaysia



B. Giri et al. (eds.), *Advances in Mushroom Research*,
https://doi.org/10.1007/978-981-95-7692-0_12

Immunomodulation, anti-inflammatory, and anticancer action are explained based on research findings. Besides, cognitive enhancement and neuroprotection as well as cardiovascular effects from mushrooms are detailed. Additionally, the metabolic and endocrine effects on the human body are described in detail. It also aims to provide researchers and food industrialists with more pathways in mushroom research, perhaps in functional food products involving mushrooms as their bioactive components.

**Keywords**

Nutraceuticals · Human health · Immunomodulation · Metabolic

## 12.1 Introduction

Mushrooms, those fascinating fleshy, spore-bearing fruiting bodies of fungi, have captivated human attention since ancient times, extending far beyond their culinary appeal (Ogidi et al. 2020; Sharifi-Rad et al. 2020). Historically, civilizations like the ancient Greeks believed mushrooms imbued strength in warriors, while the Romans revered them as "Food of the Gods," serving them only on festive occasions, and the Chinese culture, for centuries, has treasured them as an "elixir of life" and a health food (Ogidi et al. 2020). This long-standing appreciation stems from their exquisite flavor profiles and, more profoundly, their remarkable medicinal properties (Ogidi et al. 2020; Sharifi-Rad et al. 2020).

Despite their millennial existence and empirical documentation in traditional medicine, particularly in Asian countries (Bell et al. 2022; Sharifi-Rad et al. 2020), the full potential of mushrooms as food and medicine remains largely underutilized (Bell et al. 2022), especially in regions like Nigeria, where numerous wild macrofungi are yet to be properly identified and studied (Ogidi et al. 2020). However, recent advancements in medical and nutritional sciences have thrust natural products derived from both edible and nonedible mushrooms into the spotlight, as this burgeoning interest isn't surprising when you consider that mushrooms offer a unique blend of nutritional value and biologically active compounds with demonstrated health benefits.

The functional value of mushrooms is built upon a strong nutritional foundation. They are a rich source of high-quality protein containing all nine essential amino acids, dietary fiber, and essential minerals like selenium and potassium, as well as B vitamins and vitamin D (Abebaw 2020; Kumar 2015). This is complemented by their favorable low-calorie and low-fat profile, making them an ideal component of a healthy diet (Abebaw 2020). However, beyond basic nutrition, mushrooms contain functionally active compounds capable of influencing a wide range of biological pathways, which explains their therapeutic potential (Chaturvedi et al. 2018).

More than just a source of basic nutrition, mushrooms have gained considerable attention as therapeutic foods capable of preventing and managing a range of diseases such as hypertension, diabetes, hypercholesterolemia, and cancer (Kumar

2015). Their functional characteristics are primarily due to the presence of dietary fiber, especially chitin and beta-glucans (Kumar 2015; Sharifi-Rad et al. 2020). As a functional food, mushrooms possess beneficial effects for overall health and are considered an unmatched source of healthy foods and drugs (Chaturvedi et al. 2018). They are also viewed as functional food candidates for developing novel dietary supplements and therapeutics due to their limited probability of surplus dosage or toxicity and negligible side effects (Bell et al. 2022).

The estimated number of mushroom species on Earth is around 140,000, comprising approximately 22,000 identified species (Ogidi et al. 2020; Sharifi-Rad et al. 2020). Of these, about 2000 species are identified as edible (Abebaw 2020; Pandey et al. 2023), but fewer than 20 kinds appear in local markets in Nepal (Pandey et al. 2023). Only about 25 species have been intensively cultivated, and about 35 are grown on a commercial scale, while 20 are cultivated industrially (Abebaw 2020). For a long time, mushrooms were extensively used worldwide for their nutritional and medicinal features, predominantly in some Asian, Northern, and Central American countries (Sharifi-Rad et al. 2020).

The present chapter aims to provide a comprehensive account of the nutritional value, bioactive compounds, and diverse therapeutic applications of various edible and wild mushrooms. It explores their roles in preventive healthcare and immune system enhancement and their potential as antiviral, anti-inflammatory, and antineoplastic agents. Additionally, it highlights their market value as nutraceuticals and discusses future perspectives for further exploration and commercialization. The remaining subchapters aim to consolidate the current knowledge, emphasizing mushrooms with immense potential for human health and the global economy.

### 12.1.1 Beyond the Plate: Nature's Hidden Treasure

Mushrooms are macrofungi, distinguished by their fleshy, spore-bearing fruiting bodies that are visible to the naked eye. For millennia, they have held a significant place in human culture, valued not only as a source of food but also for their profound medicinal applications (Ogidi et al. 2020; Sharifi-Rad et al. 2020). Beyond their direct human use, mushrooms are integral components of terrestrial ecosystems, where they play critical roles in nutrient cycling and maintaining biodiversity (Shu et al. 2025).

Fungi represent a distinct kingdom of eukaryotic organisms, possessing complex cellular structures, including a nucleus that houses their genetic material (Ogidi et al. 2020). A key characteristic that separates them from plants is their complete lack of chlorophyll, rendering them incapable of photosynthesis (Sharifi-Rad et al. 2020). Instead of producing their own food, they are saprophytic, obtaining nutrition by secreting a wide range of extracellular enzymes to degrade complex organic matter into soluble substances, which are then absorbed (Ogidi et al. 2020). The fungal kingdom is characterized by immense biodiversity, with estimates suggesting there could be as many as 3.8 million species on Earth (Panda et al. 2022). Of this vast number, only a small fraction, approximately 120,000 species, has been formally identified (Panda et al. 2022). This massive, largely unexplored biodiversity represents a significant frontier for scientific discovery and a neglected source for new drugs.

The primary body of a mushroom fungus is the mycelium, a network of thread-like hyphae that is typically hidden within its substrate, such as soil or wood (Cateni et al. 2022). The visible part, commonly referred to as the "mushroom," is the fruit body, or basidiocarp, which is the reproductive structure (Ogidi et al. 2020). Traditional mycologists have long relied on the macroscopic and microscopic features of the fruit body, such as its shape, color, and spore characteristics, to distinguish between species. However, modern advancements in deoxyribonucleic acid (DNA) technology have proven to be powerful and essential tools for resolving taxonomic challenges and are now widely used in medicine for the rapid and accurate identification of fungi (Bell et al. 2022; Ogidi et al. 2020).

The life cycle of Basidiomycetes, the phylum to which most common mushrooms belong, is characterized by a fascinating and unique dikaryotic stage (Ogidi et al. 2020). This stage begins when mycelia from two different mating strains combine to produce a secondary mycelium that contains haploid nuclei from both parent strains. This dikaryotic state, where two separate haploid nuclei coexist within a single cell, is often the dominant phase in the mushroom's life cycle. This also allows for a unique form of mating known as di-mon mating or Buller mating, where the dikaryon retains the ability to fertilize other monokaryons it encounters (Auxier et al. 2022).

Eventually, the secondary mycelium generates a basidiocarp, where sexual reproduction culminates (Ogidi et al. 2020). Within specialized club-shaped structures called basidia, the two Haploid nuclei fuse in a process called karyogamy, giving rise to a diploid zygote. This zygote then undergoes meiosis to produce haploid nuclei, which migrate into basidiospores. These spores are then released to germinate and generate new monokaryotic hyphae, completing the cycle.

Mushrooms are indispensable to human life and vital for the environment, acting as major decomposers of organic matter, particularly in forests (Shu et al. 2025). They are key controllers of carbon and nutrient cycling, which is essential for the maintenance of biodiversity. Many mushroom species form mutually dependent symbiotic associations with most vascular plants, known as mycorrhizae. Ectomycorrhizal fungi, which include both Ascomycetes and Basidiomycetes, protect plant roots against pathogens, improve nutrient absorption from the soil, and can help remediate heavy metal contamination, thereby preventing extreme soil pollution.

Truffles, on the other hand, are a unique group of underground macrofungi typically found in arid or semi-arid ecosystems (Tejedor-Calvo and Marco 2025). They are expensive food commodities, greatly appreciated for both their culinary and medicinal properties. Ecologically, they enhance their host plant roots' capacity to resist dry spells and have been utilized by humans since the Paleolithic period, with their historical application often linked to spiritualism.

The unique biology and biochemistry of mushrooms make them a valuable resource for a wide range of applications in medicine, biotechnology, and the food industry (Panda et al. 2022). Their vast, untapped biodiversity represents a significant frontier for scientific discovery. The prospects for identifying new bioactive compounds from the thousands of underexplored species are immense, promising future innovations in pharmaceuticals, nutraceuticals, and sustainable materials (Jeitler et al. 2020).

Hence, mushrooms are a biologically fascinating and ecologically critical kingdom of organisms with profound significance for human health and culture (Ogidi et al. 2020). From their foundational role in nutrient cycling to their application as nutritious foods and sources of potent medicines, their importance cannot be overstated (Kumar 2015; Sharifi-Rad et al. 2020). The immense, largely unexplored biodiversity of fungi represents a valuable resource for future scientific discovery and sustainable use, promising novel solutions for medicine, industry, and environmental management (Panda et al. 2022).

### 12.1.2 Bioactive Powerhouses

The true hero of mushrooms lies in their vast array of bioactive compounds, known as mycochemicals (Cateni et al. 2022). These substances are the key to their extensive health-promoting properties (Cateni et al. 2022). Mycochemistry, as a distinct discipline, focuses on the isolation, structure elucidation, biosynthesis, metabolism, and biological properties of these diverse chemical substances.

Researchers are increasingly applying new molecular methods, including biomarker searches in the genome and proteome, to identify previously unknown mushrooms and detail their low-molecular-weight compounds, such as glycosphingolipids, biochrome quinones, and polyphenolic isoflavones, as well as high-molecular-weight compounds like different glucans, glycopeptides, and ribonucleoprotein complexes (Chaturvedi et al. 2018).

While the biochemical and nutrient composition of industrialized mushrooms is well-known, detailed nutritive data for many wild and underexploited edible fungi are limited or nonexistent. This gap in knowledge poses a challenge, particularly in low-income settings where these fungi serve as valuable resources for food security and income.

Nonetheless, many research studies have determined the safeguarding health outcomes of wholesome mushrooms to protect against or treat various chronic diseases. Most edible mushrooms, like most food items, have a limited probability of surplus dosage or toxicity, and along with negligible side effects, they are perfect candidates for developing novel dietary supplements and therapies.

Mushroom products are commercialized as fresh or dried, as biomass dietary supplements, or as extracts. This impacts marketing, as mushroom extracts are legally considered medicinal nutraceuticals or drugs, while biomass is typically deemed foodstuffs or dietary supplements. However, global legislation governing dietary supplements remains imprecise, with no general agreement, allowing them to be categorized as foodstuffs and/or medicinal products based on various factors.

The remarkable diversity and therapeutic potential of various mushroom species are evident in their unique nutritional profiles and extensive medicinal applications. Table 12.1 offers a comprehensive overview of several prominent edible mushrooms, detailing their macronutrient composition and highlighting their significant medicinal values, ranging from anticancer and antiviral to immunomodulatory and antiaging properties. Table 12.1 serves as a quick reference to the multifaceted benefits these fungi offer.

The efficacy of mushrooms' properties is closely linked to a rich diversity of specific classes of bioactive compounds. These mycochemicals, found in both the mushroom's fruiting body and its mycelium, work through various mechanisms to produce the health benefits observed in preclinical and clinical studies. The following sections detail the major categories of these powerful molecules.

High-molecular-weight polysaccharides, particularly beta-glucans, are among the most studied and therapeutically significant compounds found in mushrooms (Sivanesan et al. 2022). These complex carbohydrates, which consist of a glucose residue backbone with various linkages, are renowned for their potent immunomodulatory activities (Macharia et al. 2022). Rather than having direct cytotoxic effects, beta-glucans typically function as biological response modifiers, stimulating the host's innate and adaptive immune systems by activating cells like macrophages, natural killer (NK) cells, and T cells (Tacchini et al. 2023; Park 2022). This immune enhancement is a key mechanism behind their anticancer effects. Well-known examples include Lentinan from *Lentinula edodes* and Polysaccharide-K (PSK) from *Trametes versicolor*, both of which are approved as adjuvant cancer immunotherapeutics in Japan (Janardhanan et al. 2020).

In addition to these complex carbohydrates, mushrooms contain a large and diverse class of triterpenoids and steroids (Cateni et al. 2022). Triterpenoids, such as the ganoderic acids found in *Ganoderma lucidum*, have demonstrated a multitude of benefits, including anticancer, antihyperlipidemic, and neuroprotective effects (Dai et al. 2021; Janardhanan et al. 2020). Steroids, particularly ergosterol (a precursor to vitamin $D_2$), are also a key component of mushrooms. Ergosterol has been identified as a potent anti-angiogenic agent, capable of inhibiting the formation of new blood vessels that tumors need to grow (Panda et al. 2022). These compounds often exert direct cytotoxic effects on cancer cells and can modulate critical signaling pathways involved in cell survival and proliferation (Gariboldi et al. 2023).

Phenolic compounds are another key bioactive component known for their strong antioxidant and anti-inflammatory effects (Assemie and Abaya 2022). These molecules, which include flavonoids and phenolic acids, act as potent free radical scavengers and can help protect cells from oxidative damage, a key factor that contributes to the development of many chronic diseases (Cateni et al. 2022). The antioxidant properties of mushrooms are often attributed to these compounds, which can prevent lipid peroxidation and chelate metal ions, thereby reducing oxidative stress within the body (Macharia et al. 2022).

Moreover, the proteins and lectins found in mushrooms are not only of high nutritional value but also exhibit a range of pharmaceutical activities, including

**Table 12.1** Nutraceutical value of some edible mushrooms

| | | Nutritional value | | | | |
|---|---|---|---|---|---|---|
| No | Mushroom | Protein (g/100 g) | Carbohydrate (g/100 g) | Lipid (g/100 g) | Fibers (g/100 g) | Medicinal value |
| 1 | Pleurotus (oyster mushroom) | 17–42 | 37–48 | 0.5–5 | 24–31 | Anticancer, antioxidant, antitumor, antiviral, antibacterial, antidiabetic, antihypercholesterolemic, eye health, anti-arthritic, immunomodulatory, hepatoprotective, and anti-obesity |
| 2 | Ganoderma (Reishi) | 13.3 | 82.3 | 3.0 | – | Antiviral, hepatoprotective, antioxidant, antiallergenic, anticancer, hypoglycemic, immunomodulatory, hypotensive, antithrombotic, anti-inflammatory, hypocholesterolemic, antibacterial, antimutagenic, anti-osteoporotic, and antiaging |
| 3 | Lentinus (Shiitake) | 26.3 | 65.1 | 2.3 | – | Anticancer and immunomodulatory |
| 4 | Hericium (Lion's mane) | 22.3 | 57.0 | 3.5 | 7.8 | Antihypercholesterolemic |
| 5 | Flammulina (enoki) | 3.9–17.8 | 86–70.8 | 1.8–2.9 | – | Immunomodulatory and anti-asthmatic |
| 6 | Agaricus (button mushroom) | 56.3 | 37.5 | 2.7 | – | Anticancer, antidiabetic, antihypercholesterolemic, immunomodulatory, hepatoprotective, antiviral, and antimutagenic |
| 7 | Grifola (maitake) | 21.1 | 58.8 | 3.1 | 10.1 | Anticancer, antidiabetic, antihypercholesterolemic, anti-arthritic, antiviral, anti-osteoporotic, and anti-obesity |
| 8 | Cordyceps (caterpillar fungus) | 21.9 | 24.2 | 8.2 | – | Anti-asthma |
| 9 | Auricularia (wood ear mushroom) | 7.2 | 88.6 | 1.7 | – | Antiaging |
| 10 | Tremella (jelly fungus) | 4.6 | 94.8 | 0.2 | 1.4 | Anticancer and antidiabetic |

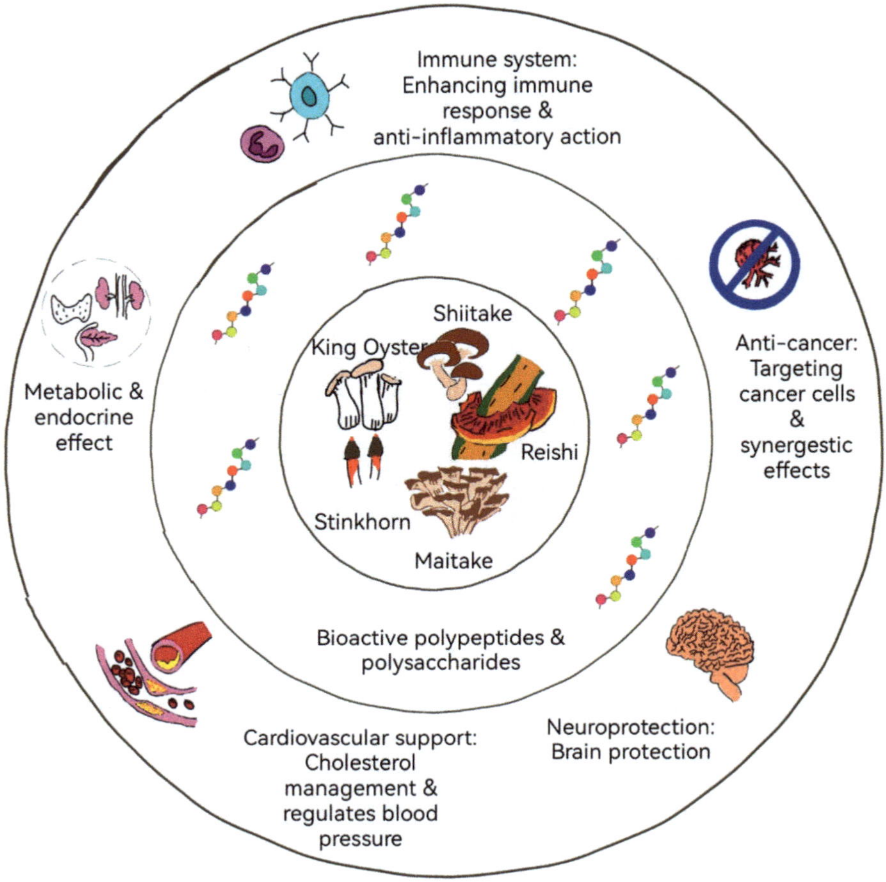


**Fig. 12.1** The role of mushrooms on human health

antitumor properties (Gariboldi et al. 2023). Lectins, a specific type of protein that binds to carbohydrates on cell surfaces, have been shown to trigger apoptosis in cancer cells, highlighting their therapeutic potential (Yan et al. 2022). This chapter delves into the contribution of mushrooms as an immune system, as an anticancer agent, as a neuroprotector, as cardiovascular support, and as metabolic and endocrine support toward better human health, as shown in Fig. 12.1.

## 12.2 Immunomodulation: Strengthening the Body's Defenses

Generally, there are a few common phyla under the Kingdom of Fungi, namely, Zygomycota, Chytridiomycota, Glomeromycota, Ascomycota (sac fungi), and Basidiomycota (club fungi). Medicinal mushrooms (MM) have been discovered mainly from the phylum Basidiomycota, followed by the phylum Ascomycota.

Club fungi reproduce externally, while Ascomycota fungi reproduce internally. Mushrooms have been widely used as medicine since ancient times. Archaeologists discovered a mummy with a mushroom as one of its possessions, believed to have been used as a laxative to treat intestinal parasite disease. The most prevalent medicinal mushrooms are from class Basidiomycota, including Reishi (*Ganoderma lucidum*), Shiitake (*Lentinula edodes* and *Lentinula polychrous*), White mushroom (*Agaricus bisporus* and *Agaricus blazei*), Maitake (*Grifola frondosa*), Turkey tail (*Coriolus versicolor*), Button mushroom (*Schizophyllum commune*), Cottony rot (*Sclerotinia sclerotiorum*), Almond mushroom (*Agaricus subrufescens*), and oyster mushrooms (*Pleurotus ostreatus*, *Pleurotus nebrodensis*, *Pleurotus sajor-caju*, and *Pleurotus citrinopileatus*). Meanwhile, examples of medicinal mushrooms from the Ascomycota phylum include Common Morel (*Morchella esculenta*), Black Morel (*Morchella conica*), Caterpillar fungus, Himalaya Viagra (*Ophiocordyceps sinensis*), Candlestick fungus (*Xylaria hypoxylon*), and Dead Moll's Fingers (*Xylaria nigripes*) (Zhao et al. 2020).

The mushroom fruiting body primarily contains 90% water and approximately 10%–40% protein and amino acids consisting of glutamine, glutamic acids, aspartic acids, valine, and leucine. Moreover, mushrooms contain carbohydrates (3%–28%), such as glycogen, chitin, and trehalose; fats (approximately 2%–8%), including palmitic, oleic, as well as linoleic fatty acids; fibers (3%–32%); and ash (8%–10%), which is mainly made up of beneficial minerals such as magnesium, iron, and zinc, and are a source of vitamins and antioxidants, including vitamin D, tocopherol, thiamine, and riboflavin. The chemical composition of mushroom products is majorly influenced by harvesting technique, storage conditions, and growth features such as mycelium development and substrate dependency on the medium as a nutrient provider, like decaying wood, sawdust, manure, and soil with organic matter, as well as environmental conditions, including temperature, humidity, pH, light, and carbon dioxide.

Medicinal mushrooms (MMs) can be defined as any macroscopic fungi that are capable of preventing and treating diseases as well as promoting balanced health through their bioactive compounds obtained from their extracts or mushroom powders of different parts, such as mycelium, which have been utilized as therapeutic agents, such as immunomodulators. In general, immunomodulators can be classified into three categories, which are immunostimulants, immunosuppressants, and immunoadjuvants (Zhao et al. 2020). To date, most immunomodulator compounds are synthetic or semi-synthetic; thus, the application of mushrooms as dietary supplements or health foods may offer natural immunomodulators with no or little toxicity. The current synthetic immunomodulator drugs reported have adverse side effects. The huge annual growth rate of natural immunomodulators worldwide shows high demand for these natural compounds as new therapeutic medicine in regulating the human immune system, including enhanced immune response, thus improving body recovery due to increased resistance to infectious diseases (Zhao et al. 2020).

### 12.2.1 Enhancing Immune Responses

Scientists have been utilizing immune stimulant compounds present abundantly in medicinal mushrooms, such as β-D-glucan, which is considered harmless but capable of triggering an immune response. Based on the theory of trained immunity (TRIM), the immune protection of an individual against other pathogens can be increased through exposure to one pathogen. Researchers suggest the immune response can be improved after innate immune cells go through an initial stimulus where mitochondrial, metabolic, and epigenetic reprogramming occurs, resulting in innate immune cells such as macrophages, neutrophils, natural killer (NK) cells, and dendritic cells having memory called trained immunity (TRIM). TRIM inducers such as β-D-glucan and BCG vaccines could provide protection against secondary heterologous stimuli, including bacterial and viral infections. Figure 12.2 shows the trained immunity (TRIM) principle.

Besides polysaccharide β-D-glucans, among familiar bioactive compounds extracted from medicinal mushrooms, MMs with immunomodulatory properties are polysaccharides with or without modification on the side chain, such as polysaccharopeptides and polysaccharide proteins. Polysaccharopeptide (PSP) can be isolated from the mushroom *T. versicolor*, while polysaccharide-protein can be found in mushrooms such as *Tricholoma giganteum* and *Macrocybe gigantea.*

The Syk Pathway (Spleen Tyrosine Kinase Pathway) briefly explains the mechanism of β-D-glucans in enhancing immune response. Firstly, β-D-glucans, which are present in mushroom cell walls, bind to Dectin-1 receptors of the innate immune

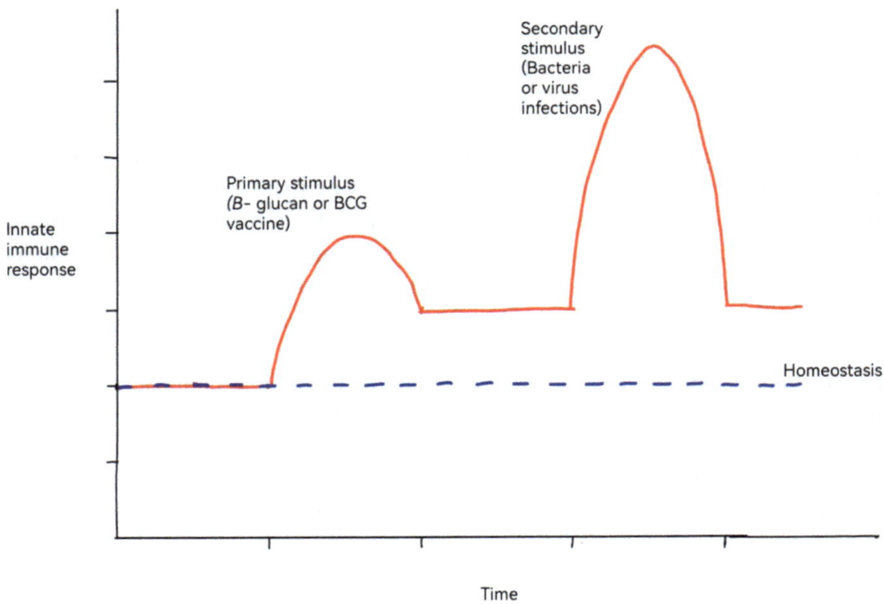


**Fig. 12.2** The Trained Immunity (TRIM) principle

system (macrophages, dendritic cells). The binding of β-D-glucan and the innate immune cell later activates the intracellular signal, the Syk enzyme (a tyrosine kinase), present inside the immune cell. The Syk enzyme activates phagocytosis, ROS (Reactive Oxygen Species), and NF-κB (Nuclear Factor kappa-light-chain-enhancer of activated B cells).

ROS is capable of killing pathogens. In general, after β-D-glucans bind with Dectin-1, it activates the enzyme complex NADPH oxidase. NADPH oxidase transfers electrons to the oxygen molecule, causing it to transform into superoxide. Superoxide dismutase (SOD) converts superoxide into hydrogen peroxide (a more powerful ROS). Under the Fenton reaction, Iron, $Fe^{2+,}$ breaks down hydrogen peroxide into hydroxyl radicals, which are considered extremely reactive and damaging to proteins, DNA, and membranes of invading pathogens. Besides, in neutrophils, the enzyme myeloperoxidase (MPO) facilitates the production of hypochlorous acid (HOCl) by the combination of both hydrogen peroxide and chloride ions ($Cl^-$). Hypochlorous acid (HOCl) is a very effective molecule to kill pathogens (bacteria and fungi).

NF-κB (Nuclear Factor kappa-light-chain-enhancer of activated B cells) is a downstream target of Syk pathways. The activation of Syk disrupts inhibitor IκB, releasing NF-κB. NF-κB moves from the cytoplasm into the nucleus. In the nucleus, it binds to DNA and activates immune-related genes such as cytokines, chemokines, Colony-Stimulating Factors (CSFs), and adhesion molecules. It initiates inflammation as well as supports immune cell activation, recruitment, and survival. The boost of cytokine production stimulates bone marrow to produce neutrophils and monocytes (immune cells).

Lectin is a large protein (with many amino acids) attached to sugar chains. It is a crucial bioactive compound isolated from MMs and reported to have immunomodulatory effects, including activating lymphocytes, stimulating production of nitrate, facilitating the production of macrophage-activating factors, and inhibiting proliferation of HeLa, S180, SW480, U937, MCF7, K562, and HepG2 cells, as well as elevating the expressions of tumor necrosis factor (TNF)-α and interleukins. The chemical structure of lectins among MM species differs from one another. Lectins are present in many MMs, including *Agaricus bisporus*, *Agrocybe aegerita*, *Floccularia luteovirens*, *Boletus edulis*, *Boletus speciosus*, *Clitocybe nebularis*, *Ganoderma capense*, *Grifola frondosa*, *Pholiota adiposa*, *Pleurotus eous*, *Russula delica*, *Russula lepida*, and *Xylaria hypoxylon* (Zhao et al. 2020).

### 12.2.2 Anti-inflammatory Action

Bioactive compounds of MMs also induce the synthesis of anti-inflammatory cytokines, including Interleukin (IL-10) and Transforming Growth Factor Beta (TGF-β). Cytokines are small proteins released by cells, mostly immune cells, including macrophages and dendritic cells, to regulate the growth, differentiation, and activities of immune cells. It acts as an important signal that can increase or decrease the inflammation against pathogens (bacteria or viruses) or abnormal cells (cancer cells), such as tumor necrosis factor, TNF-α and interferons (IFN-γ). To

**Table 12.2** Cytokines and their roles in the immune system

| Type of cytokines | Roles in immune system | Example of cytokines |
|---|---|---|
| Interleukin (IL) | Regulates the growth, activation and differentiation of immune cells | IL-6, IL-10 |
| Interferon (IFN) | Fights viral infections and tumors, activates macrophages | IFN-α, IFN-γ |
| Tumor Necrosis Factor (TNF) | Promotes inflammation and helps kill infected, abnormal, and cancerous cells | TNF-α |
| Chemokine | Recruit the right immune cells including neutrophils, monocytes, T cells, to sites of infection, injury, or inflammation | CCL2, CXCL1, CCL2 (MCP-1), CXCL8 (IL-8) |
| Colony-Stimulating Factors (CSF) | Stimulate bone marrow to produce more white blood cells (neutrophils), macrophages and monocytes | Granulocyte Colony-Stimulating Factor, G-CSF, and Granulocyte-Macrophage Colony-Stimulating Factor, GM-CSF |

simplify, TNF-α, IL-1β, IL-6, and IFN-γ are examples of pro-inflammatory cytokines that increase the immune system's response against diseases. Cytokines IL-10 and TGF-β are examples of anti-inflammatory cytokines, which are crucial to reduce inflammation and prevent overactivity (autoimmune). Table 12.2 shows the type of cytokines produced by human immune cells and their function in the immune system.

### 12.2.3 Clinical Evidence

Medicinal mushroom (MM) cell walls are abundant in β-glucans, short- or long-chain polysaccharides with β-1,6 and β-1,3 linkages that are accountable for the branching and linear chemical structures, correspondingly. However, β-glucans originating from cereals are different, where the absence of β-1,6 linkages causes no branching structures (Cerletti et al. 2021). The clinical potential of MMs using both fungal extracts and purified β-glucans for anti-inflammatory and immunomodulatory properties has been studied on cellular or experimental models. Ayeka (2018) reported on previous research on the antitumor and immunomodulatory properties of the medicinal mushroom, *Ganoderma lucidum*. *G. lucidum* enhanced inflammatory response by increasing the production of selected chemokines such as tumor necrosis factor-alpha (TNF-α), interferon-gamma (IFN-γ), and interleukin (IL-1, IL-2, and IL-6), as well as stimulating adaptive immune cells, T cells. In addition, besides polysaccharides, polypeptides are the crucial bioactive nutraceuticals with many pharmaceutical potentials, such as assisting the host in defending against the invasion of harmful pathogens, inhibiting certain enzyme activities, and boosting the adsorption and digestion of exogenous nutrition (Bhambri et al. 2022). These findings showed that mushroom bioactive metabolites are effective as immunomodulators for both adaptive and innate immune systems.

## 12.3 Anticancer Effects: A Natural Approach to Cancer Support

For centuries, mushrooms have been used in traditional folk medicine, and their use as food and medicine was mentioned in ancient treatises as early as 3000 B.C. (Janardhanan et al. 2020). While this historical use is extensive, especially in Asian countries, modern science has only recently begun to systematically investigate the therapeutic potential of their bioactive compounds (Janardhanan et al. 2020). This research has positioned medicinal mushrooms as a significant component in the field of integrative oncology (Jeitler et al. 2020). The goal is not to replace conventional treatments but to use mushroom-derived compounds as adjuvants that can support standard cancer therapies (Panda et al. 2022).

These natural products may enhance the efficacy of treatments like chemotherapy and radiotherapy while also helping to mitigate their often-severe side effects, thereby improving a patient's overall quality of life (Dan et al. 2023). We are moving beyond folklore and into the realm of molecular biology, exploring how these natural agents can target cancer's fundamental mechanisms, from the cellular level to the complex interplay of the immune system (Janardhanan et al. 2020). This review will delve into the scientific evidence from the provided documents, examining the mechanisms by which mushrooms target cancer cells, their synergistic effects with modern treatments, and the current focus of research in this promising field.

Mushrooms represent a major, yet largely untapped, source of therapeutically useful bioactive compounds (Janardhanan et al. 2020). Their anticancer activity is not due to a single mechanism but rather a multipronged assault on tumor cells (Panda et al. 2022). These effects include directly inhibiting cancer cell growth, modulating the immune system to attack tumors, preventing the formation of new blood vessels that feed them, and targeting the resilient cancer stem cells responsible for relapses and metastasis (Tacchini et al. 2023). This chapter will delve into the scientific evidence from the provided documents, examining the mechanisms by which mushrooms target cancer cells, their synergistic effects with modern treatments, and the current focus of research in this promising field.

### 12.3.1 Targeting Cancer Cells

The direct action of mushroom compounds on cancer cells is a primary area of scientific inquiry, involving a range of strategies that disrupt the fundamental processes of cancer cell survival and proliferation. A hallmark of cancer is the ability of tumor cells to evade apoptosis, or programmed cell death (Kumar 2015), but numerous mushroom compounds have been found to reactivate this critical self-destruct pathway. For instance, not only do triterpenoids from *Ganoderma lucidum* (Reishi) and the phenolic compound hispolon from *Phellinus linteus* show proven cytotoxic activity against various tumor cell lines (Janardhanan et al. 2020), but they also induce apoptosis by altering the balance of key regulatory proteins, such as upregulating the pro-apoptotic protein Bax while downregulating the antiapoptotic protein Bcl-2 (Gariboldi et al. 2023). This activation cascade ultimately involves the release of cytochrome c from the mitochondria and the subsequent activation of caspases, which are the final executioners of cell death (Yan et al. 2022).

In addition to these well-known compounds, other extracts demonstrate similar effects; for example, silver nanoparticles synthesized using an aqueous extract of *Pleurotus ostreatus* inhibited the viability of MCF-7 breast cancer cells by up to 85.14% in vitro (Hasan and Majeed 2024). Beyond simply inducing cell death, mushroom compounds can also halt the cancer cell cycle, thereby preventing replication. Supporting this, extracts from *Fomitopsis betulina* have been shown to inhibit the mitotic index in plant model organisms, suggesting a direct impact on the processes of cell division (Nowotarska et al. 2024).

Perhaps the most well-documented anticancer effect of mushrooms is their ability to act as powerful immunomodulators (Chaturvedi et al. 2018). Polysaccharides, particularly beta-glucans, are renowned for functioning as "biological response modifiers" (BRMs) (Kumar 2015). These complex carbohydrates do not typically kill cancer cells directly but instead stimulate the host's innate and adaptive immune systems (Chaturvedi et al. 2018). They achieve this by activating key immune cells such as macrophages, natural killer (NK) cells, and T cells, which are essential for recognizing and eliminating tumors (Park 2022). This principle is so well-established that some compounds, like Lentinan from *Lentinula edodes* (Shiitake) and PSK from *Trametes versicolor*, are used as well-known polysaccharide-based prescription drugs in Japan to treat cancer (Janardhanan et al. 2020). The clinical relevance of this is clear, as studies show that extracts from *Agaricus blazei* can significantly increase the activity of NK cells in patients with gynecological cancer who are undergoing chemotherapy (Bertollo et al. 2022).

Moreover, mushrooms can combat tumor growth by interfering with its ability to expand and spread. A tumor cannot grow beyond a small size without forming its own blood supply, a process known as angiogenesis (Kumar 2015), yet several mushroom compounds have demonstrated anti-angiogenic properties that effectively starve tumors of essential nutrients. Ergosterol, a compound found in many mushrooms, including *Agaricus blazei*, has been identified as a potent anti-angiogenic agent (Panda et al. 2022). In addition to cutting off the blood supply, mushrooms can also inhibit metastasis, the deadly process by which cancer spreads to distant organs (Chaturvedi et al. 2018).

Not only do they achieve this by suppressing matrix metalloproteinases (MMPs), which are enzymes that cancer cells use to invade surrounding tissues (Tacchini et al. 2023), but methanol extracts of *Phellinus linteus* and *Lentinula edodes* have also been shown to increase the expression of E-cadherin, an adhesion molecule that keeps cells bound together, thereby suppressing the migratory potential of colorectal cancer cells (Šeklić et al. 2022).

A particularly significant challenge in cancer therapy is a small subpopulation of tumor cells known as cancer stem cells (CSCs), which are believed to drive tumor initiation, recurrence, and resistance to conventional therapies (Tacchini et al. 2023). Excitingly, emerging research indicates that mycochemicals can specifically target these resilient cells. Natural molecules from mushrooms have the potential to inhibit key CSC self-renewal pathways, including Wnt, Notch, and Hedgehog (Tacchini et al. 2023). For example, polysaccharopeptide (PSP) isolated from *Trametes versicolor* has been shown to target prostate CSCs and suppress their tumorigenicity (Tacchini et al. 2023). This ability to attack the very root of cancer makes mushrooms a particularly promising area for drug development.

While these direct and immune-mediated anticancer actions are significant on their own, the full potential of medicinal mushrooms in modern oncology is perhaps best realized when they are not used in isolation but rather in concert with conventional treatments. This synergistic relationship, where mushroom compounds can enhance the efficacy of standard therapies while reducing their toxicity, represents one of the most compelling areas of integrative cancer care.

### 12.3.2 Synergistic Effects

One of the most valuable applications of medicinal mushrooms in oncology is their use in conjunction with conventional cancer treatments (Jeitler et al. 2020), as clinical evidence strongly suggests that this integrative approach can lead to better outcomes and an improved quality of life for patients (Dan et al. 2023). This synergistic effect manifests in two critical ways: enhancing the direct efficacy of treatments and reducing their debilitating side effects.

A primary way mushrooms support modern therapy is by directly enhancing its efficacy. In fact, this approach is so well-established that several mushroom-derived products, including Polysaccharide-K (PSK) from *Trametes versicolor* and Lentinan from *Lentinula edodes*, are used as approved pharmaceutical adjuvants alongside chemotherapy in Japan and China (Janardhanan et al. 2020; Dan et al. 2023). Numerous clinical trials have demonstrated that combining these compounds with standard care can improve survival rates and enhance tumor response (Dan et al. 2023; Panda et al. 2022).

For instance, a retrospective cohort analysis of patients with gastric cancer indicated that adding PSK to adjuvant chemotherapy was associated with a significant reduction in the risk of death (Dan et al. 2023). This powerful synergy is often attributed to the mushrooms' immune-enhancing properties, which support the body's ability to fight cancer while chemotherapy directly attacks the tumor (Jeitler et al. 2020). Beyond broad immune support, some extracts may also increase the efficacy of drugs through specific cellular mechanisms; for instance, *Agaricus blazei* extract sensitizes hepatocellular carcinoma cells to doxorubicin by inhibiting NF-kB, which leads to greater intracellular accumulation of the drug (Bertollo et al. 2022).

Other than improving the direct anticancer effects of treatment, a perhaps equally important role for medicinal mushrooms lies in their ability to mitigate the harsh side effects of these therapies. While effective, chemotherapy and radiotherapy are known for causing severe side effects like nausea, bone marrow suppression, anemia, and fatigue (Panda et al. 2022). Medicinal mushrooms, however, can significantly reduce these adverse effects, thereby improving the patient's quality of life (QoL) (Jeitler et al. 2020).

Clinical studies have consistently shown that patients taking mushroom extracts experience less severe side effects (Dan et al. 2023). A clear example of this was seen in a retrospective study where breast cancer patients receiving chemotherapy who also took Active Hexose Correlated Compound (AHCC) from *Lentinula edodes* had fewer neutrophil-related events and less need for granulocyte colony stimulating factor (Dan et al. 2023). Similarly, a randomized, double-blind clinical trial found that breast cancer patients supplementing with *Agaricus sylvaticus* had improved nutritional status and a reduction in adverse effects like nausea, vomiting, and anorexia (Dan et al. 2023).

The compelling clinical evidence for these synergistic benefits underscores the importance of mushrooms in patient care. However, it also highlights critical gaps in our knowledge and points toward the future direction of scientific inquiry. To fully integrate these natural therapies into evidence-based medicine, current research is now focused on moving beyond broad observations to pinpointing specific compounds, understanding their precise mechanisms, and validating their effects through rigorous clinical trials.

### 12.3.3 Research Focus

The field of mushroom-based cancer therapy is rapidly advancing, with research focused on refining our understanding and application of these natural compounds (Panda et al. 2022). A key aspect of this evolution is the shift from studying crude extracts to isolating specific bioactive compounds and elucidating their precise molecular mechanisms (Panda et al. 2022). While high-molecular-weight polysaccharides are well-studied, there is a growing interest in low-molecular-weight compounds like triterpenes, terpenes, and steroids that can directly regulate oncogenic signaling pathways (Dai et al. 2021). A powerful example of this is antroquinonol, a compound from *Antrodia camphorata*, which inhibits colon cancer stem cell-like properties by targeting the PI3K/AKT/β-catenin signaling pathway (Park 2022). This move toward identifying specific molecules is essential for developing the standardized and potent evidence-based mushroom-derived drugs of the future (Panda et al. 2022).

However, translating these promising preclinical findings into clinical practice presents a significant challenge. Despite a large volume of encouraging preclinical data, high-quality clinical evidence in humans remains limited (Panda et al. 2022). Most existing clinical studies are early Phase I or II trials, which often involve small sample sizes and suffer from a lack of standardization in the mushroom preparations used (Sivanesan et al. 2022). Furthermore, research has been concentrated on a small number of mushroom species, primarily *Lentinula edodes*, *Coriolus (Trametes) versicolor*, and *Ganoderma lucidum* (Panda et al. 2022; Sivanesan et al. 2022), with a focus on common malignancies like breast, colorectal, and prostate cancer (Dan et al. 2023). Consequently, there is a strong consensus among researchers about the urgent need for more rigorous, large-scale, double-blind, placebo-controlled clinical trials to definitively validate these findings and establish standardized therapeutic protocols (Gariboldi et al. 2023; Panda et al. 2022).

This drive for more robust clinical data fits squarely within the modern framework of integrative oncology (Jeitler et al. 2020). This approach does not seek to replace conventional medicine but rather to combine it with evidence-based complementary therapies to treat the whole person, not just the disease (Jeitler et al. 2020). In this context, mushrooms are being evaluated as part of a comprehensive treatment plan designed to support a patient's immune function and overall health during and after the rigors of standard cancer therapy.

## 12.4 Cognitive Enhancement and Neuroprotection: Boosting Brainpower

### 12.4.1 Protecting the Brain

Medicinal mushrooms have crucial therapeutic properties such as antioxidant, anti-inflammatory, antiviral, and antibacterial, as well as neuroprotective (Bonetto et al. 2025). Recent studies have discovered bioactive compounds in certain medicinal mushrooms, including polysaccharides, phenolics, and terpenes, that play a significant role in preventing age-related and neurodegenerative diseases (ND) such as Parkinson's disease (PD), Alzheimer's (AD), multiple sclerosis (MS), Huntington's disease (HD), and amyotrophic lateral sclerosis (ALS) (Abdelmoaty et al. 2024; Susanna and Rapior 2021).

Alzheimer's disease (AD) can be defined as a continuous neurodegenerative disease characterized by neuroinflammation, death of neurons, cognitive impairment with amyloid-β (Aβ) accumulation, and hyperphosphorylation (Ban et al. 2025; Tong et al. 2023). There are a few key pathologies of AD, including the presence of amyloid plaques, which are made up of Aβ protein fragments that clump together between neurons, thus disrupting cell-to-cell communication. The disruption of cell-to-cell communication triggers the immune system to respond and causes inflammation. These sticky plaques formed years before AD symptoms appear. In addition, another key feature of AD is Neurofibrillary Tangles (NFTs). NFTs are the formation of bundles of twisted fibers inside neurons due to abnormally phosphorylated (hyperphosphorylated) tau protein, which causes adverse effects, including disruption of the internal structure of neurons and death of neurons. In a normal condition, tau protein is soluble and lightly phosphorylated and binds with microtubules to provide strong, stable, and properly assembled microtubules inside the neuron; meanwhile, in an AD condition, too many phosphate groups attached to tau protein cause it to detach from neuron microtubules and misfold. The detached tau proteins clumped together, forming Neurofibrillary Tangles (NFTs) or Paired Helical Filaments (PHFs). The structure of microtubules inside neurons became weaker and disrupted, as there is no support from tau proteins, which caused adverse effects such as cognitive decline and memory loss. The excessive ROS, or imbalance between reactive oxygen species (ROS) and the body's antioxidant defenses, creates oxidative stress in the human body, which is also responsible for cellular damage, decline in neural function, and brain death due to lipid peroxidation (Lee et al. 2025; Liuzzi et al. 2023). In addition, dysfunctional autophagy jeopardized clearance of damaged (oxidized) mitochondria and mutated protein aggregates (amyloid plaques or tau protein tangles) in neurons, which causes neurodegeneration. The combination of these mechanisms synergistically affects the decline in neuronal function and causes neuronal death, especially in certain brain areas susceptible to age-related damage, namely, the substantia nigra and the hippocampus (Bonetto et al. 2025).

Medicinal mushroom extracts such as Tiger's milk (*Lignosus rhinoceros*), Wood ear mushroom (*Auricularia polytricha*), Agarikon (*Laricifomes officinalis*), Maitake (*Grifola frondosa*), Lion's mane (*Hericium erinaceus*), Reishi, Chaga (*Inonotus obliquus*), Turkey tail, and Cordyceps have been studied and commercialized as supplements to enhance brain function (Abdelmoaty et al. 2024; Ban et al. 2025; Kittimongkolsuk et al. 2021). There are several pathways to improve and protect the brain, including antioxidant activity, anti-inflammatory activity, and neuroprotection, as well as supporting the growth of new brain cells (neurogenesis). Compound INO10, an inotodiol-rich presence in mushroom Chaga (*Inonotus obliquus*) extract, is a promising compound capable of mitigating AD pathology by significantly reducing Aβ protein accumulation (amyloid plaque) and tau protein hyperphosphorylation in the hippocampus, as well as neuroinflammation. Besides inotodiol, another bioactive compound, betulinic acid, is also found in Chaga extract, and both compounds have therapeutic properties, including anti-inflammatory properties by downregulating pro-inflammatory cytokines, namely, IL-1β, IL-6, and TNF-α (Ban et al. 2025).

## 12.5 Cardiovascular Support: Promoting Heart's Health

Cardiovascular diseases (CVDs) represent a primary global health challenge, with their incidence and progression being strongly influenced by lifestyle factors, particularly diet (Dicks and Ellinger 2020). This has intensified the search for functional foods that can support heart health. Edible mushrooms have emerged as a valuable dietary component, recognized for their rich nutritional profile and diverse array of bioactive compounds that exert protective effects on the cardiovascular system (Rauf et al. 2023).

Mushrooms are low in fat, cholesterol-free, and contain beneficial nutrients like fiber, proteins, vitamins, and minerals, making them a staple in heart-healthy dietary patterns such as the Dietary Approaches to Stop Hypertension (DASH) and Mediterranean diets (Rauf et al. 2023; Riaz et al. 2022). Their therapeutic potential stems from a complex synergy of these components, which contributes to regulating key physiological markers of cardiovascular health. Mushrooms have gained attention as cardioprotective foods due to their ability to influence key physiological factors such as lipid levels, vascular tone, and oxidative stress (Badalyan et al. 2021).

### 12.5.1 Cholesterol Management

Effective management of blood lipid profiles, particularly lowering low-density lipoprotein (LDL) cholesterol, is a cornerstone of preventing atherosclerosis (dos Reis et al. 2022). Mushrooms contribute to cholesterol management through multiple bioactive compounds that interfere with cholesterol absorption from the gut and its synthesis within the body (Riaz et al. 2022). This dual-action approach makes them a valuable component of a comprehensive dietary strategy for improving lipid profiles.

A primary contributor to this effect is the soluble dietary fiber β-glucan, which is abundant in species like *Pleurotus ostreatus* (oyster mushroom) and *Lentinula edodes* (Shiitake) (Mustafa et al. 2022; Dicks and Ellinger 2020). In the gastrointestinal tract, β-glucans form a viscous gel that traps cholesterol and bile acids, promoting their fecal excretion (Riaz et al. 2022; Dicks and Ellinger 2020). This process forces the liver to utilize more circulating cholesterol to synthesize new bile acids, thereby lowering blood cholesterol levels (Dicks and Ellinger 2020). Additionally, mushrooms are a significant source of phytosterols, most notably ergosterol, which acts as a vitamin $D_2$ precursor (Ho et al. 2020; El-Ramady et al. 2022). Because ergosterol is structurally similar to cholesterol, it competes for absorption in the intestine, which reduces the amount of dietary cholesterol entering the circulation (dos Reis et al. 2022).

Beyond these structural components, mushrooms contain specific bioactives with potent cholesterol-lowering properties. Some species, especially *Pleurotus ostreatus*, naturally produce lovastatin, a compound that inhibits HMG-CoA reductase, the rate-limiting enzyme in the cholesterol biosynthesis pathway (Mustafa et al. 2022; Kała et al. 2020). By directly curbing the body's own production of cholesterol, these mushrooms mirror the mechanism of pharmaceutical statins (Kała et al. 2020). Another important compound is eritadenine, found in Shiitake mushrooms, which has been shown to lower cholesterol levels by modulating hepatic phospholipid metabolism and inhibiting S-adenosylhomocysteine hydrolase (SAHH), an enzyme involved in homocysteine production (Rauf et al. 2023; Riaz et al. 2022). By managing lipid profiles through these varied mechanisms, mushrooms help mitigate a primary driver of atherosclerotic plaque; however, their cardiovascular benefits also extend to directly influencing blood pressure.

Mushrooms employ a multifaceted strategy to support healthy cholesterol levels, utilizing soluble fibers like β-glucans to enhance excretion, sterols to block absorption, and specific compounds like lovastatin to inhibit endogenous synthesis (Dicks and Ellinger 2020; dos Reis et al. 2022; Kała et al. 2020). By managing lipid profiles, mushrooms help mitigate the primary driver of atherosclerotic plaque. However, their cardiovascular benefits also extend to directly influencing the forces within the blood vessels, namely, blood pressure, which is another critical factor in vascular stress and CVD risk.

### 12.5.2 Blood Pressure Regulation

Hypertension, or elevated arterial blood pressure, stands as a primary modifiable risk factor for numerous cardiovascular events, including heart failure and stroke (Rauf et al. 2023). Edible mushrooms possess several antihypertensive properties that work through distinct but complementary pathways, including potent antioxidant actions and the inhibition of key regulatory enzymes (Rauf et al. 2023).

The foundation of this effect lies in the ability of mushrooms to combat chronic inflammation and oxidative stress, which are underlying causes of endothelial dysfunction and arterial stiffness (Phull et al. 2022). Mushrooms are rich in phenolic compounds and flavonoids, which possess strong antioxidant properties that protect the vascular system from the damage caused by free radicals (Podkowa et al. 2021). This protective action helps maintain the flexibility and proper function of the arteries, contributing to better blood pressure control (Badalyan et al. 2021).

In addition to their general antioxidant capacity, certain mushroom species contain bioactive compounds that directly influence blood pressure regulatory systems. *Ganoderma lucidum* (Reishi), for example, is well-known for its high content of triterpenes and bioactive peptides that exhibit angiotensin-converting enzyme (ACE) inhibitory activity (Rauf et al. 2023; Kozarski et al. 2020). ACE is a central enzyme in the renin-angiotensin system that produces the potent vasoconstrictor angiotensin II; by inhibiting this enzyme, mushroom compounds promote vasodilation and help lower blood pressure (dos Reis et al. 2022). Furthermore, the favorable mineral content of mushrooms, characterized by a high potassium-to-sodium ratio, is another key factor in blood pressure regulation (Riaz et al. 2022; Rauf et al. 2023).

Thus, mushrooms support blood pressure control by protecting the vasculature from inflammatory and oxidative damage while also directly interfering with vasoconstrictive pathways through ACE inhibition (Podkowa et al. 2021; Rauf et al. 2023). The combined effect of managing both lipid levels and blood pressure contributes significantly to reducing the overall risk of CVDs. These actions are intrinsically linked to the health of the blood vessels themselves, which mushrooms support through additional protective mechanisms.

### 12.5.3 Vascular Health

Protecting the integrity and function of the vascular system is crucial for preventing the onset and progression of atherosclerosis and thrombosis (dos Reis et al. 2022). Mushrooms contribute to vascular health by shielding the endothelial lining of blood vessels, reducing inflammation, and preventing the formation of abnormal blood clots (Badalyan et al. 2021; Mehta 2020).

A key protective agent is the unique dietary amino acid ergothioneine, a potent antioxidant for which mushrooms are the richest natural source (Podkowa et al. 2021; Uffelman et al. 2023). Ergothioneine protects endothelial cells from oxidative damage, a critical initiating event in atherosclerosis, and has been shown to inhibit monocyte binding to endothelial cells, a key step in plaque formation (Phillips et al. 2022; Mustafa et al. 2022). Mushrooms also supply essential minerals like selenium, a cofactor for powerful antioxidant enzymes like glutathione peroxidase, further defending vascular tissues against free radical damage (Podkowa et al. 2021; Riaz et al. 2022).

Beyond antioxidant defense, several mushroom species exhibit direct antithrombotic effects. Fungi such as *Ganoderma lucidum*, *Auricularia auricula-judae* (wood ear mushroom), and *Pleurotus ostreatus* contain fibrinolytic enzymes that can degrade fibrin, the primary protein in blood clots (Mehta 2020; Petraglia et al. 2022). This activity helps dissolve existing clots and prevent their formation, thereby maintaining blood fluidity and reducing the risk of thrombotic events like heart attacks and strokes (Mehta 2020). This effect is complemented by flavonoids, such as quercetin, which are found in mushrooms and possess vasorelaxant properties that help improve circulation and reduce stress on the vascular system (Rauf et al. 2023).

In essence, mushrooms support a healthy vascular system through a combination of antioxidant protection, anti-inflammatory action, and antithrombotic activity (Podkowa et al. 2021; Mehta 2020). These cellular and enzymatic mechanisms, identified largely through in vitro studies, form the basis of mushrooms' cardioprotective reputation. The next critical step is to examine how these mechanisms translate into measurable clinical outcomes in human and animal studies.

### 12.5.4 Research Evidence

The cardioprotective claims for mushrooms are substantiated by a growing body of scientific research, including human clinical trials that have demonstrated quantifiable benefits for cardiovascular health. Human trials have provided direct evidence of the lipid-lowering potential of mushrooms. For instance, a pilot clinical study on healthy adults found that three months of consuming *Antrodia cinnamomea* mycelium capsules significantly reduced total cholesterol by 5.1% ($p < 0.01$) (Chen and Chang 2020).

Another randomized controlled trial involving patients with type 2 diabetes reported that a daily 16 g supplement of Agaricus bisporus (white button mushroom) powder for six weeks significantly lowered serum total cholesterol (mean reduction of 14.46 mg/dL) and LDL cholesterol (mean reduction of 9.12 mg/dL) compared to a control group. The acute effects have also been documented; supplementing a single high-fat meal with either Shiitake or white button mushroom powder was found to significantly lower the postprandial LDL response over a six-hour period in healthy individuals (Talal et al. 2023).

Evidence for blood pressure regulation is also compelling. A notable randomized trial found that incorporating 100 grams of A. bisporus into the daily diet of pregnant women significantly reduced the incidence of gestational hypertension to 4.1%, compared to 8.2% in the placebo group ($p = 0.023$) (Sun and Niu 2020). In animal models, extracts from Ganoderma lucidum have shown a profound ability to protect the heart from chemotherapy-induced damage by mitigating oxidative stress and preserving cardiac function, as measured by cardiac injury markers like creatine kinase (CK) and lactate dehydrogenase (LDH) (Veena et al. 2022).

When compared conceptually to conventional pharmaceuticals, mushrooms offer a distinct advantage for health maintenance and disease prevention (Rauf et al. 2023). While drugs like statins and ACE inhibitors provide powerful, targeted interventions for treating existing diseases, they are often accompanied by side effects (Mustafa et al. 2022). Mushrooms, on the other hand, provide a moderate, multi-targeted effect through their diverse array of bioactive compounds (Badalyan et al. 2021). This makes them an ideal component of a preventive dietary strategy, working holistically to support long-term cardiovascular health rather than serving as an acute treatment.

## 12.6 Metabolic and Endocrine Effect: Balancing the Body

Our bodies are intricate orchestras, with countless biological processes working in harmony to maintain a state of balance, or what scientists call homeostasis. Two of the most critical sections in this orchestra are the metabolic and endocrine systems. The metabolic system is our engine, converting food into energy and building blocks, while the endocrine system is the conductor, using hormones to send messages that regulate everything from our mood and stress levels to our growth and energy use (Wang et al. 2024). When this orchestra is out of tune, perhaps due to modern lifestyle pressures, poor diet, or stress, it can lead to a cascade of health issues, including metabolic syndrome, diabetes, and hormonal imbalances (Bell et al. 2022). The global rise in these conditions has spurred a search for safe, effective interventions that can support the body's natural regulatory functions (Rahman and Akter 2022).

For centuries, traditional medicine has turned to a unique natural resource to help restore this balance: mushrooms (Bhambri et al. 2022). Far more than just a culinary ingredient, certain mushrooms are now recognized as powerful "functional foods" and a source of "nutraceuticals," food-derived compounds that provide medical or health benefits (Anusiya et al. 2021). They represent a bridge between simple nutrition and potent medicine, offering a complex array of bioactive molecules like polysaccharides, terpenoids, and phenolics that can gently and effectively nudge our internal systems back toward equilibrium (Chopra et al. 2021).

### 12.6.1 Blood Sugar Control

Maintaining stable blood sugar levels is a cornerstone of metabolic health. When we eat carbohydrates, our bodies break them down into glucose, which enters the bloodstream to be used as fuel. The hormone insulin, produced by beta cells in the pancreas, acts like a key, allowing glucose to enter our cells (Rahman and Akter

2022). In conditions like insulin resistance and type 2 diabetes, this system falters. Either the body does not produce enough insulin or the cells stop responding to it properly, leading to high blood sugar (hyperglycemia), which can damage organs over time (Antonceva and Shamtsyan 2020). Excitingly, a growing body of research, including recent human trials, shows that mushrooms can intervene in this process through several clever mechanisms, making them a powerful dietary tool for managing metabolic health (Shamim et al. 2023).

One of the most direct ways mushrooms help is by slowing down how quickly sugar enters the bloodstream after a meal. They achieve this by inhibiting key enzymes responsible for carbohydrate digestion, namely α-amylase and α-glucosidase (Anusiya et al. 2021). Think of these enzymes as tiny scissors that chop up complex starches and sugars into simple glucose molecules that are easily absorbed (Aramabašić et al. 2021). By partially blocking these scissors, mushrooms ensure that carbohydrate breakdown is a slower, more gradual process. This prevents the sharp spikes in blood sugar that are particularly dangerous for individuals with diabetes (Rahman and Akter 2022). The mushroom *Grifola frondosa* (Maitake), for instance, naturally contains an alpha-glucosidase inhibitor, a mechanism similar to some pharmaceutical antidiabetic drugs. This effect is supported by studies on various mushrooms, including *Agaricus bisporus*, which have demonstrated these inhibitory activities (Shamim et al. 2023).

This effect isn't just theoretical; it has been observed in clinical settings. A randomized controlled trial involving patients with metabolic disorders found that daily consumption of a snack made from the King oyster mushroom, *Pleurotus eryngii*, led to a significant decrease in fasting glucose levels compared to a control group (Kleftaki et al. 2022). Similarly, a randomized trial on patients with type-2 diabetes showed that supplementing their diet with hot air-dried white button mushroom (*Agaricus bisporus*) powder for 16 weeks significantly improved glycemic indices (Hashemi et al. 2022). Another study showed that incorporating oyster mushroom powder into a meal beneficially affects post-meal glucose response and increases levels of glucagon-like peptide-1 (GLP-1), a hormone that helps regulate insulin secretion and appetite, in adults with impaired glucose tolerance (Dicks et al. 2022).

Beyond slowing sugar absorption, mushrooms appear to help the body use insulin more effectively and even protect the very cells that produce it. Insulin resistance, a condition where cells become "numb" to insulin's signals, is a hallmark of type 2 diabetes (Antonceva and Shamtsyan 2020). Mushroom bioactive compounds, particularly polysaccharides like β-glucans, have been shown to improve insulin sensitivity (Antonceva and Shamtsyan 2020; Rahman and Akter 2022).

The mechanism might be related to how mushroom compounds interact with key signaling pathways, such as the PI3K/Akt pathway, which is crucial for insulin signaling (Aramabašić et al. 2021). Extracts from *Pleurotus eryngii*, for example, have demonstrated the ability to lower blood glucose by influencing pathways like AMPK phosphorylation (Ahmad and Safuan 2019).

Furthermore, the powerful antioxidant properties of mushrooms play a crucial protective role. The pancreatic β-cells that produce insulin are highly susceptible to damage from oxidative stress, which is often elevated in diabetic conditions (Arunachalam et al. 2022). Mushroom polysaccharides can mitigate this damage by scavenging free radicals and boosting the body's own antioxidant defenses. Some studies in animal models even suggest a regenerative effect. Extracts from *Pleurotus eryngii* were found to partially restore pancreatic β-cells damaged by alloxan in mice, leading to increased insulin levels (Antonceva and Shamtsyan 2020). Similar regenerative potential has been noted with *Agaricus bisporus* glycoprotein preparations.

A wide array of mushrooms has demonstrated antidiabetic potential. A systematic review of clinical trials highlighted several species with promising effects (Dicks and Ellinger 2020). Grifola frondosa (Maitake) is widely studied for its antidiabetic activity, containing α-glucan (MT- alpha-glucan) and other polysaccharides that improve insulin sensitivity and ameliorate insulin resistance (Rahman and Akter 2022).

In a study on Polycystic Ovarian Syndrome (PCOS) patients, a condition often linked with insulin resistance, Maitake extract was shown to induce ovulation, suggesting an improvement in underlying metabolic function (Ahmad and Safuan 2019). Pleurotus spp. (Oyster Mushrooms), as demonstrated in human trials, species like P. eryngii and P. ostreatus are effective at managing glucose levels (Kleftaki et al. 2022). Their β-glucans and other polysaccharides contribute to these effects (Aramabašić et al. 2021).

A systematic review reported that intake of P. ostreatus can significantly reduce fasting blood glucose in diabetic subjects (Dicks and Ellinger 2020). Agaricus bisporus (white button mushroom) clinical trials have confirmed its ability to improve glycemic control in type 2 diabetes patients (Hashemi et al. 2022). Furthermore, compounds within this mushroom have been identified as agonists of peroxisome proliferator-activated receptors (PPARs), which are key regulators of glucose and lipid metabolism (Ngoc et al. 2021). Ganoderma lucidum (Reishi), a cornerstone of traditional medicine, contains Reishi polysaccharides (GL-PS), which have been found to lower serum glucose levels, in part by facilitating insulin release and protecting pancreatic β-cells (Rahman and Akter 2022). Lentinula edodes (Shiitake) extracts have also been shown to possess hypoglycemic properties, helping to regulate blood sugar in experimental models (Rahman and Akter 2022).

Mushrooms present a multipronged approach to blood sugar management. They not only slow the entry of glucose into the bloodstream but also appear to enhance the body's ability to use insulin while protecting the vital pancreatic cells responsible for its production (Aramabašić et al. 2021). This makes them a promising, safe, and effective dietary tool for managing and potentially preventing type 2 diabetes, with clinical evidence increasingly supporting their role as a complement to conventional care (Shamim et al. 2023).

### 12.6.2 Adaptogenic Effects

The term "adaptogen" refers to a unique class of natural substances that help the body adapt to stress, whether it's physical, chemical, or biological. They work like a thermostat for our stress response system, dialing it down when it's overactive and giving it a boost when it's sluggish, all without causing major side effects. The central player in our stress response is the hypothalamic-pituitary-adrenal (HPA) axis, a complex network that, when triggered, floods the body with hormones like cortisol (Wang et al. 2024). While essential for short-term "fight or flight" situations, chronic activation of the HPA axis can lead to fatigue, burnout, and a host of health problems (Lazur et al. 2023). Several medicinal mushrooms, particularly *Ganoderma lucidum* (Reishi) and *Cordyceps sinensis*, are celebrated for their adaptogenic qualities, offering a natural way to build resilience against the pressures of modern life.

Adaptogenic mushrooms help to normalize the HPA axis, preventing the excessive cortisol production that characterizes chronic stress (Lazur et al. 2023). By modulating this central stress pathway, they can help alleviate feelings of being "wired and tired" and restore a sense of calm and balance. Research in animal models of depression, which often involves inducing chronic stress, supports this. For example, extracts from *Ganoderma lucidum* have been shown to reverse the behavioral and biochemical changes caused by unpredictable chronic mild stress (UCMS) in rats, a model designed to mimic the effects of long-term stress (Wang et al. 2024). Similarly, *Cordyceps militaris* demonstrated antidepressant-like effects in a UCMS mouse model, pointing to its ability to counteract the negative impact of chronic stress (Wang et al. 2024).

The key outcome of better stress adaptation is improved energy and stamina. Adaptogens are known for their ability to combat fatigue without being overstimulating (Yamada et al. 2025). *Cordyceps* species, in particular, have a long history of use for enhancing physical performance. They are believed to improve the body's production of adenosine triphosphate (ATP), the primary carrier of energy in cells, and enhance oxygen uptake and utilization. Animal studies have shown that polysaccharides from *Cordyceps militaris* can significantly reduce physical fatigue in forced swimming tests (Wang et al. 2024). This aligns with its traditional use by athletes and those needing to combat physical exhaustion. In a human clinical study, *Flammulina velutipes* (Enoki mushroom) extract, which contains the energy-related molecule adenosine, was found to improve symptoms of male menopause, a condition often associated with fatigue and declining energy (Yamada et al. 2025).

*Ganoderma lucidum* (Reishi): often called the "mushroom of immortality," is perhaps the most famous adaptogen. Reishi has been shown in preclinical studies to have antidepressant-like and anxiolytic-like effects, helping to calm the nervous system and build stress resilience (Wang et al. 2024). *Cordyceps sinensis*/militaris: esteemed for its ability to boost energy and fight fatigue, Cordyceps is a favorite among athletes and those recovering from illness. Animal studies confirm its antidepressant-like effects and ability to counteract chronic stress (Wang et al. 2024). *Hericium erinaceus* (Lion's Mane): while famed for its neuroprotective benefits, Lion's Mane also shows adaptogenic properties. A double-blind, placebo-controlled pilot study on young adults found that supplementation with Lion's Mane had a positive impact on stress and mood (Docherty et al. 2023). Preclinical models also show it can produce antidepressant-like effects (Wang et al. 2024). Mushrooms like Reishi, Cordyceps, and Lion's Mane act as sophisticated biological response modifiers. They help the body maintain equilibrium in the face of stress by normalizing the HPA axis and bolstering energy metabolism. This makes them valuable natural tools for managing the physiological toll of chronic stress, enhancing physical and mental resilience, and promoting a state of balanced homeostasis in high-stress modern conditions.

### 12.6.3 Hormonal Balance

The endocrine system is a delicate web of glands that produce hormones, the chemical messengers that control nearly every process in our bodies, from metabolism and growth to mood and reproductive function. When this system is out of balance, it can lead to a wide range of issues, including thyroid disorders, adrenal fatigue, and reproductive problems like Polycystic Ovarian Syndrome (PCOS) (Ahmad and Safuan 2019). The conventional approach often involves hormone replacement or drugs that target specific pathways, but these can come with side effects. Medicinal mushrooms offer a more holistic approach, working to support and modulate the body's own hormonal systems through several mechanisms.

Chronic inflammation and oxidative stress can disrupt the function of endocrine glands, impairing their ability to produce and regulate hormones (Arslan et al. 2025). Mushrooms are rich in antioxidants and anti-inflammatory compounds that can protect these sensitive tissues (Abdelkader et al. 2024). For example, ergothioneine, a potent antioxidant found in high concentrations in mushrooms, has been shown to protect reproductive cells from oxidative damage (Arslan et al. 2025). Similarly, extracts from *Pleurotus ostreatus* (Oyster mushroom) can induce a classically activated macrophage response, a key part of a healthy immune and anti-inflammatory cycle (Llauradó et al. 2021). By reducing the underlying burden of stress and inflammation, mushrooms can create a more favorable environment for the endocrine system to function optimally.

Some mushrooms contain compounds that can either mimic the body's hormones or help regulate their production and activity. Mushroom sterols, for instance, are structurally similar to human hormones and can interact with hormone pathways (Anusiya et al. 2021). This may be particularly relevant in reproductive health. A systematic review on plant and mushroom extracts for PCOS found that several, including Maitake mushrooms, showed promise in improving hormonal imbalances and promoting ovulation (Ahmad and Safuan 2019). These extracts could help reduce elevated testosterone and luteinizing hormone (LH) levels while improving menstrual regularity and pregnancy outcomes (Ahmad and Safuan 2019). In a study on male rats, cordycepin from *Cordyceps militaris* was shown to ameliorate age-related testicular dysfunction and restore normal hormone levels (Arslan et al.

2025). Furthermore, a clinical trial with Enoki mushroom extract, which is rich in adenosine, found that it restored testosterone production in impaired and fatigued mouse models and beneficially affected male menopausal symptoms in human subjects (Iguchi et al. 2023; Yamada et al. 2025). In summary, mushrooms offer a well-rounded approach to blood sugar regulation. They slow the absorption of glucose, improve the body's insulin response, and help preserve pancreatic function (Aramabašić et al. 2021).

## 12.7 Conclusion: The Future of Mushrooms in the Field of Nutraceuticals

Overall, mushrooms represent a vast, largely untapped reservoir of bioactive compounds with significant potential in the fight against various diseases. The scientific evidence strongly indicates that their benefits are multifaceted, ranging from direct cytotoxic and antiproliferative effects on tumor cells to powerful immunomodulatory activities that enhance the body's own defenses. Furthermore, their role as adjuvants in conventional therapy is one of the most promising areas, offering the potential to increase treatment efficacy while simultaneously reducing debilitating side effects and improving patient quality of life. Future research must focus on isolating and characterizing specific active compounds, standardizing extraction and formulation methods, and conducting large-scale, well-designed clinical trials to confirm efficacy and establish safe therapeutic dosages. As understanding deepens, medicinal mushrooms are poised to move from the realm of traditional remedies to become a cornerstone of evidence-based integrative oncology, offering a natural and powerful approach to disease support. Beyond heart health, mushrooms also show promise in regulating metabolic processes, particularly in blood sugar control and hormonal balance, making them a key area of interest for managing broader cardiometabolic health. With a growing body of clinical evidence, they are emerging as safe and effective dietary tools for managing, and potentially preventing, type 2 diabetes, making them a valuable complement to conventional care.

**Acknowledgment** The authors would like to thank the facilities and assistance provided by the lab staff at Universiti Teknologi MARA during the completion of this research.

**Declaration** The authors do not have any competing interests.

## References

Abdelkader MAE, Mediatrice H, Lin D, Lin Z, Aggag SA (2024) Mitigating oxidative stress and promoting cellular longevity with mushroom extracts. Foods 13(24):Article 24

Abdelmoaty MM, Kadry R, Mosley RL, Gendelman HE (2024) Neuroprotective mushrooms. Neuroimmune Pharmacol Ther 3(2):129–137

Abebaw G (2020) Review on: nutritional value and health benefits of edible mushroom. J Eng Appl Sci Technol 2(4):1–2

Ahmad F, Safuan S (2019) Assessing the effectiveness of plant extracts in polycystic ovarian syndrome: a systematic review. Malays J Med Health Sci 15(2):120–129

Antonceva E, Shamtsyan M (2020) Antidiabetical and hypoglycemic action of mushroom polysaccharides. E3S Web Conf 215:05001

Anusiya G, Gowthama PU, Yamini NV, Sivarajasekar N, Rambabu K, Bharath G, Banat F (2021) A review of the therapeutic and biological effects of edible and wild mushrooms. Bioengineered 12(2):11239–11268

Aramabašić JJ, Mihailović M, Uskoković A, Grdović N, Dinić S, Vidaković M (2021) The effects of major mushroom bioactive compounds on mechanisms that control blood glucose level. J Fungi 7(1):Article 1

Arslan NP, Albayrak S, Budak-Savas A, Hacimuftuoglu A, Orak T, Ozdemir A, Karadagoglu O, Yildirim S, Cinar-Yilmaz H, Taskin M (2025) Algal and fungal antioxidants alleviate oxidative stress-induced reproductive defects. Food Sci Nutr 13(5):e70301

Arunachalam K, Sreeja PS, Yang X (2022) The antioxidant properties of mushroom polysaccharides can potentially mitigate oxidative stress, beta-cell dysfunction and insulin resistance. Front Pharmacol 13:874474

Assemie A, Abaya G (2022) The effect of edible mushroom on health and their biochemistry. Int J Microbiol 2022:8744788

Auxier B, Czárán TL, Aanen DK (2022) Modeling the consequences of the dikaryotic life cycle of mushroom-forming fungi on genomic conflict. Elife 11:e75917

Ayeka PA (2018) Potential of mushroom compounds as immunomodulators in cancer immunotherapy: a review. Evid Based Complement Alternat Med:1–9

Badalyan SM, Barkhudaryan A, Rapior S (2021) The cardioprotective properties of Agaricomycetes mushrooms growing in the territory of Armenia: review. Int J Med Mushrooms 23(5):21–31

Ban S, Do TT, Pyo JW, Moon M, Park JT (2025) INO10, a Chaga mushroom extract, alleviates Alzheimer's disease-related pathology and cognitive deficits in 3xTg-AD mice. Int J Mol Sci 26(10):4729

Bell V, Silva CRPG, Guina J, Fernandes TH (2022) Mushrooms as future generation healthy foods. Front Nutr 9:1050099

Bertollo AG, Mingoti MED, Plissari ME, Betti G, Roman JWA, Luzardo AR, Ignácio ZM (2022) Agaricus blazei Murrill mushroom: a review on the prevention and treatment of cancer. Pharmaco Res Mod Chin Med 2:100032

Bhambri A, Srivastava M, Mahale VG, Mahale S, Karn SK (2022) Mushrooms as potential sources of active metabolites and medicines. Front Microb 13:1–13

Bonetto V, Ferraresi A, Sampò S, Isidoro C (2025) Fungal bioactive compounds as emerging therapeutic options for age-related neurodegenerative disorders. Int J Mol Sci 26(10):4800

Cateni F, Gargano ML, Procida G, Venturella G, Cirlincione F, Ferraro V (2022) Mycochemicals in wild and cultivated mushrooms: nutrition and health. Phytochem Rev 21(2):339–383

Cerletti C, Esposito S, Lacoviello L (2021) Edible mushrooms and beta-glucans: impact on human health. Nutrients 13(2195):1–13

Chaturvedi VK, Agarwal S, Gupta KK, Ramteke PW, Singh MP (2018) Medicinal mushroom: boon for therapeutic applications. 3 Biotech 8(8):334

Chen WJ, Chang FW (2020) A pilot study to assess food safety and potential cholesterol- lowering efficacy of Antrodia cinnamomea solid-state cultivated mycelium in healthy adults. Evid Based Complement Alternat Med 2020(1):5865764

Chopra H, Mishra AK, Baig AA, Mohanta TK, Mohanta YK, Baek KH (2021) Narrative review: bioactive potential of various mushrooms as the treasure of versatile therapeutic natural product. J Fungi 7(9):728

Dai R, Liu M, Nik Nabil WN, Xi Z, Xu H (2021) Mycomedicine: a unique class of natural products with potent anti-tumour bioactivities. Molecules 26(4):Article 4

Dan A, Swain R, Belonce S, Jacobs RJ, Dan A, Swain R, Belonce SE, Jacobs RJ (2023) Therapeutic effects of medicinal mushrooms on gastric, breast, and colorectal cancer: a scoping review. Cureus 15(4):e37574

Dicks L, Ellinger S (2020) Effect of the intake of oyster mushrooms (Pleurotus ostreatus) on cardiometabolic parameters—a systematic review of clinical trials. Nutrients 12(4):Article 4

Dicks L, Jakobs L, Sari M, Hambitzer R, Ludwig N, Simon MC, Stehle P, Stoffel-Wagner B, Helfrich HP, Ahlborn J, Rühl M, Hartmann B, Holst JJ, Ellinger S (2022) Fortifying a meal with oyster mushroom powder beneficially affects postprandial glucagon-like peptide-1, non-esterified free fatty acids and hunger sensation in adults with impaired glucose tolerance: a double-blind randomized controlled crossover trial. Eur J Nutr 61(2):687–701

Docherty S, Doughty FL, Smith EF (2023) The acute and chronic effects of lion's mane mushroom supplementation on cognitive function, stress and mood in young adults: a double- blind, parallel groups, pilot study. Nutrients 15(22):Article 22

dos Reis EE, Schenkel PC, Camassola M (2022) Effects of bioactive compounds from Pleurotus mushrooms on COVID-19 risk factors associated with the cardiovascular system. J Int Med 20(5):385–395

El-Ramady H, Abdalla N, Badgar K, Llanaj X, Törős G, Hajdú P, Eid Y, Prokisch J (2022) Edible mushrooms for sustainable and healthy human food: nutritional and medicinal attributes. Sustainability 14(9):4941

Gariboldi MB, Marras E, Ferrario N, Vivona V, Prini P, Vignati F, Perletti G (2023) Anti-cancer potential of edible/medicinal mushrooms in breast cancer. Int J Mol Sci 24(12):10120

Hasan AM, Majeed SMA (2024) Detection of anti-cancer activity of silver nanoparticles synthesized using aqueous mushroom extract of Pleurotus ostreatus on MCF-7 human breast cancer cell line. Iraqi J Sci:1886–1894

Hashemi YH, Hosseini SA, Zakerkish M, Cheraghian B, Alipour M (2022) The effects of hot air-dried white button mushroom powder on glycemic indices, lipid profile, inflammatory biomarkers and total antioxidant capacity in patients with type-2 diabetes mellitus: a randomized controlled trial. J Res Med Sci 27(1):49

Ho LH, Zulkifli NA, Tan TC, Ho LH, Zulkifli NA, Tan TC (2020) Edible mushroom: nutritional properties, potential nutraceutical values, and its utilisation in food product development. In: An introduction to mushroom. IntechOpen

Iguchi K, Nagashima K, Mochizuki J, Yamamoto H, Unno K, Miyoshi N (2023) Enokitake mushroom and its active component, adenosine, which restores testosterone production in impaired and fatigued mouse models. Nutrients 15(9):Article 9

Janardhanan KK, Ravikumar KS, Karuppayil SM (2020) Medicinal mushroom bioactives: potential sources for anti-cancer drug development. Int J Appl Pharm:40–45

Jeitler M, Michalsen A, Frings D, Hübner M, Fischer M, Koppold-Liebscher DA, Murthy V, Kessler CS (2020) Significance of medicinal mushrooms in integrative oncology: a narrative review. Front Pharmacol 11:580656

Kała K, Kryczyk-Poprawa A, Rzewińska A, Muszyńska B (2020) Fruiting bodies of selected edible mushrooms as a potential source of lovastatin. Eur Food Res Technol 246(4):713–722

Kittimongkolsuk P, Pattarachotanant N, Chuchawankul S, Wink M, Tencomnao T (2021) Neuroprotective effects of extracts from tiger milk mushroom Lignosus rhinocerus against glutamate-induced toxicity in HT22 hippocampal neuronal cells and neurodegenerative diseases in Caenorhabditis elegans. Biology 10(1):30

Kleftaki SA, Amerikanou C, Gioxari A, Lantzouraki DZ, Sotiroudis G, Tsiantas K, Tsiaka T, Tagkouli D, Tzavara C, Lachouvaris L, Zervakis GI, Kalogeropoulos N, Zoumpoulakis P, Kaliora AC (2022) A randomized controlled trial on Pleurotus eryngii mushrooms with antioxidant compounds and vitamin D2 in managing metabolic disorders. Antioxidants 11(11): Article 11

Kozarski MS, Klaus AS, Vunduk JĐ, Jakovljević DM, Jadranin MB, Nikšić MP (2020) Health impact of the commercially cultivated mushroom Agaricus bisporus and wild- growing mushroom Ganoderma resinaceum—a comparative overview. J Serb Chem Soc 85(6):721–735

Kumar K (2015) Role of edible mushrooms as functional foods-a review. South Asian J Food Technol Environ 1(3&4):211–218

Lazur J, Hnatyk K, Kała K, Sułkowska-Ziaja K, Muszyńska B (2023) discovering the potential mechanisms of medicinal mushrooms antidepressant activity: a review. Antioxidants 12(3): Article 3

Lee SS, Ko DH, Lee GY, Kim SY, Han SY, Park JY, Park M, Kim HM, Kim YE, Yoo YC (2025) Neuroprotective effect of mixed mushroom mycelia extract on neurotoxicity and neuroinflammation via regulation of ROS-induced oxidative stress in PC12 and BV2 cells. Cells 14(13):977

Liuzzi GM, Petraglia T, Latronico T, Crescenzi A, Rossano R (2023) Antioxidant compounds from edible mushrooms as potential candidates for treating age-related neurodegenerative diseases. Nutrients 15(8):1913

Llauradó MG, Morris-Quevedo HJ, Heykers A, Lanckacker E, Cappoen D, Delputte P, Vanden BW, Salgueiro Z, Cos P (2021) Differential induction pattern towards classically activated macrophages in response to an immunomodulatory extract from Pleurotus ostreatus mycelium. J Fungi 7(3):Article 3

Macharia JM, Zhang L, Mwangi RW, Rozmann N, Kaposztas Z, Varjas T, Sugár M, Alfatafta H, Pintér M, Bence RL (2022) Are chemical compounds in medical mushrooms potent against colorectal cancer carcinogenesis and antimicrobial growth? Cancer Cell Int 22(1):379

Mehta R (2020) Anti-thrombotic attribute of different type of mushrooms. Int J Curr Microbiol Appl Sci 9(8):815–820

Mustafa F, Chopra H, Baig AA, Avula SK, Kumari S, Mohanta TK, Saravanan M, Mishra AK, Sharma N, Mohanta YK (2022) Edible mushrooms as novel myco-therapeutics: effects on lipid level, obesity and BMI. J Fungi 8(2):Article 2

Ngoc LP, Helmus R, Ehlers AW, Swart K, Besselink H, Rijke ED, Cam HDT, Brouwer A, Burg BVD (2021) Effect-directed analysis and chemical identification of agonists of peroxisome proliferator-activated receptors in white button mushroom. Food Func 12(1):133–143

Nowotarska P, Janeczek M, Wiatrak B (2024) Mushroom against cancer: aqueous extract of Fomitopsis betulina in fight against tumors. Nutrients 16(19):Article 19

Ogidi CO, Oyetayo VO, Bamidele J (2020) Wild medicinal mushrooms: potential applications in phytomedicine and functional foods. In: An introduction to mushroom. IntechOpen

Panda SK, Sahoo G, Swain SS, Luyten W (2022) Anticancer activities of mushrooms: a neglected source for drug discovery. Pharmaceuticals 15(2):176

Pandey R, Kunwar A, Ranjit R, Koirala N (2023) Wild and cultivated mushrooms of Nepal as a source of nutrients and nutraceuticals. J Nepal Biotechnol Assoc 4(1):44–51

Park HJ (2022) Current uses of mushrooms in cancer treatment and their anticancer mechanisms. Int J Mol Sci 23(18):Article 18

Petraglia T, Latronico T, Liuzzi GM, Fanigliulo A, Crescenz A, Rossano R (2022) Edible mushrooms as source of fibrin(ogen)olytic enzymes: comparison between four cultivated species. Molecules 27(23):Article 23

Phillips JM, Ooi SL, Pak SC (2022) Health-promoting properties of medicinal mushrooms and their bioactive compounds for the COVID-19 era—an appraisal: do the pro-health claims measure up? Molecules 27(7):Article 7

Phull AR, Ahmed M, Park HJ (2022) Cordyceps militaris as a bio functional food source: pharmacological potential, anti-inflammatory actions and related molecular mechanisms. Microorganisms 10:405

Podkowa A, Kryczyk-Poprawa A, Opoka W, Muszyńska B (2021) Culinary–medicinal mushrooms: a review of organic compounds and bioelements with antioxidant activity. Eur Food Res Technol 247(3):513–533

Rahman MA, Akter R (2022) Diabetes ameliorating effect of mushrooms and ameliorating diabetes. Auctores

Rauf A, Joshi PB, Ahmad Z, Hemeg HA, Olatunde A, Naz S, Hafeez N, Simal-Gandara J (2023) Edible mushrooms as potential functional foods in amelioration of hypertension. Phytother Res 37(6):2644–2660

Riaz S, Ahmad A, Farooq R, Ahmed M, Shaheryar M, Hussain M, Riaz S, Ahmad A, Farooq R, Ahmed M, Shaheryar M, Hussain M (2022) Edible mushrooms, a sustainable source of nutrition, biochemically active compounds and its effect on human health. In: Current topics in functional food. IntechOpen

Šeklić DS, Jovanović MM, Virijević K, Grujić J, Živanović M, Marković SD (2022) Effects of edible mushrooms Phellinus linteus and Lentinus edodes methanol extracts on colorectal cancer cell lines. Bio Life Sci Forum 6(1):Article 1

Shamim MZ, Mishra AK, Kausar T, Mahanta S, Sarma B, Kumar V, Mishra PK, Panda J, Baek KH, Mohanta YK (2023) Exploring edible mushrooms for diabetes: unveiling their role in prevention and treatment. Molecules 28(6):Article 6

Sharifi-Rad J, Butnariu M, Ezzat SM, Adetunji CO, Imran M, Sobhani SR, Tufail T, Hosseinabadi T, Ramírez-Alarcón K, Martorell M, Maroyi A, Martins N (2020) Mushrooms-rich preparations on wound healing: from nutritional to medicinal attributes. Front Pharmacol 11:567518

Shu W, Zhou X, Ma Y, Liu L, Duan Y, Huang C, Li C, Li Y (2025) Effects of spent mushroom substrate incorporation depth on soil microbial dynamics, carbon-nutri cycling, and maize productivity in black soil ecosystems. Environ Technol Innov 40:104481

Sivanesan I, Muthu M, Gopal J, Oh JW (2022) Mushroom polysaccharide-assisted anticarcinogenic mycotherapy: reviewing its clinical trials. Molecules 27(13):Article 13

Sun L, Niu Z (2020) A mushroom diet reduced the risk of pregnancy-induced hypertension and macrosomia: a randomized clinical trial. Food Nutr Res 64:10-29219

Susanna MB, Rapior S (2021) Agaricomycetes mushrooms (Basidiomycota) as potential neuroprotectants. Ital J Mycol 50:30–43

Tacchini M, Sacchetti G, Guerrini A, Paganetto G (2023) Mycochemicals against cancer stem cells. Toxins 15(6):360

Talal LA, Wang H, Williams BT, Morris MJ, Horvath PJ (2023) Acute effects of white button and shiitake mushroom powder supplementation on postprandial Lipemia and Glycemia following a high-fat meal. Int J Nutr 7(2):42–56

Tejedor-Calvo E, Marco P (2025) Authenticity and market potential of truffle aroma enhancers in the food industry: a review. J Food Compos Anal 147:108066

Tong Z, Chu G, Wan C, Wang Q, Yang J, Meng Z, Du L, Yang J, Ma H (2023) Multiple metabolites derived from mushrooms and their beneficial effect on Alzheimer's diseases. Nutrients 15(12):2758

Uffelman CN, Chan NI, Davis EM, Wang Y, McGowan BS, Campbell WW (2023) An assessment of mushroom consumption on cardiometabolic disease risk factors and morbidities in humans: a systematic review. Nutrients 15(5):Article 5

Veena RK, Ajith TA, Janardhanan KK (2022) Bioactive extract of mycelia biomass of Ganoderma lucidum protects doxorubicin induced cardiomyopathy. Indian J Exp Biol 60(08):Article 08

Wang CK, Kim G, Aleksandrova LR, Panenka WJ, Barr AM (2024) A scoping review of the effects of mushroom and fungus extracts in rodent models of depression and tests of antidepressant activity. Front Pharmacol 15:1387158

Williams LM, Berthon BS, Stoodley IL, Williams EJ, Wood LG (2023) medicinal mushroom extracts from Hericium coralloides and Trametes versicolor exert differential immunomodulatory effects on immune cells from older adults in vitro. Nutrients 15(9):Article 9

Yamada S, Shirai M, Nagashima K, Mochizuki J, Ono K, Kageyama S (2025) Beneficial effects of enoki mushroom extract on male menopausal symptoms in Japanese subjects: a randomized, double-blind, placebo-controlled study. Nutrients 17(7):Article 7

Yan M, Chen Y, Li M, Wu J, Fang Z, Wang J, Liu J (2022) Coprinopsis cinerea Galectin CGL1 induces apoptosis and inhibits tumor growth in colorectal cancer cells. Int J Mol Sci 24(1):235

Zhao S, Gao Q, Rong C, Wang S, Zhao Z, Liu Y, Xu J (2020) Immunomodulatory effects of edible and medicinal mushrooms and their bioactive immunoregulatory products. J Fungi 6(269):1–37

# The Application of Mushrooms in Agriculture

# 13

K. N. Abhini, N. V. Vinod, and Sinosh Skariyachan

**Abstract**

Mushrooms, which belong to the kingdom Fungi, play a vital role in sustainable farming, which is necessary to create an ecologically sound environment. This chapter reviews the scope and application of mushrooms in agriculture with an emphasis on the roles of mushrooms in managing agricultural waste, promoting plant growth, enhancing soil health, and serving as biofertilizers and biocontrol agents. Focus is given to the use of mycorrhizal fungi in improving nutrient uptake due to the mycelia's ability to establish deep into the soil. A brief overview of mushroom-derived bioactive compounds is presented, emphasizing their application in the pharmaceutical and healthcare sectors. Mushroom cultivation makes optimal use of agricultural waste, resulting in a sustainable agricultural practice. This chapter further reviews mushroom culture-based income generation, rural expansion, and agri-entrepreneurship development. The merits and demerits of mushroom process technologies are also discussed. The chapter concludes by highlighting the emergence of new technology and advancements in genetic research that will ease future mushroom cultivation.

**Keywords**

Mushrooms · Sustainable agriculture · Waste management · Ectomycorrhiza · Mycoremediation · Biofertilizer

K. N. Abhini · N. V. Vinod · S. Skariyachan (✉)
Department of Microbiology, St. Pius X College, Kasaragod, Kerala, India
e-mail: sinoshmicro@stpius.ac.in



B. Giri et al. (eds.), *Advances in Mushroom Research*,
https://doi.org/10.1007/978-981-95-7692-0_13

## 13.1 Introduction

Nowadays, farmers face several challenges, such as climate change, drought, soil degradation, pest resistance, etc., that demand a sustainable farming strategy (Farah et al. 2025). The role of microorganisms in sustainable agriculture is remarkable, improving soil health, supplying plant nutrition, and maintaining ecosystem equilibrium (Wei et al. 2024). Microorganisms, including bacteria, fungi, actinomycetes, cyanobacteria, protozoa, and algae, play a vital part in sustainable farming. They augment soil fertility, nutrient cycling, plant growth promotion, biological pest management, bioremediation, and waste decomposition (Sabalpara and Mahatma 2019). Mushrooms, which are classified as fungi, have recently gained popularity due to their likely involvement in sustainable farming (Jayaraman et al. 2024). Edible mushrooms are highly nutritious, help to reduce malnutrition, regulate lifestyle diseases, regulate hormones, and boost immunity; moreover, they ensure food security and revenue generation (Hait et al. 2025). Mushrooms are regarded as a potential superfood owing to their remarkable nutritional profile (Kaur et al. 2022). Furthermore, these are a tremendous source of bioactive composites that find application in medicines, nutraceuticals, food supplements, etc. (Kour et al. 2021). As it is cultivated in agricultural waste material, it ensures crop residue management, improves soil health, and acts as a biofertilizer and biopesticide (Jayaraman et al. 2024). An overview of the potential applications of mushrooms in various fields is illustrated in Fig. 13.1.

This chapter deals with various types of agricultural mushrooms and their application in various areas such as biofertilizers, biocontrol and pest management, waste management and bioremediation, soil health improvement, and sustainable organic farming. It also provides a comprehensive overview of the challenges and limitations of mushroom-based technologies in agriculture.

## 13.2 Taxonomy and Types of Mushrooms

The mushrooms that fit into the kingdom, Fungi, are described as macrofungi with a characteristic body that is sufficiently massive to be recognized by the human eye and harvested by hand (Hawksworth 2001). Mushroom classification is critical since there is a widespread misconception about the safety of mushroom intake. Thus, identifying mushrooms based on edibility is quite important (Li et al. 2021). Modern classification is based on the sequencing of noncoding ribosomal DNA segments, while the traditional classification method was based on characteristics like texture, color, size, habitat, etc. (Atri et al. 2017). Mainly, there are four types of mushrooms: edible mushrooms, nonedible and pathogenic mushrooms, medicinal mushrooms, and mycorrhizal and saprophytic fungi. An overview of the taxonomy and types of mushrooms commonly available for agricultural purposes is outlined in Table 13.1.

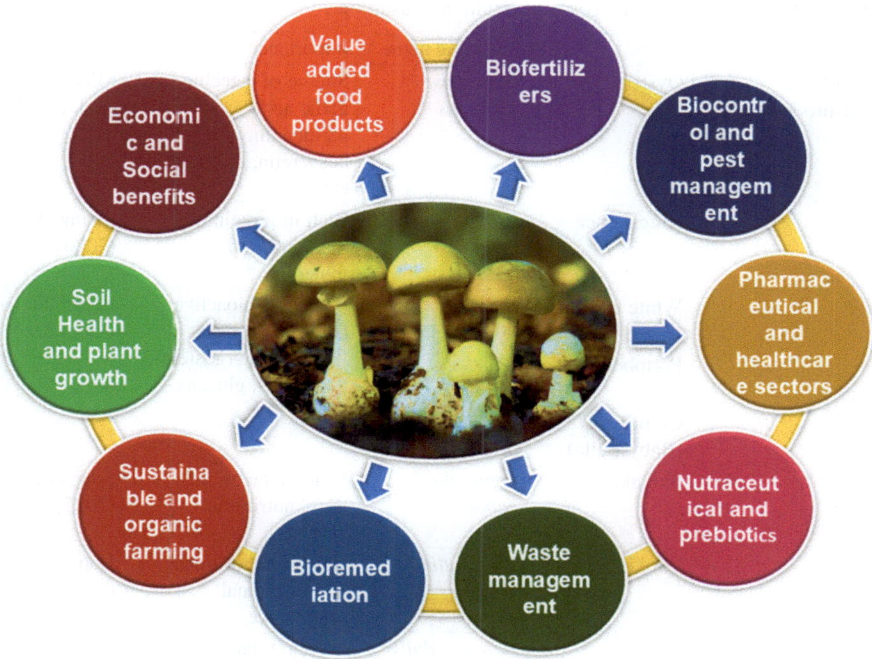


**Fig. 13.1** The potential applications of mushrooms in various fields

### 13.2.1 Edible Mushrooms

Edible mushrooms are a vast and exciting group of fungi found in various phyla, classes, orders, families, and genera depending on the type of mushroom (Zhang et al. 2021). According to investigations, there are approximately 2189 species of edible mushrooms worldwide (Li et al. 2021). A large group of edible mushrooms comes under the phylum *Basidiomycetes* (Avin et al. 2012). Some of the common edible mushrooms are *Lentinula edodes, Agaricus bisporus, Flammulina velutipes, Pleurotus ostreatus, Boletus edulis,* etc. (Mleczek et al. 2021). *Lentinula edodes* is a wood-decaying fungus that grows under dead logs' bark and develops into fruiting bodies. It is the most cultivated variety worldwide. It is familiar for its unique flavor, taste, nutritional importance, and medicinal value (Song et al. 2025). *Agaricus bisporus* has several health benefits, including increasing the bioavailability of vitamin D, promoting better gastrointestinal health, reducing the risk of cancer, and stimulating immune function. It also provides bioactive compounds like flavonoids, ergosterol, vitamin D, ergothioneine, glucans, chitin, etc. (Blumfield et al. 2020). *Flammulina velutipes,* often referred to as enoki mushrooms, is mostly regarded for its culinary and therapeutic properties (Rezaeian et al. 2023). Moreover, proteins and polysaccharides derived from *Flammulina velutipes* are used as emulsifiers in Pickering emulsions, which are used in the food, pharmaceutical, and cosmetic industries (Han et al. 2025a, b). *Pleurotus ostreatus,* also known as the

**Table 13.1** Taxonomy and types of agricultural mushrooms

| Type | | Scientific name | Importance | References |
|---|---|---|---|---|
| Edible mushroom | Oyster | *Pleurotus ostreatus* | Source of catechin, gallic acid, naringenin, hesperetin, quercetin, etc. | Mleczek et al. (2021) |
| | King oyster | *Pleurotus eryngii* | Rich in protein, fiber, vitamins, and minerals | Zhang et al. (2020) |
| | White button | *Agaricus bisporus* | Source of bioactive compounds like flavonoids, ergosterol, vitamin D, glucans, etc. | Blumfield et al. (2020) |
| | Portobello | *Agaricus bisporus* | | |
| | Cremini (Baby Bella) | *Agaricus bisporus* | | |
| | Shiitake | *Lentinula edodes* | Unique flavor and taste, nutritional benefit | Song et al. (2025) |
| | Enoki | *Flammulina velutipes* | Has both nutritional and medicinal importance | Han et al. (2025a, b) |
| | Porcini | *Boletus edulis* | Rich in vitamins, flavonoids, polysaccharides, proteins, etc. | García et al. (2021) |
| Nonedible mushroom | Fly Agaric | *Amanita muscaria* | Cycling of metallic elements | Falandysz and Treu (2019) |
| | Death cap | *Amanita phalloides* | Break down minerals, sequester more carbon | Golan et al. (2024) |
| | Destroying Angel | *Amanita virosa* | Nutrient cycling | Cai et al. (2014) |
| | Panther Cap | *Amanita pantherina* | Toxin study, soil structure improvement | Cirronis et al. (2025) |
| | Red-footed Bolete | *Erythropus* | Heavy metal accumulation | Falandysz et al. (2014) |
| | Red Cracking Bolete | *Xerocomellus chrysenteron* | Accumulation of metals | Andronikov et al. (2023) |
| | Blue-staining Slippery Jack | *Suillus tomentosus* | Exchange of nutrients with host plant | Lofgren et al. (2024) |
| Medicinal mushroom | Comb Tooth/ Coral tooth fungus | *Hericium coralloides (HC)* | Anti-inflammatory, antiviral, anti-allergic | Williams et al. (2023) |
| | Turkey tail | *Trametes versicolor* (TV) | Anti-inflammatory, antiviral, anti-allergic | Williams et al. (2023) |

(continued)

**Table 13.1** (continued)

| Type | | Scientific name | Importance | References |
|---|---|---|---|---|
| | Chaga | *Inonotus obliquus* | Antitumor and immunomodulatory | Fan et al. (2012) |
| | Maitake, ram's head, sheep's head, and hen-of-the-woods | *Grifola frondosa* | Antitumor, antidiabetic, antioxidant, immunomodulatory, anti-inflammatory, and antihyperlipidemic | Xie et al. (2025) |
| Ectomycorrhizal mushroom | King Bolete | *Boletus* | Nutrient and water absorption. Drought resistance, Soil structure, carbon cycling, and biodiversity. Forest health and regeneration, Forest resilience | Luptáková and Mihál (2020) |
| | Deceiver | *Laccaria laccata* | | |
| | Blue-staining Slippery Jack | *Suillus tomentosus* | | |
| | Fly Agaric, Death Cap, Panther Cap | *Amanita muscaria, Amanita phalloides, Amanita pantherina* | | |
| | Brittlegills | *Russula* | | |

oyster mushroom, grows on decaying wood. It is best known for its bioactive compounds like catechin, gallic acid, naringenin, hesperetin, and quercetin. It also shows antioxidant, anti-inflammatory, and anti-obesity activity (Binsaleh et al. 2025). *Boletus edulis,* the most common edible mushroom, is noted for its flavor, texture, and culinary adaptability. This mushroom is rich in health-promoting bioactive compounds like vitamins, flavonoids, polysaccharides, and proteins (García et al. 2021). Recent studies also highlight the importance of this mushroom in improving gut microbiota, which in turn reduces the risk of metabolic disorders (Zhao et al. 2025). Overall, edible mushrooms are an excellent source of nutrition, health, and environmental sustainability, in addition to being delicious. These mushrooms, whether taken fresh, dried, or as supplements, have immense promise for aiding both human health and global ecosystems (Kumar et al. 2021a, b). The major bioactive compounds present in various mushrooms are shown in Fig. 13.2.

### 13.2.2 Nonedible and Pathogenic Mushrooms

While edible mushrooms are praised for their culinary and therapeutic properties, nonedible mushrooms are generally neglected but play an important role in ecosystems, biodiversity, and scientific research. Traditional information on nonedible mushrooms plays a significant role in avoiding fatal intoxications. The most vital nonedible mushrooms are *Erythropus, Amanita muscaria, Suillus tomentosus, and*

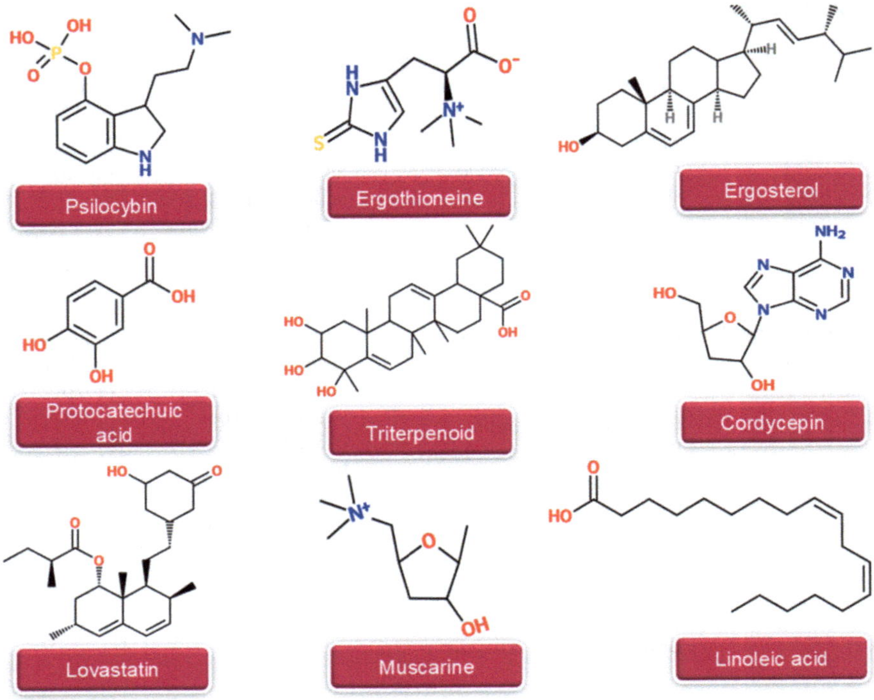


**Fig. 13.2** The structures of major bioactive compounds present in various mushrooms

*Xerocomellus chrysenteron* (Ramírez-Terrazo et al. 2021). *Erythropus* plays a vital role in bioaccumulating heavy metals, especially mercury (Falandysz et al. 2014). *Amanita muscaria* plays a significant role in the cycling of metallic elements in the ecosystem (Falandysz and Treu 2019). *Suillus tomentosus* functions as an ectomycorrhizal fungus, facilitating the exchange of nutrients with the host plant. It feeds on water and other minerals, including nitrogen and phosphorus, from the soil; in contrast, the fungus gets sugar from the host plant. Also plays a vital role in improving soil structure and carbon cycling and provides increased resistance to pathogens (Lofgren et al. 2024). Conversely, *Xerocomellus chrysenteron* plays a central role in the buildup of metals like Ag, Cd, K, P, and Rb (Andronikov et al. 2023). Like edible mushrooms, nonedible mushrooms also play a significant ecological role by participating in decomposition and nutrient cycling, improving soil health and structure, being involved in bioremediation, supporting disease suppression, and also providing biodiversity and habitat support (Ipeaiyeda et al. 2020; Elkhateeb and Daba 2022).

### 13.2.3 Medicinal Mushrooms

Medicinal mushrooms provide numerous health benefits, including anti-allergic properties, antibiotics, antifungal, antidepressant, antihyperlipidemic, antidiabetic, digestive, hepatoprotective, anti-inflammatory properties, antioxidant activity, antiviral properties, cytotoxic, immunomodulating, etc. (Venturella et al. 2021). *Hericium coralloides* (HC) and *Trametes versicolor* (TV) are two well-known medicinal mushrooms. Extracts of these mushrooms exhibited anti-inflammatory, antiviral, and anti-allergic properties (Williams et al. 2023). *Inonotus obliquus* is another medicinal mushroom with antitumor and immunomodulatory activity (Fan et al. 2012). The primary dynamic polysaccharide units in *Grifola frondosa* fruiting bodies are rich in nutrients and biological activity, including immunomodulatory, anti-inflammatory, antitumor, antidiabetic, and antioxidant properties (Xie et al. 2025). The major bioactive compounds derived from medicinal mushrooms responsible for the pharmacological activities are α- and *β*-glucans, triterpenes, erinacines and hericenones, heteroglycans, peptidoglycans, polysaccharide-K, etc. Investigations are underway to examine understudied or overlooked pharmacological characteristics (Venturella et al. 2021).

### 13.2.4 Mycorrhizal and Saprophytic Fungi

Mycorrhizal and saprophytic fungi are two important forms of fungi, having distinct ecological roles, means of nutrition, and interactions with other organisms. Mycorrhizal fungi boost plant growth by increasing nutrient and water intake, particularly phosphorus uptake from soils with high adsorption capacity. Saprophytic fungi are another important group of fungi that help in the breakdown of complex organic compounds like cellulose and lignin into simple sugars, which can provide energy sources for other organisms, including mycorrhizal fungi (Almonacid et al. 2015). Growing evidence shows that mycorrhizal fungi influence plant establishment and dispersal, control plant coexistence, and affect plant population biology and community ecology (Tedersoo et al. 2020). Mycorrhiza is classified into two types: ectomycorrhiza (EcM), which forms a mantle around the host root, and endomycorrhiza, or Arbuscular Mycorrhiza (AM), which can penetrate the roots of the host. EcM mostly belongs to Basidiomycota and Ascomycota. However, AM belongs to Glomeromycota (Tedersoo and Bahram 2019). The EcM forms a sheath around the root tip of host plants (especially woody plants) and spreads hyphae into the soil, increasing surface area, which helps in the efficient intake of water and nutrients (nitrogen and phosphorus) by plants (Lehto and Zwiazek 2011). Many mushrooms are the reproductive structure of EcM. *Boletus, Laccaria, Suillus, Amanita,* and *Russula* are examples of EcM mushrooms. Moreover, these fungi significantly influence the succession of the forest ecosystem (Luptáková and Mihál 2020). Alterations in EcM communities may disturb the nutritional equilibrium, pushing the forest ecosystem to a breaking point and adversely affecting human health. Comprehensive information on distribution and changes over time across various landscapes is essential to improve environmental models. Monitoring a specific EcM community may potentially be able to provide early indications of ecological degradation (Suz et al. 2021).

The saprophytic fungi belong to several fungal phyla that share efficient decomposing ability (Eichlerová et al. 2015). These mushrooms play a vital role in ecosystem nutrient cycling by breaking down complex substances into simpler forms. *Agaricus bisporus, Pleurotus ostreatus,* and *Volvariella volvacea* are some of the cultivable varieties of saprophytic mushrooms. The spent mushroom substrate of these mushrooms finds application as biofertilizer (Hernández et al. 2021). A recent study emphasizes the importance of saprophytic fungi in adjusting to environmental changes in arid steppes. Soil saprophytic fungal enrichment has a major impact on soil stability, physical properties, nutrients, and utility as an ecological indicator (Kang et al. 2023).

## 13.3 Mushrooms as Biofertilizers

By using microbes to solubilize nitrogen (N) and phosphorus, biofertilizers can reduce reliance on chemical fertilizers, potentially reducing climate change and enhancing soil health (Baharudin et al. 2025). A literature survey revealed that spent mushroom substrate has been used extensively as a biofertilizer. It is rich in carbon, nitrogen, and other elements, which promote healthy plant growth (Roy et al. 2015).

### 13.3.1 Role in Nutrient Cycling

Many saprophytic mushrooms play an important role in nutrient cycling by feeding on organic matter. They secrete enzymes to degrade complex polymers like lignin and cellulose and convert them into carbon, hydrogen, nitrogen, phosphorus, and other minerals. By this process, nutrition is returned to the soil (Özlü and Nicholas 2021). There are special mushrooms like *Gomphidius glutinosus*, which can essentially hyperaccumulate heavy metals. This can be successfully used to bioremediate the heavy metal-polluted area. The fungal mycelium creates a nutritional and information-sharing network on Earth and helps to construct the food web. The fungal mat is crucial to the restoration of nature because it reacts rapidly to catastrophic occurrences. They rapidly grow and restore nutrients to the soil, allowing the food chain to return to normal (Stamets 2006).

### 13.3.2 Mycorrhizal Associations and Plant Growth Promotion

Mycorrhiza is a symbiotic association between a plant root and fungi, where both benefit. It helps plants to absorb water and mineral nutrients and also provides defense against antagonists. Whereas fungi receive sugars from plants for their survival (Moora et al. 2025). These types of fungi can associate with multiple plants simultaneously and form a strong underground network that influences plant communities as well as the ecosystem (Frew et al. 2025).

#### 13.3.2.1 Arbuscular Mycorrhizal Fungi (AMF)

Nowadays, arbuscular mycorrhizal fungi (AMF) are extensively used as biofertilizers and for bioremediation. A study revealed that the application of AMF in areas of low soil health and high pathogen burden promoted the healthy growth of plants. The mycelia of AMF are well established, and they can collect all elements necessary for plant growth (Rog et al. 2025). These AMFs are also crucial in helping the plant survive harsh environments like salt and drought. Its extended hyphal network helps to absorb water during limited water availability. Additionally, they aid in absorbing scarce nutrients in dry or saline soil. Furthermore, they aid in accumulating osmo-protectants in plants, which support the maintenance of enzyme activity and cell turgor under stressful conditions (Mazumder et al. 2025).

#### 13.3.2.2 Ectomycorrhizal Fungi (EcM)

EcM fungal hyphae, in contrast to AM fungi, do not enter root cells; instead, they create a network of hyphae called Hartig's net in the root intercellular space, where the fungus and the plant exchange minerals and nutrients. EcM colonization is common in woody plants within forest ecosystems, which limits the application of EcM-based biofertilizers primarily to forestry. It helps to promote sustainable forestry and ecosystem stability. Recent studies highlighted the importance of EcM biofertilizer in forest nursery practices as they play a crucial role in supporting reforestation and afforestation efforts by improving seedling establishment and growth (Martius et al. 2025). Moreover, it plays a key role in carbon sequestration, the process of removing carbon dioxide from the atmosphere through capture and storage, in forest ecosystems (Ediriweera et al. 2025). EcM fertilizer also helps in mine site rehabilitation (Rodríguez-Uña et al. 2024). The study revealed the successful colonization, enhanced nutrient uptake, and improved growth matrices (Tüfekçi and Ortaş 2025).

### 13.3.3 Case Studies on the Application of Mushrooms as Biofertilizer

The mushroom industry generates a significant amount of waste (expired mushroom blocks and spent mushroom substrate), which presents problems for the environment and the economy. The only approach to solving this problem is to make appropriate use of these wastes. Mushrooms are a rich source of organic matter and can be used as animal feed and biofertilizer (Dowley et al. 2023). There are a few case studies presented in the following section to show the scope and application of the mushroom as a biofertilizer; the major outcomes and highlights of these studies are also shown in Table 13.2.

**Table 13.2** The beneficial applications of mushrooms as biofertilizers described in the selected case studies

| Case studies | Biofertilizer | Benefit | References |
|---|---|---|---|
| 1 | Spent mushroom substrate extract of *Pleurotus ostreatus* | Rich in micro and macro elements, reduced sugar, and phenolic compounds, it enhances the microbial community in the soil, supporting plant growth | Elsakhawy and Abd El-Rahem (2020) |
| 2 | Fermented expired mushroom blocks | Rich source of nitrogen and phosphorus. Can be used in sustainable agricultural practices | Baharudin et al. (2025) |
| 3 | Spent Mushroom Substrate from *Lentinula edodes* | High nutrient content (rich in N, P, K and Mg), ionic strength, stimulates microbial biomass, improved mineralization, and nutrient cycling | Ravlikovsky and Symochko (2020) |

#### 13.3.3.1 Case Study 1: Utilizing Spent Mushroom Substrate Extract as a Biofertilizer to Promote Rice Growth

Spent mushroom substrate (SMS) from *Pleurotus ostreatus* was air-dried in the shade for 3 days before being completely dried at 70 °C. For 2 days, the finely powdered dried SMS was soaked in water with intermittent stirring. Then, it is filtered to obtain spent mushroom substrate extract (SMSE). The chemical investigation of the extract revealed that it is rich in micro- and macroelements, a substantial amount of reducing sugars and phenols, and some bioactive compounds. The pH of the extract is near neutral, which enhances plant growth and the microbial community in the soil. The field experiment revealed that the application of SMSE accelerated the multiplication of bacterial and fungal communities. Furthermore, it enhanced the growth of cyanobacteria (*Spirulina platensis*), which plays an important role in rice fields by providing nitrogen, oxygen, and organic matter (Elsakhawy and Abd El-Rahem 2020).

#### 13.3.3.2 Case Study 2: Bokashi Fermentation for the Manufacture of Biofertilizer from Mushroom Waste

Bokashi fermentation is a fast, anaerobic composting method that uses helpful bacteria to convert organic waste into a nutrient-dense soil supplement. Unlike typical composting, fermentation is used instead of heat and air to decompose the material. The fermentation was performed under anaerobic conditions. In this study, mushroom waste was cut into small pieces and dried at 60–105 °C. Bokashi starter culture was prepared using effective microbes (EM), molasses, and mushroom waste. Dried mushroom waste was combined with the starter culture and the necessary amount of water. For 5–10 days, it is fermented anaerobically with acid. The fermentation process produced a high concentration of phosphorus (P) and nitrogen (N). These are the primary macronutrients required for the growth of plants. This suggests that mushroom waste can efficiently yield biofertilizer with high N and P content, which can be used in sustainable agriculture (Baharudin et al. 2025).

#### 13.3.3.3 Case Study 3: Spent Mushroom Substrate of Lentinula edodes as a Biofertilizer

This study compares the nutritional content, microbial biomass, and ionic strength of *Lentinula edodes* spent mushroom substrate with that of peat. The findings indicate that the SMS offers substantially greater macronutrients than peat, increases soil microbial biomass, improves nutrient cycling, and boosts microbiome activity. Therefore, SMS can be utilized to convert mushroom waste into a beneficial biofertilizer and is a viable alternative to peat (Ravlikovsky and Symochko 2020).

## 13.4 Mushrooms in Biocontrol and Pest Management

Mushrooms play a major role in pest management. A literature survey revealed that around 280 fungal species exhibit nematocidal activity. They secrete several extracellular enzymes that effectively degrade complex natural and anthropogenic substances (Rodríguez-Barrera et al. 2021).

### 13.4.1 Antagonistic Properties Against Plant Pathogens

Metabolites from mushrooms contain a variety of distinct bioactive substances, such as polysaccharides, terpenoids, steroids, phenols, and glycoprotein derivatives. These substances act as antimicrobials and can be used as biopesticides in sustainable agriculture practice (Dharani et al. 2025). Moreover, compost of spent mushroom substrate is rich in bioactive substances and also some microorganisms like bacteria, which show antagonism against pathogens. *Pseudomonas aeruginosa* and *Bacillus* sp. isolated from mushroom compost exhibited antagonistic activities against *Mycogone perniciosa*, which causes wet bubble disease (Kumar et al. 2025). The medicinal mushroom *Taiwanofungus camphoratus* is reported to possess antimicrobial activity against yeast and filamentous fungal plant pathogens and against certain pathogenic bacteria (Kim et al. 2022). A study revealed that wild mushroom extracts provide antimicrobial action against the bacterial wilt pathogen *Ralstonia solanacearum* (Erjavec et al. 2016).

### 13.4.2 Use in Integrated Pest Management (IPM)

The integrated pest control system manages pests using a variety of strategies. This strategy reduces the application of chemical pesticides, helps to protect environmental quality and biodiversity, improves crop health and productivity, and encourages long-term, cost-effective management (Ilieva et al. 2024). The mushrooms and other microbes have been successfully used in an integrated pest management strategy. The major antagonistic properties of mushrooms in integrated pest management are described in Table 13.3. In farms that grow *Agaricus bisporus* mushrooms, *Agaricus bitorquis* mushrooms may be utilized as an IPM strategy to reduce the prevalence

**Table 13.3** Major antagonistic properties of mushrooms in pest management and biocontrol

| Property | Mechanism | Example | References |
|---|---|---|---|
| Anthelmintic | Secondary metabolites act against the larval stage of gastrointestinal parasites | *Pleurotus* spp. | Rodríguez-Barrera et al. (2021) |
| Mycoparasitism | Interact with the help of haustorial cells that facilitate nutrient transfer | *Tremella* | Schoutteten et al. (2024) |
| Biopesticidal activity | Bioactive metabolite like terpenoids, phenolic compounds, flavonoids, fatty acids, and tannins acts on pests | *Lactarius deliciosus* | Volcão et al. (2022) and Abdelkhalek et al. (2025) |
| Antibiosis, anticancer, antioxidant | Bioactive metabolites responsible for the activity | *Ramaria flava* | Liu et al. (2013) |
| Enzymatic degradation | Production of nematicidal enzymes and metabolites | *Flammulina velutipes* | Ferreira et al. (2019) |
| Antifungal | Production of organic acids | *Lentinula edodes* | Kang et al. (2017) |
| Induced systemic resistance (ISR) | Production of exopolysaccharide | *Schizophyllum commune* | Yu et al. (2022) |

and stop the spread of the myceliophagous mite *M. lambi*. The extract of the mushroom *Pleurotus ostreatus* has insecticidal effects on *Macrosiphum rosae* and *Tribolium castaneum*. Furthermore, it also helps to biodegrade organophosphate pesticides in the soil (Rodríguez-Barrera et al. 2021).

## 13.5 Mushroom Cultivation on Agricultural Waste

Utilizing agricultural waste for mushroom farming is an effective and environmentally sustainable method of recycling organic byproducts while generating healthy food (Jayaraman et al. 2024). The primary goal was to safeguard and regenerate the ecosystem by converting existing agro-waste into value-added products. Various agricultural wastes, like straw, cotton waste, corncobs, cornstalks, corn gluten, sugarcane bagasse, coffee grounds, wheat bran, tea waste, pseudo-stems of bananas, legume plant residues, etc., can be used as substrates for mushroom cultivation (Grimm and Wösten 2018). The ability of *Ganoderma* mycelia to compost a range of agricultural waste was suggested by the in vitro growth of the mycelial organism on six agro-wastes: maize straw (MS), broad bean stalks (BBS), cotton stalks (CS), sugarcane bagasse (SCB), rice straw (RS), and wheat straw (WS) supplemented with wheat bran (WB) or corn gluten (CG) (Rashad et al. 2019).

The generation of *Pleurotus ostreatus* from numerous organic residues has been investigated as a sustainable substitute for agricultural waste. The study revealed that the ability of eucalyptus bark, an alternative organic residue that wood companies discard in large quantities, is an important raw material that can produce oyster mushrooms, surpassing the biological effectiveness of the conventional formula (eucalyptus sawdust), the number of fruiting bodies, and the mushroom yield. Additionally, the study indicates that lignin- and N-rich agro-industrial wastes may be useful new substrates for mushroom cultivation (Viriato et al. 2022).

### 13.5.1 Recycling Agro-waste Through Mushroom Farming

Using agricultural waste for mushroom farming also helps to recycle the waste's organic content. The mycelia help convert complex organic material in the substrate into simpler compounds. The spent mushroom substrate with these simpler molecules can be used as compost, and the nutrients are returned to the soil as soil conditioner, thereby recycling the post-mushroom substrate in an eco-friendly manner (Rashad et al. 2019). The mushroom *Pleurotus florida* (oyster mushroom) demonstrates efficient growth on the agro-wastes derived from rice and wheat. They rapidly utilize these lignin-rich materials and play an essential role in the recycling of agro-waste (Chaturvedi et al. 2021). Growing mushrooms is a scalable, environmentally responsible, and successful agro-enterprise that recycles agricultural waste. In addition to addressing environmental issues, it promotes food security, rural development, and sustainable agriculture (Tan et al. 2015).

### 13.5.2 Steps in the Production of Mushrooms

The process of producing mushrooms for human consumption or other uses is known as mushroom cultivation. Understanding which growing techniques are unique to the type of mushroom and the system being used is crucial. Multiple steps are involved in mushroom cultivation, starting with spawn preparation, moving on to substrate inoculation, spawn running, and finally fructification. Keeping things clean at all times is essential to avoid contamination and guarantee a good harvest (Silva et al. 2024).

#### 13.5.2.1 Mushroom Spawn

Spawn is the primary inoculum for mushroom cultivation. It is a mushroom mycelium grown in sterile grains for inoculation purposes. Nowadays, liquid spawn is also available as inoculum (Silva et al. 2024).

#### 13.5.2.2 Substrate Preparation

An easily available suitable substrate can be selected for mushroom cultivation. The substrate is to be sterilized using the thermal treatment method. Either the pasteurization or the sterilization method can be employed to sterilize the substrate. After sterilization, the water content of the substrate should be between 50% and 75%. The optimum pH should be in the range of 4.0–7.0. The initial pH can be adjusted using $Ca_2CO_3$ (Silva et al. 2024).

#### 13.5.2.3 Inoculation

After cooling the substrate, inoculate it with spawn. Around 5% (weight of substrate) of the seed is added to the substrate in a polythene bag. The polythene bag with substrate inoculated with mushroom seed was incubated in the dark at 24–28 °C. In the initial phase of mycelial colonization, it produces degradative enzymes like laccase and Mn peroxidase to break down the organic compounds in the substrate, making it ideal for mycelial colonization. The polythene bag can be shifted from the dark room to the fruiting room once mushroom primordia appear in the substrate (Silva et al. 2024).

#### 13.5.2.4 Fructification

The condition required for fruiting varies according to the type of mushroom cultivated. Ideal conditions required for cultivation are a temperature range from 12 to 25 °C, relative humidity of 85–95%, low $CO_2$ (<1000 ppm), moisture of 55–75%, pH of 6.5–7.5, and indirect light. Certain species, like *A. bisporus* (white button mushroom), require a casing layer over colonized mycelia to stimulate the mushroom basidioma. Once the mushroom is ready, the handpicking method can be used to harvest the mushroom. Around 3–5 harvests can be done from a mushroom bed in 7–15 days. Mushrooms are highly perishable after harvest, losing moisture and firmness and becoming vulnerable to enzymatic browning. Long-term storage and long-distance transportation are not appropriate for mushrooms. As a result, they cannot be stored at room temperature for longer than 24 h and must be sold fresh within a short time. Despite being one of the most popular preservation methods for extending product shelf life, refrigeration has limitations when it comes to preventing mushroom browning (Silva et al. 2024).

### 13.5.3 Economic and Environmental Benefits

Mushroom cultivation has shown economic viability and ecological importance through its potential to effectively use, enhance, and bio-transform agro-industrial waste. It offers financial benefits for small-scale farmers, entrepreneurs, and rural communities. It requires small capital compared to other agricultural businesses. Yield per unit area is high, which makes this very economical. Value addition offers additional income. Export-oriented cultivation of medicinal mushrooms can bring in foreign exchange. It recycles the agricultural waste, making this process sustainable and environmentally friendly (Gao et al. 2023). As it is a nutrient-rich food, it contributes to local food security. Some mushroom species can also be used for several additional purposes, such as the bioremediation and biodegradation of dangerous substances. This method converts low-value garbage into food with high nutritional value, supporting the circular economy (Silva et al. 2024).

## 13.6 Mushrooms in Waste Management and Bioremediation

### 13.6.1 Concepts and Applications of Mushroom Bioremediation

The spent mushroom compost from the mushroom industry was successfully used to remediate the mine soil contaminated with Cd, Pb, and Cu (Frutos et al. 2017). Spent mushroom substrates (*A. bisporus, P. eryngii,* and *P. ostreatus*) were effectively used to remove petroleum hydrocarbons from contaminated soil. The metagenomic analysis revealed the participation of a variety of bacteria and fungi in the successful remediation of petroleum hydrocarbons (Mayans et al. 2024). According to a study, spent mushroom compost can be fruitfully used to remediate heavy metal-contaminated soil. The biotransformation of arsenic was successfully carried out by the microbial community present in the spent mushroom compost (Dabrowska et al. 2021). A study revealed the efficiency of the mushroom *Pleurotus eryngii* in remediating aflatoxin B1, which is a major contaminant of food commodities (Branà et al. 2017). In a study, *Agaricus bisporus* substrate was well used to remediate the multi-polluted soil with Pb and polycyclic aromatic hydrocarbons (García-Delgado et al. 2015). Furthermore, EcM plays a vital role in bioremediation and acts as a long-distance biofilter under metal-stressed conditions. Some accumulate and immobilize heavy metals in the hyphal mantle (Iordache et al. 2009).

### 13.6.2 Principles and Applications of Mycoremediation

Mycoremediation is a technique that uses fungal biomass to remove heavy metals and other contaminants through biological processes such as breakdown, absorption, bioaccumulation, and bioconversion (Mohamadhasani and Rahimi 2022). The major mechanisms of mushroom-mediated bioremediation are shown in Fig. 13.3. The fungal mycelium disseminates throughout the contaminated area and secretes a variety of acids and extracellular enzymes that aid in breaking down complex organic materials into simpler nontoxic counterparts (Dinakarkumar et al. 2024). However, some mushrooms take up and accumulate heavy metals and poisons in their biomass, making them poisonous (Ab Rhaman et al. 2021). Its extended mycelial networks, synergistic relationships, adaptability, and sustainability make it a better candidate for bioremediation (Harms et al. 2011). Research shows it effectively absorbs harmful pollutants like heavy metals, dioxins, pesticides, phenols, chlorophenols, PCBs, PAHs, and effluents from pulp and paper mills (Kumar and Kaur 2019). Among other things, mycoremediation can be used effectively to disinfect soil, treat industrial and agricultural waste, purify wastewater, biodegrade hazardous waste, break down plastic and synthetic polymers, and reclaim mining sites. Thus, mycoremediation is proposed as a greener, more cost-effective, and efficient solution to address global pollution challenges. (Kumar et al. 2021a, b).

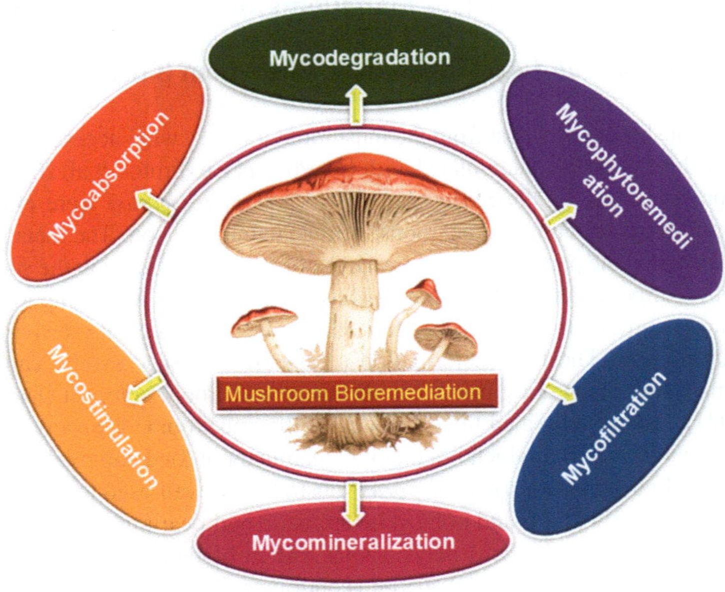


**Fig. 13.3** The major mechanisms of the applications of mushroom-mediated bioremediation

**Table 13.4** Application of mushrooms in bioremediation of heavy metals based on the selected case studies and potential outcome of the studies

| Case study | Mushroom | Outcome | References |
|---|---|---|---|
| 1 | *Pleurotus* sp. | Increased level of catalase and superoxide dismutase that enhances the accumulation of heavy metals | Mohamadhasani and Rahimi (2022) |
| 2 | *Pleurotus eryngii* | Heavy metal remediation and exhibits antioxidant properties | Goligar et al. (2023) |
| 3 | *Agaricus bisporus* | Accumulates different elements found in flotation tailings | Budzyńska et al. (2022) |

### 13.6.3 Case Studies of Mycoremediation as a Successful Practice

Several case studies comprehensively elucidate the potential applications of mushrooms as bioremediating agents. Some of the successful case studies that utilized mushrooms as mycoremediants are discussed, and their outcome is illustrated in Table 13.4.

#### 13.6.3.1 Case Study 1: Mycoremediation of Heavy Metals by Pleurotus Sp.

In this investigation, mushrooms were grown in various concentrations of cobalt, copper, and nickel. The findings showed that heavy metals increased catalase (CAT) activity and superoxide dismutase (SOD) levels. Successful accumulation of heavy metals occurred in the mycelium (Mohamadhasani and Rahimi 2022).

#### 13.6.3.2 Case Study 2: Mycoremediation of Industrial Wastewater by Pleurotus Eryngii

In this research, *Pleurotus eryngii* was cultivated in media containing concentrations of lead and cadmium at 150 mg/L, 250 mg/L, and 350 mg/L for 30 days, and the heavy metal levels in the mycelia were examined. The analysis revealed an increase in the levels of phenols, flavonoids, and DPPH compounds in the mycelia, which contribute to the mushroom's strong antioxidant properties (Goligar et al. 2023).

#### 13.6.3.3 Case Study 3: Using Agaricus Bisporus in Flotation Tailings Mycoremediation

In this investigation, *Agaricus bisporus* was cultivated on compost mixed with flotation tailings in various proportions (1%, 5%, 10%, 15%, and 20%). The *A. bisporus* strain evaluated in this research successfully accumulated specific elements (Ag, B, Cu, Ga, Ge, Rh, Au, In, Ir, K, Mg, Pd, Pt, Re, Ru, Sb, Te, and Se). This method appears to be the most effective for substrates contaminated with certain elements, which are absorbed most efficiently by the fruiting bodies. This study indicated that identifying the most appropriate fungal species to target a certain contaminant is critical for mycoremediation (Budzyńska et al. 2022).

### 13.6.4 Dye Degradation by Mushrooms

Dyes, extensively employed in sectors such as textiles, leather, paper, and cosmetics, can behave as environmental pollutants when released into water bodies or soil unless properly treated. The fungal community plays a crucial role in degrading these dyes. Azo dyes, anthraquinone dyes, and triphenylmethane dyes are major classes of dye pollutants. The mushroom *Phanerochaete chrysosporium* biodegrades the diazo dye Reactive Black 5 (RB5) that has been immobilized on sunflower seed shells and nylon sponge cubes. The sensitive soil microbe *Sinorhizobium meliloti* was used to evaluate RB5 solutions' toxicity before and after fungal treatment. The lethal dye RB5 was transformed into a harmless substance by *Phanerochaete* chrysosporium (Enayatizamir et al. 2011). A unique indigenous strain, *P. chrysosporium* IBL-03, with great manganese peroxidase (MnP) activity, was employed to remove the active textile dye's color, Drimarine Blue K2R, which is widely used in textile units. It was found that the main enzyme responsible for *P. chrysosporium's* decolorization of Drimarine Blue K2RL dye was manganese peroxidase (560 IU/mL). According to the dye adsorption studies, the dye that had initially been adsorbed on fungal mats later disappeared, most likely due to the fungus's release of MnP during secondary metabolism (Noreen et al. 2011). Dye-decolorizing peroxidases (DyPs) are a new class of heme peroxidases capable of oxidizing resistant chemicals. The recombinant GlDyP2 protein (derived from *Ganoderma lucidum*) can oxidize two lignin substrates, guaiacol and 2,6-dimethylphenol (DMP), in addition to the standard peroxidase substrate ABTS. Additionally, GlDyP2 detoxified and decolored several dyes, such as Malachite Green, Methyl Orange, Trypan Blue, Reactive Black 5, Reactive Brilliant Blue

X-BR, and Reactive Blue 19. As a result, GlDyP2 has a great deal of potential for detoxifying water that has been contaminated by dyes (Liu et al. 2024). Synthetic dyes Congo red and Remazol Brilliant Blue R were decolored by the lignocellulolytic enzymes amylase, cellulase, laccase, and xylanase that were derived from *Pleurotus ostreatus*, *P. eryngii*, and *P. cornucopiae* wasted mushroom compost (SMC) (Lim et al. 2013). Spent mushroom substrate (SMS) from white button mushrooms exhibits considerable potential for the treatment of aqueous effluents contaminated with textile dyes. Research has demonstrated both the enzymatic and nonenzymatic capabilities of SMS in effectively eliminating dyes such as crystal violet, Malachite Green, Remazol Brilliant Blue R, orange G, and rose Bengal (van Brenk et al. 2024).

The basis and mechanism of mushroom-based technologies in various fields are shown in Table 13.5. The various processes and potential benefits of edible mushroom-based technologies in the present industry are shown in Fig. 13.4.

## 13.7 Soil Health and Mushroom Activity

### 13.7.1 Impact on Soil Microbial Diversity and Structure

Microbial life is a part of the vast and ever-changing soil ecosystem (Sokol et al. 2022). Fungi, especially mushrooms, are very important in forming microbial networks due to their ability to decompose organic matter, establish symbiotic relationships, and produce secondary metabolites. Interaction between microbial communities improves soil health and nutrient cycling, which supports ecosystem stability and biodiversity (Ogwu et al. 2025). The cultivation of *P. impudicus* increased the richness of soil microbes, which clarified the differences in composition and structure of soil (Wang et al. 2024). Additionally, the Spent Mushroom Substrate (SMS) compost, with microbial diversity, improves soil multifunctionality, which supports an increase in plant biomass (Balakrishnan et al. 2024). A study revealed that intercropping with *P. ostreatus* significantly boosts the levels of total nitrogen, phosphorus, and available phosphorus in the soil (Yang et al. 2024a, b). Because of their complex metabolic activities and mycelial networks, mushrooms are essential for controlling the variety and organization of soil microbes. By increasing microbial variety, establishing functional linkages, and influencing the diversity of community members, they improve soil life and ecological robustness. Understanding and employing the relationships between microbes in soil can result in more environmentally friendly farming and forestry activities (Khan et al. 2024).

### 13.7.2 Organic Matter Decomposition and Humus Formation

Mushrooms belong to saprophytic fungi, which feed on dead organic substances by releasing extracellular enzymes and break down the complex organic material into

**Table 13.5** The scope and applications of mushrooms in various sectors of biotechnology

| Sectors | Application | Mechanism | References |
|---|---|---|---|
| Health and Pharma | Medicinal compounds Immunomodulation | Mushroom-derived biologically active substances exhibit effective antibacterial, anti-tumor, antidiabetic, anticancer, and antioxidant properties. Boosts immunity and reduces inflammation | Chugh et al. (2022) and Xie et al. (2025) |
| Food | Edible mushrooms, functional foods, and meat substitutes | Used for consumption (fresh), used as supplements in soups (powders), and used as meat substitutes by vegetarians | Rezaeian et al. (2023) and Narayanan et al. (2023) |
| Agriculture | Biofertilizer, soil conditioner, biopesticide | SMS improves soil fertility and microbial activity, and extracts exhibit insecticidal activity | Baharudin et al. (2025) and Rodríguez-Barrera et al. (2021) |
| Animal Husbandry | Feed additives | SMS used as feed additives | Dowley et al. (2023) |
| Cosmetics | Skin care | Extracts used in skin care products | Han et al. (2025a, b) |
| Industry | Enzyme production, bioplastic, and packaging | Cellulase, laccase, biofuel, textile, and paper industries. Mycelium-based bioplastic | Lim et al. (2013), El Enshasy et al. (2019), Shao et al. (2024), and Verma et al. (2023) |
| Environmental | Mycoremediation waste recycling | Degrade heavy metals, dyes, and hydrocarbons. Application of agricultural waste in mushroom production and SMS is used as a biofertilizer | Mohamadhasani and Rahimi (2022), van Brenk et al. (2024), and Mayans et al. (2024) |
| Nanobiotechnology | Mushroom-derived nanomaterials | Drug delivery, antimicrobial coating, and biosensors | Srivastava et al. (2025) |

simpler molecules and later convert it into humic substances (Collado et al. 2018). Humus is formed by the almost complete decomposition of dead leaves and other organic matter. Lignin transformation is one of the major pathways carried out in the formation of humus, along with microbial metabolites and polymerization. The ability of mushrooms to break down lignocellulosic materials makes them an excellent choice for the breakdown of organic waste, the cycling of nutrients, and humus formation in an ecosystem (Jarial et al. 2024). Cow dung combined with

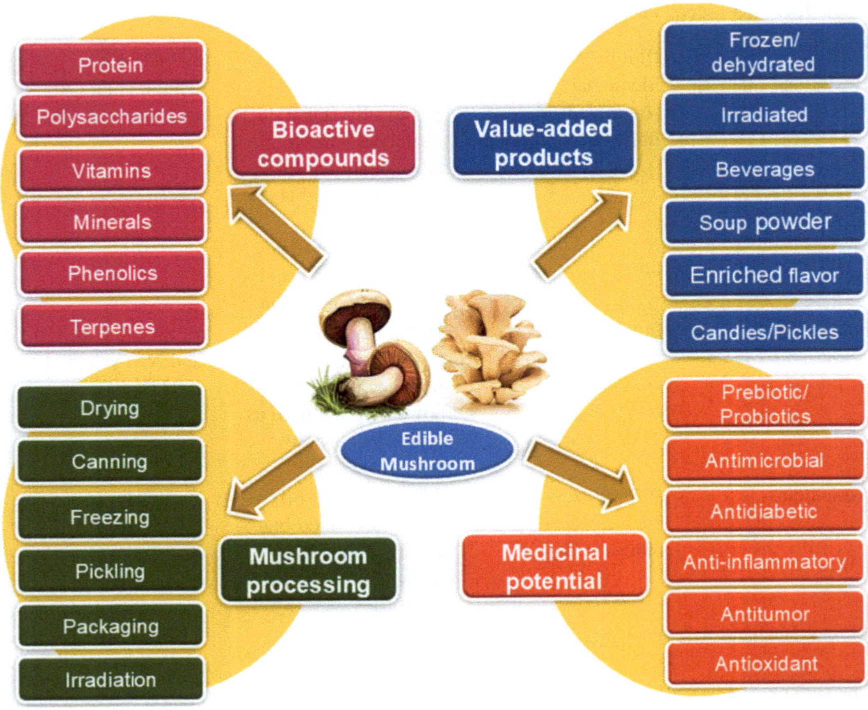


**Fig. 13.4** Various processes, potential applications, and benefits of mushroom-based technologies in the current industry

composted spent mushroom substrate aided in the creation of humus (Yang et al. 2024a, b). The humus helps to enhance soil fertility, increases the cation exchange capacity of the soil, improves soil aggregation, accelerates water retention, and helps to control soil erosion. Moreover, humus is rich in microbial diversity, which supports plant health and nutrient cycling (Piccolo and Drosos 2025).

## 13.8 Mushrooms in Sustainable Organic Farming

### 13.8.1 Compatibility with Organic Farming Practices

Organic farming is a method of farming that uses natural inputs that support soil health and biodiversity and ensure environmental sustainability. Genetically modified organisms (GMOs) and chemical fertilizers and pesticides are not taken into account in organic farming since they can alter the soil microbial community, resulting in a loss of soil health and ecological balance (Gamage et al. 2023). Mushroom cultivation exactly fits into the principles of organic farming that support soil, plant, animal, and human health. Mushroom farming effectively utilizes agricultural waste and decomposes it into simpler molecules that support soil health.

Mushroom cultivation can be included in integrated farming, which ensures soil health, biodiversity, food security, and environmental sustainability (Singh et al. 2020).

### 13.8.2 Contribution to Sustainable Agriculture Systems

Growing mushrooms contributes significantly to the management of agricultural waste in addition to producing nutritious food. Long-term soil fertility and health are maintained, pest attacks are decreased, and soil nutrient balance is achieved (Mahari et al. 2020). Growing mushrooms as part of an integrated agricultural system contributes to the sequestration of organic carbon, a crucial component in climate change mitigation. The microbial community also contributes to the stability of ecosystems. It significantly reduces climate change and promotes sustainable agricultural systems (Dou et al. 2025). Finding a balance between human activities and mushroom habitats is important for long-term coexistence (Ogwu et al. 2025).

## 13.9 Economic and Social Impact

### 13.9.1 Income Generation for Rural Communities

With increased awareness of the need for wholesome, high-quality food, the demand for nutritious foods, especially mushrooms, is rapidly increasing and will only expand as the world's population grows (Joseph et al. 2024). Mushrooms have a balanced nutritional profile with low calories and low fat, digestible carbohydrates, minerals, vitamins, fiber, and proteins (Grimm and Wösten 2018). Furthermore, mushrooms may help avoid lifestyle disorders, including diabetes, obesity, and cancer. This pharmacological importance also demands mushroom production (Bandara et al. 2021). Mushroom cultivation is the most suitable agro-based activity for rural areas. It requires minimal space with minimal investment and ensures effective utilization of agricultural waste. Moreover, it can be a reliable source of income for the rural community (Thakur 2020). It supports self-employment to sustain the rural economy and also helps to achieve sustainable developmental goals (Jayaraman et al. 2024). The economic and social relevance of mushroom-based technologies is highlighted in Table 13.6.

### 13.9.2 Role in Agri-entrepreneurship and Livelihood

In many countries, mushroom cultivation promotes agri-entrepreneurship that supports the livelihood of that particular place. This helps to generate income because it is a low-investment, high-return product with high demand (Joseph et al. 2024). The emergence of agri-entrepreneurs creates job opportunities in that particular area. Furthermore, it helps in women's and youth empowerment. Value

**Table 13.6** Economic and social impact of mushroom-based technologies

| Feature | Impact | References |
|---|---|---|
| Revenue generation | Low investment for income generation | Silva et al. (2024) |
| Employment creation | Job creation in cultivation, marketing, value-added product creation, supply chain, etc. | Sharma et al. (2025) |
| Women empowerment | Job for women in rural areas | Joseph et al. (2024) |
| Youth empowerment | Low capital investment is attractive for young agri-entrepreneurs | Sadiq et al. (2024) |
| Rural development | Economic generation in rural areas | Thakur (2020) |
| Waste utilization | SMS can be used for composting and transforming byproducts into valuable things | Silva et al. (2024) |
| Nutrition security | Nutrition-rich (protein, fiber, minerals, and vitamins) food | Galeh et al. (2025) |
| Medicinal value | Bioactive compounds act against various diseases | Lawal et al. (2025) |
| Mycoremediation | Degrades heavy metals, xenobiotics, pesticides, dyes, hydrocarbons, etc. | Yadav et al. (2021) |

addition of these mushrooms can further increase the agribusiness scope (Sadiq et al. 2024). Growing mushrooms in rural areas provides a sustainable way to address food security, land scarcity, and climate change challenges (Schilla et al. 2025).

## 13.10 Challenges and Limitations

Although growing mushrooms offers many benefits, there are also numerous drawbacks and restrictions. It is necessary to have a controlled climate, and technological know-how is another issue. Its short shelf life and delicate characteristics necessitate value addition in marketing.

### 13.10.1 Climatic, Technological, and Market Constraints

The climatic factors, such as high temperatures, extreme humidity, and restricted air flow, will all have an impact on mushroom development. It encourages the easy spread of pests and illnesses (Karavani et al. 2018). Thus, appropriate climatic conditions are required for mushroom cultivation. Another problem is acquiring technological competence in mushroom cultivation. Inadequate climate-controlled infrastructure, access to quality spawn and substrates, limited knowledge in scientific culture procedures, and poor post-harvest technologies like drying, storage, packing, etc., are all challenges (Okigbo and Anuagasi 2021). Marketing is another barrier due to its perishability. Fresh mushrooms are perishable, limiting transit and shelf life. Small growers face weak market linkages and low-price realization. Consumer knowledge and demand in rural regions are minimal. Processing and value-added units are limited (Sangeeta et al. 2024).

### 13.10.2 Issues with Large-Scale Application and Acceptance

Growing mushrooms on a large scale presents many challenges. The scaling-up requires controlled conditions (temperature, humidity), a high contamination risk, and trained staff and technical skills. Insufficient infrastructure, including climate control units and incubation laboratories, as well as cold chain, storage, and processing units (Derme et al. 2024). Incomplete or poor-quality spawn reduces production and relies on limited, concentrated spawners (Borah et al. 2019). Weak supply chains, limited customer demand in certain locations, and the lack of coordinated wholesale and export channels are all contributing factors to price volatility (Huang and Huatuco 2016). Mushroom enterprises face high capital expenditure on automation and climate control, as well as limited financial support, funding, and insurance options (Sangeeta et al. 2024). Mushrooms are generally not considered in conventional agricultural policies, with limited government support for expansion and subsidy schemes. Some places have a negative image of forest-based foods as dangerous, and a lack of knowledge among people leads to limited culinary use and consumer misunderstanding (Qwarse et al. 2021).

## 13.11 Future Perspectives and Innovations

Growing consumer preferences, sustainability requirements, and technological advancements all present exciting opportunities for the mushroom farming industry in the future (Rathod 2023).

### 13.11.1 Emerging Technologies in Mushroom-Based Agriculture

Mushroom-based biomaterials constitute a promising future for biomaterial science with the potential to transform the industry and contribute to a more sustainable future. Multidisciplinary cooperation is needed to ensure that mushroom-derived biomaterials are safe, biocompatible, and ethically accepted (Srivastava et al. 2025). An innovative method of combating agricultural pests and diseases is provided by biopesticides made from mushrooms (Omoregie et al. 2025). The emergence of the integration of controlled environment agriculture (CEA) technology is a progressive advancement in mushroom cultivation. Furthermore, the implementation of Internet of Things (IoT)-based environment control and monitoring is another progress in this sector (Cheng et al. 2023). Campaigns for education and awareness were made possible by the acceptance of new mushroom varieties and their use, particularly in functional foods and meat substitutes. Establishing a balance between consumer demands, environmental care, and the need for sustainable food production will define the future of mushroom farming (Rathod 2023).

### 13.11.2 Genetic Research and Strain Improvement

Using molecular markers like ESTs, RAPIDs, RFLPs, AFLPs, SSRIs, SCARs, microarrays, and others, it is possible to analyze any morphological, physiological, or developmental process where genetic variants exist (via mapping and profiling) with no prior knowledge. In addition, using microarray technology, it is feasible to identify genes that are activated or deactivated during development or in response to environmental changes. Additionally, molecular markers are a useful tool for identifying the quality traits of a particular strain or variety of mushrooms, which may then be applied to an existing variety to create a better strain with the desired quality attribute. Protoplasm fusion, marker-assisted selection (MAS), backcrossing tactics, electroporation and biolistic techniques, and the application of quantitative trait loci (QTLs) are all offered as innovative methods for mushroom farming (Kamal et al. 2022).

## 13.12 Conclusion

The importance of mushrooms in promoting sustainable farming methods is becoming more widely acknowledged. Beyond the production of food, they serve vital purposes like improving soil fertility, encouraging plant growth through symbiotic relationships, recycling agricultural waste, and providing natural substitutes for chemical pesticides and fertilizers. The ecological and financial significance of mushrooms in agricultural systems is highlighted by their mycorrhizal relationships, biocontrol potential, and composting abilities. Incorporating mushrooms into agricultural practices offers a promising way to increase productivity while preserving environmental health as global agriculture looks for more sustainable and regenerative methods. To fully realize the potential of mushrooms in revolutionizing contemporary agriculture, more research, implementation of technological advancements, and farmer education are required.

**Declaration** The authors declare no conflict of interest.

## References

Ab Rhaman SMS, Naher L, Siddiquee S (2021) Mushroom quality related with various substrates' bioaccumulation and translocation of heavy metals. J Fungi 8(1):42

Abdelkhalek ST, Zia S, Abdel-Haleem DR, Moussa MA, Izah SC, Wang MQ (2025) Exploring the biopesticidal potential of edible mushrooms. In: Bioactive compounds in edible mushrooms: sustainability and health applications. Springer Nature Switzerland, Cham, pp 1–54

Almonacid L, Fuentes A, Ortiz J, Salas C, Garcia-Romera I, Ocampo J, Arriagada C (2015) Effect of mixing soil saprophytic fungi with organic residues on the response of Solanum lycopersicum to arbuscular mycorrhizal fungi. Soil Use Manag 31(1):155–164

Andronikov AV, Andronikova IE, Sebek O, Martinkova E, Stepanova M (2023) Accumulation and within-mushroom distribution of elements in red cracking bolete (Xerocomellus chrysenteron) collected over the extended period from compositionally contrasting substrates. Environ Monit Assess 195(10):1157

Atri NS, Kaur M, Sharma S (2017) Characterization of lamellate mushrooms—an appraisal. In: Developments in fungal biology and applied mycology, pp 471–500. https://doi.org/10.1007/978-981-10-4768-8_24

Avin FA, Bhassu S, Shin TY, Sabaratnam V (2012) Molecular classification and phylogenetic relationships of selected edible Basidiomycetes species. Mol Biol Rep 39:7355–7364

Baharudin AS, Zainol N, Aziz NH (2025) Parameters evaluation for biofertilizer production from mushroom waste through design of experiment. J Environ Manage 378:124734

Balakrishnan K, Krishnaa D, Balakrishnan G, Manickam M, Abdulkader AM, Dharumadurai D (2024) Association of bacterial communities with psychedelic mushroom and soil as revealed in 16S rRNA gene sequencing. Appl Biochem Biotechnol 196(5):2566–2590

Bandara AR, Lian CK, Xu J, Mortimer PE (2021) Mushroom as a means of sustainable rural development in the Chin State, Myanmar. Circ Agric Syst 1(1):1–6

Binsaleh NK, Bazaid AS, Barnawi H, Alharbi B, Alsolami A, Babsail AY et al (2025) Chemical characterization, anticancer, antioxidant and anti-obesity activities with molecular docking studies of Pleurotus ostreatus biomass exposed to moist heat. J Food Meas Charact 19(4): 2476–2495

Blumfield M, Abbott K, Duve E, Cassettari T, Marshall S, Fayet-Moore F (2020) Examining the health effects and bioactive components in Agaricus bisporus mushrooms: a scoping review. J Nutr Biochem 84:108453

Borah TR, Singh AR, Paul P, Talang H, Kumar B, Hazarika S (2019) Spawn production and mushroom cultivation technology. ICAR Research Complex for NEH Region, 46

Branà MT, Cimmarusti MT, Haidukowski M, Logrieco AF, Altomare C (2017) Bioremediation of aflatoxin B1-contaminated maize by king oyster mushroom (Pleurotus eryngii). PLoS One 12(8):e0182574

Budzyńska S, Siwulski M, Budka A, Kalač P, Niedzielski P, Gąsecka M, Mleczek M (2022) Mycoremediation of flotation tailings with Agaricus bisporus. J Fungi 8(8):883

Cai Q, Tulloss RE, Tang LP, Tolgor B, Zhang P, Chen ZH, Yang ZL (2014) Multi-locus phylogeny of lethal amanitas: implications for species diversity and historical biogeography. BMC Evol Biol 14(1):143

Chaturvedi VK, Yadav N, Rai NK, Bohara RA, Rai SN, Aleya L, Singh MP (2021) Two birds with one stone: oyster mushroom mediated bimetallic Au-Pt nanoparticles for agro-waste management and anticancer activity. Environ Sci Pollut Res 28:13761–13775

Chong JL, Chew KW, Peter AP, Ting HY, Show PL (2023) Internet of things (IoT)-based environmental monitoring and control system for home-based mushroom cultivation. Biosensors 13(1):98

Chugh RM, Mittal P, Mp N, Arora T, Bhattacharya T, Chopra H et al (2022) Fungal mushrooms: a natural compound with therapeutic applications. Front Pharmacol 13:925387

Cirronis M, Giampreti A, Eleftheriou G, Gallo M, Contessa G, Faraoni L et al (2025) Coma and confirmed epileptiform activity induced by Amanita pantherina poisoning. Toxicon 254:108208

Collado S, Oulego P, Suárez-Iglesias O, Díaz M (2018) Biodegradation of dissolved humic substances by fungi. Appl Microbiol Biotechnol 102:3497–3511

Dabrowska M, Debiec-Andrzejewska K, Andrunik M, Bajda T, Drewniak L (2021) The biotransformation of arsenic by spent mushroom compost–an effective bioremediation agent. Ecotoxicol Environ Saf 213:112054

Derme T, Schwarze FW, Dillenburger B (2024) Understanding the role of controlled environments for producing mycelium-bound composites: advancing circular practices for integrating biotechnology into the construction industry. Global Chall 8(7):2300197

Dharani S, Thiribhuvanamala G, Angappan K, Swarnakumari N, Jeya Sundara Sharmila D, Manikanda Boopathi N (2025) Mushroom-derived antimicrobials–a source of eco-friendly antimicrobial compounds against phytopathogens. J Plant Pathol 12(4):1–15

Dinakarkumar Y, Gnanasekaran R, Reddy GK, Vasu V, Balamurugan P, Murali G (2024) Fungal bioremediation: an overview of the mechanisms, applications and future perspectives. Environ Chem Ecotoxicol 6:293–302

Dou T, Zhang K, Shi X, Liu W, Yu F, Liu D (2025) Crop–mushroom rotation: a comprehensive review of its multifaceted impacts on soil quality, agricultural sustainability, and ecosystem health. Agronomy 15(3):563

Dowley A, Sweeney T, Conway E, Vigors S, Ryan MT, Yadav S et al (2023) The effects of dietary supplementation with mushroom or selenium enriched mushroom powders on the growth performance and intestinal health of post-weaned pigs. J Anim Sci Biotechnol 14(1):12

Ediriweera AN, Lu W, Perez Moreno J, Kalamulla R, Mayadunna N, Pelewatta A et al (2025) Ectomycorrhizal fungal symbiosis on plant nutrient acquisition in tropical ecosystems. N Z J Bot 63(5):1–24

Eichlerová I, Homolka L, Žifčáková L, Lisá L, Dobiášová P, Baldrian P (2015) Enzymatic systems involved in decomposition reflects the ecology and taxonomy of saprotrophic fungi. Fungal Ecol 13:10–22

El Enshasy H, Agouillal F, Mat Z, Malek RA, Hanapi SZ, Leng OM et al (2019) Pleurotus ostreatus: a biofactory for lignin-degrading enzymes of diverse industrial applications. In: Recent advancement in white biotechnology through fungi: volume 3: perspective for sustainable environments. Springer International Publishing, Cham, pp 101–152

Elkhateeb WA, Daba GM (2022) The wild non edible mushrooms, what should we know so far. Int J Adv Biochem Res 6(1):43–50

Elsakhawy T, Abd El-Rahem WT (2020) Evaluation of spent mushroom substrate extract as a biofertilizer for growth improvement of rice (Oryza sativa L). Egypt J Soil Sci 60(1):31–42

Enayatizamir N, Tabandeh F, Rodríguez-Couto S, Yakhchali B, Alikhani HA, Mohammadi L (2011) Biodegradation pathway and detoxification of the diazo dye Reactive Black 5 by Phanerochaete chrysosporium. Bioresour Technol 102(22):10359–10362

Erjavec J, Ravnikar M, Brzin J, Grebenc T, Blejec A, Gosak MŽ et al (2016) Antibacterial activity of wild mushroom extracts on bacterial wilt pathogen Ralstonia solanacearum. Plant Dis 100(2): 453–464

Falandysz J, Treu R (2019) Amanita muscaria: bio-concentration and bio-indicative potential for metallic elements. Environ Earth Sci 78(24):722

Falandysz J, Dryżałowska A, Saba M, Wang J, Zhang D (2014) Mercury in the fairy-ring of Gymnopus erythropus (Pers.) and Marasmius dryophilus (Bull.) P. Karst. mushrooms from the Gongga Mountain, Eastern Tibetan Plateau. Ecotoxicol Environ Saf 104:18–22

Fan L, Ding S, Ai L, Deng K (2012) Antitumor and immunomodulatory activity of water-soluble polysaccharide from Inonotus obliquus. Carbohydr Polym 90(2):870–874

Farah AA, Mohamed MA, Musse OSH, Nor BA (2025) The multifaceted impact of climate change on agricultural productivity: a systematic literature review of SCOPUS-indexed studies (2015–2024). Discov Sustain 6(1):1–27

Ferreira JM, Carreira DN, Braga FR, Soares FEDF (2019) First report of the nematicidal activity of Flammulina velutipes, its spent mushroom compost and metabolites. 3 Biotech 9(11):410

Frew A, Varga S, Klein T (2025) Mycorrhizal networks: understanding hidden complexity. Funct Ecol 39(6):1322–1327

Frutos I, García-Delgado C, Cala V, Gárate A, Eymar E (2017) The use of spent mushroom compost to enhance the ability of Atriplex halimus to phytoremediate contaminated mine soils. Environ Technol 38(9):1075–1084

Galeh MS, Raeini MGN, Taki M (2025) From production to market; enhancing food security through efficient mushroom distribution: a study of energy and economic evaluations. Results Eng 26(10):105442

Gamage A, Gangahagedara R, Gamage J, Jayasinghe N, Kodikara N, Suraweera P, Merah O (2023) Role of organic farming for achieving sustainability in agriculture. Farming System 1(1):100005

Gao Y, Wu Z, Li W, Sun H, Chai Y, Li T et al (2023) Expanding the valorization of waste mushroom substrates in agricultural production: progress and challenges. Environ Sci Pollut Res 30(2):2355–2373

García MM, Paula VB, Olloqui ND, García DF, Combarros-Fuertes P, Estevinho LM et al (2021) Effect of different cooking methods on the total phenolic content, antioxidant activity and sensory properties of wild Boletus edulis mushroom. Int J Gastron Food Sci 26:100416

García-Delgado C, Yunta F, Eymar E (2015) Bioremediation of multi-polluted soil by spent mushroom (Agaricus bisporus) substrate: polycyclic aromatic hydrocarbons degradation and Pb availability. J Hazard Mater 300:281–288

Golan J, Wang YW, Adams CA, Cross H, Elmore H, Gardes M et al (2024) Death caps (Amanita phalloides) frequently establish from sexual spores, but individuals can grow large and live for more than a decade in invaded forests. New Phytol 242(4):1753–1770

Goligar N, Saadatmand S, Khavarinejad RA (2023) Mycoremediation of lead and cadmium by lignocellulosic enzymes of Pleurotus eryngii. AMB Express 13(1):127

Grimm D, Wösten HA (2018) Mushroom cultivation in the circular economy. Appl Microbiol Biotechnol 102:7795–7803

Hait M, Bhardwaj AK, Kashyap NK (2025) Role of edible mushroom in addressing malnutrition and food security. In: Bioactive compounds in edible mushrooms: sustainability and health applications. Springer Nature Switzerland, Cham, pp 1–40

Han J, Kawauchi M, Terauchi Y, Tsuji K, Yoshimi A, Tanaka C et al (2025a) Physiological function of hydrophobin Hydph16 in cell wall formation in agaricomycete Pleurotus ostreatus. Fungal Genet Biol 176:103943

Han X, Niu X, Xu M, Feng R, Han Q, Liu B et al (2025b) Flammulina velutipes protein-Flammulina velutipes soluble polysaccharide-tea polyphenols particles stabilized Pickering emulsions for the delivery of β-carotene. Int J Biol Macromol 285:138299

Harms H, Schlosser D, Wick LY (2011) Untapped potential: exploiting fungi in bioremediation of hazardous chemicals. Nat Rev Microbiol 9(3):177–192

Hawksworth DL (2001) Mushrooms: the extent of the unexplored potential. Int J Med Mushrooms 3(4):5

Hernández D, Ros M, Carmona F, Saez-Tovar JA, Pascual JA (2021) Composting spent mushroom substrate from Agaricus bisporus and Pleurotus ostreatus production as a growing media component for baby leaf lettuce cultivation under Pythium irregulare biotic stress. Horticulturae 7(2):13

Huang TY, Huatuco LH (2016) Agri-food supply chain for mitigation of volatilities in the role of intermediary: a case study of a mushroom trading company in Taiwan. In: Sustainable design and manufacturing 2016. Springer International Publishing, Switzerland, pp 447–454

Ilieva T, Karova A, Ivanova M (2024) Sustainable agriculture through integrated pest management: strategies for effective implementation. In: Conference on sustainable urban mobility. Springer Nature Switzerland, Cham, pp 511–520

Iordache V, Gherghel F, Kothe E (2009) Assessing the effect of disturbances on ectomycorrhiza diversity. Int J Environ Res Public Health 6(2):414–432

Ipeaiyeda AR, Adenipekun CO, Oluwole O (2020) Bioremediation potential of Ganoderma lucidum (Curt: Fr) P. Karsten to remove toxic metals from abandoned battery slag dumpsite soil and immobilisation of metal absorbed fungi in bricks. Cogent Environ Sci 6(1):1847400

Jarial RS, Jarial K, Bhatia JN (2024) Comprehensive review on oyster mushroom species (Agaricomycetes): morphology, nutrition, cultivation and future aspects. Heliyon 10(5):e26539

Jayaraman S, Yadav B, Dalal RC, Naorem A, Sinha NK, Rao CS et al (2024) Mushroom farming: a review focusing on soil health, nutritional security and environmental sustainability. Farming Syst 2(3):100098

Joseph B, Krishnakumar NM, Ceasar SA, Devassy SM (2024) Role of rural women in achieving United Nations sustainable development goals through mushroom agri-entrepreneurship: mushroom cultivation as a sustainable livelihood option for rural women. In: The role of female leaders in achieving the sustainable development goals. IGI Global, Hershey, pp 214–231

Kamal S, Barh A, Sharma K, Sharma VP (2022) Mushroom biology and advances. In: Agricultural biotechnology: latest research and trends. Springer Nature Singapore, Singapore, pp 661–688

Kang DS, Min KJ, Kwak AM, Lee SY, Kang HW (2017) Defense response and suppression of Phytophthora blight disease of pepper by water extract from spent mushroom substrate of Lentinula edodes. Plant Pathol J 33(3):264

Kang P, Pan Y, Ran Y, Li W, Shao M, Zhang Y et al (2023) Soil saprophytic fungi could be used as an important ecological indicator for land management in desert steppe. Ecol Indic 150:110224

Karavani A, De Cáceres M, de Aragón JM, Bonet JA, de-Miguel S (2018) Effect of climatic and soil moisture conditions on mushroom productivity and related ecosystem services in Mediterranean pine stands facing climate change. Agric For Meteorol 248:432–440

Kaur K, Sharma R, Srivastava I, Kaur S, Mehrotra R (2022) Edible mushrooms: nature's superfood for health and wellbeing. Int J Innov Multidiscip Res 1(1):40–53

Khan M, Khan T, Tabassum B, Hashim M (2024) Microbiome-driven soil fertility: understanding symbiotic relationships. In: Progress in soil microbiome research. Springer Nature Switzerland, Cham, pp 77–115

Kim JH, Tam CC, Chan KL, Mahoney N, Cheng LW, Friedman M, Land KM (2022) Antimicrobial efficacy of edible mushroom extracts: assessment of fungal resistance. Appl Sci 12(9):4591

Kour H, Kour S, Sharma Y, Singh S, Sharma I, Kour D, Yadav AN (2021) Bioprospecting of industrially important mushrooms. In: Industrially important fungi for sustainable development: volume 2: bioprospecting for biomolecules, pp 679–716. https://doi.org/10.1007/978-3-030-85603-8_20

Kumar R, Kaur A (2019) Oil spill removal by mycoremediation. In: Microbial action on hydrocarbons. Springer Singapore, Singapore, pp 505–526

Kumar A, Yadav AN, Mondal R, Kour D, Subrahmanyam G, Shabnam AA et al (2021a) Mycoremediation: a mechanistic understanding of contaminants alleviation from natural environment and future prospect. Chemosphere 284:131325

Kumar K, Mehra R, Guiné RP, Lima MJ, Kumar N, Kaushik R et al (2021b) Edible mushrooms: a comprehensive review on bioactive compounds with health benefits and processing aspects. Foods 10(12):2996

Kumar A, Verma R, Sharma VP, Kumar S, Kamal S, Goyanka J et al (2025) Antifungal, growth-promoting, and lignocellulose degrading-bacteria from compost and casing soil of button mushroom (Agaricus bisporus). Environ Sci Pollut Res 32(15):9747–9758

Lawal EE, Iseghohi F, Rabiu AS, Osi AR, Oyewole OA, Adetunji CO et al (2025) Application of mushrooms in the management of cardiovascular diseases. In: Mushroom biotechnology for improved agriculture and human health, pp 221–238. https://doi.org/10.1002/9781394212699.ch10

Lehto T, Zwiazek JJ (2011) Ectomycorrhizas and water relations of trees: a review. Mycorrhiza 21(2):71–90

Li H, Tian Y, Menolli N Jr, Ye L, Karunarathna SC, Perez-Moreno J et al (2021) Reviewing the world's edible mushroom species: a new evidence-based classification system. Compr Rev Food Sci Food Saf 20(2):1982–2014

Lim SH, Lee YH, Kang HW (2013) Efficient recovery of lignocellulolytic enzymes of spent mushroom compost from oyster mushrooms, Pleurotus spp., and potential use in dye decolorization. Mycobiology 41(4):214–220

Liu K, Wang J, Zhao L, Wang Q (2013) Anticancer, antioxidant and antibiotic activities of mushroom Ramaria flava. Food Chem Toxicol 58:375–380

Liu D, Diao W, Chen H, Qi X, Fang H, Yu X et al (2024) Heterologous expression and characterization of a dye-decolorizing peroxidase from Ganoderma lucidum, and its application in decolorization and detoxification of different types of dyes. World J Microbiol Biotechnol 40(10):303

Lofgren L, Nguyen NH, Kennedy PG, Pérez-Pazos E, Fletcher J, Liao HL et al (2024) Suillus: an emerging model for the study of ectomycorrhizal ecology and evolution. New Phytol 242(4): 1448–1475

Luptáková E, Mihál I (2020) Dynamics of ectomycorrhizal mycobiota (Basidiomycota) communities on a former agricultural land (West Carpathians). Mycol Prog 19(9):845–857

Mahari WAW, Peng W, Nam WL, Yang H, Lee XY, Lee YK et al (2020) A review on valorization of oyster mushroom and waste generated in the mushroom cultivation industry. J Hazard Mater 400:123156

Martius LR, Fielding D, Brown P, Milligan K, Turner M, Taylor AF (2025) Nucleation in temperate woodland regeneration: dual mycorrhizal Salix facilitate ectomycorrhizal tree establishment. Oikos 2025(4):e10974

Mayans B, Antón-Herrero R, García-Delgado C, Delgado-Moreno L, Guirado M, Pérez-Esteban J et al (2024) Bioremediation of petroleum hydrocarbons polluted soil by spent mushroom substrates: microbiological structure and functionality. J Hazard Mater 473:134650

Mazumder S, Bhattacharya D, Lahiri D, Nag M (2025) Rhizobacteria and arbuscular mycorrhizal fungi (AMF) community in growth management and mitigating stress in millets: a plant-soil microbe symbiotic relationship. Curr Microbiol 82(6):242

Mleczek M, Budka A, Siwulski M, Mleczek P, Budzyńska S, Proch J et al (2021) A comparison of toxic and essential elements in edible wild and cultivated mushroom species. Eur Food Res Technol 247:1249–1262

Mohamadhasani F, Rahimi M (2022) Growth response and mycoremediation of heavy metals by fungus Pleurotus sp. Sci Rep 12(1):19947

Moora M, Davison J, Kohout P, Zobel M (2025) The importance of the plant mycorrhizal collaboration niche across scales. Nat Rev Biodivers 1(4):1–12

Narayanan S, de Mores AR, Cohen L, Anwar MM, Lazar F, Hicklen R et al (2023) Medicinal mushroom supplements in cancer: a systematic review of clinical studies. Curr Oncol Rep 25(6): 569–587

Noreen R, Asgher M, Bhatti HN, Batool S, Asad MJ (2011) Phanerochaete chrysosporium IBL-03 secretes high titers of manganese peroxidase during decolorization of Drimarine Blue K2RL textile dye. Environ Technol 32(11):1239–1246

Ogwu MC, Richard G, Izah SC, Alimba CG, Wangboje OM (2025) Environmental roles of edible mushrooms. In: Bioactive compounds in edible mushrooms: sustainability and health applications. Springer Nature Switzerland, Cham, pp 1–24

Okigbo RC, Anuagasi C (2021) Diseases affecting mushrooms in Africa. J Food Technol Nutr Sci 129:2–10

Omoregie IP, Oyewole OA, Buba AB, Alhassan AM, Adetunji CO (2025) Agricultural applications of novel mushroom-based nanopesticide. In: Mushroom biotechnology for improved agriculture and human health, pp 45–69. https://doi.org/10.1002/9781394212699.ch3

Özlü D, Nicholas P (2021) Architecture of reforestation: mycelium as a new building material and design of the fibrous woven scaffolds. In: Construction, energy, environment and sustainability international conference. Itecons, University of Coimbra, Portugal

Piccolo A, Drosos M (2025) The essential role of humified organic matter in preserving soil health. Chem Biol Technol Agric 12(1):21

Qwarse M, Moshi M, Mihale MJ, Marealle AI, Sempombe J, Mugoyela V (2021) Knowledge on utilization of wild mushrooms by the local communities in the Selous-Niassa Corridor in Ruvuma Region, Tanzania. J Yeast Fungal Res 12(1):8–19

Ramírez-Terrazo A, Adriana Montoya E, Garibay-Orijel R, Caballero-Nieto J, Kong-Luz A, Méndez-Espinoza C (2021) Breaking the paradigms of residual categories and neglectable importance of non-used resources: the "vital" traditional knowledge of non-edible mushrooms and their substantive cultural significance. J Ethnobiol Ethnomed 17:1–18

Rashad FM, El Kattan MH, Fathy HM, Abd El-Fattah DA, El Tohamy M, Farahat AA (2019) Recycling of agro-wastes for Ganoderma lucidum mushroom production and Ganoderma post mushroom substrate as soil amendment. Waste Manag 88:147–159
Rathod MG (2023) Mushroom farming: exploring varieties, cultivation strategies, and endless possibilities. Res Rev Biotechnol Biosci 10:1–11
Ravlikovsky A, Symochko L (2020) Potential use of spent mushroom substrate of Lentinula edodes as a biofertilizer. Int J Ecosyst Ecol Sci 10(3):527
Rezaeian S, Pourianfar HR, Dowom SA (2023) Comparative assessment of β-glucan composition and structural characterization in wild and commercial Enoki mushrooms, Flammulina velutipes. J Food Compos Anal 120:105286
Rodríguez-Barrera TM, Téllez-Téllez M, Sánchez JE, Castañeda-Ramirez GS, Acosta-Urdapilleta-M, Bautista-Garfias CR, Aguilar-Marcelino L (2021) Edible mushrooms of the genus Pleurotus as biocontrol agents of parasites of importance for livestock. Scientia Fungorum 52:e1375
Rodríguez-Uña A, Salcedo I, Rodríguez-Echeverría S, Moreno-Mateos D (2024) Century-long recovery of mycorrhizal interactions in European beech forests after mining. Plant Soil 504(1): 895–909
Rog I, van der Heijden MG, Bender SF, Boussageon R, Lambach A, Schlaeppi K et al (2025) Mycorrhizal inoculation success depends on soil health and crop productivity. FEMS Microbiol Lett 372:fnaf031
Roy S, Barman S, Chakraborty U, Chakraborty B (2015) Evaluation of spent mushroom substrate as biofertilizer for growth improvement of Capsicum annuum L. J Appl Biol Biotechnol 3(3): 022–027
Sabalpara AN, Mahatma L (2019) Role of microbes in sustainable agriculture. In: Natural resource management: ecological perspectives. Springer International Publishing, Cham, pp 147–163
Sadiq MS, Singh IP, Ahmad MM, Sani BS (2024) Mushroom farming for youth empowerment: a high-value crop with minimal land requirement. ShodhPrabandhan: J Manag Stud 1(1):65–75
Sangeeta S, Sharma D, Ramniwas S, Mugabi R, Uddin J, Nayik GA (2024) Revolutionizing mushroom processing: innovative techniques and technologies. Food Chem X 23:101774
Schilla F, Mumladze G, Soukupova J, Smutka L (2025) Economic analysis of Agaricus bisporus mushrooms production and the perspective of sharing economy. Front Sustain Food Systems 8: 1415291
Schoutteten N, Begerow D, Verbeken A (2024) Mycoparasitism in Basidiomycota. In: Haelewaters D (ed) Biodiversity and evolution of fungal parasites and pathogens. Elsevier, Germany. In press. Preprint available on Authorea. https://doi.org/10.22541/au.171386938.80462578/v1
Shao G, Xu D, Xu Z, Jin Y, Wu F, Yang N, Xu X (2024) Green and sustainable biomaterials: edible bioplastic films from mushroom mycelium. Food Hydrocolloids 146:109289
Sharma A, Chauhan R, Sharma SK, Devi N, Thakur V, Verma S (2025) Mushrooming livelihoods: a socio-economic analysis of Solan district's mushroom Agripreneurs. Int J Econ Plants 12:1–8
Silva M, Ramos AC, Lidon FJ, Reboredo FH, Gonçalves EM (2024) Pre-and postharvest strategies for Pleurotus ostreatus mushroom in a circular economy approach. Foods 13(10):1464
Singh C, Pathak P, Chaudhary N, Rathi A, Dehariya P, Vyas D (2020) Mushrooms and mushroom composts in integrated farm management. Res J Agric Sci 11:1436–1443
Sokol NW, Slessarev E, Marschmann GL, Nicolas A, Blazewicz SJ, Brodie EL et al (2022) Life and death in the soil microbiome: how ecological processes influence biogeochemistry. Nat Rev Microbiol 20(7):415–430
Song X, Shang X, Zhang M, Yu H, Zhang D, Tan Q, Song C (2025) Cultivation methods and biology of Lentinula edodes. Appl Microbiol Biotechnol 109(1):1–15
Srivastava S, Mathur P, Prakash P, Falletta E, Katha U, Pagani A et al (2025) Mushroom-derived innovations: sustainable biomaterials for biomedical engineering. Biomed Mater Devices 3(1): 381–395
Stamets P (2006) Can mushrooms help save the world? Explore 2(2):152–161
Suz LM, Bidartondo MI, van der Linde S, Kuyper TW (2021) Ectomycorrhizas and tipping points in forest ecosystems. New Phytol 231(5):1700–1707

Tan WC, Kuppusamy UR, Phan CW, Tan YS, Raman J, Anuar AM, Sabaratnam V (2015) Ganoderma neo-japonicum Imazeki revisited: domestication study and antioxidant properties of its basidiocarps and mycelia. Sci Rep 5(1):12515

Tedersoo L, Bahram M (2019) Mycorrhizal types differ in ecophysiology and alter plant nutrition and soil processes. Biol Rev 94(5):1857–1880

Tedersoo L, Bahram M, Zobel M (2020) How mycorrhizal associations drive plant population and community biology. Science 367(6480):eaba1223

Thakur MP (2020) Advances in mushroom production: key to food, nutritional and employment security: a review. Indian Phytopathol 73:377–395

Tüfekçi S, Ortaş İ (2025) Contribution of indigenous ectomycorrhizal inoculation to the growth and nutrition of containizered Cedrus libani seedlings under nursery conditions. For Sci 71:681–696

van Brenk B, Kruidhof L, Kemperman AJ, van der Meer WG, Wösten HA (2024) Discoloration of textile dyes by spent mushroom substrate of Agaricus bisporus. Bioresour Technol 402:130807

Venturella G, Ferraro V, Cirlincione F, Gargano ML (2021) Medicinal mushrooms: bioactive compounds, use, and clinical trials. Int J Mol Sci 22(2):634

Verma N, Jujjavarapu SE, Mahapatra C (2023) Green sustainable biocomposites: substitute to plastics with innovative fungal mycelium based biomaterial. J Environ Chem Eng 11(5):110396

Viriato V, Mäkelä MR, Kowalczyk JE, Ballarin CS, Loiola PP, Andrade MC (2022) Organic residues from agricultural and forest companies in Brazil as useful substrates for cultivation of the edible mushroom Pleurotus ostreatus. Lett Appl Microbiol 74(1):44–52

Volcão LM, Halicki PCB, Christ-Ribeiro A, Ramos DF, Badiale-Furlong E, Andreazza R et al (2022) Mushroom extract of Lactarius deliciosus (L.) Sf. Gray as biopesticide: antifungal activity and toxicological analysis. J Toxic Environ Health A 85(2):43–55

Wang Y, Xu J, Yuan Q, Guo L, Zheng G, Xiao C et al (2024) Composition and diversity of soil microbial communities change by introducing Phallus impudicus into a Gastrodia elata Bl.-based soil. BMC Microbiol 24(1):204

Wei X, Xie B, Wan C, Song R, Zhong W, Xin S, Song K (2024) Enhancing soil health and plant growth through microbial fertilizers: mechanisms, benefits, and sustainable agricultural practices. Agronomy 14(3):609

Williams LM, Berthon BS, Stoodley IL, Williams EJ, Wood LG (2023) Medicinal mushroom extracts from Hericium coralloides and Trametes versicolor exert differential immunomodulatory effects on immune cells from older adults in vitro. Nutrients 15(9):2227

Xie D, Zhang R, Huang J, Fei Z, Wang L, Zhao J et al (2025) Efficient production, structural characterization and bioactivity of an extracellular polysaccharide from Grifola frondosa endophytic Burkholderia sp. Int J Biol Macromol 309:143090

Yadav P, Rai SN, Mishra V, Singh MP (2021) Mycoremediation of environmental pollutants: a review with special emphasis on mushrooms. Environ Sustain 4(9):1–14

Yang F, Wang M, Zhao L, Fan B, Sun N, Liu J et al (2024a) The role of cattle manure-driven polysaccharide precursors in humus formation during composting of spent mushroom substrate. Front Microbiol 15:1375808

Yang Z, Qu J, Qiao L, Jiang M, Zou X, Cao W (2024b) Tea and Pleurotus ostreatus intercropping modulates structure of soil and root microbial communities. Sci Rep 14(1):11295

Yu HL, Kim JY, Lim SH, Kang HW (2022) The defense response of pepper (Capsicum annuum L.) induced by exopolysaccharide from Schizophyllum commune. Physiol Mol Plant Pathol 118:101810

Zhang B, Li Y, Zhang F, Linhardt RJ, Zeng G, Zhang A (2020) Extraction, structure and bioactivities of the polysaccharides from Pleurotus eryngii: a review. Int J Biol Macromol 150:1342–1347

Zhang Y, Mo M, Yang L, Mi F, Cao Y, Liu C et al (2021) Exploring the species diversity of edible mushrooms in Yunnan, Southwestern China, by DNA barcoding. J Fungi 7(4):310

Zhao M, Liu Z, Geng Y, Lv X, Xu J, Zhao X et al (2025) Role of a low-molecular-weight polysaccharide from Boletus edulis Bull: Fr. in modulating gut microbiota and metabolic disorders. Int J Biol Macromol 309:142789

# The Role of Mushrooms in Breaking Down Pollutants and Keeping the Environment Clean

14

Zarimah Mohd Hanafiah, Anggita Rahmi Hafsari, Wan Hanna Melini Wan Mohtar, Tengku Nilam Baizura Tengku Ibrahim, Nur Azalina Suzianti Feisal, and Wan Abd Al Qadr Imad Wan-Mohtar

Z. M. Hanafiah
Department of Civil Engineering, Faculty of Engineering and Build Environment, Universiti Kebangsaan Malaysia (UKM), Bangi, Malaysia

Functional Omics and Bioprocess Development Laboratory, Institute of Biological Sciences, Faculty of Science, Universiti Malaya, Kuala Lumpur, Malaysia

A. R. Hafsari
Functional Omics and Bioprocess Development Laboratory, Institute of Biological Sciences, Faculty of Science, Universiti Malaya, Kuala Lumpur, Malaysia

Department of Biology, Faculty of Science, Universitas Islam Negeri (UIN) Sunan Gunung Djati Bandung, Bandung, West Java, Indonesia

W. H. M. W. Mohtar
Department of Civil Engineering, Faculty of Engineering and Build Environment, Universiti Kebangsaan Malaysia (UKM), Bangi, Malaysia

T. N. B. T. Ibrahim
Department of Environmental Health, Faculty of Health Sciences, Universiti Teknologi MARA Cawangan Pulau Pinang, Kampus Bertam, Penang, Malaysia

Occupational Health and Safety Risk Management (OHSeRM) Research Initiative Group, Universiti Teknologi MARA, Cawangan Pulau Pinang, Kampus Bertam, Kepala Batas, Pulau Pinang, Malaysia

N. A. S. Feisal
Department of Diagnostic and Allied Health Science, Faculty of Health and Life Sciences, Management and Science University, Shah Alam, Selangor, Malaysia

W. A. A. Q. I. Wan-Mohtar (✉)
Functional Omics and Bioprocess Development Laboratory, Institute of Biological Sciences, Faculty of Science, Universiti Malaya, Kuala Lumpur, Malaysia
e-mail: qadyr@um.edu.my

 
B. Giri et al. (eds.), *Advances in Mushroom Research*,
https://doi.org/10.1007/978-981-95-7692-0_14

**Abstract**

Mushrooms, particularly white-rot fungi such as *Pleurotus ostreatus*, *Trametes versicolor* and *Ganoderma lucidum*, possess diverse enzymatic systems such as laccases, manganese peroxidases, and lignin peroxidases that enable them to decompose a wide range of persistent organic pollutants, including hydrocarbons, pesticides, synthetic dyes, and emerging contaminants. Though mushrooms effectively transform or immobilise contaminants in soil and aquatic systems through enzymatic biodegradation, biosorption, and bioaccumulation, mycoremediation offers a cost-effective, eco-friendly alternative to conventional methods that are often expensive and generate secondary waste. The chapter discusses successful applications of mushrooms in degrading oil spills, remediating heavy metal-contaminated soils, treating agricultural residues, and removing colourants from industrial effluents. Additionally, the emerging role of fungi in addressing pollutants such as microplastics, nanomaterials, and endocrine-disrupting compounds is explored, showcasing the adaptability of fungal systems in complex contamination scenarios. Fungi not only break down pollutants but also enhance soil health and promote ecosystem recovery. Technological innovations such as enzyme immobilisation, microbial consortia, and artificial intelligence-assisted bioprocesses are further expanding the scope and efficiency of fungal remediation strategies. The chapter emphasises the need for interdisciplinary approaches to optimise fungal performance in real-world settings, considering variables like nutrient balance, redox conditions, and pollutant complexity. As global environmental challenges escalate, mushrooms offer a scalable and sustainable solution with dual benefits: pollution mitigation and potential co-products such as edible or medicinal biomass. Ultimately, this chapter advocates for the integration of fungi into environmental policy and practice, presenting them as vital allies in the pursuit of a cleaner, healthier, and more resilient planet.

**Keywords**

Mycoremediation · Enzymatic biodegradation · Environmental pollutants · Fungal biotechnology · Biosorption · Sustainable remediation

## 14.1 Role of Mushrooms in the Environment

Environmental pollution has emerged as a growing concern for both ecological systems and global public health. Industries, agriculture, and households consistently release highly detrimental pollutants into their environment. In response, various strategies have been developed to mitigate the chaotic and damaging effects of these pollutants (Shetty et al. 2023). Hazardous contaminants currently pose a severe environmental challenge, with both natural processes and human economic activities releasing a vast array of pollutants into the atmosphere. This extensive range includes various compound classes like polycyclic aromatic hydrocarbons (PAHs), synthetic dyes, detergents, chlorinated compounds, plastics, dioxins, and pharmaceutical compounds, as well as their complex mixtures such as creosote and oil. The leakage and accidental discharge of these contaminants are particularly problematic when the natural microbiome's biodegradation capacity is insufficient to remove or neutralise them. Microbial metal remediation itself is a complex process, influenced by factors such as cell physiology, microbial cell wall composition, and physicochemical parameters like temperature, pH, metal concentration, ionic strength, and time (Mayans et al. 2024). Microorganisms can remove metals from aqueous solutions through a process known as biosorption. Recent research has specifically termed this phenomenon, when carried out by fungi, as mycosorption (Tang et al. 2024).

Mushrooms belong to the phylum *Basidiomycetes*, a vast and varied group that, along with ascomycetes, forms the kingdom Dikarya, often referred to as the 'higher fungi' (Rathore et al. 2019). These fungi possess a distinctive, macroscopic body that can grow either above ground (epigeous) or underground (hypogeous), making them large enough to be seen and harvested by hand. Most edible mushroom species fall within the class Agaricomycetes, which produces various basidiocarps, including familiar forms like gilled mushrooms, puffballs, and bracket, coral, jelly, and crust fungi (Zhao 2023). Globally, approximately 12,000 mushroom species exist. Among these, around 2000 are considered edible, with 35 species being commercially cultivated worldwide (Jayaraman et al. 2024). Additionally, mushrooms are known to possess extensive enzyme complexes, allowing them to thrive on lignocellulosic substrates rich in lignin, cellulose, hemicellulose, and pectin (Kumla et al. 2020). Mushrooms, frequently seen only as a culinary item, are increasingly recognised as vital ecological players with significant potential to address urgent environmental problems. Beyond their role as food, these remarkable organisms are now being explored for their capacity to revolutionise environmental solutions, including pollution remediation, soil restoration, carbon sequestration, and waste management (Shin et al. 2025).

Biological approaches within industrial and environmental biotechnology are driving the creation of 'clean technologies'. These innovations prioritise maximising production while simultaneously minimising waste and transforming any generated waste into valuable resources. Moreover, these clean technologies leverage biological methods for waste remediation (Wibisono et al. 2025). Mycoremediation, for instance, is a prominent biological technique that utilises mushrooms to remove pollutants from the environment. Mushrooms are equipped with specialised enzymatic systems capable of degrading diverse pollutants, making them applicable for a wide range of applications. These enzymes (Fig. 14.1) include laccase, lignin peroxidase (LiP), versatile peroxidase (VP), manganese peroxidase (MnP), and various phenoloxidases (Gałązka et al. 2025). Among mushroom-forming fungi, wood-decaying species have long captivated researchers due to their diverse potential, from culinary to remediating environmental pollutants (Hyde et al. 2024). Mycelium, a vast network of thread-like hyphae, permeates soil and organic matter, efficiently absorbing nutrients and decomposing materials. While often likened to a

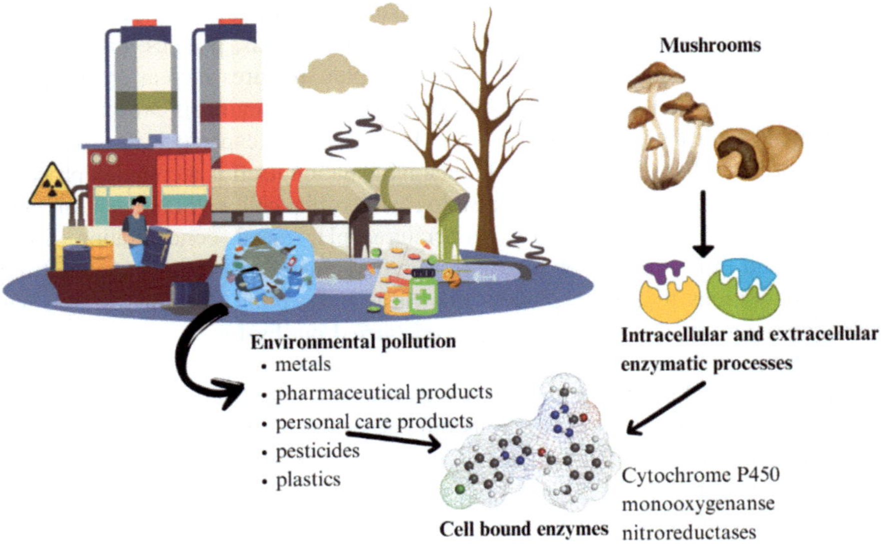


**Fig. 14.1** Overview of environmental pollutants remediated using fungal (mushroom) species via intracellular (cell-bound enzymes) and extracellular enzymatic processes

mushroom's roots, it functions as nature's potent decomposer, breaking down complex organic matter and returning vital nutrients to the soil (Yang et al. 2021).

Traditional heavy metal removal methods, including precipitation, filtration, ion exchange, carbon adsorption, evaporation, membrane technology, reverse osmosis preconcentration, redox, electrowinning, chelation, wastewater coagulation, and electrochemical processes, have consistently proven ineffective (Shofia et al. 2025). While various physical and chemical methods exist for removing and degrading stubborn and harmful pollutants from soil and water, they are often costly, generate toxic by-products, and prove ineffective against highly toxic chemicals present in low concentrations (Vidu et al. 2020). Extensive mycelial networks of mushrooms can detoxify harmful substances by secreting enzymes that break down pollutants into less toxic forms. For instance, oyster mushrooms (*Pleurotus ostreatus*) are notably effective at degrading petroleum products in oil spills, with their mycelium capable of converting hydrocarbons into organic compounds. This process offers a sustainable approach to environmental clean-up (Akpasi et al. 2023). Beyond petroleum, mushrooms also demonstrate significant potential in degrading heavy metals such as mercury, lead, and cadmium. Through the absorption and neutralisation capabilities of mycelium, contamination in soil and water can be reduced, thereby fostering healthier ecosystems (Sar et al. 2024). Mushrooms can break down and metabolise a broad spectrum of inorganic and organic pollutants, thanks to their diverse enzymatic machinery, both inside and outside their cells, and their capacity to excrete acids (Aditya et al. 2024).

Mycoremediation offers an economical, eco-friendly, and effective solution to the escalating problem of soil and water pollution. Mushrooms are ideal candidates for pollutant remediation due to their robust growth, extensive hyphal networks, production of versatile extracellular ligninolytic enzymes, high surface area-to-volume ratio, resistance to heavy metals, adaptability to fluctuating pH and temperature, and the presence of metal-binding proteins (Akhtar and Mannan 2020). This approach can be applied directly on-site (in situ) to remediate pollutants like dyes, herbicides, and pharmaceutical drugs from various industries. Alternatively, it can also be employed within bioreactors, which are systems designed with controlled physicochemical conditions to promote optimal microbial growth (Khatua et al. 2024). For instance, mushrooms' capacity to break down petroleum hydrocarbons, including alkanes, aromatic compounds, and those containing nitrogen, sulphur, and oxygen, underscores their significant potential in bioremediation (Alaidaroos 2023).

This chapter explores the roles of mushrooms in breaking down pollutants and keeping the environment clean. It details how cost-effective and sustainable mycoremediation strategies, including biosorption, bioaccumulation, and biodegradation, leverage fungi's remarkable ability to dismantle a wide range of contaminants, specifically, the pesticides, cyanotoxins, hazardous pharmaceutical compounds, and microplastics. Furthermore, it delves into recent technological advancements that boost this environmental clean-up, such as utilising microbial consortia, transcriptomics, metagenomics, and microbial engineering to improve the remediation of pollutants. The discussion also covers the intricate fungal mechanisms of action and how enzyme engineering can enhance the degradation potential of key enzymes. Ultimately, this chapter underscores the critical role of fungi in detoxifying and transforming pollutants, offering valuable insights for researchers and policymakers to implement mycoremediation for a truly effective and sustainable environment.

## 14.2 Mechanisms of Pollutant Degradation by Mushrooms

Mushroom-forming fungi (particularly white-rot fungi, which belong to the ligninolytic Basidiomycetes group) play an instrumental role in bioremediation, the process of breaking down or removing various kinds of pollutants from the environment through some of the most innovative chemical methods (Vaksmaa et al. 2023). Such organisms owe their potent remediated abilities to several processes: (1) the biosorption and bioaccumulation processes which sequester pollutants (both metals and organics) through fungal biomass, (2) the secretion of some ligninolytic oxidoreductases like peroxidases and laccases which oxidatively degrade complex organic molecules, (3) the destruction of some organic contaminants such as chemicals and fuels into smaller and less harmful materials and (4) modification or destruction of pollutants through intracellular biotransformation and detoxification processes (Fig. 14.2). These mechanisms will be described in the following sections and explained at a molecular and biochemical level.

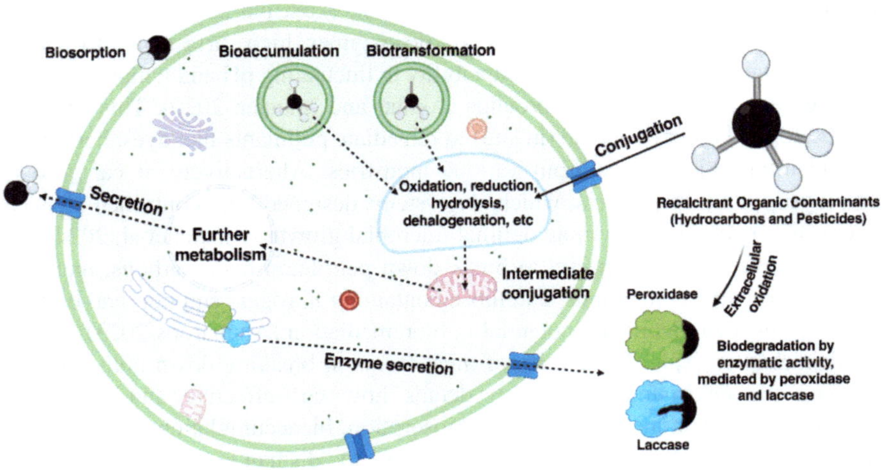


**Fig. 14.2** The overview of the mechanism of mushroom-forming fungi in degrading pollutants

### 14.2.1 Enzymatic Degradation Via Ligninolytic Enzymes (Peroxidases and Laccases)

The enzymatic breakdown, using peroxidases and laccases, is a biological process that utilises specific enzymes to dismantle complex organic materials and break them down into simpler ones. Khajehzadeh et al. (2024) have highlighted that fungi play a crucial role in environmental restoration through the ligninolytic degradation of complex organic pollutants, utilising peroxidases and laccases. Architecturally complex organic pollutants, such as lignin, pose a challenge to environmental fungi. Fungi have recently developed a method to break down lignin in specific species, such as white-rot mushrooms. This method employs the secretion of high-redox-potential enzymes, such as lignin peroxidase, manganese peroxidase, versatile peroxidase, and laccase, which is often referred to as the ligninolytic system (Hammel 1995; Lankinen et al. 2005; Vaksmaa et al. 2023). These enzymes function as non-specific oxidoreductases that are capable of interacting with complex and resistant molecular structures upon contact with a pollutant (often a large aromatic or phenolic compound). Meanwhile, the enzymes catalyse one-electron oxidation reactions that generate reactive radical intermediates, leading to the cleavage of chemical bonds in the pollutant's structure (Kumar and Chandra 2020). White-rot Basidiomycetes thus leverage their lignin-degrading machinery to break down diverse xenobiotics; for example, *Phanerochaete* and *Trametes* species have been shown to mineralise or depolymerise persistent organic compounds via these extracellular enzymes (Civzele et al. 2023; Haddadin et al. 2002; Suryadi et al. 2022; Tuisel et al. 1990). Notably, ligninolytic fungi often achieve only partial degradation of highly stable pollutants (Wang et al. 2024). However, even partial oxidation is sufficient to transform stable pollutants into smaller, more labile metabolites, which then undergo degradation (Naidu et al. 2017).

### 14.2.2 Peroxidases

LiP and MnP, which belong to the class of fungal peroxidases, are among those enzymes that have been suggested to help tackle the most recalcitrant pollutants (Ramirez et al. 2010). Such enzymes can overcome the oxidative degradation of organic compounds by utilising hydrogen peroxide as a co-substrate, with redox potentials far exceeding those of most known enzymes. Lignin peroxidase, for example, can oxidise and cleave robust polycyclic aromatic hydrocarbons by abstracting an electron to form cation radicals (Varela et al. 2000). On the other hand, manganese peroxidase (MnP) catalyses the oxidation of $Mn^{2+}$ to $Mn^{3+}$, a process that can occur with a single, non-complex, non-substrate system. The $Mn^{3+}$ ions, which are already oxidised and chelated, can oxidise other pollutants that LiP is unable to attack directly (Bilal et al. 2023; Praveen et al. 2012). Through Mn3 complex-mediated lipid peroxidation, MnP also generates radicals, such as alkoxyl radicals, in lipids, which co-oxidise the previously mentioned hydrophobic organic pollutants (Ayala et al. 2014). Some mushrooms, such as *Pleurotus* spp., produce versatile peroxidase (VP), which possesses features of both LiP and MnP and oxidises high-redox-potential aromatics and phenolic compounds by direct electron abstraction or via $Mn^{3+}$ intermediates (Sánchez-Alejandro et al. 2016). The collective action of these peroxidases triggers a cascade of oxidative reactions that can fragment large pollutant molecules into smaller pieces or add hydrophilic functional groups, rendering the toxic compounds non-toxic and facilitating further microbial degradation (Twala et al. 2020).

### 14.2.3 Laccases

Laccase is an enzyme which, together with peroxidases, is produced by several mushrooms. It belongs to the multicopper oxidases. It serves in the degradation of pollutants. Laccases perform the two-step oxidation of phenolic or aromatic compounds to yield phenoxy or aryl radicals from pollutants. Laccases catalyse the four-electron reduction of molecular oxygen to water (Tamayo-Ramos et al. 2012; Yun et al. 2025). Subsequently, these radicals can undergo numerous spontaneous reactions, such as bond rearrangements, fragmentation, or coupling. Laccases are regarded as 'green' biocatalysts, as the main by-product of their oxidation reactions is water (Shraddha et al. 2011). Laccases possess a wide range of substrate specificity, as they oxidise not only phenolic compounds but also anilines, polyphenols, various dyes, and other xenobiotics (Viswanath et al. 2014).

Laccases from fungi, such as *Trametes versicolor* or *P. ostreatus*, possess some of the highest redox potentials among laccases, reaching 0.7–0.8 V. Thus, they are non-phenolic aromatic pollutants in the presence of a small-molecule mediator (El-Batal et al. 2015). The use of redox mediators (ABTS and some phenolic compounds) enhances laccase activity by donating electrons to bulky or insoluble pollutants (those too large or too distant from the laccase active site to bind) and is capable of binding to the laccase active site (Mani et al. 2018). Using laccase-mediator systems, mushrooms can degrade dyes and pharmaceuticals as well as other persistent emerging pollutants, including endocrine disruptors. As noted by Nesma et al. (2023), the products created by laccase are prone to undergoing coupling reactions to yield higher-order oligomers (e.g. dimers, trimers) or, conversely, they may undergo cleavage into smaller structures.

Laccase can polymerise toxic compounds into larger, seemingly innocuous aggregates, which are akin to humic substances (Gałązka et al. 2023). In other instances, laccase catalyses ring cleavage or restructures the compound into less harmful forms. Furthermore, laccases are distinct because they do not require $H_2O_2$; instead, they utilise $O_2$. Laccases are often found in high concentrations, which renders them favourable in bioremediation settings, as they are readily available and secreted in large quantities (Singh et al. 2025).

Through the combined action of these enzymes, mushrooms can effectively attack a wide variety of pollutants. They have been reported to degrade persistent organic pollutants, including synthetic dyes, chlorinated phenols, nitroaromatics, and polycyclic aromatic hydrocarbons (PAHs), among others (Garcia-Delgado et al. 2015; Sahithya et al. 2022). The enzymatic oxidation typically transforms these pollutants into more hydrophilic and reactive compounds (e.g. quinones, alcohols and organic acids), which either mineralise further or become amenable to metabolism by other organisms (El-Najjar et al. 2011; Hu et al. 2021). This extracellular biodegradation is a crucial first step in fungal remediation: it 'unlocks' very large or insoluble pollutant molecules that cannot enter microbial cells on their own. For example, ligninolytic *Phanerochaete* fungi have been shown to oxidatively cleave high-molecular-weight PAHs into smaller aromatic metabolites (Pozdnyakova 2012), and laccase-producing fungi can transform endocrine-disrupting chemicals (like bisphenols and pharmaceuticals) into less active dimers or quinones (Hu et al. 2016). However, the complete mineralisation to $CO_2$ may not always be achieved by these enzymes alone because it often comes with the end-products being partially oxidised compounds. Nonetheless, these products are generally less toxic and more biodegradable than the parent pollutants, highlighting the indispensable role of fungal enzymes in initiating pollutant breakdown.

### 14.2.4 Biosorption and Bioaccumulation

In addition to active enzymatic degradation, mushrooms can remove pollutants from the environment through biosorption and bioaccumulation processes. Fungal mycelium has a unique cell wall composition, rich in glycoproteins, polysaccharides (such as chitin and glucans), and other polymers, which provide a range of functional groups capable of binding contaminants (Colosimo et al. 2021; Hyde et al. 2024; Khan et al. 2019). Hydroxyl, carboxyl, amine, thiol, and phosphoryl groups, densely present on fungal cell walls, act as natural chelators for metal ions and interact with organic molecules via adsorption. This means even dead fungal biomass (which has no metabolic activity) can serve as an efficient biosorbent for pollutants (Ahmed et al. 2016). Indeed, many studies utilise inactivated or dried mushroom mycelia to remove heavy metals and dyes from wastewater, taking advantage of surface binding without concerns of toxicity to living cells (Low et al. 2008). Using non-living biomass has practical advantages: it obviates the need to maintain growth conditions (pH, nutrients, etc.), allows for easy storage and avoids the risk of pollutants being metabolised into unknown by-products (Ahmed et al. 2016). Live fungi, on the other hand, can both adsorb pollutants on their surfaces and actively uptake them into cells, a combination that can yield very high removal efficiencies (Legorreta-Castañeda et al. 2020).

Biosorption is particularly crucial for heavy metal pollutants and persistent inorganic contaminants, which cannot be enzymatically degraded into innocuous components (Murugesan et al. 2009). Metal ions, such as lead, cadmium, mercury, and arsenic, will instead be immobilised by fungal biomass (Mohd Hanafiah et al. 2024). Fungal biosorption has proven effective even at low metal concentrations and under environmentally varied conditions (Hanafiah et al. 2025). Mushrooms growing in contaminated soils or waters often accumulate astonishing levels of metals in their fruiting bodies and mycelia, reflecting this strong sorptive capacity (Murtić et al. 2024). Heavy metals are sometimes referred to as '*conservative pollutants*' because they do not break down; therefore, sequestration is the primary remediation strategy for these pollutants. Fungal cell walls act as ion-exchange systems: for example, carboxyl groups (from glucuronic acids in glucans or amino acids in glycoproteins) can bind cationic metal species via coordination bonds; similarly, phosphate or sulfhydryl groups can immobilise metals by forming insoluble complexes (Dinakarkumar et al. 2024). As a result, waste fungal biomass from industrial processes has been repurposed as a low-cost biosorbent to clean up metal-bearing effluents (Legorreta-Castañeda et al. 2020; Muraleedharan et al. 1994; Puchana-Rosero et al. 2017).

Not only inorganic pollutants but also organic contaminants can undergo initial adsorption to fungal biomass. Hydrophobic organic chemicals (e.g. polycyclic aromatics, chlorinated pesticides) often have low bioavailability in water. Fungi tackle this by first partitioning these molecules onto their mycelial surfaces or into fungal pellets. A clear example is the degradation of certain persistent pesticides by the white-rot fungus *T. versicolor,* especially when introduced into a pesticide-spiked liquid, the fungal mycelial pellets rapidly adsorbed the hydrophobic pesticides (organochlorine, organophosphate, and pyrethroid types) onto their surface within hours. This physical sequestration concentrates the pollutants close to the fungal enzymatic machinery, effectively bioconcentrating the compounds and making them more accessible for subsequent biodegradation. In the *T. versicolor* studies, over 90% of each pesticide was eliminated within 14 days, with adsorption being the initial dominant removal mechanism, followed by slow enzymatic breakdown of the bound pesticides. Such synergy between biosorption and biodegradation is a recurring theme in fungal remediation: the fungus first attenuates the immediate toxicity or mobility of the pollutant by immobilising it, and then gradually decomposes it through metabolic activity. This two-step mechanism is beneficial for sparingly soluble pollutants, such as petroleum hydrocarbons and organochlorines, where fungi act as a living filter that traps and degrades contaminants over time.

### 14.2.5 Decomposition of Complex Organic Pollutants (Hydrocarbons and Pesticides)

Mushrooms have demonstrated the ability to decompose organic pollutants, including long-chain hydrocarbons, PAHs, and various synthetic pesticides. These compounds are typically resistant to breakdown and can persist in soils and waters for years. Fungi's prowess in degrading such molecules stems from the combined action of their extracellular enzyme arsenal and unique secondary metabolism. Petroleum hydrocarbons, which comprise mixtures of aliphatic (alkane) chains, aromatic rings, and heteroatom-containing compounds, are a prime example. Wood-decay mushrooms, such as *Phanerochaete chrysosporium* and *T. versicolor*, have been reported to degrade components of crude oil and diesel fuel. The enzymes LiP, MnP, and laccase can oxidise high-molecular-weight PAHs (having four or more fused rings) that few bacteria can attack. Early landmark studies showed that *P. chrysosporium* cultures could oxidatively transform PAHs such as benzo[a]pyrene and anthracene, leading to ring fission and accumulation of smaller aromatic acids and $CO_2$. The mechanism involves initial extracellular oxidation of the PAH by LiP (producing radical cations and unstable dihydrodiols) and MnP (via lipid peroxides), effectively cleaving the inert aromatic rings. Subsequent steps may include spontaneous rearrangements of the oxidised PAH or further metabolism by the fungus or coexisting microbes, yielding fewer toxic compounds. Although complete mineralisation of PAHs by a single fungus is rare, white-rot species often achieve significant partial degradation, reducing both the mass and toxicity of these pollutants. For instance, after fungal treatment, one might detect oxidised metabolites (e.g. quinones, phthalic acid derivatives), which are far more biodegradable and can be taken up by bacteria for final mineralisation. This collaborative degradation, where fungi perform the initial 'hard' oxidation and bacteria or other fungi complete the process, is a common scenario in soil bioremediation.

Pesticides represent another class of complex pollutants that mushrooms can decompose. Modern pesticides include organochlorine insecticides (e.g. DDT, lindane), organophosphate insecticides (e.g. parathion, chlorpyrifos), synthetic pyrethroids, carbamate insecticides, herbicides such as atrazine and glyphosate, and many others—all designed to be biologically active and often environmentally persistent. Remarkably, many fungi (including Basidiomycetes and Ascomycetes) can degrade these compounds, often employing pathways analogous to their lignin-degrading processes. White-rot mushrooms are among the most studied pesticide degraders because non-specific oxidative enzymes readily attack the complex structures of pesticides. For example, *Phanerochaete*, *Trametes*, *Pleurotus*, and *Ganoderma* species have demonstrated the ability to break down notoriously stable organochlorines, such as DDT and lindane, by cleaving their aromatic rings or dechlorinating them into less harmful metabolites. Soil fungi like *Aspergillus* and *Penicillium* species (which are not ligninolytic in the strict sense) also contribute to pesticide degradation, often via co-metabolic processes or more substrate-specific enzymes (like esterases or amidases for organophosphates and carbamates). A comprehensive survey has identified numerous genera, including *Trametes*, *Ganoderma*, *Aspergillus*, *Fusarium*, *Cladosporium*, *Rhizopus*, *Penicillium*, *Phlebia*, and *Mortierella*, with species capable of pesticide biodegradation. Each may attack a different range of chemicals, but collectively they cover a broad spectrum of pesticide chemistries.

A representative case is the degradation of three hydrophobic pesticides (the organophosphate chlorpyrifos, the organochlorine dicofol, and the pyrethroid cypermethrin) by *T. versicolor*. In a controlled study, *T. versicolor* was able to remove 95% of chlorpyrifos, 88% of dicofol, and 93% of cypermethrin from a liquid medium within 14 days. The process, as mentioned, began with the rapid adsorption of the pesticides onto fungal biomass, and over the 2 weeks, fungal enzymes progressively broke down the compounds. Metabolite analysis revealed specific degradation reactions: hydrolysis of certain functional groups, dichlorination (removal of chlorine substituents from the organochlorine structure), and oxidation of hydrocarbon moieties were all taking place, depending on the pesticide's structure. For instance, chlorpyrifos (an organophosphate with an ester linkage) can be hydrolysed by fungal enzymes into 3,5,6-trichloro-2-pyridinol. At the same time, dicofol (a DDT derivative) can undergo dichlorination to form less chlorinated intermediates, and cypermethrin (a pyrethroid ester) can be hydrolysed and oxidised to carboxylic acids. The involvement of ligninolytic enzymes is evidenced by the observation that some pesticides induce higher enzyme activities. For example, the herbicide diuron stimulates increased laccase activity in *G. lucidum*, suggesting that the fungus 'ramps up' its oxidative enzyme production in response to the pollutant challenge. Enhanced enzyme expression in the presence of pollutants is a common phenomenon, reflecting a form of co-metabolic induction or stress response that fungi use to fortify their degradative arsenal.

It is worth noting that intracellular enzymes also play a role in the decomposition of certain complex organics (a point expanded in the next section). Some pesticide molecules, after partial breakdown or uptake, are further metabolised inside fungal cells. For example, the fungal degradation of the persistent insecticide fipronil (a phenylpyrazole) was shown to proceed via intracellular cytochrome P450 enzymes in *T. versicolor*, yielding metabolites that were hydroxylated and then conjugated to sugar moieties. Likewise, the breakdown of endosulfan (a chlorinated cyclodiene insecticide) by *P. chrysosporium* can follow two routes, which are a hydrolytic pathway yielding endosulfan diol and sulphate, or an oxidative pathway via P450 that forms endosulfan ether, and blocking P450 (with an inhibitor) diverts the process entirely to the hydrolytic route, which shows that fungi have multiple, possibly redundant, means to attack a pollutant, toggling between extracellular and intracellular chemistry.

Another strategy to improve the degradation of tough pollutants is the use of microbial consortia. Recent research highlights that mixed fungal communities, or fungi in partnership with bacteria, can often degrade complex pollutant mixtures more completely than any single strain alone. In soil microcosms, a consortium of white-rot fungi achieved a 44% removal of a mixture of organochlorine pesticides, outperforming individual fungal isolates, which achieved only 26%–34% removal in other studies; pairing a ligninolytic fungus with a specialised bacterium led to synergistic degradation of herbicides and PAHs, where the fungus's enzymes performed initial oxidation and the bacterium's enzymes mineralised the intermediate products. Such synergistic interactions are an active area of research, as they mimic natural soil communities where fungi and bacteria work in tandem to decompose pollutants. The implication is that mushroom fungi not only directly break down pollutants through their mechanisms but also facilitate broader biodegradation by producing metabolites that other microbes can utilise or by physically altering the environment (e.g. increasing pollutant bioavailability) to the benefit of the community.

### 14.2.6 Biodegradation and Biotransformation

Besides extracellular degradation, mushrooms carry out several detoxification and biotransformation processes within their cells to further detoxify and process pollutants. When a pollutant (or its oxidised derivative) enters a fungal cell, it encounters metabolic machinery that, to some extent, resembles the xenobiotic metabolism systems of liver cells in higher animals. An illustration of intracellular systems is the cytochrome P450 monooxygenase pathway. Many fungi, particularly Basidiomycete mushrooms, have an exceptional amount of diversity in P450 enzymes. For instance, white-rot fungus *P. chrysosporium* contains approximately 150 different P450-encoding genes (Subramanian and Yadav 2009). These P450s are hemethiolate enzymes that catalyse *monooxygenation* reactions, typically inserting one atom of oxygen into an organic substrate ($RH + O_2 + NADPH \rightarrow ROH + H_2O$). This mechanism is ideal for transforming hydrophobic organic pollutants into more polar, reactive metabolites. In fungi, P450s are known to oxidise a range of contaminants: aliphatic and aromatic hydrocarbons (hydroxylating them to alcohols or epoxides), halogenated compounds (often initiating dehalogenation), nitroaromatics (via reduction or hydroxylation), and more.

Empirical studies highlight the role of P450s in the degradation of fungal pollutants. As mentioned, *T. versicolor* uses a P450-dependent pathway to metabolise fipronil, converting it to *hydroxylated* derivatives, which were subsequently found to be conjugated to glucose (glycosides) in the medium. This indicates a two-step detoxification: P450 introduces a hydroxyl group (Phase I metabolism), and then a glycosyltransferase adds a glucose moiety (Phase II metabolism), making the molecule more water-soluble and less toxic. Similarly, *P. chrysosporium* was shown to degrade a suite of neonicotinoid insecticides (acetamiprid, imidacloprid, clothianidin and thiacloprid) via at least two distinct P450 enzymes, each targeting different positions on these molecules. When one pathway was inhibited, alternative routes (such as hydrolysis by non-P450 enzymes) partially compensated, but overall degradation was diminished, confirming that P450-catalysed oxidation was a key step. The importance of P450s is further supported by transcriptomic studies, which show that exposure of fungi to PAHs or pesticides often induces the expression of specific P450 monooxygenases, along with other detoxification enzymes. In white-rot fungi, researchers observed that genes for certain P450s, epoxide hydrolases, and glutathione S-transferases are upregulated when the fungus is degrading PAHs, suggesting a coordinated metabolic response to handle toxic intermediates.

Following Phase I oxidation (e.g. hydroxylation by P450s or the formation of epoxides), fungi employ Phase II conjugation enzymes to detoxify and eliminate the resulting products. In the fungal context, common conjugation reactions include glucuronidation or glycosylation (the attachment of sugar molecules, such as glucose or xylose, to a pollutant metabolite) and glutathione conjugation (the addition of the tripeptide glutathione to electrophilic centres). Both pathways significantly increase the solubility of the compounds and reduce their biological activity. For instance, the hydroxylated fipronil in *T. versicolor* was found as a glycoside, indicating that it had been tagged with some sugar by a UDP-glycosyltransferase for easier excretion. Glutathione S-transferases (GSTs) in fungi can similarly attach glutathione to reactive epoxides or quinones derived from pollutant degradation, neutralising their reactivity (Antón-Herrero et al. 2023). A comprehensive study on PAH metabolism in a white-rot fungus revealed that, after initial P450 activation of PAHs to epoxides, the epoxide hydrolase and GST genes were induced, resulting in the formation of dihydrodiols and glutathione conjugates as downstream products. These multi-step pathways (hydroxylation → epoxide formation → dihydrodiol → conjugation) mirror the detoxification sequence in many organisms, underscoring an evolutionary convergence in dealing with harmful organics.

Through such biotransformation, mushrooms effectively detoxify pollutants; the resultant conjugated metabolites are typically less toxic and can be either sequestered in vacuoles or exported out of the cell. In some cases, fungi excrete these metabolites into the environment, where they may adsorb to soil or be further degraded by other microbes. In other cases, the metabolites remain bound to fungal cell components (for example, quinones might react with fungal melanin or cell wall polymers, becoming part of the fungal biomass). Either way, the bioavailability and ecological risk of the pollutant are significantly reduced. Importantly, these intracellular processes enable fungi to survive otherwise lethal toxins by chemically neutralising them. A classic illustration is fungal tolerance to the pesticide lindane (hexachlorocyclohexane): *Phanerochaete* was found to metabolise lindane via both a cytochrome P450-mediated pathway (yielding chlorinated phenols) and a glutathione-dependent dichlorination pathway; blocking the P450 route forced the fungus to rely on the slower glutathione route (Vaksmaa et al. 2023). The fungus could thus cope with the toxin by channelling it through alternative detoxification routes.

While a significant portion of research on fungal biotransformation predominantly focuses on organic compounds, it is noteworthy that certain inorganic contaminants may also undergo chemical modifications through fungal metabolic processes (Dabrowska et al. 2021). Importantly, specific fungal species possess the capability to perform redox transformations on heavy metals and metalloids, which facilitates their detoxification. This enzymatic reduction, driven by the fungus's internal reductases and consuming reducing equivalents like NADH, essentially converts a carcinogenic, mobile ion into an inert precipitate that can be removed from water. Similar microbial reductions have been observed for uranium (VI) to uranium (IV) and arsenate (V) to arsenide (III) (or further to elemental arsenic by certain yeasts), although these processes are less well-documented in mushrooms specifically. The key takeaway is that mushrooms not only accumulate metals via biosorption but, in some cases, can bio-transform them to mitigate toxicity (even though the element itself is not destroyed). The transformed metals often end up sequestered in fungal biomass or bound in the soil matrix, thus detoxified from an environmental perspective.

## 14.3 Applications of Mushrooms in Environmental Clean-Up

The application of mushrooms in cleaning up the environment is called mycoremediation. Certain types of fungi from mushroom species can break down and absorb various contaminants such as organic compounds, heavy metals, plastics and even hydrocarbons from soil and water. The mycelium, which is the vegetative part of the mushroom, consists of hyphae, a network of thread-like structures that spread through the soil and organic matter, absorbing nutrients and breaking down material. Mycelium naturally acts as one of the most powerful decomposers, able to break down complex organic matter and return essential nutrients to the environment. This ability has made it the subject of growing interest in environmental science and sustainability studies. Figure 14.3 presents a general overview of mycoremediation, highlighting how various mushroom species, such as *G. lucidum*, *Morchella* and *T. versicolor*, are used to treat pollutants, including hydrocarbons, dyes, heavy metals, and organic waste in both water and soil environments. The treatment involved several methods, such as the mycelium as a filter (mycofiltration) and typical bioreactors through several mechanisms of removal, including biosorption, biodegradation, bioaccumulation, and transformation within fungal cells. Thus, this section elaborates on the application of mushroom (the term 'fungi' will be used to ease the discussion) species in the environment, including water and soil.

### 14.3.1 Bioremediation of Agricultural Waste

Agricultural waste such as straw, sawdust and crop residues are often rich in lignocellulosic biomass, which is a complex material that mushrooms can effectively degrade. Some species of mushrooms, such as *Pleurotus*, are well-suited for growing on agricultural waste, making the process of mycoremediation integrated with mushroom cultivation. For instance, *P. ostreatus* (oyster mushroom), highly efficiently cultivated on softwood dust from mango and cashew trees in Sri Lanka (Pathmashini et al. 2009) and (Bhattacharjya et al. 2014) in the present work, also demonstrates promising results of *P. ostreatus* on sawdust from sawmills with different types of trees as the substrate. Spent fruiting bodies of mushrooms are

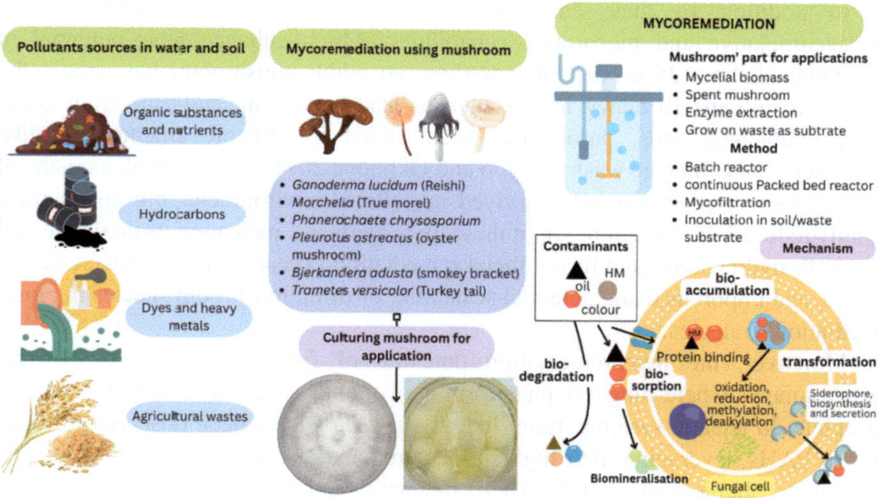


**Fig. 14.3** General overview of the application of mushrooms as bioremediation agents in water and soil environments for the removal of various contaminants

also valuable sources for mycoremediation. For instance, *P. ostreatus* has been used to degrade the compound DDT (1,1,1-trichloro-2,2-bis (4-chlorophenyl)ethane), which is a pesticide that is commonly used in agricultural industries, showing 48% of compound degradation after 28 days and 5.1% of the DDT was mineralised during 56 days of incubation on contaminated soils (Purnomo et al. 2010).

### 14.3.2 Bioremediation of Oil Spills

Hydrocarbon or oil spills in soil and water disrupt ecosystems by contaminating habitats, reducing oxygen availability, and impairing microbial and plant life. They also pose serious risks to humans' and animals' health through the accumulation of toxic compounds such as PAHs and heavy metals. Fungal species such as *P. ostreatus* and *P. chrysosporium* serve a vital function in the degradation of hydrocarbons present in polluted soils and streams. Mohamadhasani and Rahimi (2022) conducted a field trial of *P. ostreatus* and found that it can clean up petroleum oil-contaminated soils. The mycelium of the oyster mushroom could convert long-chain hydrocarbons into shorter-chain (less toxic) ones. Other types of mushrooms, such as *Phlebia acerina*, demonstrated higher benzo[a]pyrene (BaP) degradation capabilities compared with *P. chrysosporium* and *P. sordida*, achieving a degradation rate of 57.7% after 32 days of incubation under a ligninolytic condition (Zhang et al. 2023). These fungi generate ligninolytic enzymes, such as peroxidase and laccase, that decompose complicated hydrocarbons into simpler compounds. The elevated breakdown efficiency is due to the fungi's capacity to generate non-specific extracellular enzymes capable of targeting a broad range of chemical interactions inside these intricate structures.

In addition to their enzymatic versatility, many fungi exhibit strong bioaccumulative and biosorptive capabilities, which enhance their potential in mycoremediation. *Pleurotus tuber-regium*, an edible mushroom, has shown the ability to bio-absorb heavy metals such as lead, cadmium, and zinc from crude oil-contaminated soils, particularly when aided by organic amendments like fertilisers and cellulosic wastes (Ogbo and Okhuoya 2011). The study demonstrated that *P. tuber-regium* not only survived in polluted environments but also significantly reduced the concentration of these toxic elements, indicating its dual potential for both hydrocarbon and heavy metal remediation. These findings underscore the ecological importance and potential scalability of using mushroom species in bioremediation strategies.

Moreover, the mycoremediation potential of fungi is enhanced in water environments where fungal hyphae form networks that can trap and metabolise oil droplets. In aquatic systems, fungi like *T. versicolor* have been observed to break down PAHs, leveraging their high surface area and enzyme production to interact with hydrophobic contaminants. Their enzymatic toolkit includes manganese peroxidase and versatile peroxidase, which are particularly effective in oxidising recalcitrant aromatic compounds. These enzymes allow fungi to tolerate and degrade oil components even in low-nutrient or oxygen-limited environments, making them valuable tools for marine and freshwater oil spill clean-ups (Khan 2025).

Fungi offer a unique advantage in oil spill remediation through a combination of enzymatic degradation, bioaccumulation, and ecological adaptability. Their ability to function in both terrestrial and aquatic systems, coupled with low maintenance and eco-friendly operation, makes them suitable for integration into sustainable environmental management practices. Continued research, including optimisation of environmental conditions and the use of co-substrates, can further enhance the efficiency of fungal-based remediation strategies. Leveraging diverse fungal species such as *Pleurotus*, *Phlebia*, and *Phanerochaete*, researchers can develop multifaceted approaches to mitigate pollution in complex contaminated ecosystems.

### 14.3.3 Bioremediation of Heavy Metals

Certain metals occur naturally in the environment and are essential in trace amounts for the growth of plants and organisms. However, anthropogenic activities such as industrial discharges, mining, and intensive agriculture have led to elevated concentrations of both essential (e.g. Zn, Cu) and non-essential (e.g. Pd, Cd, As) metals in terrestrial and aquatic ecosystems, posing serious environmental and health risks (Oladimeji et al. 2024). These metals are persistent, non-biodegradable, and can bioaccumulate through the food chain, necessitating sustainable and effective remediation strategies.

Fungi, particularly mushroom-forming species, have shown promising potential as environmentally friendly agents for mitigating heavy metal contamination. Table 14.1 summarises several applications of fungi in heavy metal removal from the environment. Mushrooms such as *G. lucidum* have demonstrated the ability to

**Table 14.1** Fungi application in reducing heavy metals (HMs) content

| Fungi | Common name | HMs | Application and efficiency | References |
|---|---|---|---|---|
| *Pleurotus* sp. | Oyster mushroom | Cu, Ni, Co | Liquid and solid medium. *Pleurotus.* sp. more tolerant of Co than Ni and Cu | Mohamadhasani and Rahimi (2022) |
| *Morchella* | True morels | Pb | Lead-contaminated soil. The Pb content of crops was reduced by 34.83% | Wang et al. (2021) |
| *Pleurotus ostreatus* | Oyster mushroom | Cd, Cr | Cd and Cr removed 44.85%–80.36% and 14.49%–45.55%, respectively, in liquid treatment HM polluted soil with an accumulation ability of Cd (15.6 mg $kg^{-1}$) and Cr (8.9 mg $kg^{-1}$) | Li et al. (2017) |
| *Ganoderma lucidum* | Reishi or lingzhi | Pb, Zn, Ni, Cu and Cd | HM contaminated soil 45.3%–98.10% reduction | Ipeaiyeda et al. (2020) |
| *Ganoderma lucidum* | Reishi or lingzhi | Mn, Cu, Zn, Cd, Hg, Pb, and U | Contaminated sawdust with HM 60% transfer efficiency and accumulation | Tham et al. (1999) |

bioaccumulate heavy metals like lead, cadmium, and arsenic (Ipeaiyeda et al. 2020; Tham et al. 1999). Fungi are effective in bioremediation due to their secretion of organic acids (e.g. citric, oxalic, and gluconic acid), which can chelate metals and enhance their solubility or immobilisation (Dusengemungu et al. 2021). For example, *P. ostreatus* has been shown to reduce cadmium concentrations by up to 80% within 24 h under controlled laboratory conditions through intracellular accumulation and extracellular adsorption approaches (Li et al. 2017).

In addition to lowering metals' toxicity and bioavailability, this biotransformation process restricts their mobility and leaching in soil and water systems. Moreover, fungal cell walls are rich in functional groups such as carboxyl, amino, and hydroxyl groups that facilitate the passive binding of metal ions, a mechanism known as biosorption (Rashid et al. 2016). This dual action of extracellular enzyme production and cell wall biosorption enhances fungal capacity for detoxifying heavy metals from various matrices.

In soil applications, the unique morphology of fungi, particularly their extensive mycelial networks, confers an advantage in pollutant degradation. The mycelium can penetrate deeply into contaminated soils, facilitating the transport of nutrients and enzymes across large areas and enabling access to pollutants that are unreachable by bacteria (Harms et al. 2011). Additionally, fungal networks support co-metabolism, where the degradation of one compound facilitates the breakdown of others. For instance, in oil-contaminated soils, fungi such as *P. tuber-regium* not only degrade complex hydrocarbons but also immobilise co-occurring heavy metals, thereby enhancing the overall remediation efficiency (Ogbo and Okhuoya 2011).

In aquatic environments, fungal remediation is often applied through immobilised fungal biomass or purified enzymes. These systems are capable of targeting dissolved heavy metals or oil-derived pollutants. Fungi like *T. versicolor* have been immobilised on carriers for use in bioreactors or water treatment systems, effectively reducing concentrations of metals such as lead and chromium through biosorption and enzymatic degradation (Pundir et al. 2024).

### 14.3.4 Bioremediation of Dyes and Colour

Mushroom species such as *T. versicolor*, *P. ostreatus*, and *G. lucidum* are well known for their enzymatic activity and adsorption capabilities in degrading and decolourising industrial dyes. Their enzymes, including manganese peroxidase and laccase, can break down complex dye structures, thereby reducing their toxicity. For instance, *T. versicolor* has been shown to degrade up to 90% of acid dyes (Blue 62, Black 172 and Red 114) and reactive dyes (Blue 4, Orange 16, Red 120 and Black 5) within 6 days. It can also degrade 67% of Acid Orange 7 within 9 days and 51%–80% of disperse dyes (Orange 3, Red 1 and Black 1) within the same period. Moreover, *T. versicolor* has demonstrated effective decolourisation of high concentrations of Acid Blue 62 and Acid Black 172 (Yang et al. 2017). In another study, both live and dead biomass of *T. versicolor* reduced 66% of colour in tannery effluent, along with a 31%–45% reduction in chemical oxygen demand (COD) (Boujelben et al. 2022).

The reishi or lingzhi mushroom *G. lucidum* has also shown remarkable potential in removing dyes from textile effluent, up to 98% in laboratory-scale batch treatments, through a combination of adsorption, absorption, and enzymatic degradation mechanisms (Zahuri et al. 2023, 2024). The smoky bracket fungus (*Bjerkandera adusta*) was able to remove over 90% of various textile dyes, including reactive dyes (Blue 4, Red 120, Black 5, Orange 16) and acid dyes (Blue 62, Red 114, Black 172, Orange 7) within just 10–15 h using an air-lift bioreactor system (Sodaneath et al. 2017). *Mushroom Datronia* species have also shown efficacy, decolourising 86% of Remazol Brilliant Blue R and 88% of Reactive Black 5 after 160 h of treatment (Vaithanomsat et al. 2010).

White-rot fungi such as *Coriolus versicolor* (Turkey tail), *Funalia trogii* (Trog's polypore), *P. chrysosporium*, *Pleurotus florida* (white oyster mushroom), *P. ostreatus*, and *Pleurotus sajor-caju* have successfully decolourised dyes like Astrazone Black FDL, Astrazone Red FBL, and Astrazone Blue FGRL, with promising colour reduction rates of 75%–99% within 24 h (Yesilada et al. 2003). Yesilada et al. (2010) employed both live and dead pellets of *Funalia trogii*, *P. chrysosporium*, *P. ostreatus*, *P. sajor-caju*, and *Trametes versicolor* to treat various dyes. Among these were Remazol Brilliant Blue R (C.I. Reactive Blue 19), Remazol Brilliant Red 3BS (C.I. Reactive Red 239), Remazol Red BS, Remazol Turquoise Blue (C.I. Reactive Blue 21), Cibacron Red FN-3R (reactive dye), Astrazon Black (a blend of C.I. Basic Blue 159, Basic Blue 3, Basic Red 46 and Basic Yellow 28), Astrazon Blue (C.I. Basic Blue 159 and Basic Blue 3), and

Remazol Red (C.I. Basic Red 46). The removal efficiency ranged between 41% and 94%, depending on the dye and fungal species. These research findings suggest the feasibility of using repeatedly grown mycelial pellets for textile effluent treatment (Yesilada et al. 2010). Fungal pellets are reusable and can offer substantial benefits in decolourisation and detoxification when applied before or after conventional treatment methods such as the activated sludge system.

### 14.3.5 Mycofilters

Mycofiltration is a bioremediation technique mainly using the network of the mushroom mycelium as the filter to separate contaminants from water that passes through it. *Lentinus squarrosulus* is a species of edible mushroom belonging to the Polyporaceae family that has been studied by Nkechi-Nwogu et al. (2023) as the mycofiltration of soil leachate from automobile industrial areas. The experimental result shows significant water quality parameter improvement, such as pH, colour, total suspended solids and phosphate. The *P. ostreatus* mycelial network on thatching straw is also being studied by Mnkandla et al. (2024) for mycofiltration for aqueous iron (III) and imidacloprid solutions and found that 31%–94% of the contaminant was removed.

The integration of mushrooms into environmental remediation strategies, termed mycoremediation, offers a sustainable and low-cost alternative to conventional treatment technologies. Mycelium-based applications such as mycofiltration systems, fungal pellets, and substrate-amended treatments have demonstrated success in treating complex waste streams, including textile dyes, agricultural residues, and industrial pollutants. The dual benefits of pollution reduction and potential mushroom cultivation make this approach especially attractive in resource-limited and rural settings. Continued advancements in understanding fungal physiology, enzyme optimisation, and field-scale application are essential to fully harness the potential of mushrooms in global environmental clean-up efforts.

## 14.4 Mushrooms and Emerging Contaminants

The widespread presence of emerging contaminants (ECs) represents a critical frontier in environmental toxicology and remediation science. These structurally diverse, biologically active compounds, coming from pharmaceuticals, nanomaterials, personal care products, and synthetic polymers, are increasingly detected at trace levels (ng/L to μg/L), yet elicit significant ecological and physiological effects due to their potent modes of action (Li et al. 2024). In response to the limited efficacy of existing treatment systems, there is a growing call for biologically adaptive, decentralised remediation strategies (Kumar et al. 2022). Moreover, the ecotoxicity of degradation by-products remains insufficiently understood, compounding environmental risks. In response, there is a growing call for biologically adaptive, decentralised remediation systems. Mushrooms and other fungi present a promising solution through their ability to produce extracellular oxidative enzymes such as laccases and peroxidases that act on a wide range of structurally diverse pollutants. Their potential is particularly relevant in addressing the growing burden of pharmaceuticals, endocrine disruptors, microplastics, and nanomaterials in aquatic and terrestrial systems (Dinakarkumar et al. 2024; Han et al. 2025).

ECs, although present in the environment for decades, have only become routinely detectable with the advent of advanced analytical instrumentation (Noguera-Oviedo and Aga 2016). Their defining feature lies not in novelty or abundance, but in their potent biological activity at low concentrations and incomplete regulation. Many ECs, including pharmaceuticals and endocrine-disrupting compounds (EDCs), are excreted unmetabolised and resist biodegradation, leading to their pseudo-persistence in aquatic environments (Katibi et al. 2021). Compounds like carbamazepine and diclofenac have been widely reported in surface waters, sediments, and even treated drinking water across multiple continents (Patel et al. 2019). EDCs such as bisphenol A and 17α-ethinylestradiol act at nanogram to femtogram levels, disrupting hormone signalling in aquatic biota with population-level repercussions (Hense et al. 2004). Beyond these dissolved and bioactive chemicals, particulate forms of ECs, particularly microplastics and engineered nanomaterials, pose equally pressing but mechanistically distinct risks. Unlike pharmaceuticals or EDCs, these materials do not act primarily through molecular mimicry or receptor binding but through physical interactions, surface-mediated toxicity, and their ability to act as carriers. Microplastics, for example, serve as vectors for hydrophobic organic contaminants, heavy metals, and microbial pathogens, enabling their trophic transfer and accumulation across food webs. Their persistence and mobility, coupled with their capacity to concentrate pollutants on their surfaces, complicate exposure pathways and amplify ecological risks (Ziani et al. 2023). Closely related in their particulate nature, engineered nanomaterials (ENMs), such as silver nanoparticles and carbon nanotubes, present additional challenges due to their nanoscale dimensions and unique physicochemical properties. These materials exhibit toxicity profiles that are highly dependent on particle size, surface functionalisation, and environmental context, making their fate and effects difficult to predict (Dhall et al. 2024).

Pharmaceuticals and personal care products (PPCPs) epitomise the biochemical footprint of modern society. Detected in wastewater, groundwater, and even drinking water, their persistence stems from molecular designs optimised for biological activity, stability, and low-dose efficacy, traits that also hinder their removal by conventional treatment systems (Cizmas et al. 2015). Effective remediation thus requires both elimination of parent compounds and prevention of harmful transformation products, criteria that fungi are uniquely equipped to meet. Several white-rot fungi previously discussed, including *T. versicolor* and *P. chrysosporium*, have demonstrated significant extracellular enzymatic activity critical to PPCP degradation (Tu et al. 2025). These enzymes target lignin-like structures present in many pharmaceuticals, enabling fungi to act on compounds including diclofenac, carbamazepine, propranolol, and ethinylestradiol. *T. versicolor*, for example, can degrade over 90% of diclofenac within 48 h under optimal conditions enhanced by redox mediators like ABTS (Efremenko et al. 2023; Tran et al. 2010). Similarly, *P. chrysosporium* has shown success in transforming carbamazepine, resistant to bacterial systems, into less active metabolites via oxidative demethylation (Golan-Rozen et al. 2011). Compared to bacterial degradation, fungal pathways often yield detoxified, ring-cleaved products rather than reactive intermediates (Badia-Fabregat et al. 2017). However, fungal efficacy is tightly linked to environmental factors. Ligninolytic enzyme production is modulated by carbon-to-nitrogen ratios, redox conditions, and trace metals such as copper and manganese (Levin et al. 2005). Paradoxically, the nutrient-limited conditions that trigger secondary metabolism in fungi are rarely found in organic-rich wastewaters, complicating their application. Moreover, PPCP degradation is typically co-metabolic, requiring fungi to sustain enzymatic activity without prioritising biomass growth. However, many enzymatic pathways remain unresolved, and recent studies emphasise the need for genetic engineering to enhance fungal metabolic efficiency and enable their consistent use in wastewater treatment systems (Malik et al. 2023). A critical limitation lies in the fate of transformation products. While parent compounds are often removed, secondary metabolites may retain or amplify toxicity. For instance, triclosan is readily oxidised by fungal laccases, yet its dimeric by-products exhibit endocrine-disrupting potential. This underscores the need for integrated fungal treatment systems that incorporate enzymatic monitoring, by-product toxicity evaluation, and bioreactor optimisation. Although still nascent, the use of fungi as decentralised biocatalysts, particularly at PPCP point sources like hospitals or pharmaceutical plants, offers a compelling and underutilised remediation strategy.

Despite their enzymatic versatility in degrading ECs, fungi alone often face limitations in complex environmental matrices such as wastewater, sediments, and industrial effluents. These systems are characterised by chemically diverse pollutants and dynamic physical, biological, and hydraulic constraints that restrict fungal efficacy (Thirumalaivasan et al. 2024). In response, contemporary research has shifted towards integrated remediation strategies that combine fungal systems with physical adsorbents, immobilised enzymes, microbial consortia, or artificial intelligence (AI)-driven process controls. These synergistic approaches aim to overcome key drawbacks of standalone fungal treatments, such as slow degradation kinetics, limited substrate range, and operational sensitivity, while preserving their ecological advantages. Advancing such hybrid systems will require multidisciplinary efforts to enhance degradation efficiency, environmental robustness, and scalability for real-world applications.

A prominent example of such integration is the development of fungi–biochar composites, which combine the sorptive capacity of biochar with the enzymatic activity of fungi. Biochar, derived from biomass pyrolysis, offers a porous structure, abundant surface functional groups, and a high specific surface area, all of which facilitate the immobilisation of fungal hyphae and the adsorption of contaminants. This dual functionality allows pollutants to be concentrated onto the biochar surface, where they are subsequently degraded by fungal enzymes secreted in situ. Studies have shown that *T. versicolor* immobilised in CMC beads effectively biosorbs $Cu^{2+}$, $Pb^{2+}$, and $Zn^{2+}$, with stable performance across cycles and pH ranges. When applied to rotating biological contactors, it reduced total organic carbon by 70%–75% and enhanced pharmaceutical removal via advanced bio-oxidation, showing strong stability under non-sterile, continuous-flow conditions, offering a scalable solution for biological wastewater treatment (Bayramoglu 2003; Chen et al. 2023; del Álamo et al. 2018). Beyond serving as a passive scaffold, biochar actively modulates the microenvironment, enhancing redox balance, mediating electron shuttling, and sustaining laccase activity. However, the performance of such composites is highly dependent on the physicochemical characteristics of the biochar itself; for instance, high-temperature biochar may release PAHs or heavy metals that inhibit fungal growth, while low-porosity materials may impede enzyme diffusion. Therefore, the compatibility between biochar type, fungal species, and target pollutant must be carefully optimised (Mukherjee et al. 2022).

Building on the idea of decoupling fungal metabolism from fungal growth, another promising approach involves the use of immobilised enzymes. Instead of relying on live fungal cultures, this strategy isolates and anchors key enzymes, such as laccases and peroxidases, onto inert supports, forming reusable biocatalytic platforms. Immobilisation confers several advantages, such as enhancing enzyme stability across a wide range of pH and temperature conditions, preventing washout in continuous-flow systems and enabling integration into modular reactor designs such as membrane filters or cartridge systems. For example, laccases from *P. ostreatus* immobilised on silica and chitosan supports have retained over 80% of their catalytic activity after ten reuse cycles, efficiently degrading endocrine-disrupting compounds like bisphenol A and triclosan (Serbent et al. 2024). Nonetheless, challenges remain in preserving enzyme conformation and activity during immobilisation. Simple adsorption methods may result in enzyme leaching, while covalent bonding can alter active sites. Recent advances in nanoparticle-assisted immobilisation, sol-gel encapsulation, and the use of metal–organic frameworks show promise in addressing these issues, although concerns regarding cost, scalability, and regulatory acceptance persist.

While immobilised enzymes improve operational stability, they lack the metabolic diversity of whole fungal systems. This limitation can be addressed through microbial co-cultures and consortia, which exploit synergistic interactions among fungi, bacteria, and yeasts. In natural ecosystems, such interactions are fundamental to organic matter turnover, and when strategically engineered, they can broaden the range of degradable contaminants and enhance system resilience. For instance, co-cultures of fungi-bacteria of *T. versicolor* and *Pseudomonas putida* have demonstrated superior removal of ibuprofen and sulfamethoxazole, with fungi initiating oxidative cleavage and bacteria completing the mineralisation process (Espinosa-Ortiz et al. 2022). Additionally, fungal hyphae can act as conduits, so-called 'fungal highways', that facilitate bacterial transport into microporous environments, improving access to hydrophobic pollutants (Lott et al. 2025). However, maintaining stable and functional consortia poses challenges related to interspecies competition, pH drift, nutrient depletion, and quorum sensing. Innovations in synthetic ecology, including genome-scale metabolic modelling and community-level optimisation, are paving the way for more rational design of these multispecies systems.

Complementing these biological innovations, the incorporation of artificial intelligence offers a powerful tool for enhancing the performance and adaptability of fungal bioreactors. Fungal systems are inherently complex, exhibiting nonlinear responses to environmental stimuli. AI-driven models, trained on historical datasets and real-time sensor inputs, can predict key parameters such as laccase activity, fungal biomass growth, and pollutant degradation rates, enabling dynamic control of aeration, redox potential, nutrient dosing, and pH (Blessing and Olateru 2025). Algorithms such as random forests, support vector machines, and neural networks have already been applied to optimise fungal bioprocesses at the lab scale. When coupled with high-frequency monitoring tools (e.g. electrochemical sensors, online FTIR), these models can facilitate explainable, closed-loop control strategies that reduce trial-and-error experimentation and accelerate scale-up to pilot and full-scale applications. To support such data-driven control, a range of commercially available online sensors is used to continuously monitor key parameters in filamentous fungi cultivation systems (Table 14.2). These sensors provide real-time measurements of $CO_2$, dissolved $O_2$, pH, ion concentrations, and biomass viability, parameters essential for AI-assisted bioprocess control and optimisation.

Ultimately, the future of fungal remediation lies not in fungal systems alone but in their strategic integration with complementary technologies and disciplines. The convergence of ecological intelligence (fungi), material innovation (biochar, immobilisation platforms), and digital precision (AI) represents a transformative paradigm for addressing the growing complexity of EC contamination. However, realising this potential requires not only technical advancement but also interdisciplinary collaboration, lifecycle assessment, and systems-level design to ensure that these synergistic solutions are robust, scalable, and sustainable across diverse environmental contexts. Recent studies have demonstrated the potential of artificial intelligence and automation to enhance the monitoring, control, and predictive

**Table 14.2** Online sensors and analytical tools that are commercially accessible for assessing biological, chemical, and physical parameters in filamentous fungal culture systems (Wainaina and Taherzadeh 2023)

| Sensor | Remarks | Example of manufacturers |
|---|---|---|
| $CO_2$ | While other methods (e.g. RAMOS) can be employed, the fluorescence-based approach is more frequently employed. | Hach, Hamilton, Mettler Toledo |
| $O_2$ | Measurements of dissolved oxygen can be made using a Clark amperometric electrode. | Hach, Hamilton, Mettler Toledo |
| pH | A common potentiometric ion-selective glass electrode is used for the measurements. | KROHNE, Mettler Toledo, Turtle Tough |
| Temperature | Based on a thermistor, a thermometer, or a thermocouple. | High Purity New England, Parker |
| Ion-specific | For determining ions such as $NH_3^+$, $Ca^{2+}$, $K^+$, $Mg^{2+}$, $PO_4^{3-}$. | Thermo Fisher Scientific |
| Foam | Detects the presence of foam by means of a conductive loop. | Thermo Fisher Scientific |
| Pressure | Membrane pressure sensors. | High Purity New England |
| Fluorescence | Used for determining biomass viability. | iPrasense |

**Table 14.3** Showcasing examples of automation and AI implementation in bioprocesses, including fungi-based systems

| Purpose | Brief description | Automation/AI tools used and their benefits |
|---|---|---|
| Measuring fungal morphology | Morphological characterisation of *fungi* in batch cultivation | MATLAB: allowed for the estimation and morphological assessment of complete microscope slides in a single computing environment |
| Materials handling | Robotic arms are used to collect and transfer materials | Robotic systems with intelligent sensors, actuators, and the right software can monitor and control processes with high accuracy |
| Fungal cell colonies counting | Differentiating fungal cells from simple and complex backgrounds | Ilastik: Higher throughput than ImageJ and manual cell-counting methods |
| Growth of fungi measure | During the formation of cellulase, the growth of fungi in the culture media containing insoluble components was evaluated | Multiphase Artificial Neural Network (MANN): Reliably predicted fungal biomass concentration with tolerable deviation from empirical data |
| Optimising the activity of enzyme | The effect of process variables on the generation of xylanase using fungi in submerged fermentation (SmF) was assessed | Lazy learning model: better than backpropagation neural networks and multiple linear regression in terms of fit to input/output data |
| Estimating water activity (aw) | An algorithm to calculate the mass of condensed water droplets and fungal biomass during solid-state fermentation (SSF) has been proposed | MATLAB: Good agreement between proposed model and experimental data |

Adapted from Wainaina and Taherzadeh (2023)

capabilities of fungal bioreactors. Table 14.3 outlines key examples of AI implementation in fungal systems, ranging from morphological analysis to enzyme activity prediction and material handling.

The rise of pharmaceuticals, emerging contaminants, personal care products, endocrine disruptors, nanomaterials, and microplastics demands an ecological rethinking of remediation strategies, one that transcends pollutant-by-pollutant solutions and embraces multifunctional, adaptive biological systems. Within this evolving landscape, mushrooms, once overlooked as passive recyclers of lignocellulosic waste, are now emerging as versatile biocatalysts and ecological engineers, uniquely equipped to confront the biochemical excesses of the Anthropocene. Together, these advancements underscore the transformative potential of fungi when strategically integrated with modern materials, modular bioprocessing, and intelligent control systems. As fungal-based remediation moves from conceptual frameworks towards operational deployment, a deeper understanding of system integration, performance validation, and environmental compatibility will be essential. The following section explores the key challenges, technological breakthroughs, and future directions shaping this emerging frontier.

## 14.5 Conclusion

Mushroom fungi have emerged as highly effective agents in addressing environmental pollution through mechanisms such as biodegradation, biosorption, bioaccumulation, and enzymatic transformation. Their ability to produce versatile enzymes like laccase, lignin peroxidase, and manganese peroxidase allows them to break down persistent organic pollutants, including petroleum hydrocarbons, synthetic dyes, pesticides, and pharmaceutical residues. Additionally, their robust mycelial networks facilitate the immobilisation and removal of heavy metals from soil and water. The integration of mushrooms into environmental management, termed mycoremediation, offers an eco-friendly, cost-effective, and scalable alternative to conventional remediation technologies that often generate toxic by-products and require intensive resources.

Beyond their biochemical and ecological advantages, fungi are adaptable to a wide range of environmental conditions and can be applied through various methods, such as mycofiltration systems, bioreactors, and immobilised enzyme platforms. As environmental pollution becomes more complex, especially with the rise of emerging contaminants like microplastics and nanomaterials, fungi offer a sustainable biological solution with potential for further innovation. Future research should focus on enhancing the efficacy of fungal remediation through genetic engineering, consortia-based approaches, and integration with advanced technologies such as artificial intelligence and smart bioprocess control. Collectively, this chapter underscores the vital role of mushrooms not only as decomposers in nature but also as essential allies in building a cleaner, safer, and more sustainable environment.

**Acknowledgement** Thank you to the Universiti Malaya Partnership Grant 'Formulation of Lampam Sungai Sustainable Fish Feed Using Petronas Waste (MG002-2025)'.

**AI Declaration** The author(s) used AI-based tools (e.g. Grammarly and ChatGPT) solely for language editing and proofreading. The final content is the responsibility of the author(s).

## References

Aditya, Neeraj, Jarial RS, Jarial K, Bhatia JN (2024) Comprehensive review on oyster mushroom species (Agaricomycetes): morphology, nutrition, cultivation and future aspects. Heliyon 10(5): e26539. https://doi.org/10.1016/j.heliyon.2024.e26539

Ahmed AS, Mohammed-Ridha MJ, Raoof NN (2016) Kinetic, thermodynamic, and equilibrium biosorption of Pb(II), Cu(II), and Ni(II) using dead mushroom biomass under batch experiment. Biorem J 20(3):252–261. https://doi.org/10.1080/10889868.2016.1212811

Akhtar N, Mannan MA (2020) Mycoremediation: expunging environmental pollutants. Biotechnol Rep 26:e00452. https://doi.org/10.1016/j.btre.2020.e00452

Akpasi SO, Anekwe IMS, Tetteh EK, Amune UO, Shoyiga HO, Mahlangu TP, Kiambi SL (2023) Mycoremediation as a potentially promising technology: current status and prospects—a review. Appl Sci 13(8):4978. https://doi.org/10.3390/app13084978

Alaidaroos BA (2023) Advancing eco-sustainable bioremediation for hydrocarbon contaminants: challenges and solutions. Processes 11(10):3036. https://doi.org/10.3390/pr11103036

Antón-Herrero R, Chicca I, García-Delgado C, Crognale S, Lelli D, Gargarello RM, Herrero J, Fischer A, Thannberger L, Eymar E, Petruccioli M, D'Annibale A (2023) Main factors determining the scale-up effectiveness of mycoremediation for the decontamination of aliphatic hydrocarbons in soil. J Fungi 9(12):1205. https://doi.org/10.3390/jof9121205

Ayala A, Muñoz MF, Argüelles S (2014) Lipid peroxidation: production, metabolism, and signaling mechanisms of malondialdehyde and 4-hydroxy-2-nonenal. Oxidative Med Cell Longev 2014:360438. https://doi.org/10.1155/2014/360438

Badia-Fabregat M, Lucas D, Tuomivirta T, Fritze H, Pennanen T, Rodríguez-Mozaz S, Barceló D, Caminal G, Vicent T (2017) Study of the effect of the bacterial and fungal communities present in real wastewater effluents on the performance of fungal treatments. Sci Total Environ 579: 366–377. https://doi.org/10.1016/j.scitotenv.2016.11.088

Bayramoglu G (2003) Biosorption of heavy metal ions on immobilized white-rot fungus *Trametes versicolor*. J Hazard Mater 101(3):285–300. https://doi.org/10.1016/S0304-3894(03)00178-X

Bhattacharjya DK, Paul RK, Miah MN, Ahmed KU (2014) Effect of different saw dust substrates on the growth and yield of oyster mushroom (*Pleurotus ostreatus*). IOSR J Agric Vet Sci 7(2): 38–46. https://doi.org/10.9790/2380-07233846

Bilal M, Zdarta J, Jesionowski T, Iqbal HMN (2023) Manganese peroxidases as robust biocatalytic tool – an overview of sources, immobilization, and biotechnological applications. Int J Biol Macromol 234:123531. https://doi.org/10.1016/j.ijbiomac.2023.123531

Blessing A-A, Olateru K (2025) AI-driven optimization of bioremediation strategies for river pollution: a comprehensive review and future directions. Front Microbiol 16:1504254. https://doi.org/10.3389/fmicb.2025.1504254

Boujelben R, Ellouze M, Tóran MJ, Blánquez P, Sayadi S (2022) Mycoremediation of Tunisian tannery wastewater under non-sterile conditions using *Trametes versicolor*: live and dead biomasses. Biomass Convers Biorefin 14:299–312. https://doi.org/10.1007/s13399-022-02328-0

Chen X, Tang L, Wu K, Mo Y, Tang Q, Li G, Zhu Y (2023) Combined contribution of biochar and introduced AM fungi on lead stability and microbial community in polluted agricultural soil. Front Microbiol 14:1284321. https://doi.org/10.3389/fmicb.2023.1284321

Civzele A, Stipniece-Jekimova AA, Mezule L (2023) Fungal ligninolytic enzymes and their application in biomass lignin pretreatment. J Fungi 9(7):780. https://doi.org/10.3390/jof9070780

Cizmas L, Sharma VK, Gray CM, McDonald TJ (2015) Pharmaceuticals and personal care products in waters: occurrence, toxicity, and risk. Environ Chem Lett 13(4):381–394. https://doi.org/10.1007/s10311-015-0524-4

Colosimo R, Mulet-Cabero A-I, Cross KL, Haider K, Edwards CH, Warren FJ, Finnigan TJA, Wilde PJ (2021) β-glucan release from fungal and plant cell walls after simulated gastrointestinal digestion. J Funct Foods 83:104543. https://doi.org/10.1016/j.jff.2021.104543

Dabrowska M, Debiec-Andrzejewska K, Andrunik M, Bajda T, Drewniak L (2021) The biotransformation of arsenic by spent mushroom compost? An effective bioremediation agent. Ecotoxicol Environ Saf 213:112054. https://doi.org/10.1016/j.ecoenv.2021.112054

del Álamo AC, Pariente MI, Vasiliadou I, Padrino B, Puyol D, Molina R, Martínez F (2018) Removal of pharmaceutical compounds from urban wastewater by an advanced bio-oxidation process based on fungi *Trametes versicolor* immobilized in a continuous RBC system. Environ Sci Pollut Res 25(35):34884–34892. https://doi.org/10.1007/s11356-017-1053-4

Dhall S, Nigam A, Harshavardhan M, Mukherjee A, Srivastava P (2024) A comprehensive overview of methods involved in nanomaterial production and waste disposal from research labs and industries and existing regulatory guidelines for handling engineered nanomaterials. Environ Chem Ecotoxicol 6:269–282. https://doi.org/10.1016/j.enceco.2024.06.002

Dinakarkumar Y, Ramakrishnan G, Gujjula KR, Vasu V, Balamurugan P, Murali G (2024) Fungal bioremediation: an overview of the mechanisms, applications and future perspectives. Environ Chem Ecotoxicol 6:293–302. https://doi.org/10.1016/j.enceco.2024.07.002

Dusengemungu L, Kasali G, Gwanama C, Mubemba B (2021) Overview of fungal bioleaching of metals. Environ Adv 5:100083. https://doi.org/10.1016/j.envadv.2021.100083

Efremenko E, Stepanov N, Senko O, Maslova O, Lyagin I, Aslanli A (2023) Progressive biocatalysts for the treatment of aqueous systems containing pharmaceutical pollutants. Life 13(3):841. https://doi.org/10.3390/life13030841

El-Batal AI, Elkenawy NM, Yassin AS, Amin MA (2015) Laccase production by *Pleurotus ostreatus* and its application in synthesis of gold nanoparticles. Biotechnol Rep 5(1):31–39. https://doi.org/10.1016/j.btre.2014.11.001

El-Najjar N, Gali-Muhtasib H, Ketola RA, Vuorela P, Urtti A, Vuorela H (2011) The chemical and biological activities of quinones: overview and implications in analytical detection. Phytochem Rev 10(3):353–370. https://doi.org/10.1007/s11101-011-9209-1

Espinosa-Ortiz EJ, Rene ER, Gerlach R (2022) Potential use of fungal-bacterial co-cultures for the removal of organic pollutants. Crit Rev Biotechnol 42(3):361–383. https://doi.org/10.1080/07388551.2021.1940831

Gałązka A, Jankiewicz U, Szczepkowski A (2023) Biochemical characteristics of laccases and their practical application in the removal of xenobiotics from water. Appl Sci 13(7):4394. https://doi.org/10.3390/app13074394

Gałązka A, Jankiewicz U, Orzechowski S (2025) The role of ligninolytic enzymes in sustainable agriculture: applications and challenges. Agronomy 15(2):451. https://doi.org/10.3390/agronomy15020451

Garcia-Delgado C, Yunta F, Eymar E (2015) Bioremediation of multi-polluted soil by spent mushroom (*Agaricus bisporus*) substrate: polycyclic aromatic hydrocarbons degradation and Pb availability. J Hazard Mater 300:281–288. https://doi.org/10.1016/j.jhazmat.2015.07.008

Golan-Rozen N, Chefetz B, Ben-Ari J, Geva J, Hadar Y (2011) Transformation of the recalcitrant pharmaceutical compound carbamazepine by *Pleurotus ostreatus*: role of cytochrome P450 monooxygenase and manganese peroxidase. Environ Sci Technol 45(16):6800–6805. https://doi.org/10.1021/es200298t

Haddadin MS, Al-Natour R, Al-Qsous S, Robinson RK (2002) Bio-degradation of lignin in olive pomace by freshly-isolated species of Basidiomycete. Bioresour Technol 82(2):131–137. https://doi.org/10.1016/S0960-8524(01)00171-7

Hammel KE (1995) Mechanisms for polycyclic aromatic hydrocarbon degradation by ligninolytic fungi. Environ Health Perspect 103(5):41–43. https://doi.org/10.1289/ehp.95103s441

Han C, Liu Z, Teng Y, Zhang H, Li Z, Xu T, Yang Y, Teng F (2025) Careful attention to the eco-toxicity and secondary water pollution of the degradation by-products. Environ Res 271:121091. https://doi.org/10.1016/j.envres.2025.121091

Hanafiah ZM, Hafsari AR, Elango M, Ilham Z, Doni F, Wan Mohtar WHM, Gafforov Y, Wong YJ, Abdul Maulud KN, Ismail NH, Putera Mohd Yusof MY, Wan-Mohtar WAAQI (2025) Removal of cyanobacterial harmful algal blooms (HABs) from contaminated local park lake using *Ganoderma lucidum* mycelial pellets. Heliyon 11:e41205. https://doi.org/10.1016/j.heliyon.2024.e41205

Harms H, Schlosser D, Wick LY (2011) Untapped potential: exploiting fungi in bioremediation of hazardous chemicals. Nat Rev Microbiol 9(3):177–192. https://doi.org/10.1038/nrmicro2519

Hense BA, Severin GF, Welzl G, Schramm K-W (2004) Effects of 17-ethinylestradiol on zoo- and phytoplankton in lentic microcosms. Anal Bioanal Chem 378(3):716–724. https://doi.org/10.1007/s00216-003-2379-y

Hu Y-H, Zhuang J-X, Yu F, Cui Y, Yu W-W, Yan C-L, Chen Q-X (2016) Inhibitory effects of cefotaxime on the activity of mushroom tyrosinase. J Biosci Bioeng 121(4):385–389. https://doi.org/10.1016/j.jbiosc.2015.08.005

Hu Z, Shi J, Yang H, Huang J, Sheng F (2021) How organic substances promote the chemical oxidative degradation of pollutants: a mini review. Sustainability 13(19):10993. https://doi.org/10.3390/su131910993

Hyde KD, Baldrian P, Chen Y, Thilini Chethana KW, De Hoog S, Doilom M, de Farias ARG, Gonçalves MFM, Gonkhom D, Gui H, Hilário S, Hu Y, Jayawardena RS, Khyaju S, Kirk PM, Kohout P, Luangharn T, Maharachchikumbura SSN, Manawasinghe IS, Walker A (2024) Current trends, limitations and future research in the fungi? Fungal Divers 125(1):1–71. https://doi.org/10.1007/s13225-023-00532-5

Ipeaiyeda AR, Adenipekun CO, Oluwole O (2020) Bioremediation potential of *Ganoderma lucidum* (Curt:Fr) P. Karsten to remove toxic metals from abandoned battery slag dumpsite soil and immobilisation of metal absorbed fungi in bricks. Cogent Environ Sci 6(1):1847400. https://doi.org/10.1080/23311843.2020.1847400

Jayaraman S, Yadav B, Dalal RC, Naorem A, Sinha NK, Srinivasa Rao C, Dang YP, Patra AK, Datta SP, Subba Rao A (2024) Mushroom farming: a review focusing on soil health, nutritional security and environmental sustainability. Farm Syst 2(3):100098. https://doi.org/10.1016/j.farsys.2024.100098

Katibi KK, Yunos KF, Che Man H, Aris AZ, Mohd Nor MZ, Azis RS, Umar AM (2021) Contemporary techniques for remediating endocrine-disrupting compounds in various water sources: advances in treatment methods and their limitations. Polymers 13(19):3229. https://doi.org/10.3390/polym13193229

Khajehzadeh M, Ghobad-Nejhad M, Moghimi H, Soorki AA, Dai Y-C, Si J (2024) Degradation of high concentrations of anthracene using white-rot wood-inhabiting fungi and investigation of enzyme activities. Mycobiology 52(5):298–305. https://doi.org/10.1080/12298093.2024.2409485

Khan MF (2025) Fungi for sustainable pharmaceutical remediation: enzymatic innovations, challenges, and applications—a review. Processes 13(4):1034. https://doi.org/10.3390/pr13041034

Khan I, Aftab M, Shakir S (2019) Mycoremediation of heavy metal (Cd and Cr)—polluted soil through indigenous metallotolerant fungal isolates. Environ Monit Assess 191(385):1–11

Khatua S, Simal-Gandara J, Acharya K (2024) Myco-remediation of plastic pollution: current knowledge and future prospects. Biodegradation 35(3):249–279. https://doi.org/10.1007/s10532-023-10053-2

Kumar A, Chandra R (2020) Ligninolytic enzymes and its mechanisms for degradation of lignocellulosic waste in environment. Heliyon 6(2):e03170. https://doi.org/10.1016/j.heliyon.2020.e03170

Kumar M, Ngasepam J, Dhangar K, Mahlknecht J, Manna S (2022) Critical review on negative emerging contaminant removal efficiency of wastewater treatment systems: concept, consistency and consequences. Bioresour Technol 352:127054. https://doi.org/10.1016/j.biortech.2022.127054

Kumla J, Suwannarach N, Sujarit K, Penkhrue W, Kakumyan P, Jatuwong K, Vadthanarat S, Lumyong S (2020) Cultivation of mushrooms and their lignocellulolytic enzyme production through the utilization of agro-industrial waste. Molecules 25(12):2811. https://doi.org/10.3390/molecules25122811

Lankinen P, Hildén K, Aro N, Salkinoja-Salonen M, Hatakka A (2005) Manganese peroxidase of *Agaricus bisporus*: grain bran-promoted production and gene characterization. Appl Microbiol Biotechnol 66(4):401–407. https://doi.org/10.1007/s00253-004-1731-2

Legorreta-Castañeda AJ, Lucho-Constantino CA, Beltrán-Hernández RI, Coronel-Olivares C, Vázquez-Rodríguez GA (2020) Biosorption of water pollutants by fungal pellets. Water 12(4):1155. https://doi.org/10.3390/W12041155

Levin L, Forchiassin F, Viale A (2005) Ligninolytic enzyme production and dye decolorization by *Trametes trogii*: application of the Plackett–Burman experimental design to evaluate nutritional requirements. Process Biochem 40(3–4):1381–1387. https://doi.org/10.1016/j.procbio.2004.06.005

Li X, Wang Y, Pan Y, Yu H, Zhang X, Shen Y, Jiao S, Wu K, La G, Yuan Y, Zhang S (2017) Mechanisms of Cd and Cr removal and tolerance by macrofungus *Pleurotus ostreatus* HAU-2. J Hazard Mater 330:1–8. https://doi.org/10.1016/j.jhazmat.2017.01.047

Li X, Shen X, Jiang W, Xi Y, Li S (2024) Comprehensive review of emerging contaminants: Detection technologies, environmental impact, and management strategies. Ecotoxicology and Environmental Safety 278:116420

Lott C, Kabasci S, Merrettig-Bruns U, Weber M (2025) (Bio-)degradation of plastic in the environment. In: Springer handbook of circular plastics economy. Springer, Cham, pp 773–792. https://doi.org/10.1007/978-3-031-66209-6_39

Low BT, Ting YP, Deng S (2008) Surface modification of *Penicillium chrysogenum* mycelium for enhanced anionic dye removal. Chem Eng J 141(1–3):9–17. https://doi.org/10.1016/j.cej.2007.10.004

Malik S, Bora J, Nag S, Sinha S, Mondal S, Rustagi S, Hazra R, Kumar H, Rajput VD, Minkina T, Sadier NS, Almutary AG (2023) Fungal-based remediation in the treatment of anthropogenic activities and pharmaceutical-pollutant-contaminated wastewater. Water 15(12):2262. https://doi.org/10.3390/w15122262

Mani P, Kumar VTF, Keshavarz T, Sainathan Chandra T, Kyazze G (2018) The role of natural laccase redox mediators in simultaneous dye decolorization and power production in microbial fuel cells. Energies 11(12):1–12. https://doi.org/10.3390/en11123455

Mayans B, Antón-Herrero R, García-Delgado C, Delgado-Moreno L, Guirado M, Pérez-Esteban J, Escolástico C, Eymar E (2024) Bioremediation of petroleum hydrocarbons polluted soil by spent mushroom substrates: microbiological structure and functionality. J Hazard Mater 473:134650. https://doi.org/10.1016/j.jhazmat.2024.134650

Mnkandla SM, Mosoabisane MFT, Basopo N, Otomo PV (2024) Mycofiltration of aqueous iron (III) and imidacloprid solutions, and the effects of the filtrates on selected biomarkers of the freshwater snail *Helisoma duryi*. Arch Environ Contam Toxicol 86(2):187–197. https://doi.org/10.1007/s00244-023-01049-3

Mohamadhasani F, Rahimi M (2022) Growth response and mycoremediation of heavy metals by fungus *Pleurotus* sp. Sci Rep 12(1):19947. https://doi.org/10.1038/s41598-022-24349-5

Mohd Hanafiah Z, Wan Mohtar WHM, Wan-Mohtar WAAQI, Sultana A, Rohani R, Indarto A, Mundher Z, Sharil S, Sabariah T, Abdul B (2024) Removal of pharmaceutical compounds and toxicology study in wastewater using Malaysian fungal *Ganoderma lucidum*. Chemosphere 358:142209. https://doi.org/10.1016/j.chemosphere.2024.142209

Mukherjee S, Sarkar B, Aralappanavar VK, Mukhopadhyay R, Basak BB, Srivastava P, Marchut-Mikołajczyk O, Bhatnagar A, Semple KT, Bolan N (2022) Biochar-microorganism interactions for organic pollutant remediation: challenges and perspectives. Environ Pollut 308:119609. https://doi.org/10.1016/j.envpol.2022.119609

Muraleedharan TR, Venkobachar C, Leela I (1994) Investigations of fungal fruiting bodies as biosorbents for the removal of heavy-metals from industrial processing streams. Sep Sci Technol 29(14):1893–1903. https://doi.org/10.1080/01496399408002178

Murtić S, Karić L, Sinanović ĆZ, Hasanbegović A, Avdić J, Šerbo A, Hadžić A (2024) Heavy metals accumulation in the oyster mushroom basidiomes cultivated on different substrates. Acta Mycol 59:1–7. https://doi.org/10.5586/am/191601

Murugesan K, Yang I-H, Kim Y-M, Jeon J-R, Chang Y-S (2009) Enhanced transformation of malachite green by laccase of Ganoderma lucidum in the presence of natural phenolic compounds. Appl Microbiol Biotechnol 82(2):341–350. https://doi.org/10.1007/s00253-008-1819-1

Naidu Y, Siddiqui Y, Rafii MY, Saud HM, Idris AS (2017) Investigating the effect of white-rot hymenomycetes biodegradation on basal stem rot infected oil palm wood blocks: biochemical and anatomical characterization. Ind Crop Prod 108:872–882. https://doi.org/10.1016/j.indcrop.2017.08.064

Nesma ASA, Thara SS, Soni KB, Sindura KP, Krishnapriya PJP (2023) Expression profiling of laccase and β-glucan synthase genes in *Pleurotus ostreatus* during different developmental stages. Mol Biol Rep 50(9):7205–7213. https://doi.org/10.1007/s11033-023-08556-5

Nkechi-Nwogu CG, Chukwu SB, Eke PC (2023) Mycofiltration of soil leachate using *Lentinus squarrosulus* for bioremediation of soils from auto-mobile mechanic leachate soils of Ikoku Market, Port Harcourt. FUDMA J Agric Agric Technol 9(1):36–40. https://doi.org/10.33003/jaat.2023.0901.05

Noguera-Oviedo K, Aga DS (2016) Lessons learned from more than two decades of research on emerging contaminants in the environment. J Hazard Mater 316:242–251. https://doi.org/10.1016/j.jhazmat.2016.04.058

Ogbo EM, Okhuoya JA (2011) Bio-absorption of some heavy metals by *Pleurotus tuber-regium* Fr. Singer (an edible mushroom) from crude oil polluted soils amended with fertilizers and cellulosic wastes. Int J Soil Sci 6(1):34–48. https://doi.org/10.3923/ijss.2011.34.48

Oladimeji TE, Oyedemi M, Emetere ME, Agboola O, Adeoye JB, Odunlami OA (2024) Review on the impact of heavy metals from industrial wastewater effluent and removal technologies. Heliyon 10(23):e40370. https://doi.org/10.1016/j.heliyon.2024.e40370

Patel M, Kumar R, Kishor K, Mlsna T, Pittman CU, Mohan D (2019) Pharmaceuticals of emerging concern in aquatic systems: chemistry, occurrence, effects, and removal methods. Chem Rev 119(6):3510–3673. https://doi.org/10.1021/ACS.CHEMREV.8B00299

Pathmashini L, Arulnandhy V, Wijeratnam RW (2009) Cultivation of oyster mushroom (*Pleurotus ostreatus*) on sawdust. Ceylon J Sci Biol Sci 37(2):177–182. https://doi.org/10.4038/cjsbs.v37i2.505

Pozdnyakova NN (2012) Involvement of the ligninolytic system of white-rot and litter-decomposing fungi in the degradation of polycyclic aromatic hydrocarbons. Biotechnol Res Int 2012:1–20. https://doi.org/10.1155/2012/243217

Praveen K, Usha KY, Viswanath B, Reddy BR (2012) Kinetic properties of manganese peroxidase from the mushroom Stereum ostrea and its ability to decolorize dyes. J Microbiol Biotechnol 22(11):1540–1548. https://doi.org/10.4014/jmb.1112.12011

Puchana-Rosero MJ, Lima EC, Ortiz-Monsalve S, Mella B, da Costa D, Poll E, Gutterres M (2017) Fungal biomass as biosorbent for the removal of Acid Blue 161 dye in aqueous solution. Environ Sci Pollut Res 24(4):4200–4209. https://doi.org/10.1007/s11356-016-8153-4

Pundir A, Thakur MS, Prakash S, Kumari N, Sharma N, Parameswari E, He Z, Nam S, Thakur M, Puri S, Puranik S, Kumar S, Madhu, Kumar M (2024) Fungi as versatile biocatalytic tool for treatment of textile wastewater effluents. Environ Sci Eur 36(1):185. https://doi.org/10.1186/s12302-024-01007-3

Purnomo AS, Mori T, Kamei I, Nishii T, Kondo R (2010) Application of mushroom waste medium from Pleurotus ostreatus for bioremediation of DDT-contaminated soil. Int Biodeterior Biodegradation 64(5):397–402. https://doi.org/10.1016/j.ibiod.2010.04.007

Ramirez DA, Munoz SV, Atehortua L, Michel FC Jr (2010) Effects of different wavelengths of light on lignin peroxidase production by the white-rot fungi *Phanerochaete chrysosporium* grown in submerged cultures. Bioresour Technol 101(23):9213–9220. https://doi.org/10.1016/j.biortech.2010.06.114

Rashid A, Bhatti HN, Iqbal M, Noreen S (2016) Fungal biomass composite with bentonite efficiency for nickel and zinc adsorption: a mechanistic study. Ecol Eng 91:459–471. https://doi.org/10.1016/j.ecoleng.2016.03.014

Rathore H, Prasad S, Kapri M, Tiwari A, Sharma S (2019) Medicinal importance of mushroom mycelium: mechanisms and applications. J Funct Foods 56:182–193. https://doi.org/10.1016/j.jff.2019.03.016

Sahithya K, Mouli T, Biswas A, Scorlet MT (2022) Remediation potential of mushrooms and their spent substrate against environmental contaminants: an overview. Biocatal Agric Biotechnol 42:102323. https://doi.org/10.1016/j.bcab.2022.102323

Sánchez-Alejandro F, Juarez-Moreno K, Baratto MC, Basosi R, Vazquez-Duhalt R (2016) Tryptophan-surface modification of versatile peroxidase from *Bjerkandera adusta* enhances its catalytic performance. J Mol Catal B Enzym 124:45–51. https://doi.org/10.1016/j.molcatb.2015.12.001

Sar T, Marchlewicz A, Harirchi S, Mantzouridou FT, Hosoglu MI, Akbas MY, Hellwig C, Taherzadeh MJ (2024) Resource recovery and treatment of wastewaters using filamentous fungi. Sci Total Environ 951:175752. https://doi.org/10.1016/j.scitotenv.2024.175752

Serbent MP, Magario I, Saux C (2024) Immobilizing white-rot fungi laccase: toward bio-derived supports as a circular economy approach in organochlorine removal. Biotechnol Bioeng 121(2):434–455. https://doi.org/10.1002/bit.28591

Shetty SS, Deepthi D, Harshitha S, Sonkusare S, Naik PB, Kumari NS, Madhyastha H (2023) Environmental pollutants and their effects on human health. Heliyon 9(9):e19496. https://doi.org/10.1016/j.heliyon.2023.e19496

Shin H-J, Ro H-S, Kawauchi M, Honda Y (2025) Review on mushroom mycelium-based products and their production process: from upstream to downstream. Bioresour Bioprocess 12(1):3. https://doi.org/10.1186/s40643-024-00836-7

Shofia SI, Vickram AS, Saravanan A, Deivayanai VC, Yaashikaa PR (2025) Sustainable separation technologies for heavy metal removal from wastewater: an upgraded review of physicochemical methods and its advancements. Sustain Chem Environ 10:100264. https://doi.org/10.1016/j.scenv.2025.100264

Shraddha, Shekher R, Sehgal S, Kamthania M, Kumar A (2011) Laccase: microbial sources, production, purification, and potential biotechnological applications. Enzyme Res 2011(1): 217861. https://doi.org/10.4061/2011/217861

Singh AK, Abellanas-Perez P, de Andrades D, Cornet I, Fernandez-Lafuente R, Bilal M (2025) Laccase-based biocatalytic systems application in sustainable degradation of pharmaceutically active contaminants. J Hazard Mater 485:136803. https://doi.org/10.1016/j.jhazmat.2024.136803

Sodaneath H, Lee JI, Yang SO, Jung H, Ryu HW, Cho KS (2017) Decolorization of textile dyes in an air-lift bioreactor inoculated with Bjerkandera adusta OBR105. J Environ Sci Health 52(11): 1099–1111. https://doi.org/10.1080/10934529.2017.1340753

Subramanian V, Yadav JS (2009) Role of P450 monooxygenases in the degradation of the endocrine-disrupting chemical nonylphenol by the white rot fungus *Phanerochaete chrysosporium*. Appl Environ Microbiol 75(17):5570–5580. https://doi.org/10.1128/AEM.02942-08

Suryadi H, Judono JJ, Putri MR, Eclessia AD, Ulhaq JM, Agustina DN, Sumiati T (2022) Biodelignification of lignocellulose using ligninolytic enzymes from white-rot fungi. Heliyon 8(2):e08865. https://doi.org/10.1016/j.heliyon.2022.e08865

Tamayo-Ramos JA, van Berkel WJH, de Graaff LH (2012) Biocatalytic potential of laccase-like multicopper oxidases from *Aspergillus niger*. Microb Cell Factories 11:165. https://doi.org/10.1186/1475-2859-11-165

Tang H, Xiang G, Xiao W, Yang Z, Zhao B (2024) Microbial mediated remediation of heavy metals toxicity: mechanisms and future prospects. Front Plant Sci 15:1420408. https://doi.org/10.3389/fpls.2024.1420408

Tham LX, Matsuhashi S, Kume T (1999) Responses of *Ganoderma lucidum* to heavy metals. Mycoscience 40(2):209–213. https://doi.org/10.1007/BF02464301

Thirumalaivasan N, Gnanasekaran L, Kumar S, Durvasulu R, Sundaram T, Rajendran S, Nangan S, Kanagaraj K (2024) Utilization of fungal and bacterial bioremediation techniques for the treatment of toxic waste and biowaste. Front Mater 11:1416445. https://doi.org/10.3389/fmats.2024.1416445

Tran NH, Urase T, Kusakabe O (2010) Biodegradation characteristics of pharmaceutical substances by whole fungal culture *Trametes versicolor* and its laccase. J Water Environ Technol 8(2): 125–140. https://doi.org/10.2965/jwet.2010.125

Tu Y, Han L, Ding X, Zhu X, Xu G, Xie X, Zhou Q, Shuang C, Li A (2025) Degradation of pharmaceuticals and personal care products (PPCPs) in municipal wastewater by white rot fungi: neutral environment exploitation and degradation mechanisms. Chem Eng J 507:160794. https://doi.org/10.1016/j.cej.2025.160794

Tuisel H, Sinclair R, Bumpus JA, Ashbaugh W, Brock BJ, Aust SD (1990) Lignin peroxidase H2 from *Phanerochaete chrysosporium*: purification, characterization and stability to temperature and pH. Arch Biochem Biophys 279(1):158–166. https://doi.org/10.1016/0003-9861(90)90476-F

Twala PP, Mitema A, Baburam C, Feto NA (2020) Breakthroughs in the discovery and use of different peroxidase isoforms of microbial origin. AIMS Microbiol 6(3):330–349. https://doi.org/10.3934/microbiol.2020020

Vaithanomsat P, Apiwatanapiwat W, Petchoy O, Chedchant J (2010) Decolorization of reactive dye by white-rot fungus *Datronia* sp. KAPI0039. Kasetsart J Nat Sci 44(5):879–890

Vaksmaa A, Guerrero-cruz S, Ghosh P, Zeghal E, Hernando-Morales V, Niemann H (2023) Role of fungi in bioremediation of emerging pollutants. Front Mar Sci 10:1–21. https://doi.org/10.3389/fmars.2023.1070905

Varela E, Martínez AT, Martínez MJ (2000) Southern blot screening for lignin peroxidase and aryl-alcohol oxidase genes in 30 fungal species. J Biotechnol 83(3):245–251. https://doi.org/10.1016/S0168-1656(00)00323-0

Vidu R, Matei E, Predescu AM, Alhalaili B, Pantilimon C, Tarcea C, Predescu C (2020) Removal of heavy metals from wastewaters: a challenge from current treatment methods to nanotechnology applications. Toxics 8(4):101. https://doi.org/10.3390/toxics8040101

Viswanath B, Rajesh B, Janardhan A, Kumar AP, Narasimha G (2014) Fungal laccases and their applications in bioremediation. Enzyme Res 2014:163242. https://doi.org/10.1155/2014/163242

Wainaina S, Taherzadeh MJ (2023) Automation and artificial intelligence in filamentous fungi-based bioprocesses: a review. Bioresour Technol 369:128421. https://doi.org/10.1016/j.biortech.2022.128421

Wang Y, Tan R, Zhou L, Lian J, Wu X, He R, Yan F, He X, Zhu W (2021) Heavy metal fixation of lead-contaminated soil using Morchella mycelium. Environ Pollut 289:117829. https://doi.org/10.1016/j.envpol.2021.117829

Wang P, Feng X, Lv Z, Liu J, Teng Q, Chen T, Liu Q (2024) Temporal dynamics of lignin degradation in *Quercus acutissima* sawdust during *Ganoderma lucidum* cultivation. Int J Biol Macromol 268(P1):131686. https://doi.org/10.1016/j.ijbiomac.2024.131686

Wibisono DAS, Saw C-Y, Wu T-Y, Chau C-F (2025) Advancing industrial food byproduct management: strategies, technologies, and progress in waste reduction. Processes 13(1):84. https://doi.org/10.3390/pr13010084

Yang S-O, Sodaneath H, Lee J-I, Jung H, Choi J-H, Ryu HW, Cho K-S (2017) Decolorization of acid, disperse and reactive dyes by *Trametes versicolor* CBR43. J Environ Sci Health A 52(9):862–872. https://doi.org/10.1080/10934529.2017.1316164

Yang L, Park D, Qin Z (2021) Material function of mycelium-based bio-composite: a review. Front Mater 8:737377. https://doi.org/10.3389/fmats.2021.737377

Yesilada O, Asma D, Cing S (2003) Decolorization of textile dyes by fungal pellets. Process Biochem 38(6):933–938. https://doi.org/10.1016/S0032-9592(02)00197-8

Yesilada O, Yildirim SC, Birhanli E, Apohan E, Asma D, Kuru F (2010) The evaluation of pre-grown mycelial pellets in decolorization of textile dyes during repeated batch process. World J Microbiol Biotechnol 26(1):33–39. https://doi.org/10.1007/s11274-009-0138-8

Yun J, Yang Q, Liu G (2025) Mechanisms of lignin degradation and persistent free radical formation under light or thermal exposure. Cell Rep Sustain 2(1):100267. https://doi.org/10.1016/j.crsus.2024.100267

Zahuri AA, Abdul Patah MF, Kamarulzaman Y, Hashim NH, Thirumoorthi T, Wan Mohtar WHM, Mohd Hanafiah Z, Amir Z, Wan-Mohtar WAAQI (2023) Decolourisation of real industrial and synthetic textile dye wastewater using activated dolomite. Water 15(6):1172. https://doi.org/10.3390/w15061172

Zahuri AA, Wan Mohtar WHM, Hanafiah ZM, Abdul Patah MF, Show PL, Gafforov Y, Wan-Mohtar WAAQI (2024) Mycoremediation of industrial textile wastewater using *Ganoderma lucidum* pellets and activated dolomite in batch bioreactor. Mol Biotechnol. https://doi.org/10.1007/s12033-023-01035-z

Zhang W, Li Q, Wang J, Wang Z, Zhan H, Yu X, Zheng Y, Xiao T, Zhou LW (2023) Biodegradation of benzo[a]pyrene by a white-rot fungus *Phlebia acerina*: surfactant-enhanced degradation and possible genes involved. J Fungi 9(10):978. https://doi.org/10.3390/jof9100978

Zhao R-L (2023) Edible and medicinal macrofungi. J Fungi 9(9):908. https://doi.org/10.3390/jof9090908

Ziani K, Ioniță-Mîndrican C-B, Mititelu M, Neacșu SM, Negrei C, Moroșan E, Drăgănescu D, Preda O-T (2023) Microplastics: a real global threat for environment and food safety: a state of the art review. Nutrients 15(3):617. https://doi.org/10.3390/nu15030617

# Eco-friendly Leather from Mushrooms

# 15

Tamilarasi Sambu Periyasamy,
Leena Catherine Anthonimuthu Francis Xavier, Jegadeesh Raman,
Suresh Balasubramanian, and Hariprasath Lakshmanan

**Abstract**

The rising environmental concerns linked to the traditional leather-making process, like overconsumption of water, pollution with chemicals, and the killing of animals for their skins, have convinced the manufacturers to look for more earth-friendly replacements. What is even more interesting is that one new product is leather coming from mushrooms, which is often called "mycelium leather." Mycelium, which is the root part of the fungus can be cultivated in labs to produce a strong, soft, and compostable material that has similar properties to leather in terms of feel and strength. A big difference between animal leather and mushroom leather is that the latter does not require that much land, water, and time and does not bring along the moral dilemmas associated with cattle breeding. Moreover, the production does not involve the toxic leather tanning processes common in the synthetic or animal skin industry. Recent advancements in material science and biotechnology can help companies to make the scalability, durability, and beauty of mushroom leather a lot better, making it suitable for use

Tamilarasi Sambu Periyasamy and Leena Catherine Anthonimuthu Francis Xavier contributed equally with all other contributors.

T. S. Periyasamy · H. Lakshmanan (✉)
Division of Biochemistry, School of Life Sciences, Ooty Campus, JSS Academy of Higher Education and Research, Longwood, Ooty, The Nilgiris, Tamil Nadu, India
e-mail: hariprasath80@jssuni.edu.in

L. C. A. F. Xavier · S. Balasubramanian
Division of Microbiology, School of Life Sciences, Ooty Campus, JSS Academy of Higher Education and Research, Longwood, Ooty, The Nilgiris, Tamil Nadu, India

J. Raman
Agricultural Microbiology Division, National Institute of Agricultural Sciences, Rural Development Administration, Wanju, Jeollabuk, Republic of Korea

 
B. Giri et al. (eds.), *Advances in Mushroom Research*,
https://doi.org/10.1007/978-981-95-7692-0_15

in fashion, furniture, and automotive industries. Many new companies and big international brands are spending money on both the research and the commercialization of mushroom leather, making it a strong competitor in the industry of sustainable materials. There are still some hurdles to overcome, such as mass production, cost, and customer acceptance. Mushroom leather, in general, is a game-changing move toward green innovation that provides a fit of performance, sustainability, and ethical accountability.

**Keywords**

Mushroom leather · Mycelium · Eco-friendly materials · Sustainable alternatives · Biodegradable leather

## 15.1 Introduction

The international leather industry has been associated with significant environmental and ethical concerns for many years. International knowledge of environmental protection and animal treatment has become larger in recent times, as has the demand for vegan alternatives and a cut in plastic usage. There is a drive to substitute animal, fungal, and plant materials (Shin et al. 2025). Conventional leather production is highly dependent on animal farming, which leads to deforestation, water wastage, methane production, and biodiversity loss (Daddi et al. 2024). Mycelium, the vegetative fungal network of fibers, is quickly becoming a viable, sustainable substrate to produce leather-like biomaterials. In contrast to animal hide (highly associated with greenhouse gas emissions, land use, water usage, and toxic tanning substances) and most synthetic "vegan" leathers (usually plastic-based and nonbiodegradable) (Wang 2022), mycelium is cultivable from agricultural side-streams on comparatively low energy inputs and results in a predominantly biological, biodegradable matrix with inherent fibrous texture (Xu et al. 2025; Alemu et al. 2022).

The attractiveness of mycelial leather is due to biology and process versatility (Raman et al. 2022). Various fungal species, particularly most Basidiomycetes like Ganoderma, Trametes, and Pleurotus, create hyphae with different diameters, branching, and density properties, directly affecting the mechanical behavior, surface texture, and water interaction of harvest mats. Choosing strains and adapting growth conditions, substrate type, incubation time, and fermentation mode, etc., allow designers to control end-product properties (Elsacker et al. 2023; Jones et al. 2018a). Several growth strategies, such as solid surface growth, liquid-surface (static) culture, and submerged bioreactor production of fungal mass followed by mat formation, have been developed. Solid surface growth tends to provide more rapid mat formation and improved structure, and liquid methods can expedite harvest and scaling, and submerged fermentation can provide large biomass but calls for extra shaping or post-processing (Crawford et al. 2024). Patents and recent publications catalog these processes and their industrial variations (Stelzer et al. 2021).

Post-processing lies at the heart of transforming fragile mycelial mats into tough, leather-like sheets. Post-treatments documented in the literature are plasticizers (glycerol), crosslinking reagents (phenolics/tannins, salts), controlled drying and heat-pressing, and tanning or coating processes to enhance tensile strength, abrasion resistance, and water repellency (Jones et al. 2021). Optimization of post-treatment is a critical factor determining matching performance with intended applications like bags, footwear uppers, and upholstery (Manan et al. 2022; Ariyani et al. 2024). Green claims regarding mycelial leathers are promising but require lifecycle examination. Some operations and companies seem low-impact since they transform waste feedstocks into biomaterials and do not involve livestock emissions; however, some scale-up practices (e.g., $CO_2$-enriched growth chambers, post-treatments that are solvent-intensive) can increase carbon and toxicity footprints (Volk et al. 2024; Livne et al. 2022). Recent patent and lifecycle assessments require transparency and LCA comparisons based on standardized standards against animal and PU/TPU leathers (Kowalska et al. 2013).

Mechanical performance continues to be an issue in the key materials science. Pure mycelial sheeting tends to be lower in tensile strength and toughness compared with animal hide or industrially engineered synthetics (Benetti et al. 2025). Various approaches to enhance mechanical properties include multi-layer growth (re-growing hyphae in various directions), composite methods (mycelium grown using woven fiber scaffolds), chemical crosslinking, heat-press densification, and blended formulations (Saez et al. 2024). Studies indicate that by combining attentive strain selection, controlled cultivation, and aimed post-treatment processes, one can obtain materials suitable for limited to moderate structural requirements. Scalability and reproducibility are live engineering challenges. The production of uniform mats at commercial scales necessitates constant spawn, aseptic manufacturing, oxygen/$CO_2$-controlled regimes, and either a constant or well-standardized batch process. Scale engineering is now a leading commercial frontier, as evidenced by the intense private-sector activity in the patent literature dedicated to growth chamber design, substrate treatment, and automated harvesting (Ariyani et al. 2024).

Apart from performance and extent, regulatory and safety issues must be confronted. Genetic modification of strains used carries over bioactive fungal metabolites, occupational spore exposure, and chemical post-treatment fate (dyes, fixatives); these are all issues mentioned in recent critiques. Authors suggest extensive testing of toxicity, biodegradation, and supply-chain transparency prior to mass commercial use (Jones et al. 2021). The product and design possibilities for leather made from mushrooms may be useful for fashion accessories and home textiles, low-load footwear and car interiors, particularly where biodegradability and low ecological impact are promoted as key values. Designers are working with natural pigments, embossing, and mixed assemblies (mycelium + textile backing) to balance form and function. Branding and consumer acceptance will be the key drivers to adopting the mycelium-based leather products (Mallon 2024). Overall, mycelium-based mushroom leather is a rapidly converging field of microbiology, materials science, design, and supply-chain engineering.

This book chapter has transitioned from proof-of-concept to patents, startups, and peer-reviewed papers. The remaining gaps are in standardized performance metrics, open lifecycle assessments, and scalable low-impact post-processing. The ongoing interdisciplinary efforts merging strain engineering, fermentation design, harmless chemistry for finishing, and transparent lifecycle reporting will convey whether mushroom leather is headed for the masses as a sustainable leather replacement.

## 15.2 Fungi Used for Leather Production

The sustainable materials transition in the global fashion and textile industries is in its early stages. This transition has been influenced by heightened environmental awareness, consumer awareness of ethical issues associated with conventional leather production, and the associated priorities of the global economy. In this regard, mycelium-based leather, formed by creating a leather-like substance from a fungal network, serves as a carbon-neutral, biodegradable, cruelty-free substitute for traditional leather (Jones et al. 2021). Advancements in technology within biotechnology and materials science have improved scalability and durability as a commercially viable textile substitute for leather. Consequently, mycelium leather would be regarded as a groundbreaking technology or innovation for the eco-friendly elements of fashion, automotive interiors, and luxury products (Williams et al. 2022).

### 15.2.1 Mycelium: The Substrate

Mycelium is a huge, interconnected matrix of tubular cells, called hyphae. In nature, this network will secrete enzymes to decompose organic matter and is essentially the earth's recycler (Song et al. 2025; Abdulhusein and Kadim 2024). To create material, this growth is harnessed by inoculating fungi on a substrate, usually agricultural waste such as sawdust, straw, or corn husks, that has a stable carbon-based nutrient source. This inoculated substrate is then incubated in a controlled environment (Wijayarathna et al. 2021; Ariyani et al. 2024). The mycelium grows, consumes nutrients, and binds the substrate particles into a thick three-dimensional mat. Once harvested, dried, and processed (often through compression and some other type of tanning), this mat becomes a durable bio-leather. The features of this leather, such as thickness, pliability, tensile strength, and textile properties, are determined by three key factors, including the fungal species, composition of substrates, and the growth conditions (Ugbede et al. 2023). Commonly used fungal species include *Agaricus bisporus*, *Fomes fomentarius*, *Ganoderma lucidum*, *Phellinus ellipsoideus*, and *Pleurotus ostreatus* (Raman et al. 2022; Abdulhusein and Kadim 2024).

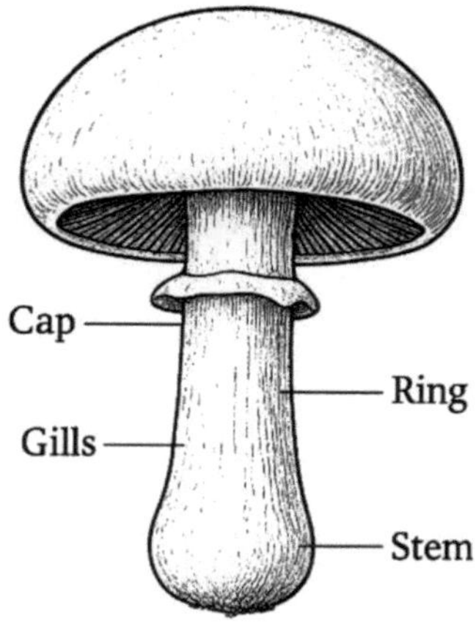


**Fig. 15.1** *Agaricus bisporus*

## 15.2.2 *Agaricus bisporus*

*Agaricus bisporus*, or button mushroom, is the most consumed and cultivated edible mushroom species worldwide (Chatterjee et al. 2023) (Fig. 15.1). As a member of the family *Agaricaceae* fungi, it can be grown and harvested, and the same mushrooms can be eaten from existing extraction with endless variability. Commercially, it is grown largely in controlled situations where the environmental conditions are most suited to growing on composted organic substrate like straw, chicken manure, and gypsum (Zhang et al. 2024). It exhibits three different forms: the immature button stage, where the cap is small and has not separated; the brown-cremini stage; and the fully mature portobello with an open, dark brown cap and rich meat-like texture (Lin and Sun 2019). Nutritionally, this is a food source high in protein, dietary fiber, B vitamins, selenium, and antioxidants and low in calories and fat. Its application in bio-leather is truly novel and revolves around the concept of waste valorization (Ramos et al. 2019).

The residual material from *A. bisporus* (Spent Mushroom Substrate, or SMS) is thrown away after the mushroom fruit has been harvested. This agricultural waste, abundant in mycelium, is collected, processed (cleaned, pressed, and dried), and becomes a heavy, flat sheet. All the SMS that goes into the process is turning the compost into a material that is heavy, strong, coarse, natural, and generally very strong, more amenable to use in rigid accessories, packaging, or job site panels rather than soft, flexible clothing. This has the potential to create a unique model for a circular economy by upcycling agricultural waste into a sustainable material again (Baars et al. 2024).

## 15.2.3 *Fomes fomentarius*

*Fomes fomentarius*, also known as the tinder fungus or hoof fungus, is not only a biological novelty, but also a species of great historical and developing industrial

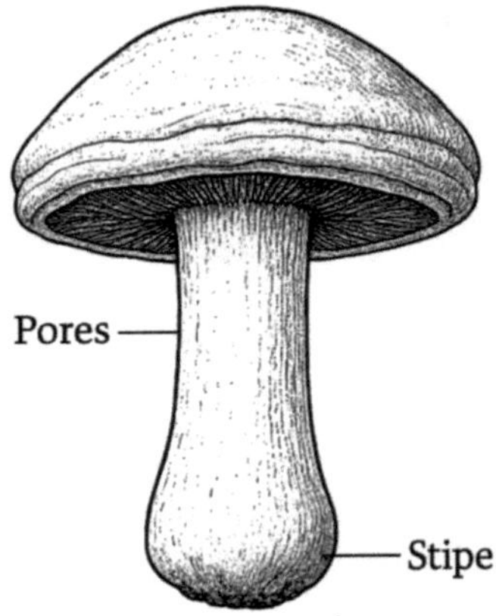


**Fig. 15.2** *Fomes fomentarius*

importance with regard to the sustainable production of bio-leather (Fig. 15.2). *Fomes fomentarius* is usually found growing like a shelved hoof on the rotting bark of birch and beech trees deep within Northern forests. For thousands of years, humans have utilized the fungus for many purposes, and in its most famous application, it was intended as tinder since the dried inner tissue of the fungus's fruiting body, or amadou, would easily catch fire from a spark. However, amadou was also changed into a strong, felt-like, durable textile for clothing accessories; this technique's historic context has been renewed through scientific innovations, positioning *Fomes fomentarius* as the prototype for fungal bio-leather development (Chen et al. 2021).

Fomes fomentarius exhibits several characteristics, as a species and as a fungus, that are favorable for bio-leather due to its complicated biology. It is important to note that *F. fomentarius* differs from many fungi harvested for use in mycelium-based materials since those rely solely on the vegetative mycelial network grown on substrates (Kalitukha et al. 2023). *F. fomentarius* has been harvested primarily for the sterile tissue of its fruiting body: amadou. Amadou is comprised of a dense and convoluted conglomerate of hyphae that naturally has resilient, fibrous, and light qualities (Müller et al. 2021). When harvested, dried, and processed by soaking, pounding, stretching, etc., amadou can take on flexible, leather-like properties. Historically, harvesters have used manual labor to process amadou, but its properties can now be further developed and standardized through more controlled laboratory processes, while still maintaining those same properties and a high-quality end product (Xia et al. 2021).

### 15.2.4 *Ganoderma lucidum*

*Ganoderma lucidum*, or the *reishi* mushroom, has been considered for medicinal purposes in traditional medicine for centuries (Fig. 15.3). But beyond its medicinal qualities, this fungus is now being developed into a sustainable material, particularly

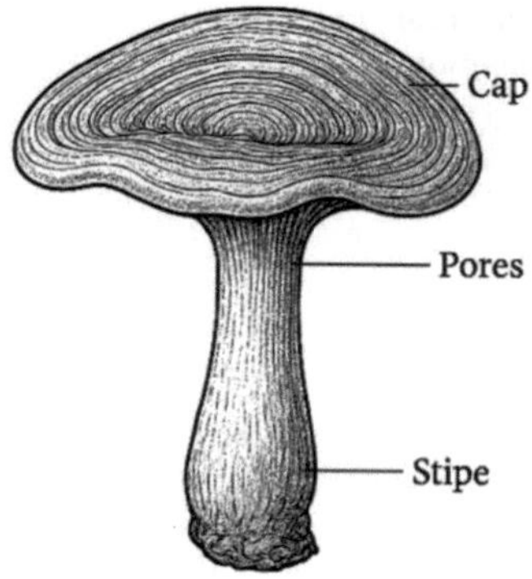


**Fig. 15.3** *Ganoderma lucidum*

bio-leather (Chang et al. 2006). Not only does *G. lucidum* bio-leather offer a sustainable solution to the waste created by the fashion and manufacturing industries by being both environmentally sustainable and fully biodegradable, but also, like *G. lucidum*, the bio-leather does not come from an animal like traditional leather and is an ethical alternative to synthetic leathers. The biological properties of *G. lucidum* make this fungus an ideal candidate to make a high-quality material that may replicate some degree of the durability, quality, touch, and flexibility of traditional leather (Hennicke et al. 2016).

*G. lucidum's* robust and dense mycelial network is one of the most attractive attributes of this organism as a source of bio-leather (Porter et al. 2023). The mycelial network allows for a very rapid and consistent growth; therefore, it is possible to produce bio-leather sheets in large pieces that are all consistent in composition. When woven and combined into a sheet, *G. lucidum's* mycelium exhibits favorable characteristics that are desirable in a robust vertical or textile (Solanilla-Duque et al. 2021). Depending on the substrate and processing methods used, *G. lucidum* can replicate many leather textures ranging from completely silky smooth surfaces to soft suede textures. Its intrinsic porosity allows for natural dyes to buffer, providing many design options where chemicals are not involved (Luangharn et al. 2019; Susel et al. 2021).

Companies like MycoWorks have successfully commercialized *G. lucidum* into bio-leather specifically for high-end applications. Through promotion, *G. lucidius* is marketed as a luxury good able to compete with high-end animal leathers in the textile and leather markets. MycoWorks and their patented technologies, like Fine Mycelium™, allow for precise engineering of the mycelium to increase density, strength, and uniform growth (Williams et al. 2022).

### 15.2.5 *Phellinus ellipsoideus*

*P. ellipsoideus* is a white-rot fungus, indicating it can produce large quantities of biomass quickly, and produce decomposer enzymes that can decompose wood lignin and cellulose (Fig. 15.4). The fruiting bodies of *P. ellipsoideus* can be enormous,

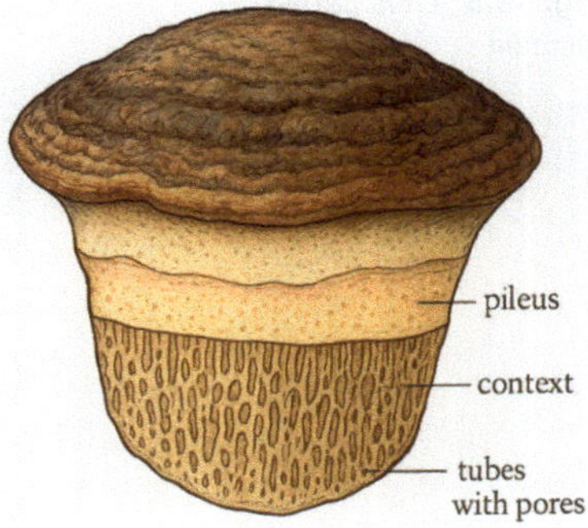


**Fig. 15.4** *Phellinus elllipsoideus*

with some measuring as much as ten or more meters in length and weighing hundreds of kilograms, reflecting the ability of this fungus to produce biomass at very high rates when conditions are right (Thanh et al. 2017). *P. ellipsoideus* grows through a dense and rapidly colonizing mycelial network that can propagate largely on lignocellulosic substrates, including wood chips, sawdust, and agricultural residue. This growth aspect is important for our *P. ellipsoideus* bio-leather process, for as it grows and colonizes the substrate, it binds the substrate together into a compact structure, where the mycelium acts as a binder and, in the densification of the material, creates a material that could be processed into leather-like sheets.

*Phellinus ellipsoideus* is not yet commercially available, but this experimental fungus has a surprising possibility for bio-leather made from it, based on its impressive ability to generate large masses of biomass—as evidenced by its record-breaking fruit bodies. The process would utilize the fungi's strong mycelium grown on either wood chips or agricultural waste, which consumes and binds this organic matter into a thick, cohesive mat. This mat would be processed after harvest by some combination of heat-pressing, drying, and possibly natural tanning to form a sheet material with properties specifically associated with leather (Bustillos et al. 2020). Likely, the bio-leather would have some significant thickness, a robust durability, and a natural, organic texture from the woody substrate, which probably would not be right for supple clothing but would be good for anything with a structure like bags, shoes, or panels for architecture.

### 15.2.6 *Pleurotus ostreatus*

*Pleurotus ostreatus*, commonly known as the oyster mushroom, is an excellent organism to produce bio-leather (Fig. 15.5). They have a unique growth habit that results in highly branched and web-like mycelium that grows quickly and completely integrates a range of agricultural wastes, including straw, coffee grounds, and sawdust. As the fungus grows, it integrates these materials into a mass that is solid and in sheet form with its mycelium. At harvest, this mycelium biomass can be

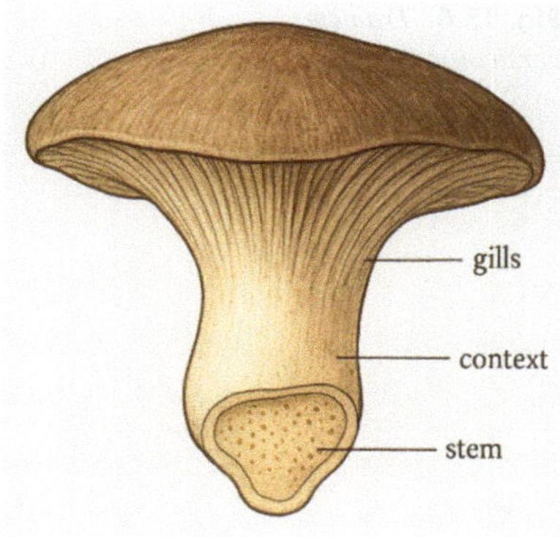


**Fig. 15.5** *Pleurotus ostreatus*

harvested, heat-pressed together, and dried after the fungus is heat-killed. It is a very bio-leather at this point (El-Ramady et al. 2022).

Bio-leather made with oyster mushroom mycelium will typically be light in weight and flexible with a soft touch like soft suede. It has been popular for some fashion accessories, apparel, and upholstery. The production of bio-leather using oyster mushroom mycelium is an incredibly sustainable and completely biodegradable product that generates less carbon impact to produce since it converts low-value waste (Melanouri et al. 2021). Currently, bio-leather is not as rigid as other leathers that have been derived from fungi like Ganoderma, but the research continues! Exploration of substrate blends and post-processing continues to improve the function of strength, water resistance, and function. *Pleurotus ostreatus* bio-leather is an excellent, low-cost, accessible, and scalable product for material development that is aligned with a growing market for ethical and environmentally sustainable solutions (Raman et al. 2020).

### 15.2.7 *Trametes versicolor*

*Trametes versicolor* (turkey tail) is valuable in the production of robust, fibrous mycelial mats that can be processed into environmentally friendly leather alternatives (Fig. 15.6). The mycelium of *Trametes versicolor* gives rise to a dense interwoven structure with high binding ability, which is ideal for making flexible sheets. With proper treatment, these sheets are used in the production of handbags, purses, wallets, and belts, providing a green alternative to animal and synthetic leathers (Karthikeyan et al. 2024; Yang et al. 2017). The fungus's natural enzymatic activity not only provides strong materials but also enhances the compression and surface treatment's structural integrity. Mycelium composites from T. versicolor have been shown by researchers to be resistant to everyday use (Kniep et al. 2024). Therefore, they can be used in consumer products like handbags and designer accessories. The composite material provides mechanical strength while still being biodegradable; thus, it is a very good green alternative for fashion and lifestyle

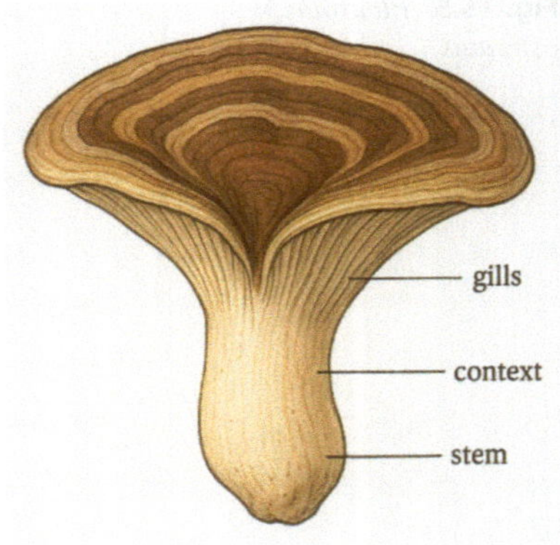


**Fig. 15.6** *Trametes versicolor*

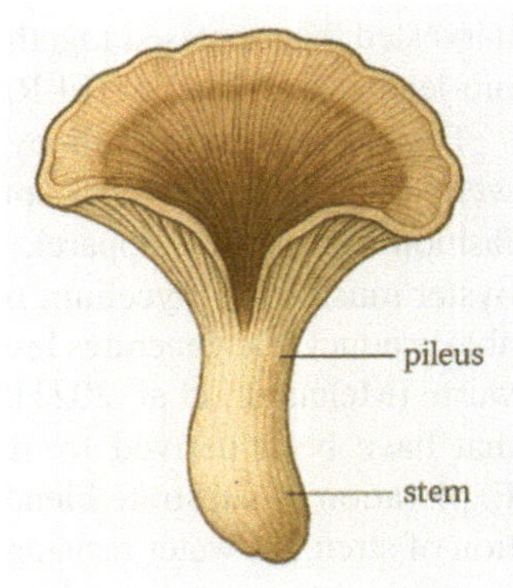


**Fig. 15.7** *Schizophyllum*

products (Pawłowicz et al. 2024). T. versicolor leathers are being considered for applications in carpeting, lining of shoes, and eco-friendly packaging, besides being used in fashion. Fungal leather startups draw attention to the visual appeal of *T. versicolor* products since they can be dyed, embossed, or treated to resemble conventional leather. With such features, the material can be viewed as a significant contributor to the sustainable handbag production and circular economy initiatives (Shen et al. 2024).

### 15.2.8 *Schizophyllum*

The *split gill fungus* (*Schizophyllum* commune) has been the subject of extensive studies because of its ability to produce a polysaccharide called schizophyllan that improves mycelium mat strength and flexibility (Fig. 15.7). The thick mycelial growth of S. commune forms delicate yet strong sheets, which can be converted into sustainable leather materials. The scientists observed that S. commune could be cultivated on agricultural waste that was about to be thrown away, thus lessening the waste and making the production of biomaterials affordable (d'Errico et al. 2025).

Interestingly, the leather-like sheets made from S. commune have also been evaluated for fashion items, mainly handbags, wallets, and other accessories. The material's lightness and strength give it the perfect characteristics for such products; on the other hand, finishing indeed allows the material to be embossed, dyed, and textured to look like leather (Benetti et al. 2025). Besides, its ability to break down naturally is a major plus for the product when compared with the petroleum-based synthetics that are usually applied in accessories and lifestyle items, as these are not degradable. The fashion world is not the only one interested in S. commune mycelium composites; they are also being tested for shoe parts, upholstery, and even durable packaging. The fungus, along with its durability and water-repellency features, provides the mission of giving functional use to luxury and daily products alike. Such versatility has resulted in the industries' willingness to pilot-scale produce new material innovations such as eco-leather products, thereby placing S. commune at the center of attention as a species worth investing in for the purpose of upscaling sustainable material innovation (Meyer et al. 2020; Jiang 2022).

### 15.2.9 *Fomitella fraxinea*

*Fomitella fraxinea*, a polypore fungus and brown-rot type, appeals in the production of bio-based material due to its mycelium networks, which are very dense and highly homogeneous. The by-product of its cultivation on lignocellulosic residues is mats of high structural integrity that can be used as biofabrics. There are recent studies that show *F. fraxinea* mycelium is among the most stable and durable as compared to other white-rot fungi that are less rigid; hence, the use of this fungus in eco-leather is very promising (Benetti et al. 2025; Raman et al. 2022). The resulting leather-like texture from *F. fraxinea* sheets is usable for footwear, handbags, and upholstery materials. The resistance and toughness of its mycelium composites make them suitable for items that go through repeated wear and environmental stresses. Its property of thick, layered mat development also allows for the customization of material thickness according to product specifications, from soft purses to stiffened shoe soles (Whabi et al. 2024). Apart from fashion, the use of *F. fraxinea*-based bio-composites is being investigated for automotive interior components, furniture upholstery, and sustainable architectural panels. The fungus's brown-rot metabolism further assists in the production of composites that imitate some of the natural leather's tactile and textural properties and also have the feature of being biodegradable. Thus, *F. fraxinea* is seen as the future species in the quest for sustainable mycelium-based leather substitutes (Sharma et al. 2024; Kumar et al. 2025).

### 15.2.10 *Pleurotus djamor*

*Pleurotus djamor*, referred to as the pink oyster mushroom, is a fungus species with great potential to produce sustainable materials. The mycelium of the fungus develops into a network of fibrous mats that are interconnected, flexible, strong, and light, making them suitable for bio-based leather alternatives (Crawford et al. 2024). *P. djamor* is faster in occupying agricultural residues compared to other Pleurotus species, thus allowing the low-cost and scalable production of mycelium composites. Researchers have been focusing on *P. djamor* for making products such as handbags, wallets, footwear lining, and foam packages, where durability and appearance are very important. Also, the natural pigmentation of *P. djamor* helps to lessen the dye consumption, therefore making the processing eco-friendlier. The mycelium-based composites of P. djamor have a very light weight but high tensile strength, which makes it very attractive for use in fashion fabrics and green accessories (Kapoor et al. 2024). *P. djamor* mycelium is being used outside of fashion for furniture upholstery, acoustic panels, and interior design material, combining functionality and biodegradability. The feature of being compatible with bio-based polymers not only adds extra thermal and water resistance but also enlarges the area of application for long-lasting consumer products. To sum up, *P. djamor* demonstrates how fungal diversity can be exploited to achieve the circular economy's objectives and to develop new material solutions for sustainable leather (Gutarowska et al. 2024).

### 15.2.11 *Lentinula edodes*

*Lentinula edodes*, or the shiitake mushroom, is among the most cultivated edible fungi globally, but its mycelium has also attracted interest as a biomaterial for alternative leathers. The compact, interlaced mycelial mats of *L. edodes* possess tensile strength, flexibility, and permeability, properties that are necessary for the development of sustainable leather-like composites (Shin et al. 2025). Besides, its ability to be adapted to lignocellulosic agri-residues renders it an inexpensive substrate for sustainable production systems. *L. edodes* mycelium is used as a leather substitute in manufacturing handbags, wallets, belts, and shoe linings where strength and beauty are required (Baars et al. 2024). Its fibrous microstructure from the natural material provides toughness, with surface coatings enhancing water repellency and suppleness to facilitate comparison with animal leather. Other startups and research initiatives are also using *L. edodes* in upholstery materials and fashion accessories, focusing on biodegradability and a lower environmental footprint relative to polyurethane-based synthetic leathers (Singh et al. 2025). In addition, the combination of *L. edodes* mycelium with biodegradable polymers improves its performance in terms of heat stability and elasticity, so it is appropriate for high-performance markets, including furniture, automotive interiors, and green textiles. As consumers increasingly seek to enhance sustainable awareness, *L. edodes* is not only valued for its edible and medicinal properties but also for its circular economy use, providing an alternative, renewable, and biodegradable material to traditional leather (Lu et al. 2025).

### 15.2.12 *Flammulina velutipes*

*Flammulina velutipes*, the enoki mushroom, is a broadly cultivated edible mushroom, but its mycelium has been newly investigated as a renewable raw material for bio-based alternatives to leather. The close and elastic mycelial web of *F. velutipes* can be grown on numerous lignocellulosic substrates to produce mats that have the properties of animal leather in strength, softness, and breathability (Shin et al. 2025). Its potential for growth under controlled fermentation conditions also qualifies it for scalable biomanufacturing of leather-like materials. The mycelium of *F. velutipes* has been explored for manufacturing fashion accessories such as handbags, wallets, belts, and shoes (Bhatia and Katoch 2022). Scientists have demonstrated that following chemical and mechanical treatment, composites of *F. velutipes* mycelium have enhanced durability, water resistance, and elasticity, with improved performance in consumer products. Also, surface engineering allows for tailor-made textures and colors, facilitating their incorporation into luxury fashion and upholstery products (Wang 2022). Outside fashion, *F. velutipes*-derived mycelium leather is also being considered for sustainable packaging, furniture covers, and automotive interior applications, pointing toward its application in a circular economy. With respect to synthetic leather made from petroleum and not biodegradable, *F. velutipes* composites show lower carbon imprints and compostability, which are concomitant with international sustainability objectives (Angelova et al. 2021; Im et al. 2025). Its double worthiness as a food and material-generating fungus further solidifies its place as a pivot in emerging bio-based industries.

### 15.2.13 *Cerioporus squamosus*

*Cerioporus squamosus*, also commonly referred to as Saddle, is a basidiomycete fungus that creates large fruiting bodies in hardwoods. New studies have emphasized its mycelial growth potential as a strong potential candidate for bio-based materials, such as eco-friendly leather-like composites (Iordache et al. 2023). The mycelium of this fungus produces strong interlaced hyphae and thus forms thick mats that are suitable for flexible materials due to the mechanical strength they provide. In this respect, the fungus *C. squamosus* can be considered quite a good competitor compared to its other fungal peers; it will thus be more convenient to use this mushroom in large-scale bio-fabrication. The mycelium of *C. squamosus* has already been tested in the production of such fashion items as bags, wallets, and belts, where not only the elongation at break but also the surface finish is of utmost importance (Baars et al. 2024). When the material is subjected to treatment with eco-friendly adhesives and alternatives to tannins, its durability and water resistance are greatly enhanced, making it a practical replacement for synthetic leathers. The aesthetic of *C. squamosus* mycelium has also opened doors for different product finishes that are in demand by luxury and sustainable fashion markets. Besides accessories, the researchers are looking at its application in upholstery, shoes, and interior decorative items as well. Because of the high rate of growth and the generation of a large amount of biomass, it has been proposed that it could be used to produce biodegradable composites and light structural materials. *C. squamosus* leather-like materials have notably been associated with a lower carbon footprint as compared to petroleum-derived synthetics, thus aligning with circular economy and sustainability goals (Iordache et al. 2023).

### 15.2.14 *Bjerkandera*

*Bjerkandera* species, especially *Bjerkandera adusta*, are ligninolytic white-rot fungi that are known for lignin degradation and efficient growth on lignocellulosic substrates. They are great candidates for biomaterial production, such as leather-like materials, due to their strong mycelial colonization and tight mat development. Mechanical stability and flexibility are introduced to the interwoven hyphal networks of *Bjerkandera*, features important to produce alternatives to animal leather (Wang 2022). Experimental research has demonstrated that *Bjerkandera*-based mycelium composites can be converted into sustainable leather alternatives with post-processing methods like pressing, tanning with natural agents, and bio-based coatings. These products have been applied in the manufacture of wallets, belts, and bags, where durability and aesthetic surface finish are crucial. Due to its enzymatic system, *Bjerkandera* also ensures biodegradability, separating it from petroleum-based synthetic leathers (Elkhateeb et al. 2022).

Their biodegradable, lightweight, yet strong properties make them ideal as sustainable alternatives to plastics and polyurethane across many industries. Notably, lifecycle assessments indicate that mycelium composites from plants such as *Bjerkandera* offer reduced greenhouse gas emissions compared to synthetic leather production (Gržetić and Banjanac 2025), consistent with worldwide sustainability and circular economy targets.

### 15.2.15 *Cerrena unicolor*

*Cerrena unicolor*, a white-rot basidiomycete, has been of interest for sustainable leather-like material production because it possesses a mycelial network that grows rapidly and a lignin-degrading enzymatic system that is highly efficient. The high-density and homogeneous mats of *C. unicolor* render it ideal for creating flexible, resistant, and biodegradable composites that can be used as environmentally friendly substitutes for animal leather and synthetic polymers. It has been shown by recent research that *C. unicolor* mycelium can be converted into fashion accessory materials, footwear materials, and upholstery materials that take advantage of its strength and surface quality, without the high environmental load associated with traditional leather processing (Madusanka et al. 2024).

## 15.3 Processing of Fungi for Leather Production

Processing of fungi transforms a raw fungal mat into viable material, aids in stabilization and preservation, and activates the chemical bonding of the biomass (Shin et al. 2025). The processing involves (Fig. 15.8):

(i) Pre-treatment
(ii) Chemical treatment
(iii) Physical treatment
(iv) Post treatment

### 15.3.1 Pre-treatment

The pre-treatment process treats the fungal tissue by eliminating unwanted biomolecules and altering its chemical structure to allow for chemical crosslinking and other processing steps that enhance durability and leather-like characteristics (Vandelook et al. 2021) (Fig. 15.9). The process also stabilizes the tissue during drying to ensure that it does not become brittle and increases its resistance to deterioration and mechanical stresses by microorganisms during tissue processing and in later usage (Meyer et al. 2021).

***Hydration and selection*** The precursor fungal tissue is treated with lipids, moisture, or hydration agents, including glycerol or sorbitol, to increase its water content to provide flexibility. The tissue may also be sectioned to sizes that are appropriate (Elsacker et al. 2023).

**Fig. 15.8** Processing of fungi for leather production

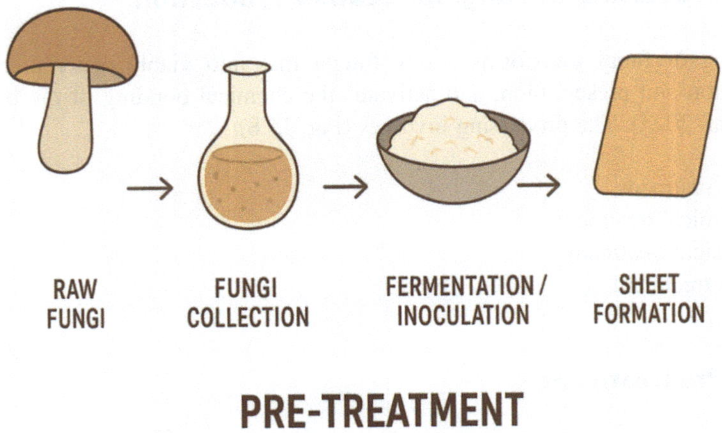


**Fig. 15.9** Pre-treatment processing of mushroom leather production

**Procedure**

- Immerse the harvested mycelial tissue in a lipid-containing solution, moisture, or hydrating agents (glycerol or sorbitol).
- Incubate for 1–24 h under room temperature conditions to saturate and maximize pliancy.
- With sterile tools, cut or trim the hydrated tissue into desired sizes or shapes for future applications.

### 15.3.2 Chemical Treatment

The hydrated tissue is next subjected to chemical treatments such as sodium hydroxide and acetic acid or isopropanol, ethanol, or methanol for timeframes from seconds to 6 months. Chemical treatments will remove soluble components of the extracellular matrix, including carbohydrates and proteins, as well as denature proteins (Amobonye et al. 2023) (Fig. 15.10).

**Preparation of Chemical Solution**

- Prepare the solution of sodium hydroxide (NaOH), acetic acid, or alcohol (such as isopropanol, ethanol, or methanol).
- Concentrations and pH are a function of the desired treatment severity (common NaOH concentration range is 1–10%).

**Immersion**

- You need to completely immerse the sectioned fungal tissue in the prepared solution.

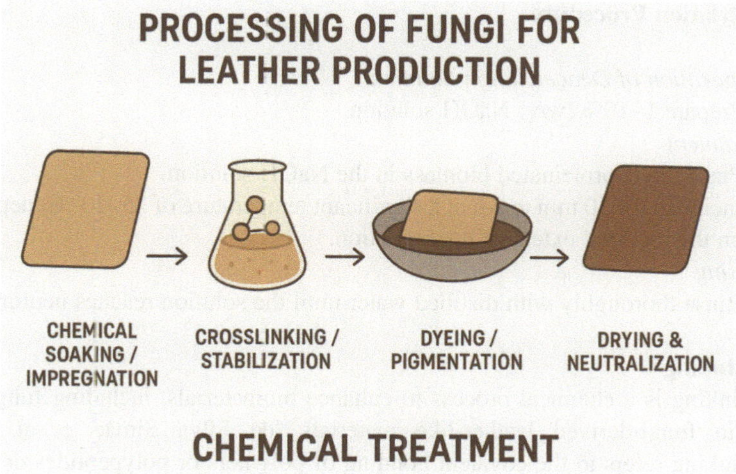


**Fig. 15.10** Chemical treatment of the processing of mushroom leather production

- Keep the tissue immersed at room temperature or at (slightly) elevated temperatures (25–30 °C) for 5 s to 6 months, depending on the degree of treatment you want to apply.

**Rinsing**

- After immersion, rinse the tissue with water or a neutralizing solution to remove residual chemicals.

**Deproteination and Deacetylation**

Chemical treatment will also deacetylate the chitin within the fungal cell walls, remove extracellular matrix components, and denature proteins, creating more reactive sites available for further crosslinking (Benetti et al. 2025; Bustillos et al. 2020).

**Deproteination Procedure**

1. *Preparation of Deproteination Solution*
   - Prepare an ethanol, isopropanol, or sodium hydroxide (NaOH) aqueous solution (1–4% w/v).
2. *Treatment*
   - Place the fungal biomass in the prepared solution.
   - Incubate for seconds to hours, depending on biomass thickness and the amount of protein to be removed.
   - Gently stir to allow for even treatment.
3. *Rinsing*
   - Rinse the biomass thoroughly with distilled water until it attains neutral pH (pH 7), so excess chemicals are removed.

**Deacetylation Procedure**

1. *Preparation of Deacetylation Solution*
   - Prepare 1–10% (w/v) NaOH solution.
2. *Treatment*
   - Place the deproteinated biomass in the NaOH solution.
   - Incubate for 30 min to 3 h at a significant temperature of 25–50 °C, depending on the required extent of deacetylation.
3. *Rinsing*
   - Rinse thoroughly with distilled water until the solution reaches neutral pH 7.

**Crosslinking**

Crosslinking is a chemical process to enhance biomaterials, including fungal biomass in fungi-derived leather-like materials (da Silva Simão et al. 2020). Crosslinking refers to the covalent bonding of polymers or polypeptides or epoxy-like polymers, which improves biomaterial mechanical properties, abuse tolerance, and biological durability (Wijayarathna et al. 2021; Gandia et al. 2021). For fungi-based biopolymer materials, crosslinking generally occurs with the chitin fibrils and other polymer structures in the mycelium. Common crosslink agents include genipin, adipic acid, or phenol that react with amino or hydroxyl groups in the polymer to create stable covalent bonds (Amobonye et al. 2023; Sundar et al. 2010).

1. *Preparation and Chemical Treatment:*
   - The dried fungal biomass or mycelium mats are chemically infused by either vacuum infusion or soaking in solutions with crosslinking agents such as genipin, adipic acid, or phenol. The time frame for chemical infusion can vary from a few seconds to several months, depending on the intended degree of crosslinking.
   - The material may undergo other chemical treatments, such as deacetylation or some removal of extracellular matrix, in order to perform better crosslinking and gain more bonding sites.
2. *Activation of the Fibrils*
   - The chemical treatment denatures proteins or deacetylates chitin so that functional groups like amines or hydroxyls on the fibrils are exposed and available for crosslinking.
3. *Reaction with Crosslinking Agents*
   - The material is placed or infused with the crosslinking agent. The cross linker + will react covalently with the functional groups on the chitin fibrils or the other polymers and join them with stable bonds (i.e., amide bonds).
4. *Fixation*
   - The product, once cross-linked, may be further subjected to processes like heating during heat pressing (either hot or cold) to decrease thickness and further fix the structure of the material.
   - The material could then be dried by various methods, including convection ovens, freeze-drying, air drying, and conductive drying to fix the structure.

### 15.3.3 Physical Treatment

Physical processes are important for transforming raw fungal mycelium into leathery materials with enhanced durability, flexibility, and mechanical stiffness. Following cultivation, mycelial mats are generally dried, pressed, and hot-pressed to evaporate excess water and form a flat sheet structure (Jenitta et al. 2013) (Fig. 15.11). These treatments not only improve the density and resistance to microbial decay but also assure the presence of the fibrous matrix of chitin and glucans, which gives fungal leather its mechanical strength (Jones et al. 2021; Orlita 2004). Another benefit from physical compression is the smoothness and surface texture enhancement, which factors in the compatibility of using it in fashion and upholstery applications (Narayanan 2024; Khambhaty 2020). Physical treatments on a sophisticated level, such as freeze-drying, heat-curing, and mechanical stretching, are also utilized to customize the final product's porosity, elasticity, and toughness (Haneef et al. 2017). Freeze-drying helps in keeping the structural integrity and lightweight properties, while hot pressing at controlled temperatures does the opposite by making the material tougher through partial denaturation of fungal proteins (Abdulhusein and Kadim 2024). Furthermore, the application of both shear processing and ultrasonic treatment has been successful in enhancing the regularity of the fungal leather and in reducing its brittleness so that the eco-friendly fungi-leather has similar characteristics to conventional animal leather (Arman et al. 2023). Thus, physical processing is the necessary first step in the process of producing sustainable, customizable, and functional leather substitutes from fungi.

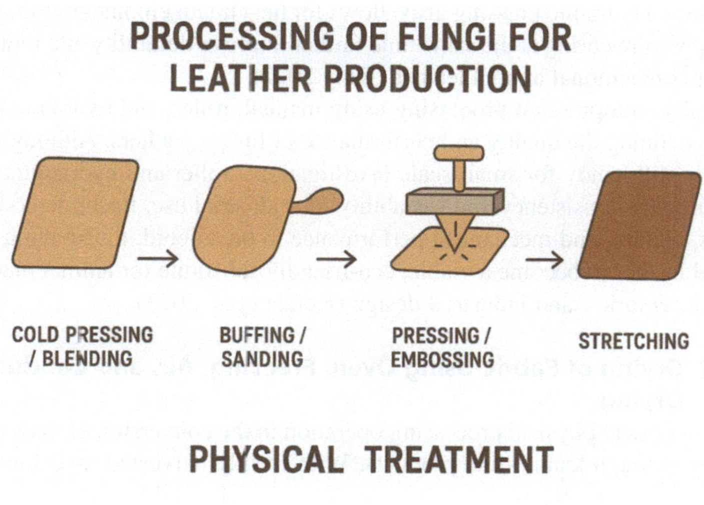


**Fig. 15.11** Physical treatment of the processing of mushroom leather production

#### 15.3.3.1 Compression Using Rollers, Manual Press, or Hydraulic Press

Compression is a fundamental treatment in the process of making fungal leather, which physically and literally has a positive effect on density, mechanical properties, and surface characteristics. The porous structure of the fungal mycelium mats formed at the end of mycelium cultivation, accompanied by air pockets, leads to loss of mechanical performance if not treated (Kefale et al. 2023). Compression with rollers or pressing tools removes this porosity and aligns the fibrous structures, yielding higher tensile strengths and elasticity (Dabeta 2018). Furthermore, the processing step improves the dimensional stability required in fashion, upholstery, and industrial applications where leather has to go through mechanical stress. Manual pressing is likely the simplest technique used for fungal mat compression. The mats are compressed to the desired uniform thickness and firmness by using weighted plates or small manual presses (Elkhateeb et al. 2022). Although it is a non-costly method and it can be used in small-scale production, manual pressing is not conducive to full production, lacks accuracy, and can be unreliable if an attempt is made to keep a certain pressure over a large area. Despite all these drawbacks, it is still an unavoidable step in the laboratory-scale fungal leather production, especially in the prototyping part. Roller compression is, on the other hand, a more precise method for fungal leather processing. Mycelium sheets passing through the roller machines ensure that a uniform pressure is applied and the processing is continual, thus making it suitable for the textile and paper industries (Barrera Torres et al. 2025). Roller methods give more room for flexibility, and a continuous rolling surface is provided, which is perfect for dyeing and finishing treatments. In addition to this, the rolling process promotes the bonding of hyphal networks, thus providing strength while preserving the breathability of the fungal leather. Hydraulic systems make it possible to minutely adjust pressures to regulate sheet thickness, density, and texture (Abraham 2023). The process has been found to greatly enhance tensile strength, water resistance, and resistance to tearing relative to manual or roller compression. Hydraulic pressing also allows for heat treatment integration, whereby heat and pressure curing at the same time further improve durability and replicate the texture of conventional animal leather (Jiang 2022).

Generally, compression processing using manual, roller, and hydraulic means is central to defining the quality and performance of fungal leathers. Although manual pressing is still handy for small-scale investigations, roller and hydraulic compression offer better consistency and scalability for industrial use. Each method enables thickness, texture, and mechanical performance to be tailored, thus paving the way for fungal leather to become a viable, eco-friendly substitute for animal materials in fashion, accessories, and industrial design (Kefale et al. 2023).

#### 15.3.3.2 Drying of Fabric Using Oven, Freezing, Air, and Conductive Drying

Drying is a critical physical processing operation in the conversion of mats of fungal mycelium to tough leather-like products. With freshly harvested mycelium having high water content, effective drying assures microbial stability, avoids decomposition, and improves mechanical strength. Among the different processes, oven drying is commonly used because it is efficient and can result in uniform products. Controlled oven drying at low to moderate temperatures drives out moisture without damaging the fibrillar network of the fungal mat, which results in enhanced stiffness and dimensional stability (Song et al. 2025). Overheating causes thermal protein and polysaccharide degradation, which can deleteriously affect flexibility and color. Air drying, while being slower, is still one of the most cost-efficient and environmentally friendly methods. Gradual evaporation of moisture under atmospheric or controlled conditions without much energy use is possible using this technique. Air-dried fungal leathers tend to maintain their inherent texture and porosity and are thus well-suited for applications where breathability is a consideration (Jones et al. 2021). However, it is a time-consuming process, and the finished product is likely to exhibit heterogeneity in thickness and mechanical properties due to nonuniform drying.

Freezing and freeze-drying (lyophilization) are sophisticated techniques to maintain the three-dimensional structure of fungal mycelium under dehydration. Rapid freezing of the material and sublimation of ice under vacuum reduce structural collapse and shrinkage, leading to lightweight and porous fungal leather. Freeze-dried mats are more elastic and resilient than oven- or air-dried counterparts. Due to the high operation cost and equipment needs, however, the method is still restricted from large-scale implementation in fungal leather industries. Conductive drying, where direct heat transfer by contact with hot surfaces occurs, allows for energy-efficient and quick dehydration of fungal mats (Liu et al. 2024). The method enables even heating by conduction to control the reduction of thickness and the creation of smooth surfaces, thus enhancing flexibility and tearing resistance. Concurrently, the process is usually combined with pressing operations, where both heating and pressing are combined to further improve water resistance and mechanical properties. Conductive drying thereby provides a good balance between product quality and energy efficiency (Majib et al. 2024). Drying methods overall have a great impact on the performance, texture, and functionality of leathers based on fungi. Oven drying is the easiest and most reliable approach, while air drying is environmentally friendly but less certain. Freeze-drying gives the best quality materials with great preservation of structures, but at a higher cost, while conductive drying combines efficiency with the added benefit of mechanical strength. The drying method choice is based on the final application and the necessary compromise between cost, sustainability, and performance, which makes drying a critical factor in the fungal leather production process (Sharma et al. 2024).

### 15.3.4 Post-treatment

The mycelium leather post-treatment process has always been a way to enhance the mechanical properties, stability, and marketability of the product. Among the most recent advancements, the use of polyethylene glycol (PEG) is a dual-purpose application as a plasticizer and crosslinking agent to make a strong and hard material. A related experiment with mycelial leather from brown-rot fungi suggested that the application of 20% PEG to the samples significantly increased Young's modulus and tensile strength, thus producing a thicker and leatherier product (Raman et al. 2022). Liquid fermentation combined with freeze-thaw cycles is another new method, which improves both structural and tensile properties while using less energy for drying. Ganoderma mycelium tests produced a tensile strength of 6.22 MPa and elongation at break of 18.9%, pointing out the technique's capability of becoming a post-treatment solution (Song et al. 2025). While specific methods are being investigated, broader reviews highlight that general post-processing treatments such as tanning, coating, and mechanical finishing can enhance the durability, water resistance, and visual attractiveness of the products. A very recent 2025 comprehensive review covers these aspects and emphasizes the way in which treatments lead to improved mechanical performance and sustainability of mycelium-based leather (Benetti et al. 2025). Intellectual property reviews have also pointed out dominant market trends—such as the introduction of new plasticizers and crosslinkers, signaling the growing interest in large-scale, high-quality post-treatment technologies for fungal biomaterials (Elsacker et al. 2023).

#### 15.3.4.1 Stretching

Stretching as a post-treatment is an essential step in improving the structure and functions of fungal leather. The application of tensile strain to the material after drying or partial curing can help to orient the chitinous fibers present in the mycelial network, thereby resulting in higher tensile strength, flexibility, and dimensional stability. Besides, stretching under controlled conditions reduces the occurrence of shrinkage and warping of the material, which usually accompanies the dehydration process. Research on mycelium composites has proven that mechanical treatments like pressing or deformation greatly enhance strength and flexibility, validating the suitability of stretching as an ancillary treatment (Song et al. 2025). Besides enhancing the mechanical performance, stretching also supports surface evenness and elasticity, which are sought after for uses in fashion and textiles (Jones et al. 2021).

Studies that have mixed mycelium mats with biopolymer coatings, including polyhydroxybutyrate and polylactic acid, have shown enhanced Young's modulus and elongation potential, with implications that mechanical strain can be used to strengthen the bonding and orientation of fibers (Xu et al. 2025). Stretching fungal sheets as part of post-processing alone can also tailor thickness, make it softer, and increase flexibility, enabling fungal leather to be closer to the texture of animal-based leather. Practical uses of stretching have already materialized in experimental and small-scale industrial applications. For instance, DIY and artisanal methods tend to employ wooden frames or clamps to stretch fungal mats during drying to avoid shrinkage and to achieve more uniform shapes (Hassan 2020). When paired with plasticizers like glycerol or crosslinking reagents like polyethylene glycol, further stretching enhances elasticity and toughness to make fungal leather products fulfill consumers' demands for flexibility and resilience. Such strategies underscore the promise of stretching as an uncomplicated but effective post-treatment method for enhancing fungal leather manufacturing (Elsacker et al. 2023).

#### 15.3.4.2 Dyeing

Fungal leather dyeing generally targets even color, high fastness, and minimum effect on breathability and biodegradability. Due to the high content of mycelium cell walls in chitin/chitosan and β-glucans, numerous colorants can be immobilized by ionic bonds or low-level crosslinking, which allows for low-temperature, aqueous dyeing using dip-exhaust, padding/spray, or post-coating with pigment binders. Current research on mycelium mats indicates utilitarian pathways like bio-based pigments and acrylic "leather" coatings to produce even coloration without toxic chrome systems, and patent reviews detail spray/pad coloring and pigmented polyurethane/acrylic topcoats transposed from leather coatings to mycelial substrates (Crawford et al. 2024; Elsacker et al. 2023).

Natural pigments such as fungal pigments (e.g., melanins, azaphilones) and dyes derived from plants are increasingly preferred to be in conformance with mycelium material's sustainability profile; reviews report their chemistry and use, and current surveys promote bio-mordants (tannins, enzymes, biopolymers) over heavy-metal salts for enhancing depth and fastness. With the cationic nature of chitosan, such bio-mordant approaches are mechanistically congruent with mycelium leather, enabling anionic dye attachment without compromising compostability (Lee et al. 2014). Aside from topical dyeing, integrated coloration strategies are being developed: coating/finishing layers containing pigments are common in the mycelium-leather patent literature, and leather science documents hybrid chitosan–dye complexes that tan and color simultaneously in a single bath with exceptional rub and wash fastness chemistry that can be directly translated to chitosan-carrying mycelial mats (Jones et al. 2021). The studies conducted on grown mycelium sheets also demonstrate simple post-dyeing and painting methods that work with heat-press consolidation and thus allow for easy, low-energy large-scale processes (Kalia and Singh 2020).

#### 15.3.4.3 Drying

Drying is an indispensable step in the post-treatment of fungal leather, which serves to stabilize the mycelial matrix by water removal and growth of microbes' inhibition, thus adjusting the physical and mechanical properties of the material at the same time. The oven-drying procedure is one of the most common methods, where the regulated temperature regimes (40–80 °C) help in the removal of moisture from the sheets, though if the sheets are overheated, it may lead to brittleness and discoloration (Farrahncor et al. 2025). The air-drying method takes longer but is more environmentally friendly and maintains some flexibility; however, it may cause uneven dehydration and longer processing time. Freeze-drying, on the contrary, is very common because it keeps the porous three-dimensional network and minimizes shrinkage, giving a more leathery feel than conventional hot-air drying (Jones et al. 2021).

Recent studies demonstrate that drying by freeze-thaw cycles and freeze-drying improves tensile toughness and tensile strength of fungal mats made from Ganoderma and Pleurotus. Hot pressing or heated plate pressing, which are conductive drying techniques, not only densify the structure but also add to the stiffness and improve surface smoothness, which is favorable for further finishing and coating treatments (Xu et al. 2025). The mycelium sheets' water vapor permeability is significantly influenced by drying, which is a parameter that greatly affects the comfort and breathability of the leather for wearing. Fungal leather's mechanical properties, like elasticity and stress-strain behavior, are also highly dependent on the drying route, as humidity-sensitive biopolymers such as chitin and glucans undergo reorganization during the process of dehydration (Sharma et al. 2025).

The comparative assessments point out that drying is not merely a process of preservation but a process of operational modification that will affect the thermal conductivity, the durability, and the biodegradability of mycelium composites. It has been profoundly recommended that freeze-drying be employed as the method of drying to ensure the structural integrity of bio-based products, besides the fact that it offers support for the scaling up of the production, even if the energy costs are higher (Campos et al. 2025). Meanwhile, industrial usage still leans toward oven and air-drying methods due to their ease and lower expenses, although they can have an impact on the porosity of the materials. Newer techniques, such as hybrid drying that utilizes both air and conductive heating, have shown potential in achieving a balance between energy conservation and product quality (Attias et al. 2020). Moreover, subsequent pressing after drying further stabilizes dimensions and prevents cracking during finishing operations. Controlled drying of mycelium composites leads to thermal conductivities comparable to those of insulation foams, which is proof of the multifunctionality that drying methods can offer (Shankar et al. 2024). Thus, it is concluded that the drying methods are of paramount importance not only in the case of extending the life of the fungal leather but also in the case of altering its end-use performance in the fashion, upholstery, and eco-friendly material markets (Mallon 2024; Lu et al. 2025).

## 15.4 Type of Leather Manufactured from Fungi

Fungal leather is a biocompatible alternative to both animal and synthetic leathers. The production process involves cultivating the fungal mycelium, which is the root-like network of the fungus, on agricultural waste. The produced material is further processed, treated, and compacted to acquire the characteristics and feel of animal leather. There are different kinds of it, and they include mycelium sheets, composite fungal leather, and bio-fabricated fungal hides.

### 15.4.1 Cultivation of Mycelium for Leather Production

The escalating necessity for eco-friendly replacements for leather made from animals has resulted in a very positive response toward the production of fungal mycelium-based materials (Kanth and Hussain 2024). Mycelium, which is the vegetative underground branching structure of a fungus, can be produced in a controlled manner to get the same texture and strength as animal skin leather (Rathinamoorthy et al. 2023). While the production of mycelium leather is going on, traditional leather that utilizes livestock breeding and toxic agents for tanning is overburdened by an enormous carbon footprint, cruelty, and a deprived eco-system (Muthu and Ramchandani 2024; Jhanji 2023). The field of biotechnology and materials science has recently made it possible to alter the fungal strain, substrate, and culture conditions to produce better quality leather and strength (Haneef et al. 2017; Appels et al. 2019; Jones et al. 2017). The process of fungus cultivation for leather starts with the selection of a suitable species of fungi, usually from the genera Ganoderma, Pleurotus, and Fomes. Such organisms are chosen not only for their ability to grow into easily workable, fibrous mats that closely mimic the collagen fibers of animal skin but also for their aesthetic value (Holt et al. 2012; Antinori 2021). The usual practice is to produce by growing fungi on waste materials generated in agriculture and industry, like sawdust, corn husks, or rice bran, which not only reduces production costs but also contributes to the waste treatment and circular bioeconomy processes. The fungi are introduced into clean substrates and cultivated under specific temperature, humidity, and air conditions to guarantee rapid and uniform mycelium growth and total substrate colonization (Jones et al. 2018a, b).

After colonization, the mycelium of the fungus is subjected to special techniques that make the material suitable for leather-like applications. The mats that are obtained from the fungi are put through hot pressing, plasticization, crosslinking, and coating with natural or synthetic binders to improve their tensile strength, elasticity, and water resistance (Camere and Karana 2018; Appels et al. 2018). Genetic and metabolic engineering strategies are being explored to help mycelium produce more of the structural protein, which in turn would improve the mechanical and aesthetic qualities of the mycelium (Aiduang et al. 2024). Treatments with natural oil, biopolymer, and environment-friendly tanning agents can keep on enhancing the flexibility and strength of mycelium leather, according to recent studies (Akhter et al. 2024). The scalability of mycelium leather production is one of the areas that is drawing the most attention, and companies are moving to bioreactor-based fermentation process systems instead of classic solid-state production (Karana et al. 2018). The submerged and semi-solid fermentation processes in which mycelium is grown are being optimized to shorten the growth cycle and to get more material without lowering the quality of the product. Furthermore, a huge boost in 3D printing and bio-fabrication technologies makes it possible to customize mycelium leather into novel textures, thicknesses, and surface patterns, making it suitable for various sectors such as fashion, furniture, and automotive industries (Bonga 2024; Silvestri 2022). However, the potential of mycelium leather is not enough to answer all the challenges related to its widespread use. Future research and industrial development are expected to tackle the high cost of production, limited scalability, and variability in performance as compared to conventional leather, which are some of the barriers (Meyer et al. 2016). The commercial viability of this biomaterial will also be determined by regulatory clearances, market acceptance, and competitive pricing. On the other hand, mycelium cultivation is a big change to sustainable biomaterials, giving a sustainable and eco-friendly alternative to animal leather that is in harmony with global sustainability aims (Jinanukul et al. 2024; Sahu et al. 2025).

### 15.4.2 Processing Techniques in Fungal Leather Manufacturing

The process of making fungal leather includes growing mycelium, but it also includes exposing it to some postharvest processing methods to get the material characteristics desired. Mycelium mats are very important in the process of making fungal leather, as they lack the necessary strength, flexibility, and water resistance for functional uses (Trivedi et al. 2025; Song et al. 2025). Usually, these treatments combine biological processes with chemical alterations and mechanical treatments to improve the texture and durability and imitate the characteristics of traditional leather (Benetti et al. 2025). By fine-tuning these techniques, companies can produce fungal leather with different properties for use in fashion, upholstery, and industry (Ekegårdh and Kristersson 2021). The pressing and drying stage is one of the most crucial in the processing of fungal leather, as it removes the excess water and compresses the mycelial network into a solid sheet (Attias et al. 2020). Hot pressing not only eliminates water but also helps to align mycelial fibers, which results in increased tensile strength and lower porosity (Appels et al. 2019). Freeze-drying is sometimes used to maintain the structural uniformity, but its usage is limited on an industrial scale. Dried products are sure to be dimensionally stable and ready to receive additional treatments (Jiang 2022; Song et al. 2025).

The chemical modification process is a necessary step in fungal leather production. Chitosan, starch, and other plant polymers, which are classified as natural adhesives, are the most frequently applied (Hou 2024). In addition, plant polyphenols, enzymes, and bio-based aldehydes are among the environmentally friendly tanning agents that are used as substitutes for the chromium salts (Elsacker et al. 2019). These processes not only maintain the durability of the leather and extend its life, but they also do so without the release of harmful chemicals into the atmosphere (Jones et al. 2021). Apart from that, biopolymers such as polylactic acid or polyvinyl alcohol are mixed in to have a smoother surface finish (Liu et al. 2024). The surface finish is very important for fungal leather's cosmetic nature. Natural waxes, oils, or polyurethane-based biodegradable coatings are used to enhance the shine, water resistance, and abrasion resistance (Basak et al. 2024). By means of embossing and texturing, the look and feel of animal hides are replicated. Moreover, pigmentation and the incorporation of natural dyes allow producers to have the option of coloring the fungal leather without the use of synthetic dyes, which are a major source of pollution (Fuck et al. 2018). These stages finally pave the way for the material to be diverse and customer-friendly.

The extended processing methods include crosslinking and enzymatic processing for the additional toughness of fungal leather. Crosslinking, whether through natural or synthetic agents, bonds mycelium fibers more tightly, which in turn gives rise to better elasticity and greater resistance to tearing (Deeg et al. 2017). Enzymatic processing, in particular laccase and peroxidase, is reported to do cellular wall modification of fungal materials for the purpose of enhancing flexibility and improving biodegradability (Temporiti et al. 2022). The next generation of fungal leather is also going to utilize an incorporation of nanomaterials such as cellulose nanofibers or graphene to not only improve toughness but also add functional properties like antimicrobial activity (Antony Jose et al. 2025). The production of large amounts of fungal leather will entail processing equipment that is both economical and environmentally friendly. The combination of automated pressing, coating, and finishing technologies with bioreactor-based growth is in the process of being developed to reduce variability and improve efficiency (Sharma et al. 2025; Jones et al. 2021). The investigation into green processing reagents, new finishing technologies, and hybrid bio-fabrication methods will be an ongoing one to get to the level of conventional leather. Thus, the various processing methods affirm the view of fungal leather as a high-performance and eco-friendly substitute for animal leather in global markets.

### 15.4.3 Types of Fungal Leather Products

Mycomaterial leather, which is made from mycelium, is increasingly seen as a sustainable and futuristic choice compared to fossil and animal leathers. It is made by cultivating fungal mycelium on agricultural or industrial waste; thus, the resulting products have many different characteristics as to density, texture, and strength (Amobonye et al. 2023; Crawford et al. 2024). Mycelium leather's adaptability is such that it can be turned into a wide range of consumer and industrial products while at the same time having a very low environmental impact that is comparable to the one associated with traditional leather production (Song et al. 2025). The production of mycelium leather shoes is one of the most common uses for this type of leather, where its properties of softness, breathability, and resistance to water are very much appreciated. Sneakers, sandals, and boots made of mycelium are already in the product line of some companies, which consider them to be eco-friendly alternatives to traditional animal hides (Jones et al. 2021). It is true that the processing of this material allows for the creation of different textures and finishes, thus making it particularly attractive to those designers who want to combine fashion and sustainability in the footwear industry.

In the world of fashion, clothing, and luxury items, fungal leather is paving its way toward being an alternative to the premium hides of jackets, handbags, belts, and wallets. The reason for its wide use is that the texture of the material can be personalized; thus, designers can create smooth, suede, or embossed finishes, which not only allow design freedom but also reduce the environmental impact. High-end firms are gradually warming up to the idea of using fungal leather to meet the consumers' demand for an ethical and sustainable fashion (Battal et al. 2025). For instance, besides fashion, fungal leather has also been invented for furniture and interior design applications. Upholstery, sofas, decorative panels, and automotive interiors have started to use fungal leather due to its durability, lightweight, and aesthetic appeal (Elsacker et al. 2023). The material's adaptability allows architects and designers to introduce green biomaterials to the construction of both residential and commercial buildings. Furthermore, fungal leather is being adopted in the biodegradable packing and accessories industry as well. Its capability to be molded and strong structure make it a choice of material for manufacturing cases, covers, and eco-friendly packaging that can replace plastic and synthetic materials. This scenario puts fungal leather on the map as a new biomaterial that supports the implementation of circular economy principles and reduction of plastic pollution (Verma et al. 2023).

In addition to lifestyle items, fungal leather can also be applied in the industrial and automotive sectors, especially as a substitute for composites based on petroleum. It is the very reason that its fire-resistant and biodegradable characteristics make it the right choice for car seat covers, steering wheel grips, and aviation upholstery. The study is still going on in the direction of augmenting the leather's mechanical and thermal properties, thus eventually making the petroleum-based materials more vulnerable (Wattanavichean et al. 2025). Among other things, the production of fungal leather into different product categories is a clear sign of its acceptance as a sustainable kit in all trade sectors. Fungal leather is a win-win situation; it not only saves the earth but also makes its way through the rat race of sustainable alternatives by displacing leathers derived from animals and plastics in the uses of fashion, interiors, packaging, automotive applications, etc. It does so by supporting waste valorization and eco-protection at the same time.

#### 15.4.3.1 Sheet-Based Mycelium Leather

Mycelium leather is laced with a fungus that is purposely grown into large and flat mats that imitate the texture and softness of animal leather (Raman et al. 2022). The mats are grown on the farm waste like sawdust, straw, or corn husks, plus the whole process is sustainable and environmentally friendly (Song et al. 2025). The difference between sheet-based mycelium and conventional leather is that it can be done in bioreactors under controlled conditions with less energy and water, which not only makes it a possible substitute but also a potential competitor for various industries (Williams et al. 2022). The flexibility and strength of the fungal product leather are due to the tight hyphal networks that are formed during growth (Kniep et al. 2024). To enhance the performance, post-processing techniques such as hot pressing, drying, and sustainable tanning are applied, which not only improve longevity but also water resistance (Karunarathne et al. 2024). Research has indicated that treatment using natural binder sheet-based fungal mats can reach tensile strength and elasticity levels like those of synthetic leathers that are commercially used for footwear and apparel (Saha 2023).

Sheet-based mycelium leather has been the subject of increasing interest by the fashion industry, as it is possible to tailor the product in the entire gamut of thickness, color, and texture during the production process (Amobonye et al. 2023). Premium fashion brands and technology startups such as MycoWorks and Bolt Threads have already taken the lead with the launch of their prototypes for this material's application in the form of handbags, footwear, and jackets (Farra 2021). Its combination with natural dyes and patterning upon growth effectively reduces the need for manual finishing treatments, thus giving it an edge in the sustainable fashion industry more so (Elsacker et al. 2023). No doubt, the mycelium leather is the most eco-friendly option compared to both animal and synthetic leathers. It means no more animal slaughter, a significant reduction in the generation of greenhouse gases, and no more toxic chemicals associated with the tanning process (Gutarowska et al. 2024). Moreover, as it is a biodegradable substance, the material aligns with the principles of a circular economy and can be disposed of through safe decomposition at the end of its life cycle without adding to the heap of nonbiodegradable waste. Although the technology is still very much in the laboratories of mass commercialization, the hurdles must be overcome to give the consumer a flawless product in size and thickness, hold the performance over the duration of use, and cut down the price of production (Zhao et al. 2025). Provided that the technological advancements in bioprocessing and processing continue uninterrupted, mycelium leather will grow to become a mainstream contender with the capacity to satisfy the global market.

#### 15.4.3.2 Composite Mycelium Leather

Composite mycelium leather is created through the combination of fungal mycelium with different materials like plant fibers, polymers, or textile backings for durability and functional property enhancement (Rathinamoorthy et al. 2023). Compared to sheet-based mycelium leather that relies mainly on the continuous growth of fungi, composite leather comprehensively integrates the biological network with engineered reinforcements to yield higher tensile strength, flexibility, and water resistance. This formation of a composite allows the adjustment of physical properties and the extension of the product life, which makes composite mycelium leather more suitable for demanding applications such as footwear and furniture (Bahua et al. 2024). The growth of fungal mycelium on lignocellulosic substrates and the subsequent association of the developed biomass with natural fibers, such as hemp, jute, or cotton, and compaction with biodegradable or synthetic polymers are basically the steps that are involved in the process. Studies have shown that the natural fibers rich in cellulose not only strengthen the fungal matrix but also improve the interface between the mycelial web and the composite layers, thus increasing the durability under repeated stress (Aiduang et al. 2022). Usually, chemical treatments such as the use of plant-derived crosslinkers for controlling the hydrophobicity and abrasion resistance of the material are performed (Choi et al. 2025).

Composite mycelium leather demonstrates certain mechanical advantages over sheet-based materials. As an example, mycelium plant fiber composites have shown almost double the tensile strength compared to pure mycelium mats, besides being biodegradable and taking up less space (Liu et al. 2024). Tearing and microbial resistance have been significantly improved with the use of biopolymer matrices like polylactic acid (PLA) or polyhydroxyalkanoates (PHA), thereby prolonging the lifespan of customer applications (Scaffaro et al. 2018). These improvements are making composite mycelium leather closer to the mechanical benchmarks of animal leather. Composite mycelium leather, meanwhile, remains sustainable by providing the planet with a certain amount of ecological footprint, together with functional and long-lasting durability. Usage of renewable natural fibers lessens the need for fossil-based reinforcements, and lifecycle assessments have revealed that composite forms still possess a significantly lower carbon footprint when compared to either animal or synthetic leather alternatives (Peeters 2023). Among the startups such as MycoWorks and Ecovative, these are the companies that are taking the composite route toward scaling up their production while at the same time maintaining the same level of material performance in the textile, automotive, and furniture industries (Williams et al. 2022). Notwithstanding its advantages, it is still a challenge to produce the composite on a large scale and at the same time ensure the quality of the composite throughout the lots is uniform. The usual situations of different fiber types, fungal strains, and curing processes lead to mechanical properties being inhomogeneous, denying industrial standardization. Plus, while biodegradability and durability are research-focused areas, one can see that some reinforcements could potentially negate the environmental benefits if they are made from nonbiodegradable polymers (Benetti et al. 2025). The ongoing developments of bio-based polymer chemistry and precision fermentation are promising future composite mycelium leathers to push the limits of barriers and become an alternative to animal leather in the mass market.

#### 15.4.3.3 Bio-fabricated Mycelium Polymer Hybrids

Bio-fabricated mycelium polymer composites are an innovative and eco-friendly material that is formed by combining fungal mycelium with natural or synthetic polymers to create durable, flexible, and customizable products. Mycelium alone provides a self-assembling fiber network that is highly porous and biodegradable, but it does not have the tensile strength and the water resistance that would be required for large-scale applications. The biological matrix has been strengthened by the incorporation of polymers, thus assisting the researchers in improving mechanical stability, hydrophobicity, and toughness; therefore, the hybrid materials are considered suitable for applications such as leather substitutes, packaging, and biomedical scaffolds (Taghizadehmakoei 2024). Mycelium composites are made by incorporating polymers either during the cultivation period, when the dispersion of polymers takes place along with the fungal development, or after harvesting, when polymer-based coatings or infiltrations are applied (Taghizadehmakoei 2024). For example, the use of such biodegradable polymers as polylactic acid (PLA) and polyhydroxyalkanoates (PHA) has been effective in increasing the flexibility and water resistance of mycelium composites. At the same time, biocompatible artificial polymers like polyurethane have been utilized to manage the elasticity and surface smoothness, thus offering new possibilities in footwear, furniture, and wearable textiles (Andia and Peralta Gomez 2025).

The current summary highlights that bio-fabricated mycelium polymer hybrids not only maintain the mechanical properties of traditional leathers and plastics but are also biodegradable. For instance, mycelium PLA composites demonstrated stronger tensile strength and improved dimensional stability compared to ordinary mycelium sheets (Nicolae et al. 2023). In a different scenario, PHA-infused mycelium hybrids had better thermal stability and microbial resistance, which projected longevity of the product in humid environments. Such diversity makes polymer hybrids a great platform for moving away from petroleum-based materials in consumer products. Besides the mechanical properties, these hybrids also contribute to the circular economy and sustainability goals. Mycelium biopolymer composites are reported to have much lower life cycle greenhouse gas emissions than animal leather as well as polyurethane-based synthetic materials (Akhter et al. 2024; Kefale et al. 2025). Moreover, the modularity of fusing the growth of fungi with various polymers can offer specific functions, ranging from lightweight packaging foams to high-performance textile substitutes. Some startups and research consortia are scaling up hybrid technologies for commercial application in fashion, automotive, and interior design sectors (Rathinamoorthy et al. 2023). Even with these benefits, it is challenging to match biodegradability with durability since the incorporation of some synthetic polymers can undermine environmental advantage. Additional work is emphasizing bio-based and recyclable polymers that maintain functional performance without restricting end-of-life compostability (Wattanavichean et al. 2025). Precision fermentation, green polymer synthesis, and surface engineering developments are anticipated to further enhance the range of bio-fabricated mycelium polymer hybrids as a next-generation material alternative to animal and fossil products.

## 15.5 Eco-friendly Mushroom Leather-Like Products

Environmentally friendly mushroom leather-like products are green substitutes for animal and synthetic leather, produced using mycelium, the fungi root system. Examples of such products are wallets, handbags, belts, shoes, and jackets, which are durable with a lower environmental footprint (Fig. 15.12; Table 15.1). Compared to conventional leather, they use less material, produce less carbon, and are biodegradable. Mushroom leather is also a versatile product, which can be used by the fashion, accessories, and packaging industries. Therefore, it favors environment-friendly consumer choices as well as enabling circular economy principles.

**Fig. 15.12** Eco-friendly mushroom leather-like products

**Table 15.1** Eco-friendly mushroom leather-like products

| Product | Mushroom species (mycelium source) | Reference |
|---|---|---|
| Wallets | *Ganoderma lucidum*, *Pleurotus ostreatus* | Battal et al. (2025); Kapoor et al. (2024) |
| Handbags/purses | Like *Ganoderma lucidum*, *Pleurotus ostreatus*, *Aspergillus niger*, and *Pleurotus djamor* | Mallon (2024); Crawford et al. (2024); Al-Bader and Mohammed (2022) |
| Backpacks | *Ganoderma lucidum*, *Trametes versicolor* | Rathinamoorthy et al. (2023); Dhanda et al. (2024) |
| Footwear (shoes, boots, sneakers, sandals) | *Ganoderma lucidum*, *Trametes hirsuta*, *Pleurotus ostreatus* | Jhanji et al. (2022); Silverman et al. (2020); Bi and Crnković (2024) |
| Belts | *Pleurotus djamor*, *Fomes fomentarius* | Jhanji et al. (2022); Gandia et al. (2021) |
| Hats/caps | *Ganoderma applanatum* | Kumar et al. (2025); Kaaronen (2025) |
| Jackets and coats | *Ganoderma lucidum (reishi mushroom)*, *Pleurotus ostreatus (oyster mushroom)*, and *Fomes fomentarius* | Mallon (2024); Kapoor et al. (2024) |
| Sustainable textiles and fabrics (for clothing) | *Ganoderma lucidum (reishi mushroom)*, *Pleurotus ostreatus (oyster mushroom)*, and *Fomes fomentarius* | Wattanavichean et al. (2025); Ekegårdh and Kristersson (2021); Bi and Crnković (2024) |
| Watch straps | *Ganoderma lucidum* | Gandia et al. (2021); Jhanji et al. (2022); Jones et al. (2021) |
| Phone/laptop cases | *Ganoderma lucidum*, *Pleurotus ostreatus* | Elsacker et al. (2023); Spinnewijn (2022); Kapoor et al. (2024) |
| Packaging foams | *Pleurotus ostreatus*, *Ganoderma lucidum* | Kandel (2022); Wang (2022); Manan et al. (2022) |
| Furniture upholstery | *Ganoderma lucidum*, *Trametes versicolor* | Jones et al. (2021); Singh et al. (2025) |

### 15.5.1 Wallet

The world's transition toward sustainable materials has boosted the creation of sustainable alternatives to conventional animal leather, with fungal mycelium leading the way. Wallets, one of the most commonly used fashion accessories, are increasingly being made from mycelium-based leather due to its durability, flexibility, and close resemblance to traditional leather (Battal et al. 2025). In contrast to synthetic leather produced from petroleum-based products, mycelium leather provides for biodegradability; thus, wallets produced from this are both fashionable and eco-friendly. Wallets that have been produced using fungal mycelium are created by growing organisms like *Ganoderma lucidum* and *Pleurotus ostreatus*, which produce compact and intertwined networks of mycelia (Kapoor et al. 2024). These are then treated with drying, pressing, and in some cases, chemical or biological processes to attain the required mechanical properties and texture. The end product is tear- and bend-resistant, with the necessary qualities to withstand daily usage in wallets (Sahu et al. 2025; Rathinamoorthy et al. 2023).

The wallets also lessen dependence on animal products, which fits into the increasing need for fashion that is cruelty-free. The ecological advantages of mycelium wallets are substantial since they consume fewer resources than animal husbandry and circumvent the use of toxic chemicals in the tanning of animal hide (Liu et al. 2024). Moreover, the time taken for production is significantly less, as mycelium leather can be grown within weeks, while that of animals takes years to grow livestock. The efficiency, coupled with the fact that mycelium is biodegradable, positions wallets as a perfect sustainable option in the fashion sector (Jones et al. 2021). The mycelium wallet's commercialization has advanced rapidly, and it is with the startups and the fashionable brands that the sustainable fashion consumer market disruption is already considered. More than just a guarantee to consumers of vegan and eco-friendly products, mycelium wallets also demonstrate advancements in materials science (Pertile et al. 2025). The material's physical property development through further research would signal its acceptance by the fashion industry, and wallets would be one of the major drivers for this transition to take place, as they are the most practical and symbolic products at the same time.

### 15.5.2 Handbags and Purses

Handbags and purses rank among the most sought-after fashion accessories in the world, but their traditional manufacture with animal leather has necessitated ethical and environmental concerns (Mallon 2024). Mycelium leather has risen as a sustainable material for sustainable handbags and purses due to its strength, biodegradability, and cruelty-free production, which makes it very appealing to responsible consumers (Nithyaprakash et al. 2020). Fungi like *Ganoderma lucidum*, *Pleurotus ostreatus*, *Aspergillus niger*, and *Pleurotus djamor* are also frequently applied in handbag production since they create dense interwoven mats of mycelia like animal skin collagen fibers. These mats are physically and chemically treated to enhance mechanical properties, flexibility, and water resistance so that handbags remain both functional and beautiful. The end product has a look and feel of traditional leather and therefore can be used for upmarket as well as down-market fashion applications (Crawford et al. 2024; Al-Bader and Mohammed 2022). Some startups and research communities have previously proven the viability of handbags constructed from mycelium leather, illustrating their value in displacing animal leather within the fashion world. By using agricultural waste as substrate for the fungal growth, the process also gains greater sustainability as it valorizes waste streams (Rathinamoorthy et al. 2023). Vegan, sustainable, and ethically sourced consumer products have been gaining popularity, and with that, it is just a matter of time before the handbags and purses made of fungal mycelium will be considered the most important breakthrough in sustainable fashion innovations.

### 15.5.3 Backpacks

Backpacks have become a common accessory in daily life for students, business professionals, and tourists. However, traditional leather backpacks are still a significant factor in environmental harm. The leather production is a resource-consuming process, while fossil fuel-based plastics are used in synthetic alternatives, which are also non-sustainable. The leather made from mycelium has been an alternative for a backpack with its durability, flexibility, and biodegradability claim without the unethical issues of the animal products (Rathinamoorthy et al. 2023). This makes mycelium leather backpacks a good choice in the growing market for products related to a sustainable lifestyle and fashion. Most industrial methods for making mycelium backpacks rely on fungi such as Ganoderma lucidum and Trametes versicolor to create dense, interwoven mycelial networks that are fit for durable accessories (Dhanda et al. 2024). The material is then processed through mechanical pressing and either chemical or enzymatic processes to strengthen the tensile strength, elasticity, and water resistance, which are the vital characteristics of contemporary weight-carrying backpacks that are prone to outdoor conditions (Gao 2023). These engineered features make fungal leather backpacks equal to or even better in functional performance than traditional leather backpacks without sacrificing environmental responsibility. Various business firms and research partnerships are presently testing mycelium-based backpacks, emphasizing the prospect of such commercial use. Utilizing agricultural waste as substrates for the cultivation of fungi, the process not only produces eco-friendly materials but also encourages the practice of circular economies (Perry 2022).

### 15.5.4 Footwear (Shoes, Boots, Sneakers, Sandals)

Leather-like products from mushrooms have also been introduced as a renewable substitute for animal-based and synthetic leathers in shoemaking, such as shoes, boots, sneakers, and sandals. These mycelial biomaterials are mainly produced from mycelium, the filamentous vegetative part of fungi with a root-like structure, which is designed to be engineered to replicate the texture, durability, and flexibility of traditional leather (Jhanji et al. 2022; Silverman et al. 2020). In contrast to animal leather, mushroom leather does not involve animal killing, and as opposed to man-made materials, it minimizes reliance on petrochemical plastics, hence the environmentally friendly preference for buyers (Ekegårdh and Kristersson 2021). The application of mushroom leather as a material for footwear design solves several challenges of sustainability. Common leather making is linked with large amounts of greenhouse gas emissions, cutting of forests, and water-using tanning (Baars et al. 2024). Synthetic leather, on the other hand, comes with toxic chemicals and microplastic contamination.

Conversely, mushroom leather needs little land, water, and energy inputs and can be treated with nontoxic treatments to attain desired attributes like waterproofness, permeability, and tensile strength to be useful for functional footwear use (Bi and

Crnković 2024). Further, its biodegradability means that those shoes that get discarded from mushroom leather contribute minimally to landfill waste (Benetti et al. 2025). Some shoe companies and researchers already use mushroom leather in sneakers, sandals, and boots. For example, Adidas and Stella McCartney have collaborated with the creators of mycelium leather to develop prototype shoes with the same durability as conventional materials but with eco-friendly properties (Williams et al. 2022). Continued research aims to further increase mechanical strength, flexibility, and resistance to wear to further increase its large-scale commercial application in various categories of footwear (Ariyani et al. 2024). This makes mushroom leather shoes an exciting innovation in the world movement toward eco-friendly fashion.

### 15.5.5 Belts

Mushroom leather, created from fungi mycelium, is being utilized in fashion items like belts because of its eco-friendly and sustainable characteristics. While animal leather is produced through livestock rearing and tanning in an environmentally devastating manner, mushroom leather can be grown in controlled conditions with low water, land, and energy needs (Jhanji et al. 2022; Jhanji 2023). It thus becomes a potential material for environmentally aware consumers looking for cruelty-free options with no compromise on durability and looks (Ariyani et al. 2024). The belts produced from mushroom leather are not only biodegradable but also versatile in aesthetic appeal, with textures and finishes equivalent to leather (Gandia et al. 2021). They can be naturally dyed and treated with nontoxic applications, which increase both the belts' sustainability and value in the fashion industry (Battal et al. 2025). While mushrooms, also called fungi, are less durable than cowhide and not so suitable for heavy-duty use, the belt's design and aesthetics will surely play a great role in its acceptance as a strong, stylish, and eco-friendly choice (Xu et al. 2025).

### 15.5.6 Hats/Caps

Mushroom leather is a product made from the mycelium of certain types of fungi like Ganoderma lucidum, Pleurotus ostreatus, and Fomes fomentarius, which have excellent fibrous networks and can produce pliable bio-composites (Kumar et al. 2025). Controlled cultures can grow these fungi to produce sheets that are light and breathable, as well as suitable for making accessories like hats and caps (Kaaronen 2025). Contrary to the processing of animal leather that consumes plenty of resources or the petroleum dependency in synthetic leather, mushroom leather requires only a fraction of inputs like land, water, and energy; hence, it is an environmentally friendly and very sustainable choice for headgear (Kapoor et al. 2024). Products based on mycelium used in making hats and caps are not only soft and long-lasting but also provide natural patterns and can be colored with plant-based or organic dyes (Nithyaprakash et al. 2020). The fact that mushroom leather is biodegradable means that the products produced from such materials will not add to landfill waste, thus being in line with circular fashion principles (Vunduk and Biketova 2022). The tensile strength and weather resistance of mycelium leather are still being improved through research, thus assuring that the fungi like Ganoderma and Pleurotus will keep on being the source of quality alternatives for fashionable, sustainable headgear (Wang 2022).

### 15.5.7 Jackets and Coats

Mushroom leather-like material coats and jackets are environmentally friendly alternatives to leather made from animals and synthetic polymers. Different types of fungi, such as Ganoderma lucidum (reishi mushroom), Pleurotus ostreatus (oyster mushroom), and Fomes fomentarius, have been used in the production of mycelium-based materials, which are tough, flexible, and resilient and suitable for outerwear (Mallon 2024; Kapoor et al. 2024). The mats of mycelium used in postharvest treatment reach a leather-like texture and possess the properties of tensile strength, flexibility, and breathability that are on par with traditional leather jackets (Jhanji et al. 2022). Mycelium jackets and coats have the effect of minimizing environmental pollution associated with tanning operations and are a cruelty-free, biodegradable alternative for green fashion. In addition to sustainability, mycelium leather jackets and coats are a blend of innovation and environmental awareness. Experiments have proved that mycelium composites obtained from *Ganoderma and Pleurotus* possess adjustable surface features and durability, which enable them to endure regular wear and weather (Pertile et al. 2025). With increasing customer preferences for plant- and fungi-based textiles, fashion companies are gradually embracing mycelium leather jackets and coats as high-end, environmentally friendly materials, hence reshaping sustainable fashion trends (Rathinamoorthy et al. 2023).

### 15.5.8 Sustainable Textiles and Fabrics (for Clothing)

Mycelium-based sustainable fabrics and textiles are a revolutionary alternative to petroleum-based synthetics and animal-derived leather in the fashion industry. Fungal strains like *Ganoderma lucidum (reishi)*, *Pleurotus ostreatus (oyster mushroom)*, and *Fomes fomentarius (tinder fungus)* are widely utilized to produce leather-like sheets with mechanical properties that can be regulated (Kapoor et al. 2024). Mycelium textiles are lightweight, breathable, and biodegradable, which is beneficial for apparel and fashion uses. In comparison to traditional leather, its production involves little water and shuns poisonous tanning chemicals, thus greatly reducing the environmental impact (Jhanji et al. 2022). Mycelium cloth can also be designed to mimic suede, leather, or woven fabric so that wide creativity in fashion design is possible without sacrificing durability (Wattanavichean et al. 2025).

Mushroom-based fabric has attracted significant attention from both academia and industry due to its customizable properties, including strength, elasticity, and surface texture. In recent times, it has been revealed by studies that Ganoderma and Pleurotus mycelium composites have tensile properties resembling natural leather, while Fomes fomentarius-derived fibers have the highest strength among all fungi, which is suitable for fabric-like applications (Ekegårdh and Kristersson 2021; Bi and Crnković 2024). With the consumer base increasingly asking for no-animal, eco-friendly clothing, fashion companies and even small brands are looking for mushrooms as their new material. The great app for mycelium textiles in the future will not only be circular fashion but also a combination of aesthetics and sustainability (Kumar et al. 2025).

### 15.5.9 Watch Straps

Mushroom leather materials have come to be quite popular and are being used in the fashion accessories market, mainly for watch bands. Various fungi such as Ganoderma lucidum (reishi mushroom), Pleurotus ostreatus (oyster mushroom), and Fomes fomentarius (tinder fungus) are believed to be the very best organisms in the process of discovering the toughest, most flexible, and yet most sustainable animal leather substitutes (Jones et al. 2021). Mycelium leather watch straps take toughness, lightness, and biodegradability, and in this way become very suitable for daily use while at the same time reducing the need for traditional animal hides and synthetic materials. In contrast to traditional tanning procedures, which are chemically toxic and generate waste, fungal leather production relies on carbon-neutral cultivation on crop residues, hence supporting sustainable systems of production (Jhanji et al. 2022; Jhanji 2023).

Current research points to the fact that mycelium composites have comparable tensile strength, elasticity, and surface finishes to traditional leather, which renders them suitable for use in smaller accessories such as watch straps, where wear resistance and flexibility are essential (Gandia et al. 2021). For example, Fomes fomentarius fibers have high structural stability, whereas the Pleurotus and Ganoderma strains yield smooth, leather-like sheets that are capable of being dyed and finished for high-end uses (Basak et al. 2024). As consumers increasingly demand cruelty-free, environmentally friendly products, mushroom watch straps are being embraced by green fashion companies to meet aesthetics with eco-friendliness (Khan 2025).

### 15.5.10 Phone/Laptop Cases

Mushroom leather materials made from mycelium are gaining more attention in the production of eco-friendly, durable, flexible, and biodegradable phone and laptop covers. Mycelium composites, grown on waste substrates from agriculture, are specifically developed to provide protection from impacts, be light in weight, and have different textures, and thus, they are suitable for electronic device housing while simultaneously getting rid of plastic waste (Elsacker et al. 2023). The study of

Ganoderma and Pleurotus mycelium composites has indicated their mechanical properties are on a par with those of conventional plastics, while fibers from Fomes fomentarius provide scaffold-like structures, thus marking the advent of the use of such materials in making eco-friendly electronic device accessories (Spinnewijn 2022; Kapoor et al. 2024).

### 15.5.11 Packaging Foams

Mushroom foams for packaging, extracted from the mycelium of such fungi as Ganoderma lucidum (reishi), Pleurotus ostreatus (oyster mushroom), and Fomes fomentarius (tinder fungus), are being developed as eco-friendly replacements for polystyrene and polyurethane foams produced from petroleum. The organic foams are cultivated by growing fungal mycelium on crop waste, whereby the mycelial networks bind the particles into soft, protective composites that are suitable for packaging (Kandel 2022; Wang 2022). Researchers have suggested that the mycelium foams have the same compressive strength, thermal insulation, and water resistance as the synthetic foams but with complete biodegradability and compostability after use (Manan et al. 2022). These advances lead to less plastic waste and a lower carbon footprint, thus making mycelium foams a possible material for sustainable packaging in the electronics, food, and consumer goods sectors (Wattanavichean et al. 2025).

### 15.5.12 Furniture Upholstery

Mushroom leather-like composites, from fungal mycelium like *Ganoderma lucidum (reishi mushroom), Pleurotus ostreatus (oyster mushroom)*, and Fomes *fomentarius (tinder fungus*), are also gaining momentum as renewable alternatives for furniture upholstery. These mycelium composites can be tailored to hardness, flexibility, and hydrophobicity, rendering them as good substitutes for animal hide-based leather and synthetic vinyl, which find widespread application in sofa upholstery, chairs, and cushions (Jones et al. 2021; Singh et al. 2025). Studies have indicated that the tensile strength, elasticity, and abrasion resistance properties of mycelium composites match those of conventional upholstering materials, while being biodegradable and having lower environmental footprints (Rashdan and Ashour 2023). Mycelium upholstering also has the potential to be cultivated on crop waste substrates, reducing costs of production and embracing circular economy principles, making it an emerging green innovation for sustainable interior design (Sahu et al. 2025).

## 15.6 Sustainable Alternative for Bovine Leather

Production of bovine leather is fundamentally unsustainable and unethical due to egregious environmental and ethical impacts. At its origin, it is a co-product of the meat industry; thus, the demand supports industrial cattle farming that is considered the main cause of deforestation, losses in biodiversity, and greenhouse emissions, the latter occurring through methane (Kefale et al. 2023). Yet, it is the process of tanning that turns raw hide into very durable leather that is most damaging. It is resource-intensive, consuming large quantities of water and often using toxic chemicals, including chromium salts, which may end up in local water supplies unless perfectly handled, creating huge health hazards for tannery workers. To add on, the whole process, including the land, water needed for raising cattle, energy-intensive processes, and global shipping, emits a heavy carbon footprint. From the ethical point of view, there are major animal welfare issues since it remains associated with an industry frequently condemned for its inhumane practices. Although it is unquestionably durable, the mounting environmental and ethical issues surrounding conventional bovine leather render it a matter of concern for the conscious consumer (Saha et al. 2020).

### 15.6.1 Plant-Based Leather

Plant-based leathers, considered sustainable materials, are a huge step forward in material technology. For example, cactus leather from Nopal cactus and Pinatex from pineapple leaf waste come from plants that require little water and, importantly, make use of agricultural waste that is already being produced, thereby creating a circular economy and a potential income for farmers (Malabadi et al. 2025; Rashid et al. 2024). Moreover, mushroom leather (Mylo) grows from mycelium grown in vertical farms on organic waste, thereby creating a material that has an extremely low evolutionary footprint in just a matter of weeks (Duangsuwan et al. 2023; Kim et al. 2021). While many of these are currently combined with a polymer such as polyurethane for strength and water resistance, these materials greatly lessen the ecological damage incurred by cattle farming. They stand as a cruelty-free option that is partially biodegradable and continues to fortify its quality, granting them the higher-shelf status among cutting-edge fashion and automotive brands that want to present a more ethical and environmentally conscious material profile (Wjunow et al. 2023).

Fruit waste leather acts as a great material for circular economy practices, transforming crop waste into something valuable. The most famous fruit leather example is Pinatex—it is produced through pineapple leaf cellulose fibers (Josroy and Sundervel 2024). When the pineapple fruit is picked for food, the leaves are collected. The long fibers get pulled out via decortication, then washed and felted into a nonwoven mesh. This mesh is then coated in a polylactic acid (PLA), bio-based polymer coating to produce a tough, leather-like fabric that is versatile and also partly biodegradable (Das et al. 2022). Apple leather is created using the pomace—the skin, core, and seeds left over from the juice and compote making. This waste material is dehydrated and ground into a fine powder before mixing with polyurethane (PU) and being applied to a fabric backing (Rimantho et al. 2024). Some other innovations in this group are venturing into waste from grapes, bananas, and mangoes (De Menezes Rodrigues et al. 2023). The most important benefit of these materials is that they create value for the current supply chains without demanding more land, water, or pesticides to cultivate the raw material, thus solving the crucial problem of agricultural waste while developing a new, sustainable textile.

### 15.6.2 Bio-fabricated Leather

Lab-grown leather provides a surgical alternative to creating leather that is biologically identical to animal leather without raising and killing animals. It comprises cultivating collagen-producing cells in a bioreactor and growing actual leather tissue that may be tanned and finished in traditional ways (Absharina et al. 2025). The principal advantage is that it avoids animal cruelty, and there may be land-use and methane emission concerns from the ranching that can be completely avoided (Sahu et al. 2025). Still early in commercialization, bio-fabrication faces several other challenges with high costs of production and energy usage, representing the bridge between biotechnology and sustainable design toward a future where consumers can have the very look and feel of good leather with none of the ethical dilemmas or the bulk of its environmental impact (Whabi et al. 2024).

#### 15.6.2.1 Methods of Bio-fabrication

**Collagen Solutions**

This approach uses collagen in a liquid form for adhesives, enabling the assembly and patterning of bio-fabricated materials while surpassing conventional fabrication methods. The method starts with collagen, which is extracted and purified either from animal or nonanimal sources, and then is dissolved in an acidic solution to stop fibril formation at an early stage. The acidic solution becomes viscous, and bio-inks—whose application can be controlled and programmed to be like 3D printing—are applied to the seams or surfaces of bio-fabricated leather parts or biomaterial substrates. The crucial change happens when the solution is treated with environmental signals like temperature or pH shift. This triggers gelation, in which the collagen molecules in solution gelate and form a firm and cross-linked hydrogel. The hydrogel serves as a natural glue, providing strong bonds in lieu of petroleum-based synthetic adhesives and, in some cases, eliminating the need for stitching. This invention improves the biodegradability of the material in general and allows for designs like seamless garments and accessories, which improves manufacturing (Tewari et al. 2023).

**Electrospinning and Electrocompaction**
Using electrical forces is an electrohydrodynamic processing technique that enables collagen-based material engineering. Think of it as constructing leather but starting at the molecular level and going fine-tuned to the nano-level (Khan 2025). To begin with, fueling a collagen solution with a high voltage, such as through a syringe, spins a polymer jet. This polymer jet elongates and breaks into remarkable, often nano-scale, fibers. These fibers are gathered on a rotating drum to create a nonwoven porous mat. This mat resembles the extracellular matrix of skin; it can either be used as a scaffold for cell growth or as an independent material (Attias et al. 2020). Electrocompaction is a complementary process where an electric field is used to compress collagen molecules electrostatically and align them into dense, highly ordered fibrils or microfilaments with enhanced mechanical properties. The delicate, fibrous scaffold resulting from electrospinning is a material that results from both methods and possesses a high surface area and desirable properties. They may be tanned, dyed, and finished in the usual manner for leather products to yield a final product that replicates the traditional one (Alaneme et al. 2023; Madusanka et al. 2024).

**Cross-Linked Collagen Fibrils**
This method is a bottom-up approach to biomaterials that aims to recreate a leather-like material through the reconstitution of purified collagen. This is done by finely tuning the reconstitution of the collagen so that its intrinsic hierarchical structure is preserved. The method involves initially preparing a neutralized solution of collagen molecules. These molecules are then prompted, under specific biochemical conditions, to self-assemble into fibrils. These fibrils are the basic fibrous units responsible for tensile strength. However, the pioneering fibril network is weak and hydrophilic. To develop a strong material, the network is cross-linked to enable permanent bonds by agents such as genipin or tannins. These agents can create strong covalent bonds between the individual fibrils. With crosslinking, the material's mechanical strength, thermal stability, and resistance to enzymatic scavenging improve drastically. A final and equally important step is to lubricate the cross-linked matrix with biocompatible agents. Such biocompatible lubrication permits the fibrils to slide relative to one another, thereby imparting the required flexibility, softness, and drape to the material, as required by supple leather (Majib et al. 2024; Jo et al. 2023).

### 15.6.3 Recycled Leather

Recycled leather, frequently referred to as reconstituted leather or leather dust, is created by using excess leather pieces (i.e., scraps, shreds, and offcuts) gathered from manufacturing and binding them together with a polyurethane or latex binder onto a fibrous substructure (Liu et al. 2023). In contrast to virgin leather, which is sourced from a hide, or bio-fabricated leather, which is cell-based, recycled leather is a composite that transforms potential landfill waste into a new material (Senthil

2024). Being a type of recycled plastic, rPET is used to make leather alternatives, thereby averting the problem of pollution through waste. Plastic bottles and other plastic wastes are converted into a fine polyester fiber before being coated to form a sturdy leather-coated cloth. It does so by disrupting the landfill disposal methods and plastic dumping into ocean currents, thereby converting harmful waste into a useful long-term product. Unfortunately, this is not complete in achieving sustainability, considering that, being synthetic, it is dependent on fossil fuels and may also disperse microplastic pollution in different stages throughout its lifecycle. Neither is it biodegradable. For consumers who want to focus on minimizing waste and go for something that is entirely vegan and waterproof, rPET is a type of practical, easily available alternative that tackles a completely different form of environmental degradation (Omoloso et al. 2020).

Recycled leather is part of the circular economy because it prevents postindustrial scrap such as leather offcuts, dust, and shavings from being incinerated or sent to landfill. This process minimizes the waste footprint of the tannery and fabricator. Recycled leather also reduces the need for new animal hides and showcases benefits to land use and water consumption from livestock farming and the greenhouse gases released into the environment. When contrasted to the whole work of tanning raw hides, the making of recycled leather uses less water and energy (Stejskal et al. 2023). The main advantages of the material are cost and flexibility. It allows people to make an ethical choice with their clothing, coming in at a small fraction of both bio-fabbed and traditional leather costs. It becomes possible to hold a more accurate and closer tolerance since much of the natural grain pattern, natural color, texture, and thickness are removed, thus producing sheets with a consistent texture, color, and thickness without the imperfections that are found in real leather. Such uniformity is particularly advantageous in mass production. It can also be easily embossed with a whole array of grain patterns or other embossings, dyed to virtually any shade, and finished to individual aesthetic and functional requirements. These properties are the cause of its widespread variety of applications in the various sectors (Kodaloğlu and Kodaloğlu 2024). In fashion, it is used for items such as wallets, journals, watch straps, and nonstructural components of bags. It is widely used in the furniture industry as the upholstery for the seats and backrests of residential and contract seating and for decorative panels and inlays. In the car industry, it is being used in noncritical interior parts like door panel covers, dashboard overlays, and console coverings. It is also used for home décor items, bookbinding, and even to make special-effects wall covering to demonstrate that sustainability can be practical and versatile (Giehl et al. 2024).

## 15.7 Development of Fungi-Derived Fabrics

The innovation of fungi-based fabrics, starting with materials like Muskin, Mylo™, and Reishi™, represents a complete transition from previous sustainable textile production, from the harvesting of resources to cultivated production. This innovation begins by taking advantage of the properties of mycelium, the complex root-like structure of fungal threads that grow underneath the earth. Mycelium can be grown vertically in specialized facilities using the low-value agricultural wastes available in abundance and not on previously cultivated land (Liu et al. 2024).

Agricultural waste can include sawdust, wood chips, and straw. This waste serves as a source of nutrients, and in a cycle like fermentation, the mycelium breaks down and mingles this waste into a dense mat of intergrown tissue. The growing cycle is incredibly rapid, taking only weeks to develop a solid biomass, a far cry from the years it takes to raise a cow to leather. All these are extremely effective, with little water input, no sun, and no release of greenhouse gas emissions. The mycelial mat formed as a result is then harvested and goes through an important processing phase. It is treated in a process of heat pressing and drying, which stops the growth and gives it a pliable, flexible sheet form (Ali et al. 2021). One of the main benefits of this bio-fabrication is the greatly diminished use of toxic chemicals. In contrast to traditional leather tanning that depends heavily on chromium and other toxins, mycelium-derived materials tend to use natural, vegetable-based tanning processes or need only very mild solutions to gain stability and toughness.

The final material has a texture that is strikingly leather-like, supple, and flexible, with a natural, irregular grain that is very highly valued in luxury products. In addition, at the end of its product life, a fabric made from fungi that has not been blended with synthetic polymers is completely biodegradable and can be composted, returning nutrients to the soil and ending a circular lifecycle. This places mycelium not just as a substitute but as a root technology for a new generation of grown, rather than manufactured, materials that provide a scalable solution for the interlocked crisis of industrial poisoning, animal suffering, and depletion of resources (Vico et al. 2024).

The path of fungi-based fabric innovation is one of swift scientific progress and deliberate business development, from a specialized idea to a material being endorsed by international brands. Mycoremediation was one of the early areas of research looking into the natural ability of mycelium to form strong structures (Minh and Ngan 2021). One of the major companies in mycelium-based materials was Ecovative Design, which started off by making mycelium packaging that could replace polystyrene foam, thus altering the whole perception of mycelium as a material suitable for industry. A breakthrough in this area was not without problems regarding the growth process, which needed to be improved to produce thinner, more even, and supple mats that would have the correct density and flexibility for fashion use. Such a development would include not only careful selection of the strain but also identifying specific fungal species such as Ganoderma and Fomes, which are known to naturally have leather-like properties, and then finally refining the substrate and growth conditions, such as temperature, humidity, $CO_2$ levels (Chittenden 2023), etc.

The material development process continues with the aim to improve its performance in various aspects such as tensile strength, abrasion resistance, and water repellency, and at the same time keep its sustainable ratings intact. This is usually done with bio-based coating or making minimal changes during the growing phase instead of synthetic laminates (Kniep et al. 2024). The existing challenge is that of producing on a mass level at a price that competes with incumbent materials, a barrier that is being met by using computerized, high-volume vertical farming methods. As production increases, the emphasis is also on making full lifecycle analyses to quantify and demonstrate the low environmental impact of the material, especially in minimal water and land use (Shakir and Ahmad 2024). The creation of fungus-based fabrics is therefore an astounding example of biomimicry, showing how an understanding of biological processes can be used to generate high-performance materials that are innately consistent with planetary health and offer a new standard for the future of manufacturing and the fashion world (Jones et al. 2018a).

### 15.7.1 Applications of Fungi-Derived Fabrics

- Mycelium-based textiles can be applied in various sectors, providing a sustainable option to traditional fabrics (Sharma et al. 2024). It is in the clothing and fashion trade that their influence is most conspicuous. These are the fabrics chosen by both major fashion brands and avant-garde ones to realize the new vegan generation of luxury articles. This also pertains to shoes—such as sneakers and boots—as it is employed in uppers, accents, and even stiffeners (El-Den et al. 2024).
- As an accessory material, it has been skillfully patterned into handbags, wallets, belts, watch straps, and hats, oftentimes with a high-end, suede-like finish that speaks to the ethically minded consumer. Outside of accessories, mycelium is being studied for nonwoven apparel applications like jackets, waistcoats, and caps to bring unique textures and narratives to high-fashion lines (Abdel-Maksoud et al. 2023).
- Perhaps one of the most scalable and ready-to-implement applications would be in protective packaging. Mycelium is grown in molds specifically created for a particular product, like electronics, wine bottles, or sensitive components. In just a few days, it forms a solid block that is identical in function to expanded polystyrene (EPS) foam, except it is totally compostable at home after use. The product converts agricultural waste into a valuable protective tool, which eliminates one major waste source for the shipping and logistics industry (Bai et al. 2022).
- Additionally, the insulating and fire-retardant properties of the material are being investigated for potential use in advanced industrial applications, such as biodegradable building and automotive insulation (Zhang et al. 2025; Daget et al. 2024).

## 15.8 Properties of Fungi-Derived Leather

Fungi-leather, which is made primarily from mycelium, is light, flexible, long-lasting, breathable, and biodegradable, providing a green alternative to synthetic and animal leathers. It may be biologically engineered to replicate various textures, colors, and thicknesses while possessing environmentally friendly and nonpoisonous characteristics. Porosity, mechanical strength, and thermal insulation are the properties that make it fit for textiles, footwear, furniture, and wraps. Its adjustability is very high with physical, chemical, and biological post-treatments. Overall, it is an environmentally friendly solution that has good performance and little ecological impact (Wattanavichean et al. 2025). The different properties are mechanical, physical, chemical, thermal, and aesthetic function properties.

### 15.8.1 Mechanical Properties

Fungi-based leathers, made from mycelium, have mechanical strength that can be tuned according to the type of fungus, its growing conditions, and the treatments applied after the harvest. Mycelium sheets in their natural form usually have a tensile strength of 2–7 MPa, but through methods like hot pressing or chemical crosslinking, can be increased to over 10 MPa and are on par with natural leathers (Wattanavichean et al. 2025). The elasticity and tensile properties depend on the density and orientation of the hyphal network, which can be designed for flexibility or rigidity as per the application requirement (Kumar et al. 2025). Talking about flexibility and elongation, mycelium-based leathers display elongation at break of 10–25%, which means the material can be distorted and molded without breaking (Sierra et al. 2023). This trait, together with breathability and lightness, makes this material perfect for acquisitive uses such as shoes and apparel. Besides, the compressive strength and Young's modulus can be increased by reinforcing with natural fibers or applying polymer coatings; thus, tougher composite materials are formed (Haneef et al. 2017).

All these changes make fungal leather durable, yet it is lighter and more friendly to the environment. Tear resistance and abrasion resistance are the major mechanical parameters too; they are necessary for the product's durability. Working on fungal leather's abrasion, heat pressing, and surface treatment have been the methods used to create a smoother, denser structure with fewer damage points (Mautner 2023). Moreover, fungi-derived materials can be manufactured to have even thickness and be quite free from natural blemishes (like scars on animal hides), thus giving a more uniform mechanical property across the large-area sheets (Angelova et al. 2021). The combined mechanical characteristics of these materials make fungi-based leather an attractive, eco-friendly alternative for many sectors like fashion, furniture, and packaging.

### 15.8.2 Physical Properties

Mushroom leather, largely derived from fungal mycelium, has unique physical properties that enable it to be classified as an eco-friendly alternative to both animal and synthetic leathers. The material's strength, flexibility, and porosity are mainly determined by the species of fungus, the substrate mixture, and the treatments applied, such as pressing, chemical crosslinking, and drying. Studies have shown that mycelium from Ganoderma lucidum, Pleurotus ostreatus, and Fomes fomentarius yields leather with tensile strengths and elongation percentages reaching up to 50%, thereby being comparable to the case of natural leather for lightweight uses (Byoungmin et al. 2025). Furthermore, such material has low-density, breathable, and adjustable mechanical properties, which make it possible to replicate the smoothness and feel of animal leather when introduced with eco-friendly coatings (Javadian et al. 2020).

Mushroom leather not only conveys pleasant qualities in terms of strength and flexibility but also shows good thermal, acoustic, and water absorption characteristics that further enhance its performance in different areas. The mycelial fibrous network provides the material with natural insulation, making it light and strong, together with the factors of biodegradability and low environmental impact (Basak et al. 2024; Jeong and Shin 2023). Among processing techniques, hot pressing and chemical tanning lead to a more hydrophobic and smoother surface of the leather, thus enhancing its resistance to wear and extending the range of usability in fashion and upholstery to the point where it is common (Ko et al. 2007). Ongoing research amplified that those leathers made from mycelium are on par with traditional leathers in terms of performance, offering an eco-friendly alternative that satisfies the criteria of strength, flexibility, and sustainability (Rajapakshe et al. 2025).

### 15.8.3 Chemical Properties

Mushroom leather is a synthetic creation made of a highly interconnected chitin, β-glucans, proteins, and polysaccharide network that gives it the same structural strength and flexibility as animal collagen in leather. The chemical groups present on the surface—hydroxyl, amino, and carboxyl—make it possible to modify the surface through eco-friendly tanning, acetylation, or coating procedures to get improved hydrophobicity and resistance (Pertile et al. 2025). Moreover, the stability and resistance to microbial attacks that lead to degradation are given by the phenolic residues and the lignin-like materials that are inserted from the growth substrate. The use of bio-based resin and natural tannin in post-processing gives the product extra chemical durability while still allowing for the Indian mushroom leather to be classified as a biodegradable material, thus making it a safe and sustainable choice over conventional leathers (Plavan et al. 2019).

### 15.8.4 Thermal Properties

The mycelial structure that mainly comprises chitin, glucans, and proteins is the reason behind the superb thermal properties of mushroom leather, which offer natural insulation and a low rate of heat transfer (Tansakul and Lumyong 2008). Mycelium's porous structure gives it a low thermal conductivity, akin to that of commercial synthetic foams, thereby making it light and efficient as a thermal insulator. Moreover, the product possesses remarkable fire resistance since mycelium undergoes charring rather than melting or emitting toxic smoke, which is one of the main reasons in comparison to synthetic leathers (Owen et al. 2021). Mushroom leather can be further stabilized thermally through the application of post-processing techniques such as hot pressing, chemical crosslinking, or bio-based flame retardant incorporation and allowed to withstand moderate heat levels (Pertile et al. 2025). The characteristics of mycelium leathers make them suitable not only for the fashion industry but also for interior and packaging applications where thermal insulation and fire resistance are indispensable (Meyer et al. 2021).

### 15.8.5 Aesthetic Function Properties

Mushroom leather, along with its unique aesthetic and functional features, is suggested as a sustainable alternative to animal and synthetic leathers. The mycelial network, which is present in the mushroom leather, allows for various types of surface textures with smooth, suede-like, and fibrous finishes depending on the type of fungus and processing method used (Nithyaprakash et al. 2020). The material can be made soft, elastic, and breathable for fashion and upholstery applications while still being biodegradable (Rühse and Horvath 2024). Moreover, the material can easily absorb the natural dye and coating, allowing different color variations to be made without the use of any harmful chemicals (Javadian et al. 2020). From the functional point of view, mushroom leather joins lightweight flexibility with durability, resistance to wear, and the ability to undergo shaping and stitching operations, thus allowing its use in such products as footwear, bags, and interior design accessories (Khan et al. 2024). Such functional and aesthetic features are in favor of the consumer, and hence, mycelium leather can be positioned as a value-added material in the sustainable sectors due to its increasing popularity among consumers.

## 15.9 Market Potential for Fungi-Derived Leather

Fungal leather, or mycelium leather, has been the most talked-about and praised of all the revolutionary and green alternatives to the traditional animal and synthetic leathers due to its machine-manufactured eco-friendliness, biodegradability, and versatile material properties. Valued at over 400 billion dollars, the global leather industry is under scrutiny as environmental pollution, animal welfare, and high carbon emissions keep rising incessantly (Muthu and Ramchandani 2024). The more people go for plant-based and eco-friendly lifestyles, the more mushroom leather becomes a viable solution with an even smaller ecological footprint, which in turn will aid its entrance into the mainstream fashion and materials markets (Camere and Karana 2018). The production of fungus-based leather takes less effort than the production of leather from animals, utilizing agricultural by-products and even reducing the emission of greenhouse gases, the use of water, and the area of land taken by half (Appels et al. 2019). Unlike petrochemical-based synthetic leathers that cause microplastic pollution, mushroom leather is biodegradable and compostable, thus maintaining harmony with the principles of a circular economy (Haneef et al. 2017). These eco-friendly characteristics not only attract pro-environment consumers but also sustainable brands, making it a competitive alternative in the market.

Mushroom leather is now being used by high-fashion brands as a part of their collection, which has great commercial potential. Adidas, Stella McCartney, and Hermès are some of the leading fashion houses that have partnered with mycelium material startups Bolt Threads (Mylo™) and MycoWorks (Reishi™) to create prototypes and luxury items (van Rooyen 2024; Mallon 2024; Ekegårdh and Kristersson 2021). The collaborations are a clear indication of the mushroom leather's luxury application versatility and the trend of its acceptance among international markets. The success of such market entry for these products indicates consumer demand, as well as industry willingness to embrace fungi-based materials. The potential of the global market for fungi-based leather is fueled by growing demand in the fashion, footwear, furniture, and automotive sectors for sustainable materials. The eco-leather market is expected to expand significantly, and mushroom leather stands as a standout innovation in this market niche (Spinnewijn 2022; Muthu and Ramchandani 2024). According to forecasts by analysts, the bio-based leather market could reach multi-billion-dollar market values by 2030, with major contributions from fungi-based material due to scalability and consumer interest in cruelty-free options (Perez 2020).

Consumer behavior is significant in determining the market for mushroom leather. Growing concern for environmental damage, combined with the need for products that are animal-free as well as vegan-friendly, has widened the market base for bio-based products. Consumers belonging to the millennial and Gen Z generations, which control a large share of the luxury fashion and lifestyle market, have a special affinity for sustainable innovations, imparting mushroom leather good commercial muscle (Chalannavar et al. 2025). Mushroom leather, in its functional aspects, is a highly adaptable being, possessing properties such as strength, flexibility, and texture that can be adjusted to fit the requirements of different industries. Its capability to adapt gives it the bearability with natural leathers and synthetic ones, thus opening its market to the whole range of applications from shoes and fashion accessories to upholstery and car interiors (Elsacker et al. 2019). Moreover, the fact that one can tailor the look and mechanical properties gives it an edge as a sustainable material in the high-end category.

On the other hand, despite these pros, there are still scaling up of production and manufacturing cost competitions with the established alternative leathers that are constantly being challenged. The present situation is that the output is being restricted to small amounts because of the high R&D costs and complicated post-treatment processes. Still, the technological progress that is taking place, together with the investment from the major companies and the government policies that support the innovation of sustainable materials, is expected to overcome these obstacles and speed up the process of commercialization (Attias et al. 2020). The market of fungal-derivative leather is large enough, and its attractiveness to many industries is characterized by high growth. Its combination of environmental benefits, performance advantages, and consumer acceptance places it as an important driving force in the future of sustainable materials. The more technology advances to make it easier and cheaper to produce, the more mushroom leather will become a common material in fashion, furniture, cars, and other sectors. Thus, it will be a great help in the global transition of industries to sustainability (Jones et al. 2017).

## 15.10 Conclusion

Among the eco-friendly alternatives, mushroom leather has become a great choice, which goes on to deal with the environmental and ethical issues created by the leather industry. Its low resource consumption, complete biodegradability, and use of nontoxic processing methods make it not only a challenger but also the leader of the eco-friendly movement in the manufacturing industry. Continuous research on mycelium growth and material engineering is rapidly making mushroom leather stronger, more attractive, and longer lasting. The competition in the market, industrial interest, and consumer awareness are very quickly leading to its progress. Many of the companies started using mushroom leather, which will set the standards in the fashion, interior design, and other material-dependent industries. Ultimately, mushroom leather is a brave new world solution that combines environmental protection with contemporary convenience.

**Acknowledgments** The authors would like to thank the Head, School of Life Sciences (Ooty Campus), JSS AHER, Mysuru, for providing the facilities and suggestions in the manuscript.

**Declaration** The authors have no competing interests to declare that are relevant to the content of this review article.

## References

Abdel-Maksoud G, Nasr H, Kassem M (2023) Biopolymer applications for different treatments of the historical paper manuscripts and leather bindings: a review. Egypt J Chem. https://doi.org/10.21608/ejchem.2023.225573.8320

Abdulhusein HS, Kadim BM (2024) Antimicrobial substances and strategies to avoid bacterial and fungal effects in leather manufacturing. Kafkas Univ Fen Bilim Enst Derg 2024:1467530. https://doi.org/10.58688/kujs.1467530

Abraham SM (2023) Pineapple fibre-reinforced polyester composites. In: Polyester-based biocomposites. CRC Press, pp 25–43

Absharina D, Padri M, Veres C, Vágvölgyi C (2025) Bacterial cellulose: from biofabrication to applications in sustainable fashion and vegan leather. Fermentation 11:23. https://doi.org/10.3390/fermentation11010023

Aiduang W, Chanthaluck A, Kumla J, Jatuwong K, Srinuanpan S, Waroonkun T, Oranratmanee R, Lumyong S, Suwannarach N (2022) Amazing fungi for eco-friendly composite materials: a comprehensive review. J Fungi 8:842. https://doi.org/10.3390/jof8080842

Aiduang W, Jatuwong K, Luangharn T, Jinanukul P, Thamjaree W, Teeraphantuvat T, Waroonkun T, Lumyong S (2024) A review delving into the factors influencing mycelium-based green composites (MBCs) production and their properties for long-term sustainability targets. Biomimetics 9:337. https://doi.org/10.3390/biomimetics9060337

Akhter S, Jahan MS, Rahman ML, Ruhane TA, Ahmed M, Khan MA (2024) Revolutionizing sustainable fashion: jute–mycelium vegan leather reinforced with polyhydroxyalkanoate biopolymer crosslinking from novel bacteria. Adv Polym Technol 2024:1304800. https://doi.org/10.1155/2024/1304800

Alaneme KK, Anaele JU, Oke TM, Kareem SA, Adediran M, Ajibuwa OA, Anabaranze YO (2023) Mycelium based composites: a review of their bio-fabrication procedures, material properties and potential for green building and construction applications. Alex Eng J 83:234–250. https://doi.org/10.1016/j.aej.2023.10.012

Al-Bader SM, Mohammed NJ (2022) Fungal screening of women shoulder handbags with special reference to get rid of *Aspergillus niger* via essential oil vapours. AIP Conf Proc 2398:040016. https://doi.org/10.1063/5.0095494

Alemu D, Tafesse M, Mondal AK (2022) Mycelium-based composite: the future sustainable biomaterial. Int J Biomater 2022:8401528. https://doi.org/10.1155/2022/8401528

Ali MF, Hossain MS, Ahmed S, Chowdhury AS (2021) Fabrication and characterization of eco-friendly composite materials from natural animal fibers. Heliyon 7(5):e06954. https://doi.org/10.1016/j.heliyon.2021.e06954

Amobonye A, Lalung J, Awasthi MK, Pillai S (2023) Fungal mycelium as leather alternative: a sustainable biogenic material for the fashion industry. Sustain Mater Technol 38:e00724. https://doi.org/10.1016/j.susmat.2023.e00724

Andia A, Peralta Gomez N (2025) SynBio architecture: from lab-grown materials to bio-programmable infrastructures. Int J Archit Comput 23:856–871. https://doi.org/10.1177/14780771251352947

Angelova GV, Brazkova MS, Krastanov AI (2021) Renewable mycelium based composite–sustainable approach for lignocellulose waste recovery and alternative to synthetic materials–a review. Z Naturforsch C 76(11–12):431–442

Antinori ME (2021) Pure mycelium materials: characteristics and applications. Doctoral thesis

Antony Jose S, Cowan N, Davidson M, Godina G, Smith I, Xin J, Menezes PL (2025) A comprehensive review on cellulose nanofibers, nanomaterials, and composites: manufacturing, properties, and applications. Nanomaterials 15(5):356. https://doi.org/10.3390/nano15050356

Appels FV, Dijksterhuis J, Lukasiewicz CE, Jansen KM, Wösten HA, Krijgsheld P (2018) Hydrophobin gene deletion and environmental growth conditions impact mechanical properties of mycelium by affecting the density of the material. Sci Rep 8(1):4703. https://doi.org/10.1038/s41598-018-23171-2

Appels FV, Camere S, Montalti M, Karana E, Jansen KM, Dijksterhuis J, Krijgsheld P, Wösten HA (2019) Fabrication factors influencing mechanical, moisture- and water-related properties of mycelium-based composites. Mater Des 161:64–71. https://doi.org/10.1016/j.matdes.2018.11.027

Ariyani NR, Wijayanti SP, Putra NG, Kusumawati MB, Setiawan AAR, Isharyadi F, Widyastuti N, Djarot IN, Handayani T (2024) Mycelium-based leather potential for sustainability: a review. AIP Conf Proc 3001(1):030012. https://doi.org/10.1063/5.0184187

Arman I, Ansari KB, Danish M, Farooqi IH, Jain AK (2023) Ultrasonic-assisted feedstock disintegration for improved biogas production in anaerobic digestion: a review. Bioenergy Res 16(3):1512–1527. https://doi.org/10.1007/s12155-023-10608-4

Attias N, Danai O, Abitbol T, Tarazi E, Ezov N, Pereman I, Grobman YJ (2020) Mycelium bio-composites in industrial design and architecture: comparative review and experimental analysis. J Clean Prod 246:119037. https://doi.org/10.1016/j.jclepro.2019.119037

Baars JJ, Mishra P, Hendrickx P, van der Horst C, van Peer A (2024) Development of a circular sustainable culturing process for natural leather-like materials based on fungal mycelium. Stichting Wageningen Research, Wageningen Plant Research, Business Unit Plant Breeding, Wageningen. https://doi.org/10.18174/676633

Bahua H, Wijayanti SP, Putra AS, Ariyani NR, Isharyadi F, Nuha N, Paminto AK, Mulyono M, Djarot IN, Widyastuti N, Sitomurni AI (2024) Life cycle assessment (LCA) of leather-like materials from mycelium: Indonesian case study. Int J Life Cycle Assess 29:1916–1931. https://doi.org/10.1007/s11367-024-02351-5

Bai Z, Wang X, Zheng M, Yue O, Xie L, Zha S, Dong S, Li T, Song Y, Huang M, Liu X (2022) Leather for flexible multifunctional bio-based materials: a review. J Leather Sci Eng 4:1. https://doi.org/10.1186/s42825-022-00091-6

Barrera Torres G, Gutierrez Aguilar CM, Lozada ER, Tabares Montoya MJ, Ángel Álvarez BE, Sánchez JC, Jaramillo Carvalho JA, Santos RJ (2025) Application of post-industrial leather waste for the development of sustainable rubber composites. Polymers 17:190. https://doi.org/10.3390/polym17020190

Basak S, Shakyawar DB, Samanta KK, Kumar N, Ghosh S (2024) Sustainable flexible composite: an upcoming development in technical textile. BTRA Scan 53:3

Battal Ö, Acar M, Uzun M (2025) Fashionable sustainable leather. J Art Des Music 4:10. https://doi.org/10.55554/2785-9649.1048

Benetti B, Conti F, Dimitriadis P (2025) Mycelium-based leather: a review on post-processing treatments and material enhancements. Rigas Tehn Univ Zinatn Raksti 29:390–404. https://doi.org/10.2478/rtuect-2025-0026

Bhatia S, Katoch P (2022) Mushroom: a review on multipurpose fungi and its commercial utilization. Int J Adv Agric Sci Technol 9:66–84. https://doi.org/10.47856/ijaast.2022.v09i05.007

Bi Z, Crnković T (2024) Bio-leather: sustainable clothing fabrics made from simple media ingredients and slime mold Physarum polycephalum. J Ind Text 54:15280837241254510. https://doi.org/10.1177/15280837241254510

Bonga KB (2024) Tuning of self-growing natural composite materials using non-pathogenic fungal strains—towards 3D constructs. Doctoral thesis

Bustillos J, Loganathan A, Agrawal R, Gonzalez BA, Perez MG, Ramaswamy S, Boesl B, Agarwal A (2020) Uncovering the mechanical, thermal, and chemical characteristics of biodegradable mushroom leather with intrinsic antifungal and antibacterial properties. ACS Appl Bio Mater 3: 3145–3156. https://doi.org/10.1021/acsabm.0c00164

Byoungmin C, Yeonju K, Hoyoung L, Yoonhee A, Yang J, Jonghoon K (2025) Pretreatment conditions of mushroom mycelium as a leather substitute. https://doi.org/10.2478/9788367405829-041

Camere S, Karana E (2018) Fabricating materials from living organisms: an emerging design practice. J Clean Prod 186:570–584. https://doi.org/10.1016/j.jclepro.2018.03.081

Campos AD, Wahalathanthrige HJ, Russell S, Harrison MD, Strong PJ (2025) Conversion of soluble compounds in distillery wastewater into fungal biomass and metabolites using Australian Ganoderma isolates. J Fungi 11:432. https://doi.org/10.3390/jof11060432

Chalannavar RK, Malabadi RB, Divakar MS, Komalakshi KV, Kamble AA, Karamchand KS, Kolkar KP, Coronado KVC, Munhoz ANR (2025) Industrial *Cannabis sativa* (fiber or hemp): hemp-made leather. World J Adv Res Rev 25:2207–2218. https://doi.org/10.30574/wjarr.2025.25.2.0503

Chang MY, Tsai GJ, Houng JY (2006) Optimization of the medium composition for the submerged culture of Ganoderma lucidum by Taguchi array design and steepest ascent method. Enzyme Microb Technol 38(3-4):407-414. https://doi.org/10.1016/j.enzmictec.2005.06.011

Chatterjee S, Das A, Paul D, Chakraborty S, Choudhury P (2023) Utilization of fleshing waste of leather processing for the growth of zygomycetes: a new substrate for economical production of bio-polymer chitosan. J Environ Manag 343:118141. https://doi.org/10.1016/j.jenvman.2023.118141

Chen H, Abdullayev A, Bekheet MF, Schmidt B, Regler I, Pohl C, Vakifahmetoglu C, Czasny M, Kamm PH, Meyer V, Gurlo A, Simon U (2021) Extrusion-based additive manufacturing of fungal-based composite materials using the tinder fungus *Fomes fomentarius*. Fungal Biol Biotechnol 8:1. https://doi.org/10.1186/s40694-021-00129-0

Chittenden T (2023) Foraging for fashion's future: the use of mycelium materials and fungi intelligence in fashion design. Fash Style Pop Cult 12. https://doi.org/10.1386/fspc_00174_1
Choi SM, Lee DH, Zo SM, Sood A, Han SS (2025) Gel-based approaches to vegan leather: opportunities and challenges in mimicking leather properties. Gels 11:395. https://doi.org/10.3390/gels11060395
Crawford A, Miller SR, Branco S, Fletcher J, Stefanov D (2024) Growing mycelium leather: a paste substrate approach with post-treatments. Res Dir Biotechnol Des 2:e6. https://doi.org/10.1017/btd.2024.6
da Silva Simão R, de Moraes JO, Carciofi BAM, Laurindo JB (2020) Recent advances in the production of fruit leathers. Food Eng Rev 12(1):68-82. https://doi.org/10.1007/s12393-019-09200-4
d'Errico A, Schröpfer M, Mondschein A, Wösten HA (2025) Characterization of the surface charge and reactivity of *Schizophyllum commune* mycelium material. Colloids Surf B Biointerfaces 254:114852. https://doi.org/10.1016/j.colsurfb.2025.114852
Dabeta E (2018) Study on effect of drying techniques on area yield and quality of leather. AAU
Daddi T, Ahmad S, Albano F (2024) Fostering environmental sustainability of leather industry: environmental impact assessment and improvement prospects for leather-tanning technologies. Environ Sci Pollut Res Int 32(54):29995–30012. https://doi.org/10.1007/s11356-024-35625-4
Daget TM, Kassie BB, Abate MT, Teshome MF, Arega MM, Atalie D (2024) Eco-friendly and bio-based approaches for flame-retardant functionalization of textile materials: a review. Polym Plast Technol Mater 63:1887–1916
Das RK, Mizan A, Zohra F, Başaran B, Ahmed S (2022) Application of indigenous plant-based vegetable tanning agent extracted from *Xylocarpus granatum* in semi-chrome and chrome retanned leather production. Tekstil Konfeksiyon 32:258–264. https://doi.org/10.32710/tekstilvekonfeksiyon.1066811
De Menezes Rodrigues G, Garcia VADS, Yoshida CMP, Vanin FM, De Carvalho RA (2023) Fruit-based leathers: a comprehensive review of terminologies, composition, and quality attributes. Food Sci Eng 4:216–236. https://doi.org/10.37256/fse.4220232791
Deeg K, Gima Z, Smith A, Stoica O, Tran K (2017) Greener solutions: improving performance of mycelium-based leather. Final report to MycoWorks, pp 1–24
Dhanda V, Arsalan S, Kaushal S (2024) Revolutionizing material: the rise of bio leather as eco-friendly and sustainable approach. Int J Agron 7:121–128. https://doi.org/10.33545/2618060X.2024.v7.i11b.1954
Duangsuwan S, Junkong P, Phinyocheep P, Thanawan S, Amornsakchai T (2023) Development of green leather alternative from natural rubber and pineapple leaf fiber. Sustainability 15:15400. https://doi.org/10.3390/su152115400
Ekegårdh N, Kristersson E (2021) From fungi to vegan leather: a case study highlighting the potentials of mycelium-based leather in the textile industry and the unique features of its value chain. Accessed from https://www.diva-portal.org/smash/get/diva2:1654482/FULLTEXT01.pdf. Accessed on 13 Apr 2024.
El-Den RG, Wahab SA, Atallah H, Saleh Y, Hassabo A (2024) Environmental sustainability and innovations in the fashion industry. J Text Color Polym Sci. https://doi.org/10.21608/jtcps.2024.259736.1314
Elkhateeb WA, Galappaththi MCA, Wariss HM, Haesendonck KV, Daba GM (2022) Fungi-derived leather (mushroom leather). MycoKing 1:1–5
El-Ramady H, Abdalla N, Fawzy Z, Badgar K, Llanaj X, Törős G, Hajdú P, Eid Y, Prokisch J (2022) Green biotechnology of oyster mushroom (*Pleurotus ostreatus* L.): a sustainable strategy for myco-remediation and bio-fermentation. Sustainability 14:3667. https://doi.org/10.3390/su14063667
Elsacker E, Vandelook S, Brancart J, Peeters E, De Laet L (2019) Mechanical, physical and chemical characterisation of mycelium-based composites with different types of lignocellulosic substrates. PLoS One 14:e0213954. https://doi.org/10.1371/journal.pone.0213954
Elsacker E, Vandelook S, Peeters E (2023) Recent technological innovations in mycelium materials as leather substitutes: a patent review. Front Bioeng Biotechnol 11:1204861. https://doi.org/10.3389/fbioe.2023.1204861

Farra E (2021) Stella McCartney introduces her first garments made of Mylo, the "leather" alternative grown from mushrooms. Obtido em 3 de 3 de 2022, de Vogue: https://www.vogue.com/article/stella-mccartney-first-mylo-mushroom-leather-alternative-ready-to-wear

Farrahnoor A, Hishamuddini ADN, Yusoff H, Hyie KM, Maideen NC, Fohimi NAM, Zhou BT (2025) Influence of drying temperature in the oven on physical, morphology and mechanical properties of mycelium composite. Pertanika J Sci Technol 33:1. https://doi.org/10.47836/pjst.33.1.10

Fuck WF, Brandelli A, Gutterres M (2018) Leather dyeing with biodyes from filamentous fungi: a special review. J Am Leather Chem Assoc 113:299–310

Gandia A, Van Den Brandhof JG, Appels FV, Jones MP (2021) Flexible fungal materials: shaping the future. Trends Biotechnol 39:1321–1331. https://doi.org/10.1016/j.tibtech.2021.03.002

Gao X (2023) Wearing pineapple leaves? Consumer responses to fashion products made from repurposed materials. PhD thesis, Wageningen University and Research

Giehl D, Hansen É, Robinson LC, De Aquim PM (2024) Reusing finished leather waste to produce pigmented thermoplastic polyurethane composite. Collagen Leather 6:1. https://doi.org/10.1186/s42825-024-00149-7

Gržetić J, Banjanac K (2025) Lignin in water purification: enzymatic and adsorption-maintained processes. In: Handbook of lignin. Springer Nature, Singapore, pp 1–32. https://doi.org/10.1007/978-981-97-2664-6_63-1

Gutarowska B, Gibka D, Olejnik K, Pospiech P, Boruta T, Kapela T, Makowski K (2024) Biocomposites based on mould biomass and waste fibres for the production of agrotextiles: technology development, material characterization, and agricultural application. Materials 17:6084. https://doi.org/10.3390/ma17246084

Haneef M, Ceseracciu L, Canale C, Bayer IS, Heredia-Guerrero JA, Athanassiou A (2017) Advanced materials from fungal mycelium: fabrication and tuning of physical properties. Sci Rep 7:41292. https://doi.org/10.1038/srep41292

Hassan RRA (2020) In vitro study for use of cactus gel in enhancing the mechanical strengths of vegetable tanned leathers under accelerated aging. Int J Precis Eng Manuf 21:145–155. https://doi.org/10.1007/s12541-019-00178-x

Hennicke F, Cheikh-Ali Z, Liebisch T, Maciá-Vicente JG, Bode HB, Piepenbring M (2016) Distinguishing commercially grown *Ganoderma lucidum* from *Ganoderma lingzhi* from Europe and East Asia on the basis of morphology, molecular phylogeny, and triterpenic acid profiles. Phytochemistry 127:29–37. https://doi.org/10.1016/j.phytochem.2016.03.012

Holt GA, Mcintyre G, Flagg D, Bayer E, Wanjura JD, Pelletier MG (2012) Fungal mycelium and cotton plant materials in the manufacture of biodegradable molded packaging material: evaluation study of select blends of cotton byproducts. J Biobaased Mater Bioenergy 6:431–439. https://doi.org/10.1166/jbmb.2012.1241

Hou C (2024) Fungal living materials and the application of bacteria-fungi interaction on regulating mycelial growth: a review. Master's thesis

Im HS, Tagawa S, Jeong JS, Shin HJ (2025) Production of decolorized mushroom pulp for nonwoven cotton composite. Fibers 13:30. https://doi.org/10.3390/fib13030030

Iordache O, Cazan B, Perdum E, Tănăsescu C, Secăreanu L (2023) From fungi to biomaterials: exploiting filamentous fungi strains for sustainable applications. In: The 11th international conference TEX TEH, pp 131–138

Javadian A, Le Ferrand H, Hebel DE, Saeidi N (2020) Application of mycelium-bound composite materials in construction industry: a short review. SOJ Mater Sci Eng 7:1–9. https://doi.org/10.3929/ethz-b-000543782

Jenitta XJ, Gnanasalomi VDV, Gnanadoss JJ (2013) Treatment of leather effluents and waste using fungi. Int J Comput Algorithm 2:294–298

Jeong YH, Shin HJ (2023) Development of eco-friendly composite materials using *Trametes orientalis* mycelium and regional biomass. 버섯 27:67

Jhanji Y (2023) Mushroom and corn fibre—the green alternatives to unsustainable raw materials. In: Sustainable fibres for fashion and textile manufacturing. Woodhead Publishing, pp 129–158. https://doi.org/10.1016/B978-0-12-824052-6.00012-3

Jhanji Y, Kothari VK, Gupta D (2022) A critical review on prediction of functional & performance attributes of textiles by artificial neural network. Indian J Fibre Text Res 47(2):252–258

Jiang L (2022) Production of bioresins from fungal mycelia. In: Fungal biopolymers and biocomposites: prospects and avenues. Springer, pp 255–293. https://doi.org/10.1007/978-981-19-1000-5_13

Jinanukul P, Kumla J, Aiduang W, Thamjaree W, Oranratmanee R, Shummadtayar U, Tongtuam Y, Lumyong S, Suwannarach N, Waroonkun T (2024) Comparative evaluation of mechanical and physical properties of mycelium composite boards made from *Lentinus sajor-caju* with various ratios of corn husk and sawdust. J Fungi 10:634. https://doi.org/10.3390/jof10090634

Jo YK (2023) Natural biopolymer-based hydrogels as designer matrices for organoid cultures. Organoid 3:e13. https://doi.org/10.51335/organoid.2023.3.e13

Jones M, Huynh T, Dekiwadia C, Daver F, John S (2017) Mycelium composites: a review of engineering characteristics and growth kinetics. J Bionanosci 11:241–257

Jones M, Bhat T, Huynh T, Kandare E, Yuen R, Wang CH, John S (2018a) Waste-derived low-cost mycelium composite construction materials with improved fire safety. Fire Mater 42:816–825. https://doi.org/10.1002/fam.2637

Jones M, Bhat T, Kandare E, Thomas A, Joseph P, Dekiwadia C, Yuen R, John S, Ma J, Wang CH (2018b) Thermal degradation and fire properties of fungal mycelium and mycelium-biomass composite materials. Sci Rep 8:17583. https://doi.org/10.1038/s41598-018-36032-9

Jones M, Gandia A, John S, Bismarck A (2021) Leather-like material biofabrication using fungi. Nat Sustain 4:9–16. https://doi.org/10.1038/s41893-020-00606-1

Josroy AV, Sundervel A (2024) Unlocking the potential of tropical fruit waste to develop a biobased material: an environment-friendly approach. Text Leather Rev 7:340–351. https://doi.org/10.31881/tlr.2023.171

Kaaronen RO (2025) Living in the mycelial world. https://doi.org/10.1111/tops.70003

Kalia A, Singh S (2020) Myco-decontamination of azo dyes: nano-augmentation technologies. 3 Biotech 10:384. https://doi.org/10.1007/s13205-020-02378-z

Kalitukha L, Galiano A, Harrison F (2023) Medicinal potential of the insoluble extracted fibers isolated from the Fomes fomentarius (Agaricomycetes) fruiting bodies: A review. Int J Med Mushrooms 25(3). https://doi.org/10.1615/IntJMedMushrooms.2022047222

Kandel B (2022) Mycelium: using mushrooms for the future packaging materials in Europe. Bachelor's thesis

Kanth SV, Hussain H (2024) Capacity building of global leather sector for inclusivity and sustainable development. In: Emerging trends in leather science and technology. Springer, Singapore, pp 277–302. https://doi.org/10.1007/978-981-99-9754-1_10

Kapoor S, Verma N, Charak H, Dutta K, Gupta A, Gupta V (2024) Flexible fungal materials. In: Microbial approaches for sustainable green technologies. CRC Press, p 154

Karana E, Blauwhoff D, Hultink EJ, Camere S (2018) When the material grows: a case study on designing (with) mycelium-based materials. Int J Des 12(2):119–136

Karthikeyan MR, Raja D, Sri AA, Bilal S, Pooja C, Sindhuja KG (2024) Mycelium-based flame retardant treatments for textiles using extracts of Turkey tail, reishi, and shiitake. J Text Assoc 85:642–648. https://doi.org/10.63665/jta.v85i6.05

Karunarathne A, Nabiyeva G, Rasmussen CJ, Alkhoury K, Assem N, Bauer J, Chester SA, Khalizov AF, Gor GY (2024) Effects of humidity on mycelium-based leather. ACS Appl Bio Mater 7:6441–6450. https://doi.org/10.1021/acsabm.4c00586

Kefale GY, Kebede ZT, Birlie AA (2023) A systematic review on potential bio leather substitute for natural leather. J Eng 2023:1629174. https://doi.org/10.1155/2023/1629174

Kefale GY, Bizuneh B, Getachew L, Karthikeyan R (2025) Innovative pathways in leather industry: a comprehensive review of sustainable technologies and strategies. J Sci Ind Res 84:703–715. https://doi.org/10.56042/jsir.v84i6.10781

Khambhaty Y (2020) Applications of enzymes in leather processing. Environ Chem Lett 18:747–769. https://doi.org/10.1007/s10311-020-00971-5

Khan R (2025) Mycelium-based bioproducts: A novel material for a sustainable economy – A comprehensive review. The Microbe 8:100439. https://doi.org/10.1016/j.microb.2025.100439

Khan MR, Shakil MSR, Vorselaars B, Rahman M, Jawad IA, Ahmed W (2024) Nanotechnology in leather manufacturing. In: Micro and nanomanufacturing, vol II. Springer, Cham, pp 23–52. https://doi.org/10.1007/978-3-031-70499-4_2

Kim H, Song JE, Kim HR (2021) Comparative study on the physical entrapment of soy and mushroom proteins on the durability of bacterial cellulose bio-leather. Cellulose 28:3183–3200. https://doi.org/10.1007/s10570-021-03705-0

Kniep J, Graupner N, Reimer JJ, Müssig J (2024) Mycelium-based biomimetic composite structures as a sustainable leather alternative. Mater Today Commun 39:109100. https://doi.org/10.1016/j.mtcomm.2024.109100

Ko WC, Liu WC, Tsang YT, Hsieh CW (2007) Kinetics of winter mushrooms (Flammulina velutipes) microstructure and quality changes during thermal processing. J Food Eng 81:587–598. https://doi.org/10.1016/j.jfoodeng.2006.12.009

Kodaloğlu M, Kodaloğlu FA (2024) Environmentally-friendly recycled leather-reinforced composite: thermal and acoustic properties. Teknik Bilimler Derg 14:29–34. https://doi.org/10.35354/tbed.1484783

Kowalska M, Cordella M, Wolf O (2013) European Commission Joint Research Centre address: Joint Research Centre, Institute for Prospective Technological Studies, Edificio EXPO-C/Inca Gracilaso, 3-E 41092 Seville

Kumar P, Kapoor A, Pal DB, Raghunathan M (2025) Advanced materials from fungal mycelium: fabrication and tuning of physical properties. In: Fungal waste biomass management for energy, environment and value-added products. Springer, Cham, pp 219–240. https://doi.org/10.1007/978-3-031-82599-6_9

Lee SC, Eun CS, Kim WJ (2014) Dyeing properties of leather using red natural dyes. J Soc Leather Technol Chem 98:252–258

Lin X, Sun D (2019) Research advances in browning of button mushroom (Agaricus bisporus): affecting factors and controlling methods. Trends Food Sci Technol 90:63–75. https://doi.org/10.1016/j.tifs.2019.05.007

Liu J, Recupido F, Lama GC, Oliviero M, Verdolotti L, Lavorgna M (2023) Recent advances concerning polyurethane in leather applications: An overview of conventional and greener solutions. Collagen Leather 5(1):8. https://doi.org/10.1186/s42825-023-00116-8

Liu X, Zhang X, Wang X, Yue O, Jiang H (2024) Engineered, environmentally friendly leather-like bio-based materials. Trends Biotechnol 43:1104–1115. https://doi.org/10.1016/j.tibtech.2024.07.004

Livne A, Wösten HA, Pearlmutter D, Gal E (2022) Fungal mycelium bio-composite acts as a CO2-sink building material with low embodied energy. ACS Sustain Chem Eng 10:12099–12106. https://doi.org/10.1021/acssuschemeng.2c01314

Lu X, Brennan MA, Brennan CS (2025) Harnessing the multifaceted potential of mushrooms: sustainable development, health promotion, and industrial innovation. Int J Food Sci Technol 60:vvaf159. https://doi.org/10.1111/ijfs.15980

Luangharn T, Karunarathna SC, Mortimer PE, Hyde KD, Thongklang N, Xu J (2019) A new record of Ganoderma tropicum (Basidiomycota, Polyporales) for Thailand and first assessment of optimum conditions for mycelia production. MycoKeys 51:65–83. https://doi.org/10.3897/mycokeys.51.33513

Madusanka C, Udayanga D, Nilmini R, Rajapaksha S, Hewawasam C, Manamgoda D, Vasco-Correa J (2024) A review of recent advances in fungal mycelium based composites. Discov Mater 4:84–88. https://doi.org/10.1007/s43939-024-00084-8

Majib NM, Yaacob ND, Ting SS, Rohaizad NM, Rashidi AMA (2024) Fungal mycelium-based biofoam composite: a review in growth, properties and application. Prog Rubber Plast Recycl Technol 41:91–123. https://doi.org/10.1177/14777606241252702

Malabadi NRB, Kolkar NKP, Chalannavar NRK, Baijnath NH (2025) Plant-based leather production: an update. World J Adv Eng Technol Sci 14:31–59. https://doi.org/10.30574/wjaets.2025.14.1.0648
Mallon S (2024) Materialities of "mushroom-leather": a critical appraisal of the material dimensions and sustainability of fungus-based alternatives to leather for use in fashion and textiles. Ethnol Fennica 51:81–110
Manan S, Atta OM, Shahzad A, Ul-Islam M, Ullah MW, Yang G (2022) Applications of fungal mycelium-based functional biomaterials. In: Fungal biopolymers and biocomposites: prospects and avenues. Springer, Singapore, pp 147–168. https://doi.org/10.1007/978-981-19-1000-5_9
Mautner A (2023) Fungi-derived composite nanopapers. In: Bioprospects of macrofungi. CRC Press, pp 233–257
Melanouri E, Dedousi M, Diamantopoulou P (2021) Cultivating Pleurotus ostreatus and Pleurotus eryngii mushroom strains on agro-industrial residues in solid-state fermentation. Part I: screening for growth, endoglucanase, laccase and biomass production in the colonization phase. Carbon Resour Convers 5:61–70. https://doi.org/10.1016/j.crcon.2021.12.004
Meyer V, Andersen MR, Brakhage AA, Braus GH, Caddick MX, Cairns TC, de Vries RP, Haarmann T, Hansen K, Hertz-Fowler C, Krappmann S (2016) Current challenges of research on filamentous fungi in relation to human welfare and a sustainable bio-economy: a white paper. Fungal Biol Biotechnol 3:6. https://doi.org/10.1186/s40694-016-0024-8
Meyer V, Basenko EY, Benz JP, Braus GH, Caddick MX, Csukai M, de Vries RP, Endy D, Frisvad JC, Gunde-Cimerman N, Haarmann T (2020) Growing a circular economy with fungal biotechnology: a white paper. Fungal Biol Biotechnol 7:5. https://doi.org/10.1186/s40694-020-00095-z
Meyer M, Dietrich S, Schulz H, Mondschein A (2021) Comparison of the technical performance of leather, artificial leather, and trendy alternatives. Coatings 11:226. https://doi.org/10.3390/coatings11020226
Minh NT, Ngan HN (2021) Vegan leather: an eco-friendly material for sustainable fashion towards environmental awareness. AIP Conf Proc 2406:060019. https://doi.org/10.1063/5.0066483
Müller C, Klemm S, Fleck C (2021) Bracket fungi, natural lightweight construction materials: hierarchical microstructure and compressive behavior of Fomes fomentarius fruit bodies. Appl Phys A 127(3):178. https://doi.org/10.1007/s00339-020-04270-2
Muthu SS, Ramchandani M (2024) Vegan alternatives for leather. Springer, Heidelberg
Narayanan M (2024) Heavy metals removal potential of metal tolerant fungal biomass on leather industry effluent and assessment of treated effluent toxicity by in vitro studies. Biomass Convers Biorefin 15:23979–23988. https://doi.org/10.1007/s13399-024-04984-7
Nicolae M, Roussel V, Koelle M, Huron S, Steimle J, Teyssier M (2023) Biohybrid devices: prototyping interactive devices with growable materials. In: Proceedings of the 36th annual ACM symposium on user interface software and technology. ACM, New York, pp 1–15. https://doi.org/10.1145/3586183.3606774
Nithyaprakash V, Niveathitha S, Shanmugapriya V (2020) Designer activism strategies for sustainable leather product designs. In: Muthu SS (ed) Leather and footwear sustainability: manufacturing, supply chain, and product level issues. Springer, Singapore, pp 57–89. https://doi.org/10.1007/978-981-15-6296-9_4
Omoloso O, Mortimer K, Wise WR, Jraisat L (2020) Sustainability research in the leather industry: a critical review of progress and opportunities for future research. J Clean Prod 285:125441. https://doi.org/10.1016/j.jclepro.2020.125441
Orlita A (2004) Microbial biodeterioration of leather and its control: a review. Int Biodeterior Biodegrad 53:157–163. https://doi.org/10.1016/S0964-8305(03)00089-1
Owen MM, Achukwu EO, Arukalam IO, Muhammad M, Romli AZ (2021) Effect of processing temperatures on the thermal and mechanical properties of leather waste-ABS composites. Compos Adv Mater 30:26349833211060056. https://doi.org/10.1177/26349833211060056
Pawłowicz T, Gabrysiak KA, Wilamowski K (2024) Investigating the potential of polypore fungi as eco-friendly materials in food industry applications. Forests 15:1230. https://doi.org/10.3390/f15071230

Peeters SS (2023) Assessing modifications on mycelium-based composites and the effects on fungal degradation and material properties. Doctoral dissertation, Eindhoven University of Technology

Perez RC (2020) Bioengineering basics, biometrology, and biocivics with *Ganoderma lucidum*. Stanford University

Perry M (2022) Hortus curious: discover the world's most weird and wonderful plants and fungi. Dorling Kindersley Ltd, London

Pertile B, Fontana RC, Isoton R, Brandalise RN, Camassola M (2025) Mycelium-based products: multifunctional flexible mycomaterial, a textile alternative produced from fungi and lignocellulose waste. World J Microbiol Biotechnol 41:77. https://doi.org/10.1007/s11274-025-04470-z

Plavan V, Ishchenko O, Resnytskyi I, Liashok I, Kuchynska D (2019) Modified polysaccharides application for antiseptic leather processing. Tekhnichni Nauky Tekhnologii 18:209–214

Porter DL, Hotz EC, Uehling JK, Naleway SE (2023) A review of the material and mechanical properties of select *Ganoderma* fungi structures as a source for bioinspiration. J Mater Sci 58: 3401–3420. https://doi.org/10.1007/s10853-023-08214-y

Rajapakshe BPS, Maneesha KAV, Jayapema KKC, Navaratne S (2025) Development of value-added savory cream from button mushroom (*Agaricus bisporus*) and evaluation of chemical, physical and microbiological characteristics. J Future Foods 5:410–418. https://doi.org/10.1016/j.jfutfo.2024.07.019

Raman J, Jang K, Oh Y, Oh M, Im J, Lakshmanan H, Sabaratnam V (2020) Cultivation and nutritional value of prominent *Pleurotus* spp.: an overview. Mycobiology 49:1–14. https://doi.org/10.1080/12298093.2020.1835142

Raman J, Kim DS, Kim HS, Oh DS, Shin HJ (2022) Mycofabrication of mycelium-based leather from brown-rot fungi. J Fungi 8:317. https://doi.org/10.3390/jof8030317

Ramos M, Burgos N, Barnard A, Evans G, Preece J, Graz M, Ruthes AC, Jiménez-Quero A, Martínez-Abad A, Vilaplana F, Ngoc LP, Brouwer A, Van Der Burg B, Del Carmen Garrigós M, Jiménez A (2019) *Agaricus bisporus* and its by-products as a source of valuable extracts and bioactive compounds. Food Chem 292:176–187. https://doi.org/10.1016/j.foodchem.2019.04.035

Rashdan W, Ashour A (2023) Mycelium-based materials: an alternative for sustainable interior design. New Des Ideas 7:44–61

Rashid A, Shah A, Khan S, Adamjee MA, Shahab A (2024) Potential of Cactus Leather in Pakistan's Leather Industry: A Qualitative Analysis. Int J Soc Sci Entrep 4(3):1–16. https://doi.org/10.58651/ijsse.v4i3.298

Rathinamoorthy R, Sharmila Bharathi T, Snehaa M, Swetha C (2023) Mycelium as sustainable textile material—review on recent research and future prospective. Int J Cloth Sci Technol 35: 454–476. https://doi.org/10.1108/IJCST-01-2022-0003

Rimantho D, Chaerani L, Sundari AS (2024) Initial mechanical properties of orange peel waste as raw material for vegan leather production. Case Stud Chem Environ Eng 10:100786. https://doi.org/10.1016/j.cscee.2024.100786

Rühse V, Horvath AS (2024) Fungi matters: a comparison of contemporary fungi imagery between art, design, and architecture. In: Politics of the machines: lifelikeness & beyond, Aachen 2024. BCS Learning & Development, pp 24–31

Saez D, Hofman G, Grizmann D, Werner A, Trautz M (2024) Enhancing mycelium-based composites: anisotropic material design and mechanical performance. In: IASS 2024 symposium redefining the art of structural design. ETH, Zurich

Saha S (2023) Sustainability and development of biomaterials in textile: a review. In: Indo-Japan textile research conference. Springer Nature, Singapore, pp 83–124. https://doi.org/10.1007/978-981-97-6968-1_7

Saha N, Ngwabebhoh FA, Nguyen HT, Saha P (2020) Environmentally friendly and animal free leather: fabrication and characterization. AIP Conf Proc 2289:020049. https://doi.org/10.1063/5.0028467

Sahu S, Bhardwaj P, Singh G, Bhalla A, Arya SK (2025) A pioneering review on *Ganoderma lucidum*–derived leather: taking a step towards a cruelty-free leather manufacturing. Environ Sci Pollut Res 32:1–17. https://doi.org/10.1007/s11356-025-36130-y

Scaffaro R, Lopresti F, Marino A, Nostro A (2018) Antimicrobial additives for poly (lactic acid) materials and their applications: current state and perspectives. Appl Microbiol Biotechnol 102: 7739–7756. https://doi.org/10.1007/s00253-018-9220-1

Senthil R (2024) Regenerated products from leather industrial solid waste: future perspective and current advances. J Hazard Mater Lett 5:100112. https://doi.org/10.1016/j.hazl.2024.100112

Shakir MA, Ahmad MI (2024) Bioproduct advances: insight into failure factors in mycelium composite fabrication. Biofuels Bioprod Biorefin 18:1739–1754. https://doi.org/10.1002/bbb.2620

Shankar J, Verma P, Shah MP (eds) (2024) Microbial approaches for sustainable green technologies. CRC Press, Boca Raton

Sharma M, Verma S, Chauhan G, Arya M, Kumari A (2024) Mycelium-based biocomposites: synthesis and applications. Environ Sustain 7:265–278. https://doi.org/10.1007/s42398-024-00305-z

Sharma M, Lim LT, Kaur G (2025) Tailoring Structure-Property Relationships of Fungal Mycelium for Material Applications: A Process Engineering Approach for Pure Mycelium-Based Biomaterials. New Biotechnol. https://doi.org/10.1016/j.nbt.2025.12.006

Shen SC, Lee NA, Lockett WJ, Acuil AD, Gazdus HB, Spitzer BN, Buehler MJ (2024) Robust myco-composites: a biocomposite platform for versatile hybrid-living materials. Mater Horiz 11:1689–1703

Shin H, Ro H, Kawauchi M, Honda Y (2025) Review on mushroom mycelium-based products and their production process: from upstream to downstream. Bioresour Bioprocess 12:1. https://doi.org/10.1186/s40643-024-00836-7

Sierra LAB, Mendes-Pereira T, García GJY, Werkhaizer CQ, de Rezende JB, Rodrigues TAB, Góes-Neto A (2023) Current situation and future perspectives for the use of fungi in the biomaterial industry and proposal for a new classification of fungal-derived materials. PeerJ Mater Sci 5:e31. https://doi.org/10.7717/peerj-matsci.31

Silverman J, Cao H, Cobb K (2020) Development of mushroom mycelium composites for footwear products. Cloth Text Res J 38:119–133. https://doi.org/10.1177/0887302X19890006

Silvestri L (2022) TERRAFORMA – material investigation on the possibilities to combine natural growth of mycelium and unfired clay for novel sustainable product design applications. Master's thesis

Singh S, Singh NP, Agrawal S, Kumar A (2025) Unleashing the potential of white-rot fungi mycelium for functional biomaterials development. Discov Mater 5:96. https://doi.org/10.1007/s43939-025-00288-6

Solanilla-Duque JF, Del Rosario Salazar-Sánchez M, Herrera RR (2021) Potential of lignocellulosic residues from coconut, fique, and sugar cane as substrates for *Pleurotus* and *Ganoderma* in the development of biomaterials. Environ Qual Manag 32:343–353. https://doi.org/10.1002/tqem.21826

Song L, Liu Y, Xiao S, Yuan X, Han X (2025) Revolutionizing eco-friendly leather production: a freeze-thaw and liquid fermentation approach with fungal mycelium. J Fungi 11:326. https://doi.org/10.3390/jof11040326

Spinnewijn O (2022) Innovative leather alternatives: a short-lived trend or the future of fashion? Thesis

Stejskal J, Ngwabebhoh FA, Sáha P, Prokeš J (2023) Carbonized leather waste: a review and conductivity outlook. Polymers 15(4):1028. https://doi.org/10.3390/polym15041028

Stelzer L, Hoberg F, Bach V, Schmidt B, Pfeiffer S, Meyer V, Finkbeiner M (2021) Life cycle assessment of fungal-based composite bricks. Sustainability 13(21):11573. https://doi.org/10.3390/su132111573

Sundar VJ, Gnanamani A, Muralidharan C, Chandrababu NK, Mandal AB (2010) Recovery and utilization of proteinous wastes of leather making: a review. Rev Environ Sci Biotechnol 10(2): 151–163. https://doi.org/10.1007/s11157-010-9223-6

Susel GDE, Domínguez EMH, Fernández AGE, Cervantes JÁ, Medellín LDR, Mendoza BM (2021) Biomaterial obtenido a partir de micelio de hongo (*Ganoderma lucidum*) y residuos agrícolas. S Fla J Dev 2(3):4663–4681. https://doi.org/10.46932/sfjdv2n3-065

Taghizadehmakoei A (2024) Pavilion design based on biogenic mycelium architectural feedstock. Doctoral dissertation, University of British Columbia. https://doi.org/10.14288/1.0442062

Tansakul A, Lumyong R (2008) Thermal properties of straw mushroom. J Food Eng 87(1):91–98. https://doi.org/10.1016/j.jfoodeng.2007.11.016

Temporiti MEE, Nicola L, Nielsen E, Tosi S (2022) Fungal enzymes involved in plastics biodegradation. Microorganisms 10(6):1180. https://doi.org/10.3390/microorganisms10061180

Tewari S, Reshamwala SMS, Bhatt L, Kale RD (2023) Vegan leather: a sustainable reality or a marketing gimmick? Environ Sci Pollut Res 31(3):3361–3375. https://doi.org/10.1007/s11356-023-31491-8

Thanh NT, Tuan NN, Kuo P, Dung DM, Phuong DL, Giang DTT, Wu T, Thang TD (2017) Chemical constituents from the fruiting bodies of *Phellinus igniarius*. Nat Prod Res 32(20): 2392–2397. https://doi.org/10.1080/14786419.2017.1413572

Trivedi M, Kushwaha R, Mordina B, Jha AA, Singh V (2025) Utilizing fungal mycelium in bio-nanocomposites as a replacement for chemically synthesized lightweight polymers. In: Fungal waste biomass management for energy, environment and value-added products. Springer Nature, Cham, pp 59–87. https://doi.org/10.1007/978-3-031-82599-6_3

Ugbede AS, Abioye OP, Aransiola SA, Oyewole OA, Maddela NR, Prasad R (2023) Production, optimization and partial purification of bacterial and fungal proteases for animal skin dehairing: a sustainable development in leather-making process. Bioresour Technol Rep 24:101632. https://doi.org/10.1016/j.biteb.2023.101632

van Rooyen A (2024) Sustainable fashion tech innovation: a case study of Bolt Threads' redefinition of fashion systems, design activism, and social entrepreneurship. Discern 5(2):1–13

Vandelook S, Elsacker E, Van Wylick A, De Laet L, Peeters E (2021) Current state and future prospects of pure mycelium materials. Fungal Biol Biotechnol 8(1):20. https://doi.org/10.1186/s40694-021-00128-1

Verma N, Jujjavarapu SE, Mahapatra C (2023) Green sustainable biocomposites: substitute to plastics with innovative fungal mycelium-based biomaterial. J Environ Chem Eng 11(5): 110396. https://doi.org/10.1016/j.jece.2023.110396

Vico A, Maestre-Lopez MI, Arán-Ais F, Orgilés-Calpena E, Bertazzo M, Marhuenda-Egea FC (2024) Assessment of the biodegradability and compostability of finished leathers: analysis using spectroscopy and thermal methods. Polymers 16(13):1908. https://doi.org/10.3390/polym16131908

Volk R, Schröter M, Saeidi N, Steffl S, Javadian A, Hebel DE, Schultmann F (2024) Life cycle assessment of mycelium-based composite materials. Resour Conserv Recycl 205:107579. https://doi.org/10.1016/j.resconrec.2024.107579

Vunduk J, Biketova AY (2022) The cultural use of mushrooms. In: Edible fungi: chemical composition, nutrition and health effects. The Royal Society of Chemistry, pp 327–358. https://doi.org/10.1039/9781839167522-00327

Wang Y (2022) Evaluation of novel mycelium-based bio-foams using unexplored mycelium species and agricultural waste as packaging materials. Thesis

Wattanavichean N, Phanthuwongpakdee J, Koedrith P, Laoratanakul P, Thaithatgoon B, Somrithipol S, Kwantong P, Nuankaew S, Pinruan U, Chuaseeharonnachai C, Boonyuen N (2025) Mycelium-based breakthroughs: exploring commercialization, research, and next-gen possibilities. Circ Econ Sustain 5:1–43. https://doi.org/10.1007/s43615-025-00539-x

Whabi V, Yu B, Xu J (2024) From nature to design: tailoring pure mycelial materials for the needs of tomorrow. J Fungi 10(3):183. https://doi.org/10.3390/jof10030183

Wijayarathna EKB, Mohammadkhani G, Soufiani AM, Adolfsson KH, Ferreira JA, Hakkarainen M, Berglund L, Heinmaa I, Root A, Zamani A (2021) Fungal textile alternatives from bread waste with leather-like properties. Resour Conserv Recycl 179:106041. https://doi.org/10.1016/j.resconrec.2021.106041

Williams E, Cenian K, Golsteijn L, Morris B, Scullin ML (2022) Life cycle assessment of MycoWorks' Reishi™: the first low-carbon and biodegradable alternative leather. Environ Sci Eur 34(1):120. https://doi.org/10.1186/s12302-022-00689-x

Wjunow C, Moselewski K, Huhnen Z, Sultanova S, Sabantina L (2023) Sustainable textiles from unconventional biomaterials—cactus based. Eng Proc 37:58. https://doi.org/10.3390/ecp2023-14652

Xia Y, Jin Y, Qi J, Chen H, Chen G, Tang S (2021) Preparation of biomass carbon material based on Fomes fomentarius via alkali activation and its application for the removal of brilliant green in wastewater. Environmental Technology & Innovation 23, 101659. https://doi.org/10.1016/j.eti.2021.101659

Xu R, Wu W, Zhong Y, Zhang L, Xu H, Mao Z, Ji B (2025) High-strength and biodegradable mycelial leather materials cross-linked with dialdehyde carboxymethyl cellulose. ACS Sustain Chem Eng 13(21):8057–8067

Yang Z, Zhang F, Still B, White M, Amstislavski P (2017) Physical and mechanical properties of fungal mycelium-based biofoam. J Mater Civ Eng 29(7):04017030. https://doi.org/10.1061/(ASCE)MT.1943-5533.0001866

Zhang F, Fu G, Liu H, Wang C, Zhou J, Ngai T, Lin W (2024) Sustainable leather alternatives: high-performance and dyeable bio-based materials from fungal chitin and tannic acid. Carbohydr Polym 348:122800. https://doi.org/10.1016/j.carbpol.2024.122800

Zhang J, Yeoh GH, Kabir II (2025) Polyurethane materials for fire retardancy: synthesis, structure, properties, and applications. Fire 8(2):64. https://doi.org/10.3390/fire8020064

Zhao A, Jones MP, Weiland K, Bismarck A (2025) Shifu-inspired fungal paper yarns. Adv Sci 12:e11975. https://doi.org/10.1002/advs.202511975

# Exploring Uncharted Applications of Mushrooms for Textile and Packaging

# 16

Pooja Bhadoriya, Aditi Virmani, and Bhoomika Rajeev Gupta

**Abstract**

Mushrooms, long known for their nutritional, medicinal, and ecological roles, are now evolving as sustainable bio-resources in material science. The filamentous network of fungal hyphae has emerged as a sustainable alternative to petrochemical-derived textiles and packaging materials. In the textile sector, mycelium-based matrices have shown significant promise as substitutes for nonwoven fabrics and leather analogues. Similarly, in the packaging industry, mycelium-derived foams and films have been developed as biodegradable replacements for synthetic polymers like polystyrene and polyethylene. Mushroom mycelium represents a promising edge toward sustainability and is likely to allay the habit of using nonrenewable polymers. This chapter explores the architectural properties of mushroom mycelium as a potential biomaterial for packaging and textiles, details of their production process using biodegradable waste materials, applications, and their industrial adaptations available in the market. The chapter examines challenges and hurdles in the extended use of this biomaterial, such as durability, waterproofing, production scalability, and cost efficiency, and suggests solutions to these obstacles by making advancements in fungal biotechnology, substrate engineering, and hybrid composites. We have also discussed the advanced applications of mushroom mycelium in space exploration, construction, and furniture, boosting the circular economy and carbon neutrality. Thus, by using engineering methodologies and the biological basis of mushroom growth, mycelium-based textiles and packaging can transform into ecologically aligned and industrially viable alternatives.

P. Bhadoriya (✉) · A. Virmani · B. R. Gupta
Shaheed Rajguru College of Applied Sciences for Women, University of Delhi, Vasundhara Enclave, Delhi, India



B. Giri et al. (eds.), *Advances in Mushroom Research*,
https://doi.org/10.1007/978-981-95-7692-0_16

**Keywords**

Myco-leather · Myco-packaging · Mushroom in textiles · Fungal biotechnology · Sustainable packaging · Biomaterial

## 16.1 Introduction

Over the past several years, mushrooms have transcended familiar roles in food and medicine to emerge as some of the strongest players of sustainable innovation. Previously valued only for their ecological, nutritional, and medicinal applications, they are now emerging to be realized as revolutionizing bio-resources in material science, particularly in the textile and packaging fields. In particular, the mushroom mycelium, as a filamentous vegetative structure of fungi, has increasingly attracted attention as an abundant and renewable matrix material for the fabrication of biodegradable materials. Mushroom-based mycelium is now the focal point of a revolution, with scientists and innovators harnessing its properties to make sustainable alternatives for packaging and textiles. These two sectors are significant for their environmental footprint but have never been targeted before by fungal biotechnology (Attias et al. 2019; Jones et al. 2020b).

The textile and packaging industries are two of the largest contributors to global pollution, consuming huge quantities of nonrenewable resources and producing millions of tons of waste annually. The fashion industry alone is accountable for around 10% of global carbon emissions and produces more than 92 million tons of waste annually (UNE and Notten 2020; UNEP 2022). Similarly, single-use plastic packaging accounts for almost 40% of global plastic production, much of which contributes to landfill waste and sea waste (Ellen MacArthur Foundation 2016). As environmental concerns continue to rise, the need for sustainable practices has never been more urgent. From climate change and pollution to overconsumption of natural resources, the status quo of industrial practices is no longer viable. The pressing need for environmentally friendly alternatives has motivated scientists and innovators to look for new biomaterials. Mycelium, the vegetative part of fungi, is one such prospect. Its potential is immense, as it is a natural, versatile, and resilient material having excellent end-of-life performance, thus enabling the creation of next-generation materials that challenge the dominance of petroleum-based plastics and animal-derived leathers. This chapter explores the new and fast-emerging field of mushroom application in textiles and packaging, with a focus on mycelium-based material as a new sustainable alternative. The chapter discusses the biological principles of mycelium growth, material properties, and production processes. We also discuss commercial uses and future directions while at the same time dealing with current limitations in scalability and durability. With society moving toward circular economies and carbon neutrality, fungi are offering solutions that go beyond simple decomposition as they provide regenerative pathways to solve some of our most pressing industrial problems (Elsacker et al. 2021).

## 16.2 The Science of Mycelium: A Living and Sustainable Biomaterial

The fruiting body of mushrooms has been the main center of attention in scientific, culinary, and commercial terms for hundreds of years. From traditional Chinese medicine to modern gourmet cooking and medical research, the surface cap of the mushrooms is considered the most important part of the organism. But the mushroom's underground network, or mycelium, was largely ignored, considered a kind of substrate-propping scaffold or parasite precursor of fruiting. More recently, however, mycelium has emerged as a potential game-changer in terms of sustainability. It has been hailed for its green potential, from biodegradable packaging and regenerating textiles to environmentally friendly building materials, and even when it comes to remediating pollutants (Haneef et al. 2017; Jones et al. 2020b; Elsacker et al. 2021). Mushroom mycelium has a massive network and important ecological roles; hence, it can be an ecological instrument in addressing such problems. Mushroom mycelium is the complex of fungus vegetative cells (hyphae) that act as the root system of a fungus (Fig. 2.1). It is often hidden in the substrate woven of hyphae and serves as the nutrient-absorbing and exploratory vegetative body of fungi in soil or on decaying plant or animal material (called the substrate) responsible for nutrient uptake, reproduction, and interactions with the environment. It is widely distributed in nature, present in forests as well as arable lands, playing an important role in the ecosystem by functioning as an anaerobic decomposer of organic matter and a communal collector of energy (Jennings 1984).

The mycelium of a mushroom shows a distinctive pattern of structural, biochemical, and functional traits, which give it adaptability for use in textile and packaging material. Its wall is composed of chitin and chitosan, which give rigidity, tensile strength, and stability like natural polymers such as cellulose and keratin (Fig. 16.1). These ingredients are necessary to create hard and durable things (Mitrano et al. 2021). Another significant polysaccharide, β-glucans, improve both extensity and cohesion, providing flexibility to the polymeric-like structure of the mycelial network and therefore, an inherent characteristic to the textile and soft packaging applicability (Haneef et al. 2017). Furthermore, mycelium produces enzymes and structural proteins that facilitate the attachment and regulate the degradation of substrate, enabling it to colonize the insides and/or outsides of lignocellulosic materials.

Mycelium can colonize and digest a diverse range of lignocellulosic substrates—from agricultural waste like wheat straw or sawdust and waste husks—converting them into a dense matrix of fibers. Additionally, the physicochemical properties of mycelium depend on the form of hyphal colonization developed over a substrate. This serves as an important morphological identification tool for the selection of appropriate species of mushrooms as bio-resources (Alaneme et al. 2023). In fungi, the arrangement of hyphae is categorized into monomitic, dimitic, and trimitic hyphae that have generative, skeletal, and binding hyphae, respectively (Fig. 2.1) (Porter and Naleway 2022). Monomitic species such as *Agaricus bisporus* have only generative hyphae and are usually found in softer substrates. Dimiticis is described

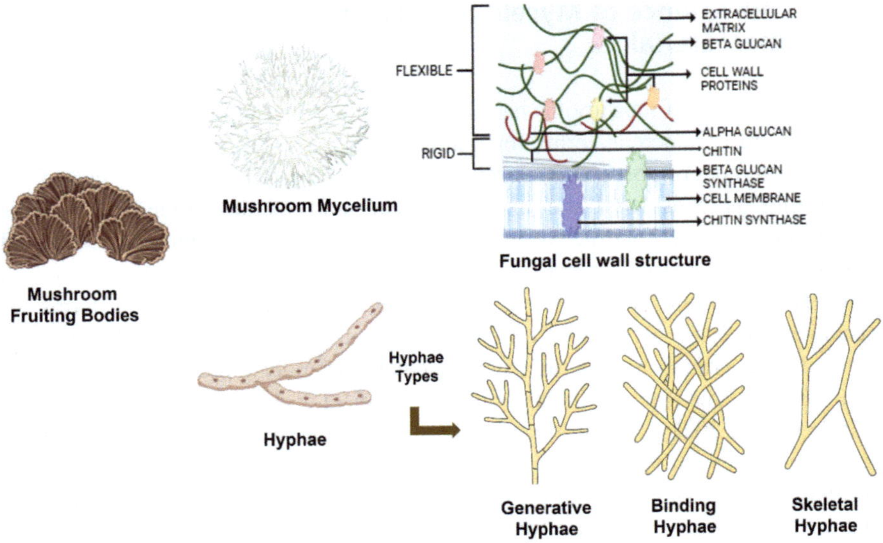


**Fig. 16.1** Structural, biochemical, and functional traits of mushroom mycelium. Schematic diagram showing fruiting bodies of mushrooms, underlying mycelium network, individual hyphae, and three hyphal types—generative, binding, and skeletal. The figure also depicts the morphology of the cell wall with distinct flexible and rigid layers constituted of extracellular matrix, β-glucans, α-glucan, chitin, cell wall proteins, and membrane-associated enzymes such as chitin synthase and β-glucan synthase. (Modified after Shin et al. 2025)

when a combination of generative and skeletal hyphae or generative and binding hyphae is present, as in *Grifola frondosa*. *Trametes versicolor* basidiomycetes contain all three generative, skeletal, and binding hyphae. The arrangement of these hyphae is one of the reasons for further structural reinforcement that allows dimitic and trimitic fungi to become efficient lignocellulose colonizers, e.g., wood. Therefore, these features of dimitic and trimitic fungi endow them with tolerance to environmental stress, which is strongly correlated with their ecological importance in forest ecosystems as well as in industrial processes, such as biodegradation and bioremediation.

Due to the strength of trimitic hyphal structures, *Ganoderma* species are currently in demand for companies that engage in the business of mycelium-based products (Porter and Naleway 2022). The dimitic or trimitic fungal matrix exhibits qualities like porosity, lightweight structure, and enhanced compressive strength. These qualities can all be tailored according to the requirement through varying growth conditions, substrate selection, and strain improvement by genetic engineering (Girometta et al. 2019; Shin et al. 2025). This matrix can be molded into desired shapes, offering versatility in design and function. The biodegradability and energy efficiency of mycelium make it a possible solution to artificial polymers and leather and, hence support the extended use of mycelium as green packaging material, sustainable clothing, and accessories connecting biology with industrial revolutions.

## 16.3 Mycelium-Based Sustainable Packaging Materials

The move to adopt sustainable packaging alternatives represents one of the most urgent objectives across industries globally, as an increasing urgency to address plastic waste and to minimize the environmental impact of packaging materials takes center stage. Mushroom-based packaging represents a novel and sustainable route that harnesses the biological characteristics of mycelium to produce a green, strong, and lightweight material alternative to plastic and foam that cradles products for shipping (Ashok et al. 2016).

### 16.3.1 Mycelium Packaging Production Process

In the production of myco-packaging or mushroom-based packaging, fungal mycelium is cultivated on agricultural waste such as straw, sawdust, hemp hurds, or other lignocellulosic wastes that act as a source of nutrients and as a structural framework. Figure 16.2 depicts the process for making mycelium-based packaging material. The procedure has the following phases.

#### 16.3.1.1 Substrate Making

The substrate, originating from any type of agricultural waste, is pasteurized or sterilized first. This step helps in eliminating competing microflora and ensuring an ideal environment for the growth of the desired mushroom species.

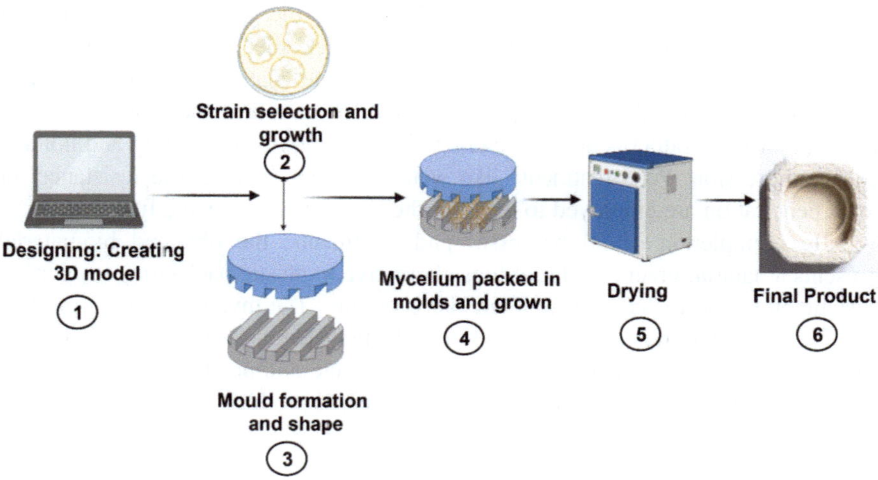


**Fig. 16.2** Steps in the production process of packaging material from mushroom mycelium. The figure shows the stepwise fabrication of mycelium-based packaging. It begins with digital design and 3D model creation, followed by mushroom strain selection, formation of molds, growth of mycelium within molds, and finally drying the mycelium-colonized mold to obtain sustainable, biodegradable packaging material

#### 16.3.1.2 Inoculation and Growth

The prepared substrate is then inoculated with the desired mushroom species/strain, like *Pleurotus, Ganoderma*, or *Trametes*. Mushroom mycelium starts colonizing the substrate under suitable conditions of temperature (25–35 °C), moisture (60–65%), oxygen, and carbon-di-oxide levels. Darkness is generally preferred for incubation to inhibit fruiting body formation while promoting rapid mycelial expansion (Islam et al. 2018).

#### 16.3.1.3 Molding and Sculpting

Molding gives shape to the mycelium growth. Molding can be achieved by keeping the seeded substrate into the molds of different shapes, like bottles, containers, or blocks. As the mycelium grows, it attaches firmly to the shape like a natural adhesive and fills the cavity, and so it is sculpted accordingly. This process of molding and sculpting, along with mycelium growth, can span from days to even weeks, as per mushroom species and substrate used (Elsacker et al. 2019).

#### 16.3.1.4 Drying

After the completion of growth, the sculpted mycelium mold is kept for drying in a sterile environment. Drying can be done in ovens at temperatures varying between 60 and 65 °C or in air at room temperature. Drying helps to preserve the shape and structure of mycelium mold and prevent further growth (Jones et al. 2020b).

The colonization of its growth substrate by mycelium functions as natural glue, binding cohesively and forming a material that is firm, lightweight, and can be tailored as needed. Shaping of the material can be achieved during the growth period, as discussed above, or else it can be done post-mycelial complete growth. In post-molding, first, the mycelium is dried to halt the growth. The removal of moisture content from the biomaterial expands shelf life, increases dimensional stability, and makes it lightweight. Afterwards, the material is sculpted using different types of molds as per requirements to provide a firm shape and look. The finishing of the product is done by adding techniques like laser cutting and hot and cold compression. Post-treatments like water-repellent coating, fire resistance, or resin penetration are employed to increase their application-specific functioning.

After completion of all processes and treatments, the obtained biomaterial presents a natural, green, nonhazardous alternative to plastic packaging, supporting sustainability and the concept of a circular economy. The myco-packaging production is done with utmost caution. Right from the first step, i.e., the choice of the right substrate, to the final finishing technique, each stage is crucial in determining the quality, biodegradability, and stability of the finished product.

### 16.3.2 Applications and Industry Adoption

A range of companies across various industries have adopted mushroom-based packaging as an eco-friendly alternative to plastic and styrofoam (Table 16.1).

**Table 16.1** List of companies using mushroom-based packaging materials

| Company name and location | Applications | Mushroom species used | Key products | References |
|---|---|---|---|---|
| Ecovative Design, USA | Sustainable packaging, MycoLeather | Oyster (*Pleurotus* spp.) | MycoComposite™, Forager™ leather | Ecovative (2023) |
| Nature's Fynd, USA | Mycelium-based Food & Materials | *Fusarium* strain (not traditional) | Sustainable protein & materials | Nature's Fynd. (2025) |
| BioFab, Australia | Sustainable packaging | Oyster (*Pleurotus* spp.) | Biodegradable mycelium packaging | Biofab (2025) |
| Grown.bio, Netherlands | Mycelium packaging & furniture | Oyster (*Pleurotus* spp.) | Custom mycelium packaging | Hugohek (2025) |
| Paradise Packaging Co., USA | Eco-friendly packaging | Oyster (*Pleurotus* spp.) | Compostable protective packaging | Paradise packaging Pvt. ltd. (2023) |
| Dell, USA | Sustainable packaging | Oyster (*Pleurotus* spp.) | Mycelium-based protective packaging for electronics | Dell (2024) |
| IKEA, Sweden | Eco-friendly packaging & furniture | Oyster (*Pleurotus* spp.) | Biodegradable mycelium packaging (Styrofoam alternative) | Barrett (2020) |
| Biohm, United Kingdom | Construction & insulation | Oysters & Other Fungi | Mycelium-based insulation panels, biobased building materials | Biohm (2016) |

Made from mycelium, this material is grown into custom molds, creating strong yet biodegradable packaging that decomposes naturally within weeks.

Pioneered by companies like Ecovative, Mushroom® Packaging, and MycoWorks, this innovation is now being embraced by major brands such as Dell, IKEA, and Lush to reduce their carbon footprint. Dell Technologies, for example, transitioned to the use of mycelium-based packaging as a substitute for expanded polystyrene (EPS) to transport laptops and servers, effectively curtailing plastic waste. IKEA has also considered using mycelium-based packaging for small products like glassware and kitchen accessories as part of their sustainable design (IKEA 2021). In the food and beverage industry, while mycelium packaging is not as prevalent because of its water vulnerability, innovations like coated mycelium trays are being experimented with for usage in single-use food containers, coffee cup sleeves, and takeaway containers. Start-ups like Paradigm Shift and Magical Mushroom Company in the UK have been at the forefront of applying mycelium to packaging wine bottles and cosmetics, with biodegradable inserts replacing traditional inserts (Magical Mushroom Company 2023). Apart from packaging, mycelium-based material applications have also transitioned into the building industry due to their high insulation properties. Products like wall panels and soundproofing boards are gaining attention as part of circular and sustainable building materials. These diverse applications across industries highlight the growing potential of mycelium as a versatile and eco-friendly alternative.

## 16.4 Mushrooms in the Textile Industry

The global textile industry ranks among the most polluting sectors, contributing a great scale toward environmental degradation. The production process here is water-demanding, and near about 200 L of water is needed to produce 1 kilogram of fabric. The textile dyeing and finishing industry creates about 70–80% of wastewater, which contains heavy metals and toxic synthetic colorants, and there is a release of microplastics and microfibers in a substantial amount (Williams et al. 2022; Liu et al. 2024). Amid the pressing need for sustainable alternatives, fungal-based materials, particularly mycelium, are emerging as innovative, novel, and earth-friendly alternatives. Mycelium composites and bio-leathers are contenders as biodegradable, low-carbon substitutes for conventional textiles, offering a promising solution for the fashion and textile industry (Rathinamoorthy et al. 2023). Mushrooms and fungal materials are now researched not only as food or medicine but also as fashionable, functional, and contemporary materials, such as self-cultivated mycelium sheets, fungal leather, and colorable bio-textiles (Shankar et al. 2024).

### 16.4.1 Fungal Mycelia as a Vegan Leather Substitute

The production process of conventional leather poses numerous environmental, ethical, and human health concerns. This is so because the process involves extremely high water usage and contributes to pollution through the release of toxic tanning chemicals such as chromium and formaldehyde (Jones et al. 2023; Nguyen 2024). The livestock industry, which supplies hides for leather, is a significant source of greenhouse gas emissions and drives deforestation to make space for cattle grazing. Animals used for leather often suffer inhumane treatment on factory farms, enduring confinement, painful procedures, and cruel handling during transport and slaughter. Moreover, tannery workers, particularly in developing countries, are frequently exposed to hazardous chemicals, putting them at risk for serious health issues while working in unsafe and exploitative conditions. Its production is also highly resource-intensive, leading to land degradation, water shortages, and the accumulation of large volumes of solid waste (Kanagaraj et al. 2006; Doe 2023).

One of the most inventive and sustainable advancements in biomaterials is mycelium leather, a fungal-derived alternative to animal skin and synthetic leathers. This emerging bio-based material closely mimics the texture, flexibility, and durability of traditional leather without bypassing the ethical and environmental issues of its conventional forms. The biomaterial leverages the complex three-dimensional networks of fibers formed by mushroom mycelium, grown in controlled environmental conditions, to generate sheet materials that closely resemble traditional leather varieties (Appels et al. 2018; Jones et al. 2020a).

## 16.4.2 Production Process of Mycelium Leather

The production process of mycelium-based leather consists of various strictly controlled procedures—from the choice of substrate to post-processing—each of which affects the final texture, durability, and functionality of the final product. Figure 16.3 shows the process for leather production with mycelium material. It typically consists of the following steps.

### 16.4.2.1 Substrate Preparation

The appropriate lignocellulosic substrates are selected according to the mushroom species that serve as the nutritional and structural base for mycelial growth. Frequently used substrates are sawdust, wheat bran, corn husks, straw, hemp hurds, and other agro-industrial derivatives (Appels et al. 2018; Elsacker et al. 2021). These substrates are normally sterilized or pasteurized to get rid of microbial contaminants and other competitors, ensuring a clean environment for mycelial colonization.

### 16.4.2.2 Inoculation and Growth

The substrate after sterilization is inoculated with selected mushroom species. Post-inoculation, the substrate is kept in trays or molds, designed to guide the form and

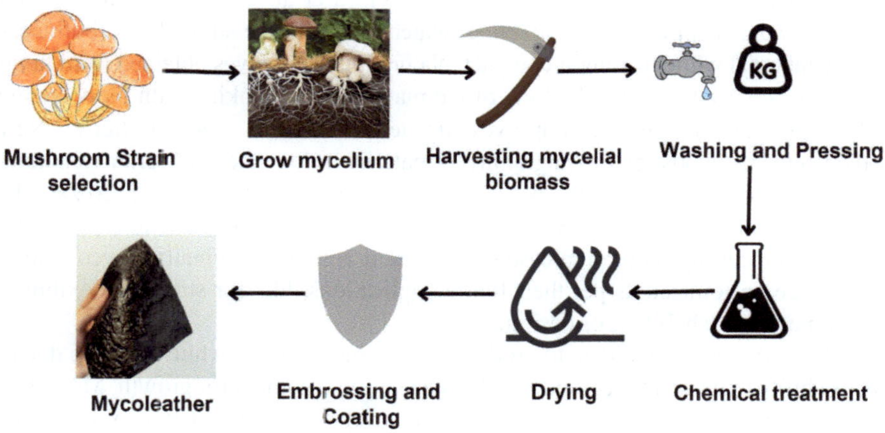


**Fig. 16.3** Steps in the production process of leather from mushroom mycelium. The schematic figure illustrates a stepwise fabrication of mycelium-based leather. It begins with mushroom strain selection and its cultivation, followed by harvesting of mycelial biomass, washing, and pressing. After these chemical treatments are done, they are followed by drying. Finishing steps involve embossing and coating to obtain a sustainable leather-like material

dimensions of the developing mycelial mat. The mycelium starts colonizing the substrate under appropriate conditions of temperature (25–30 °C), relative humidity (60–70%), and limited illumination to restrict fruiting (Appels et al. 2018). The gaseous exchange of oxygen and carbon-di-oxide should also be optimized to favor the morphological features of mycelium and its thickness. The period for complete colonization spans from 5 to 14 days, depending on the physiochemical nature of the substrate and the mushroom species.

#### 16.4.2.3 Mycelial Mat Formation

The mushroom mycelium grows and expands its fibers in and around the substrate, binding to it cohesively, giving rise to a mycelial mat. The density, borders, and alignment of this mat can be varied by altering the ingredients in the substrate, airflow, and the shape of the mold used (Attias et al. 2019).

#### 16.4.2.4 Harvesting and Inactivation

The harvesting of the mycelial mat is done once the desired dimensions and density are obtained. Harvesting is done by drying the mycelial mat in an aseptic environment in a hot air oven at 50–70 °C. Drying the mat kills the inoculated mushroom species, stabilizes it, and removes moisture (Haneef et al. 2017; Elsacker et al. 2021).

#### 16.4.2.5 Post-Processing and Functionalization

The final step in the mycelial-leather production process is to enhance its appearance, usefulness, and functionality. For this, different types of treatments are done to process the mycelial mat, which include surface treatments, hot pressing, coloring, and biochemical alterations. The surface of the myco-mycelial mat is smeared with plant oils or natural wax, which makes it waterproof, while tanning with natural agents renders flexibility and durability (Crawford et al. 2024). Heat, along with compression, is applied in hot pressing, which enhances the strength and bulkiness of the material while making it compact. Natural pigments/dyes obtained from fungi or plants are used to uplift the look of the material. Crosslinking with biopolymers like chitosan or alginate improves its tear resistance and elasticity. New advancements in post-processing of myco-material include the application of natural water-repellent substances like beeswax, shellac, or resins (Jones et al. 2020a). To obtain a texture like the conventional leather texture, the myco-material is processed using laser cutting and compressing, like grain and suede. Finally, myco-leather sheets are customized as per their future applications, like for strength, flexibility, tenderness, stretchability, or design.

The most viable candidate to produce leather substitutes is white-rot fungi due to their ability to inherently colonize and utilize various biomass for growth. More than 69 different species of mushrooms are used in myco-leather production, of which the major ones include species such as *Pleurotus ostreatus, Trametes versicolor, Ganoderma lucidum,* and *Agaricus bisporus*, which have significant potential for leather substitute production (Amobonye et al. 2023). Other than these, *Fomitopsis iberica, Fomitopsis pinicola, Stereum hirsutum*, and *Schizophyllum commune* are also used to some extent. Among all, *Ganoderma lucidum*, *Pleurotus ostreatus*,

*Fomitopsis pinicola*, and *Trametes versicolor* are known for forming robust, fibrous, and cohesive hyphal networks, proving to be the best selection for leather production (Haneef et al. 2017; Islam et al. 2018). These fungal mycelia adapt their natural fibrous structure in a way that resembles the fibrous units found in animal skins. When used as composite materials, they act as support layers for coatings applied on top of them. This structure not only strengthens the material but also adds plasticity to the overall product. Myco-leather obtained from these mushroom species is sustainable in terms of the circular economy and minimal waste production (Sydor et al. 2022). Regarding the pathogenicity of these mushroom species, they have been found to have negligible risk to process and cultivate with minimal requirements of confinement (Jones et al. 2017).

### 16.4.3 Applications and Industry Adoption

The myco-leather has gone from laboratory-scale innovations to real-life applications. Different enterprises are using this sustainable biomaterial in the fashion world (Table 16.2). Companies like Bolt Threads and MycoWorks (USA) have introduced and patented mycelium leather named Mylo™ and Reishi™, respectively, for making fashion accessories. The material is gaining traction while replacing conventional leather forms. Elite brands like Stella McCartney, Hermes, and Adidas have teamed up to incorporate mycelium/fungal leather into shoes, handbags, and accessories (Attias et al. 2019; Jones et al. 2020a, b; Manan et al. 2021).

**Table 16.2** List of companies adopting mushroom leather and textiles in the fashion industries

| Company name and location | Applications | Mushroom species used | Key products | References |
|---|---|---|---|---|
| MycoWorks, USA | Mycelium leather | Reishi (*Ganoderma lucidum*) | Fine mycelium leather | Mycoworks (2023) |
| Bolt Threads, USA | Mushroom leather | Reishi (*Ganoderma lucidum*) | Mushroom leather | Bolt Threads (2023) |
| MOGU, Italy | Mycelium textiles and acoustic panels | Oyster (*Pleurotus_spp.*) | Mycelium based flooring, soundproofing | Mogu (2023) |
| NEFFA, Netherlands | Fashion and leather alternative | Reishi (*Ganoderma lucidum*) | Mycelium based textiles | NEFFA (2023) |
| MuSkin, Italy | Mushroom mycelium-based fabrics | Phellinus ellipsoideus | Fabrics, leather, paper,etc | Life Materials (2025) |
| Ecovative, USA | Mushroom leather (forage) | Reishi (*Ganoderma lucidum*) | Air mycelium mushroom leather | Ecovative (2023) |

### 16.4.4 Dyes from Mushrooms

In recent years, the global textile and dye industry has been in the spotlight for its ecological impact. Traditional petrochemical-based synthetic colorants are toxic and nonbiodegradable and generate serious water pollution and cause environmental damage (Bechtold and Mussak 2009). Conventional dyeing processes are often water-intensive and rely on harsh chemicals, contributing significantly to environmental pollution. The use of fungal dyes, derived from various species of mushrooms, lichens, and molds, offers a promising solution with their rich spectrum of natural colors, biodegradability, and often lower toxicity. Moreover, mushroom-based dyes/colorants are UV stable and exhibit antimicrobial properties (Venil et al. 2020). Notable mushroom genera known for producing abundant dyes include *Trametes, Cortinarius, Omphalotus, Pycnoporus,* and *Dermocybe* (Benedict et al. 2020). The distinct and rich coloration of mushroom species is due to various secondary metabolites produced by them (Gill and Steglich 1987). Mushroom pigments are known to have an ecological role in their life cycle, as they protect mushroom species from UV radiation and pathogens and also safeguard mushroom spores. Species such as *Cortinarius semisanguineus, Phaeolusschweinitzii, Hydnellumpeckii*, and *Inonotushispidus* yield bioactive colorants such as anthraquinones, melanins, and pulvinic acids.

The extraction of pigments from mushrooms is usually done in solvents like water, ethanol, methanol, and acetone, depending on their biochemical nature. The production amount can be scaled up by optimizing conditions like temperature, pH, and fermentation time (Zhang et al. 2021; Villalobos-Pezos et al. 2024). However, methods employed for pigment extraction are not sustainable, but scientists are working in this area to have environmentally friendly extraction procedures, such as the usage of ultrasound technology or supercritical $CO_2$ (Sirohi et al. 2024). The ethanol or water-soluble pigments are easily transferred to natural fibers through traditional mordanting processes. For instance, the phenoxazinone dye known as cinnabarin has been employed in imparting color to cotton fabrics and has an antibacterial effect (Masae et al. 2020). Similarly, carminic-red anthraquinones obtained from *Cortinarius sanguineus* mushrooms have been used to successfully color silk and wool using alum as a mordant (Herrala et al. 2022).

Recent research has been focused on the identification of novel pigment-producing fungal species with superior colorfastness and color gamut, the optimization of extraction processes for industrial acceptability, and the evaluation of the green sustainability of fungal dye production and use. Collectively, this work emphasizes the transformative power of fungal biotechnology for the creation of ecologically benign and aesthetically pleasing alternatives to textile coloration (Venil et al. 2020).

### 16.4.5 Mycelium-Blended Clothing

The clothing made from mycelium and mushroom fibers represents an innovative leap in sustainable textiles. The root-like structure of fungi is bio-fabricated into a versatile fabric. When mixed with natural fibers like hemp, jute, cotton, or flax, it develops into strong mycelium-fiber composites, which encompass superior mechanical, thermal, and eco-friendly characteristics (Islam et al. 2018; Appels et al. 2019; Jones et al. 2020a). Mycelium acts as a natural adhesive by colonizing lignocellulosic substrates via enzymatic degradation, intertwining with fiber-strengthened materials to create composite structures using self-assembly (Elsacker et al. 2019). Performance properties of the composites, like strength, porosity, and aesthetics, can be engineered through effective choice of mushroom species like *Ganoderma lucidum* or *Pleurotus ostreatus* and optimization of substrate constituents. The synergistic incorporation of natural fibers is responsible for the biodegradability and mechanical strength of the resulting material, while their properties, like fiber orientation, density, and pretreatment methods, influence the degree of mycelial colonization and composite strength (Haneef et al. 2017).

The production process involves growing mycelium cells on organic matter in controlled environments, which then form a 3D network. It starts with substrate preparation, inoculation, controlled growth, molding, and heat-activated deactivation—amounting to an energy-saving, chemical-free process that is suitable for environmentally friendly manufacture (Appels et al. 2018). The resulting mycelium-fiber composites are of low weight (60–200 kg/m$^3$), mechanically strong (with compressive strength up to 1.2 MPa), thermally insulated, and acoustically absorbent, with additional benefits such as complete biodegradability and intrinsic fire resistance due to chitin content (Haneef et al. 2017; Islam et al. 2018; Elsacker et al. 2021).

Fabrics made of mycelium are heralded for their sustainability, offering a vegan, biodegradable, and often carbon-negative alternative to conventional materials like leather. It possesses beneficial properties such as durability, flexibility, and natural resistance to water, fire, and microbes, making it suitable for a wide range of apparel, from seamless clothing to high-fashion accessories, ushering in a new era of eco-conscious fashion. Certain technologies, like those from MycoTEX, can even create seamless garments directly from these fungi, eliminating much of the waste associated with traditional textile production. The method significantly reduces water consumption and energy use by avoiding steps like spinning, weaving, and extensive sewing.

## 16.5 Challenges and Limitations of Mushroom-Based Biomaterials

Even with its potential, the broad application of mushroom-based materials faces numerous challenges.

### 16.5.1 Packaging

Despite the promising sustainability and biodegradability of mushroom-based packaging, several technical and environmental challenges limit its scalability and industrial application. Using synthetic plastic-made molds for shaping mycelium and moving raw material from places remains a costly environmental price to pay while making this sustainable biomaterial. Though these concerns can be alleviated by the usage of molds made from recycled plastic through on-the-spot 3D printing and obtaining substrate from local agricultural residual material, they still need to be considered. Making mycelium-based packaging is a laborious process, as only growing mycelium up to complete colonization of substrate and transforming it into compact composites takes 2 to 3 weeks, as compared to their synthetic counterparts, rendering it unproductive (Jones et al. 2020a). Maintenance of critical parameters of temperature, humidity, light exposure, and aseptic environment for optimal growth and yield poses commercial challenges, particularly when these parameters are fluctuating among different mushroom species. Furthermore, myco-packaging exhibits limited stretchability and water resistance compared to conventional plastics. Although chemical treatments can enhance these properties, they may adversely affect the material's biodegradability (Islam et al. 2018). The uniformity of the product at each production cycle is difficult to sustain owing to the usage of agricultural waste generated from mixed sources (Appels et al. 2019). Lastly, the high production amount, regulatory obstacles, and emerging myco-material technology are challenges to large-scale commercial operations (Haneef et al. 2017). To explore the complete potential of myco-material as a viable option to plastic material, all these multidimensional hurdles need to be tackled through research on optimizing substrate composition, technology to control environmental parameters, and making a hybrid material.

### 16.5.2 Textiles

The use of mushroom mycelium as a sustainable alternative to animal-derived leather and in the synthetic fabric industry has garnered a considerable amount of attention and is acknowledged well, but still, numerous hindrances prevent its extensive utilization. One of the prime concerns related to their mechanical performance in the textile industry is with respect to their flexibility, brittleness, and shelf life, which is found to be poor for long-term and high-strain usage. Furthermore, untreated mycelium can easily absorb water, resulting in swelling or structural degradation, so moisture sensitivity is still a major disadvantage. Water resistance can be improved through coatings, but the environmental benefits are hampered because coatings frequently decrease the material's biodegradability and add synthetic elements. Scalability and consistency present yet another significant obstacle. Because fungal growth is organic, batches can differ greatly in texture, thickness, and durability, making industrial processing and quality control more difficult (Rathinamoorthy et al. 2023). Likewise, the difficulties in dyeing, finishing, and attaining consistent colorfastness limit the aesthetic and functional customization of mycelium textiles. Most applications remain specialized or experimental due to consumer perception and high production costs, which further limit commercial viability. Mycelium textiles will mainly remain limited to eco-conscious or luxury design markets unless there is additional innovation in cultivation methods and post-processing (Kim and Ruedy 2019).

## 16.6 Futuristic Implications of Mycelium-Based Innovations

Engineered mycelium-based products have taken prevalence in modern times; thus, they are now being explored across diverse sectors such as construction, furniture, and even sectors like aerospace, etc.. These futuristic applications can help in shifting the focus from the increasing usage of plastic and other damaging substances to more sustainable approaches, thus yielding a better, safer, and greener environment.

### 16.6.1 Packaging

Innovations include making mycelium-derived films and sheets, which can be grown as thin, flexible layers resembling plastic wrap or parchment and serve as alternatives to synthetic films used in food wrappers, inner packaging layers, or sachets. These films are grown on flat surfaces or substrates and can be tuned for transparency, water resistance, and strength through controlled culturing and post-treatment techniques (Smith 2025). Innovations are also expanding into real-time active packaging systems. Where living or semi-living mycelial networks are embedded into packaging materials that can respond to environmental signals such as changes in humidity, temperature, or the presence of certain gases, thus preventing food contamination and acting as biosensors (Schultz et al. 2024).

### 16.6.2 Textile Industry

Mycelium composites are the integration of mycelium into natural fiber substrates such as hemp, cotton, and jute. Such composite fabrics are ideal for medical textiles, biodegradable packaging cloth, and acoustic interior panels. Myco-fibers include the extraction and spinning of fungal cell wall components such as chitin and β-glucans into myco-fibers. These myco-fibers can then be spun into yarns, woven, or blended with conventional fibers, offering a fully textile-compatible form of mycelium (Williams et al. 2022).

Biofilm coatings involve the use of mycelium-derived biofilms in coating technologies, which are applied to traditional fabrics as a functional finish. It is seen that these coatings can provide antimicrobial protection, UV resistance, thermal regulation, and moisture management without adding toxic chemicals or compromising the hand-feel of the base fabric. Some companies like Bolt Threads and MycoWorks are in the process of incorporating these systems into their production line (Kaiser et al. 2023).

### 16.6.3 Myco-Architecture for Space Exploration

NASA is actively exploring fungi as a sustainable material platform through its Myco-Architecture Off Planet initiative, which envisions using mycelium-based structures for future extraterrestrial habitats on the Moon and Mars (Silverstein 2020). This initiative aims at growing habitats and equipment in situ instead of transporting conventional building materials from Earth by utilizing mycelium's natural ability to self-assemble into robust, lightweight, and biodegradable structures. A key innovation under this program involves developing hybrid living systems that integrate mycelium with photosynthetic organisms such as cyanobacteria to establish a closed-loop bioregenerative life-support architecture. In this model, cyanobacteria perform photosynthesis, producing oxygen and sugars. Mycelium consumes these outputs as a growth substrate and converts them into robust structural materials (NASA 2021). The fungal mycelium converts them to structural material, which is fire-retardant and sound. Once fully grown, these mycelium-based components can be dried to become inert, stable composites capable of withstanding the harsh conditions of space environments (Doyle 2024). This bio-fabrication strategy aligns with long-duration space missions by promoting resource efficiency, in-situ manufacturing, and reduced launch mass, making it a promising solution for future off-world habitation (Silverstein 2020).

### 16.6.4 Construction

Mycelium-based materials are emerging as sustainable alternatives to traditional building materials like plastics, foams, and even concrete in the construction industry. It includes the production of bricks made from mycelium, which is done by growing fungal networks within agricultural or cellulose-based substrates, which are naturally self-binding, requiring no synthetic adhesives (Global Construction 2025). When dried, the resulting blocks are lightweight, structurally stable, and highly customizable—offering circular, carbon-negative alternatives to concrete, gypsum, or petroleum-based insulation foams.

### 16.6.5 Furniture

Furniture making has traditionally relied on conventional materials such as plastics, foams, and processed wood products. However, these are now being increasingly replaced by mycelium-based materials, which can be molded into custom geometries using biodegradable substrates (Bitting et al. 2022). Mycelium offers a unique combination of mechanical integrity, lightweight structure, thermal insulation, and inherent fire resistance, making it a viable alternative to traditional raw materials. This shift also eliminates the need for synthetic adhesives or resins, contributing to more sustainable and eco-friendly furniture design (Growing Mycelium 2024).

## 16.7 Conclusion

Mushroom-mycelium-based material offers a revolutionary and significant step toward sustainability and innovation, providing an eco-friendly solution to many industrial products that are degrading and polluting the environment. The unique architecture and biochemistry of mycelium have rewritten their roles beyond food and medicine in the fields of textiles, packaging, construction, and many more to add. The advancement in fungal biotechnology in the textile and packaging fields indicates the paradigm shift toward a circular and bio-based economy, valorizing different agricultural wastes into useful commodities. Mycelium-based leather, fibers, and foam composites are biodegradable, versatile, durable, energy efficient, and thus strongly support sustainability goals to minimize industrial waste generation, reduce carbon footprint, and contribute to climate resilience. While noteworthy progress has been achieved in transferring the mycelium-based sustainable technology from test bed to industry, hurdles still come in the way of scaling production and improving performance and cost efficiency. All these hurdles can be conquered by association among scientists, engineers, policymakers, designers, and entrepreneurs. Revolutionizing mycelium-based material science requires inputs from the scientific world, funding, and partnership between industry and academia, as it holds immense potential for a more sustainable and greener future.

**Acknowledgments** The authors would like to thank the management of Shaheed Rajguru College of Applied Sciences for Women, University of Delhi, for providing facilities for carrying out the present study.

**Declaration** The authors declare that no generative AI tools or AI-assisted technologies have been used during the writing process of this book chapter.

## References

Alaneme KK, Anaele JU, Oke TM, Kareem SA, Adediran M, Ajibuwa OA, Anabaranze YO (2023) Mycelium based composites: a review of their bio-fabrication procedures, material properties and potential for green building and construction applications. Alex Eng J 83:234–250

Amobonye A, Lalung J, Awasthi MK, Pillai S (2023) Fungal mycelium as leather alternative: a sustainable biogenic material for the fashion industry. Sustain Mater Technol 38:e00724. https://doi.org/10.1016/j.susmat.2023.e00724

Appels FVW, Dijksterhuis J, Lukasiewicz CE, Wösten HAB (2018) Mycelium materials: the growing prospects of fungal biomaterials. Fungal Biol Biotechnol 5:5. https://doi.org/10.1186/s40694-018-0053-1

Appels FVW, Camere S, Montalti M, Karana E, Jansen KMB, Dijksterhuis J, Wösten HAB (2019) Fabrication factors influencing mechanical, moisture- and water-related properties of mycelium-based composites. Mater Des 161:64–71. https://doi.org/10.1016/j.matdes.2018.11.027

Ashok A, Rejeesh C, Renjith R (2016) Biodegradable polymers for sustainable packaging applications: a review. Int J Biobased Plast 1(11)

Attias N, Danai O, Tarazi E, Zelinger E (2019) Mycelium bio-composites in industrial design and architecture: comparative review and experimental analysis. J Clean Prod 246:119037. https://doi.org/10.1016/j.jclepro.2019.119037

Barrett A (2020) IKEA replaces Styrofoam with mushroom bioplastics. Bioplastics News, 7 February. https://bioplasticsnews.com/2020/02/07/ikea-styrofoam-mushroom-ioplastics/

Bechtold T, Mussak R (2009) Handbook of natural colorants. Wiley, Chichester

Benedict C, Biedermann PHW, Müller C (2020) Fungal secondary metabolites as eco-friendly dyes. Fungal Biol Rev 34:100–117

Biofab (2025). Biofabtech.com. Biofab. http://biofabtech.com/

Biohm (2016) BIOHM: Mycelium insulation (#Future Of Home). https://www.biohm.co.uk/mycelium

Bitting S, Derme T, Lee J, Van Mele T, Dillenburger B, Block P (2022) Challenges and opportunities in scaling up architectural applications of mycelium-based materials with digital fabrication. Biomimetics 7(2):44. https://doi.org/10.3390/biomimetics7020044

Bolt Threads (2023) Mycelium Inspired. https://myceliuminspired.com/companies/bolt-threads

Crawford A, Miller SR, Branco S, Fletcher J, Stefanov D (2024) Growing mycelium leather: a paste substrate approach with post-treatments. Res Dir Biotechnol Des 2:e6

Dell (2024) Sustainable technology: eco-friendly & energy-efficient devices. https://www.dell.com/en-us/lp/dt/sustainable-devices#Sustainable-Materials

Doe J (2023) Mycelium-based materials: an alternative for sustainable interior design. New Des Ideas 7(1):48

Doyle T (2024) NASA advances research to grow habitats in space from fungi. NASA News Release, 26 June. https://www.nasa.gov/news-release/nasa-advances-research-to-grow-habitats-in-space-from-fungi/

Ecovative Design (2023) Mycelium technology platform and packaging applications. https://ecovativedesign.com

Ellen MacArthur Foundation (2016) The new plastics economy: rethinking the future of plastics. Ellen MacArthur Foundation, Cowes

Elsacker E, Vandelook S, Brancart J, Peeters E, De Laet L, Van Wylick A (2019) Mechanical, physical and chemical characterisation of mycelium-based composites with different types of lignocellulosic substrates. PLoS One 14(7):e0213954. https://doi.org/10.1371/journal.pone.0213954

Elsacker E, Vandelook S, Brancart J, Peeters E, De Laet L (2021) Mechanical, physical and chemical characterization of mycelium-based composites with different types of lignocellulosic substrates. PLoS One 16(2):e0245994. https://doi.org/10.1371/journal.pone.0245994

Gill M, Steglich W (1987) Pigments of fungi (macromycetes). Prog Chem Org Nat Prod 51:1–317

Girometta C, Picco AM, Baiguera RM et al (2019) Physico-mechanical and thermodynamic properties of mycelium-based biocomposites: a review. Sustainability 11(1):281

Global Construction (2025) Living mycelium in sustainable construction materials: a fungal revolution. ArchUp, 7 June. https://archup.net/living-mycelium-in-sustainable-building-materials/

Growing Mycelium (2024) Materials Assemble. https://materialsassemble.com/the-craft/growing-mycelium/

Haneef M, Ceseracciu L, Canale C, Bayer IS, Heredia-Guerrero JA, Athanassiou A (2017) Advanced materials from fungal mycelium: fabrication and tuning of physical properties. Sci Rep 7:41292. https://doi.org/10.1038/srep41292

Herrala M, Yli-Öyrä J, de Albuquerque AF, de Farias NO, Morales DA, Räisänen R, Freeman HS, Umbuzeiro GA, Rysä J (2022) Waterless dyeing and in vitro toxicological properties of biocolorants from Cortinarius sanguineus. J Fungi 8(11):1129

Hugohek (2025) Grown-design: beautiful products with fungus and biomass. Grown.bio. https://www.grown.bio/

IKEA (2021) IKEA explores packaging made from fungi. IKEA Sustainability Report. https://www.ikea.com

Islam MR, Tudryn G, Bucinell R, Schadler L, Picu RC (2018) Morphology and mechanics of fungal mycelium. Sci Rep 8:7531. https://doi.org/10.1038/s41598-018-25807-2

Jennings DH (1984) The ecology and physiology of the fungal mycelium. Cambridge University Press, Cambridge
Jones M, Huynh T, Dekiwadia C, Daver F, John S (2017) Mycelium composites: a review of engineering characteristics and growth kinetics. J Bionanosci 11(4):241–257. https://doi.org/10.1007/s42235-020-0045-6
Jones M, Bhat T, Kandare E, Yuen R, Wang CH, John S, Ma J (2020a) Waste-derived low-cost mycelium composite construction materials with improved fire safety. Fire Mater 44(2):194–203. https://doi.org/10.1002/fam.2771
Jones M, Bhat T, Wang CH, John S, Moinuddin KA (2020b) Engineered mycelium composite construction materials from fungal biorefineries: a critical review. Mater Today Sustain 7:100028. https://doi.org/10.1016/j.mtsust.2020.100028
Jones E, Smith A, Lee B (2023) The environmental impact of traditional leather production. J Sustain Mater 15(2):123–135
Kaiser R, Scott J, Dade-Robertson M, Elsacker E (2023) Knit-myelium hybrids: designing and scaling up with bio fabrication systems based on mycocrete injection in 3D knitted tubular formworks. In: Heinzel T, Dumitrescu D, Tomico O, Robertson S (eds) Proc textile intersections Conf 2023. September, London, UK, pp 20–23
Kanagaraj J, Velappan KC, Chandra Babu NK, Sadulla S (2006) Solid wastes generation in the leather industry and its utilization for cleaner environment—a review. J Sci Ind Res 65(7):541–548
Kim Y, Ruedy D (2019) Mushroom packages: an Ecovative approach in packaging industry. In: Filho WL (ed) Handbook of engaged sustainability. Springer, Cham, pp 1–25. https://doi.org/10.1007/978-3-319-53121-2_27-1
Life Materials (2025) MuSkin: cruelty-free and oil-free vegan leatherette alternative to animal-based leather. https://lifematerials.eu/shop/muskin-cruelty-free-and-oil-free-vegan%20leatherette-alternative-to-animal-based-leather/
Liu Y, Chen J, Duan D, Zhang Z, Liu C, Cai W, Zhao Z (2024) Environmental impacts and biological technologies toward sustainable treatment of textile dyeing wastewater: a review. Sustainability 16(24)
Magical Mushroom Company (2023) Our products: sustainable packaging solutions. https://www.magicalmushroom.com
Manan S, Ullah MW, Ul-Islam M, Atta OM, Yang G (2021) Synthesis and applications of fungal mycelium-based advanced functional materials. J Bioresour Bioprod 6(1):1–10
Masae M, Saithong P, Kongsong P, Wattanasen S (2020) UV protection and antibacterial properties of cotton fabrics dyed. Sci Eng Connect 43(2):203–212
Mitrano DM, Rentsch D, Nowack B (2021) Sustainable biomaterials: trends and knowledge gaps. Environ Sci Technol 55(1):75–82
Mogu (2023) Mycelium Inspired. https://myceliuminspired.com/companies/mogu
Mycoworks (2023) Mycelium Inspired. https://myceliuminspired.com/companies/mycoworks
NASA (2021) Myco-architecture: Building with fungi. https://www.nasa.gov/directorates/spacetech/myco-architecture-building-with-fungi/
Nature's Fynd (2025) Fungi-based foods for optimists. https://www.naturesfynd.com/
NEFFA (2023) Mycelium Inspired. https://myceliuminspired.com/companies/mycotex
Nguyen T (2024) Eco-efficiency measures in the leather industry: an educational case. J Account Manag Inf Syst 23(2):365–380
Paradise Packaging Pvt. Ltd (2023) Paradise Packaging. https://www.paradisepackaging.net/
Porter DL, Naleway SE (2022) Hyphal systems and their effect on the mechanical properties of fungal sporocarps. Acta Biomater 145:272–282
Rathinamoorthy R, Sharmila Bharathi T, Snehaa M, Swetha C (2023) Mycelium as sustainable textile material—review on recent research and future prospective. Int J Clothing Sci Technol 35(3):454–476

Schultz N, Fazli A, Piros S, Barranco-Origel Y, DeLa CP, Schneider DY (2024) Characterization of mycelium biocomposites under simulated weathering conditions. ACS Appl Bio Mater 7(12): 8408–8422. https://doi.org/10.1021/acsabm.4c01192
Shankar MP, Hamza A, Khalad A, Shanthi G, Kuppireddy S, Kumar DS (2024) Engineering mushroom mycelium for a greener built environment: advancements in mycelium-based biocomposites and bioleather. Food Biosci:105577
Shin HJ, Ro HS, Kawauchi M, Honda Y (2025) Review on mushroom mycelium-based products and their production process: from upstream to downstream. Bioresour Bioprocess 12(1):3
Silverstein K (2020) NASA's myco-architecture project explores fungus-based buildings on Mars. NASA Centennial Challenges
Sirohi R, Negi T, Rawat N, Sagar NA, Sindhu R, Tarafdar A (2024) Emerging technologies for the extraction of bioactives from mushroom waste. J Food Sci Technol 61(6):1069–1082
Smith C (2025) Clean the Sky. https://www.cleanthesky.com/innovation/mycelium-packaging
Sydor M, Bonenberg A, Doczekalska B, Cofta G (2022) Mycelium-based composites in art, architecture, and interior design: a review. Polymers 14(1):145. https://doi.org/10.3390/polym14010145
UNEP, Notten P (2020) Sustainability and circularity in the textile value chain: global stocktaking. UNEP, Nairobi
United Nations Environment Programme (2022) Sustainability and circularity in the textile value chain: global stocktaking. https://www.unep.org
Venil CK, Velmurugan P, Dufossé L, Renuka Devi P, Veera Ravi A (2020) Fungal pigments: potential coloring compounds for wide ranging applications in textile dyeing. J Fungi 6(2):68. https://doi.org/10.3390/jof6020068
Villalobos-Pezos M, Muñoz-Fariña O, Ah-Hen KS, Garrido-Figueroa MF, García-Figueroa O, González-Esparza A, Bastías-Montes JM (2024) Optimization of phenolic content extraction and effects of drying treatments on physicochemical characteristics and antioxidant properties of edible mushroom Pleurotus ostreatus (Jacq.) P. Kumm (oyster mushroom). Antioxidants 13(12): 1581
Williams E, Cenian K, Golsteijn L, Morris B, Scullin ML (2022) Life cycle assessment of MycoWorks' Reishi™: the first low-carbon and biodegradable alternative leather. Environ Sci Eur 34(1):120
Zhang F, Xue F, Xu H, Yuan Y, Wu X, Zhang J, Fu J (2021) Optimization of solid-state fermentation extraction of Inonotus hispidus fruiting body melanin. Foods 10(12):2893

MIX
Papier aus verantwortungsvollen Quellen
Paper from responsible sources
FSC® C105338

If you have any concerns about our products,
you can contact us on
**ProductSafety@springernature.com**

In case Publisher is established outside the EU,
the EU authorized representative is:
**Springer Nature Customer Service Center GmbH**
**Europaplatz 3, 69115 Heidelberg, Germany**

Printed by Libri Plureos GmbH
in Hamburg, Germany